Life Study

A Textbook of Biology

The cover photograph shows a hive bee visiting a sage flower. The anthers are brushing the bee's abdomen and depositing pollen.

The photograph opposite shows a pair of Canada geese with their brood.

The drawings are by the author, whose copyright they are unless otherwise stated, and whose permission should be sought before they are reproduced or adapted in other publications.

The full-colour illustrations on pages 188, 200–1 and 204–5 are by Pamela Haddon.

Designer: Richard Glyn Jones

British Library Cataloguing in Publication Data

Mackean, D. G.
Life Study
1. Biology
I. Title
574 QH308.7

ISBN 0–7195–3861–0
(trade cased edition)

ISBN 0–7195–3783–5
(school limp edition)

First published 1981
Reprinted 1982, 1984, 1986, 1988,
1990 with revisions, 1992,
1994, 1995, 1996, 1998, 1999

Printed in Hong Kong by
Colorcraft Ltd

John Murray

Life Study

A Textbook of Biology

D. G. Mackean

CONTENTS

SECTION ONE
Principles of Biology

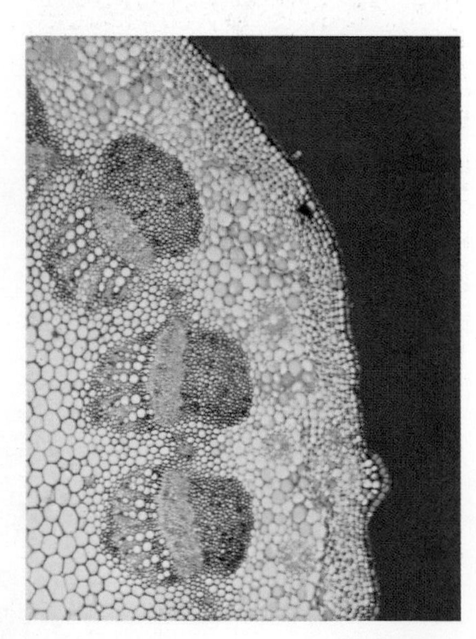

SECTION TWO
Flowering Plants

SECTION THREE
Human Physiology

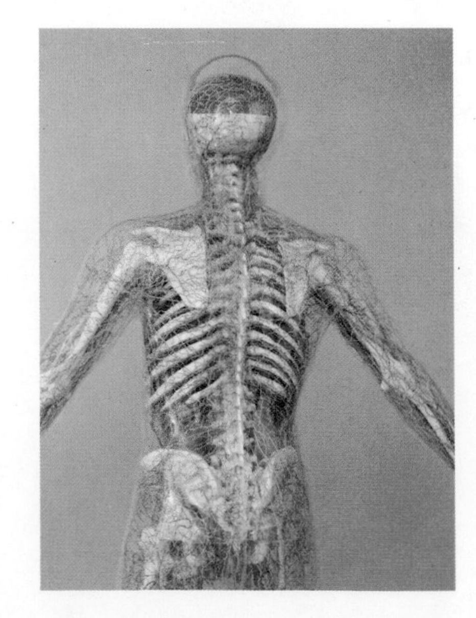

SECTION FOUR
Genetics and Evolution

SECTION FIVE
Characteristics of Living Things

SECTION SIX
Health

To the student

This book contains most of the information you need to study biology for GCSE. The aim of the book is to ensure that whatever syllabus or teaching plan you follow, the facts are easy to find, understand and learn.

Life Study is therefore meant to be used as a reference book—that is, you look up the information when you need it rather than trying to read through the book from start to finish. You may need to read a whole chapter at a time, in which case the list of contents on the previous two pages will direct you to it. For gathering information from various chapters you will need to use the index on pages 261 to 266. In the index, the page numbers printed in bold type show where the subject is introduced or most fully explained.

The experiments described are only a small part of the practical work you may do and they are included mainly for revision purposes (the instructions may not be detailed enough to follow in the laboratory). For more detailed instructions you are recommended to use *Class Experiments in Biology* or *Experimental Work in Biology* (details below).

The questions in the text are meant to test your understanding of what you have read in the chapter. Usually, they do not ask you just to reproduce what the text says, but to use the information in a slightly different way.

Certain scientific terms are used without explanation. You can look up any that are not familiar in the glossary on pages 258 to 260.

Author's acknowledgements

I am very grateful to Mary Ashton, Margaret Mackean, John Barker and Pat Lowry for reading the manuscript and for their valued comments. I should like to thank all those people who searched for and submitted photographs; individual credits are given on page 266. I am also indebted to Pamela Haddon for her paintings of plants and animals in Chapters 23 and 24.

Practical work

D. G. Mackean, *Experimental Work in Biology*, Combined Edition (John Murray, 1982).
150 tried and tested experiments. Detailed instructions for students, and questions to test their understanding of experimental design and interpretation of results. Teachers' book available.

D. G. Mackean, C. Worsley and P. Worsley, *Class Experiments in Biology* (John Murray, 1982).
40 experiments selected from *Experimental Work in Biology*, with photocopiable worksheets for the students. Teachers' book available.

Biology Resource Pack for GCSE

Photocopiable questions for self-assessment, class discussion, manipulation and interpretation of data. Also extensive lists of videos, slides, software, teaching packs, wall charts and practical work. (John Murray, 1990)

SECTION ONE

Principles of Biology

1 Cells and tissues

CELL STRUCTURE

Almost every part of a plant or an animal is made of very small structures called **cells**. One way to examine the cells of a plant or animal is to cut very thin slices, called **sections** (Figure 1), and study them under the microscope.

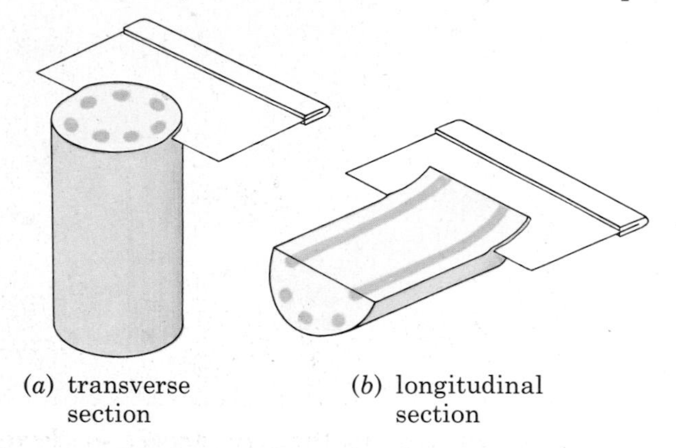

Figure 1 Cutting sections of a plant stem

Sections of one cell thickness can be cut and mounted on a glass slide. Light will pass through them and the details of the cells can be seen.

The photograph in Figure 4 shows a greatly magnified section through the tip of a plant shoot. You can see the stem, part of two leaves at the sides and two small leaves at the tip. These structures are all seen to be made up of box-like cells. Figure 2 shows what three of these cells might look like when cut open.

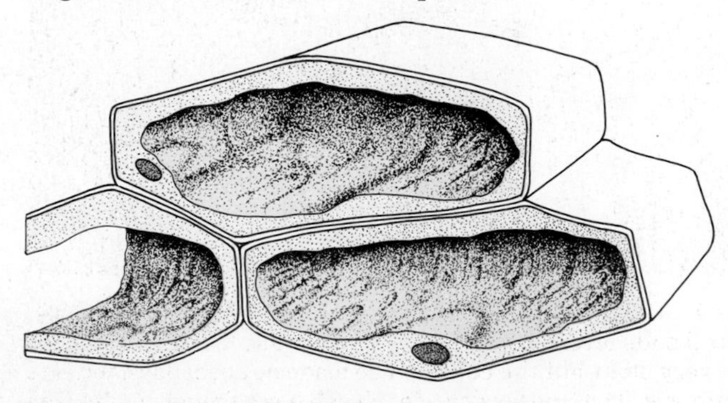

Figure 2 Three plant cells cut open

Animal cells need much greater magnifications to be seen clearly. Those shown in Figure 6 are of kidney tubules magnified about 800 times to make the details clear enough.

Thousands of cells packed together in an orderly way make up the **tissues** of living organisms: the skin, bone or muscle tissue of an animal, or the leaf and stem tissue of a plant.

Cells may be very different shapes and sizes. Muscle cells and nerve cells, for example, may be very long (see Figure 13*e* on page 6). However, *all* cells have a **cell membrane**, enclosing the living substance, called **cytoplasm**, which contains a **nucleus** (Figure 3).

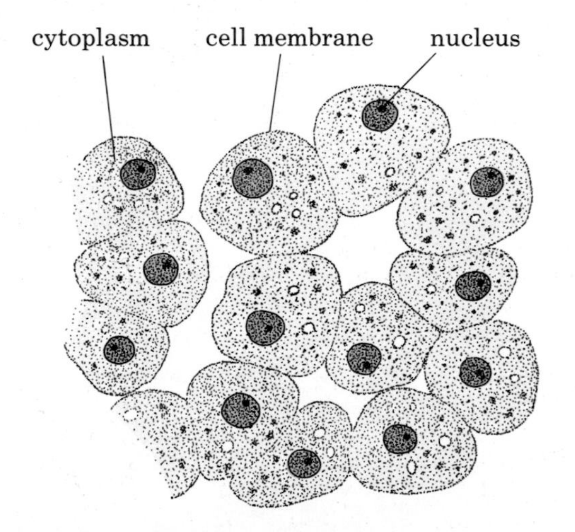

Figure 3 A group of animal cells

Cell membrane. This is a very thin, flexible layer round the cell. Although it is very thin, it keeps the cell contents from escaping and mixing with the surroundings. The cell membrane also has control over the substances which are allowed to enter or leave the cell.

Cytoplasm. Seen under an ordinary microscope, this looks like a thick liquid with particles in it. Cytoplasm contains **enzymes** (page 12), which control the chemical changes taking place in it. There are also a number of tiny structures which play a part in building up new cytoplasm, producing substances to be used

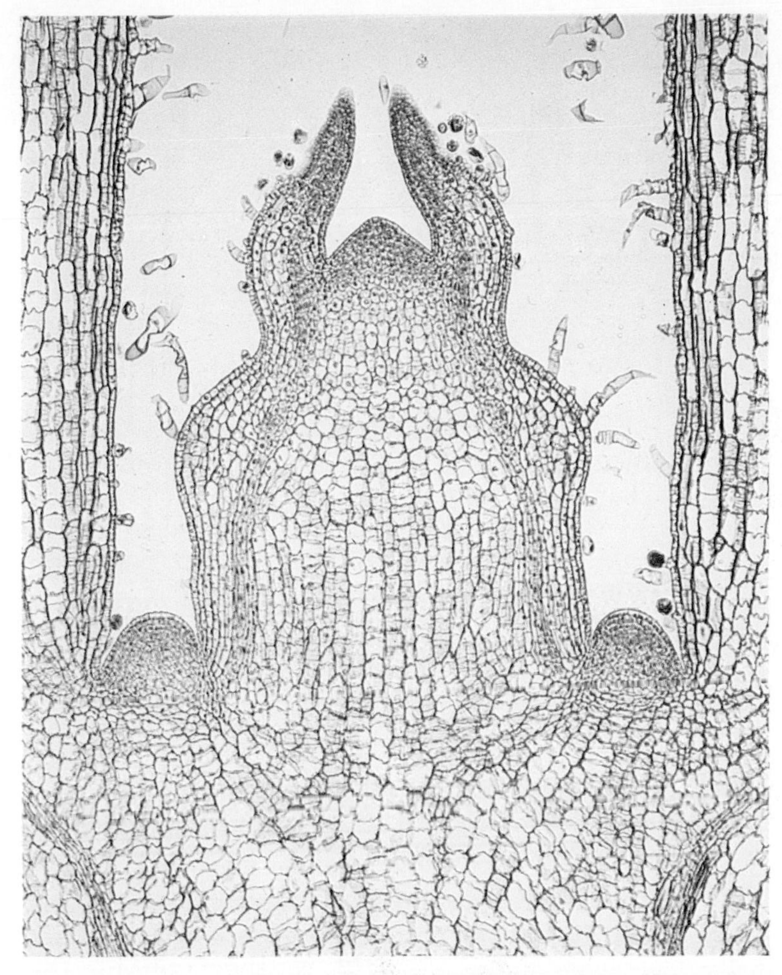

Figure 4 Longitudinal section through the tip of a plant shoot (×80). The cell walls are stained blue and the nuclei red. The two smallest leaves and the three buds look pink. This is because they contain a large number of small, dividing cells, and so there are many nuclei present.

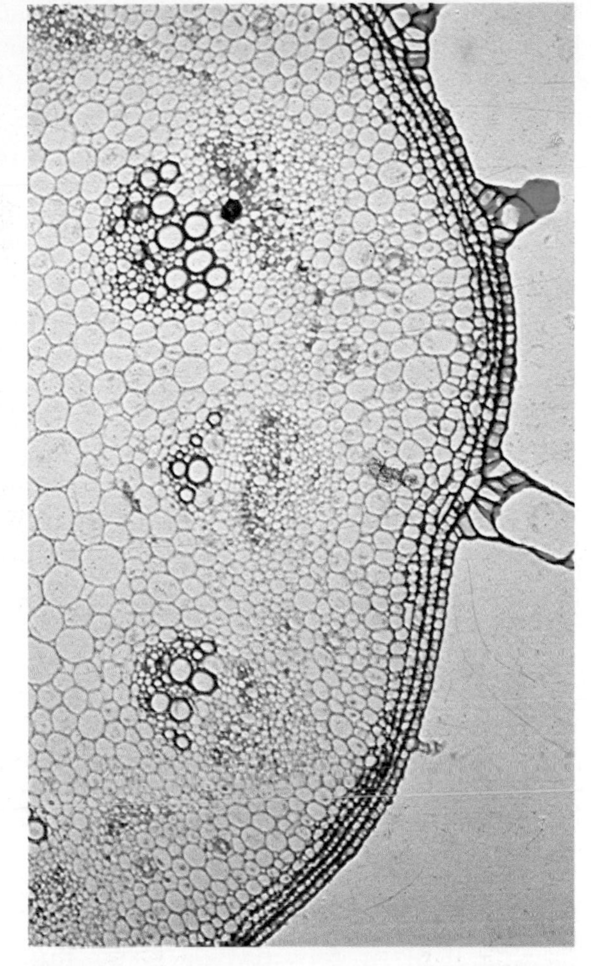

Figure 5 Section through part of a sunflower stem (×60). Although the cells forming the stem are all similar in structure, they vary a great deal in shape and size. The three structures in the stem are veins.

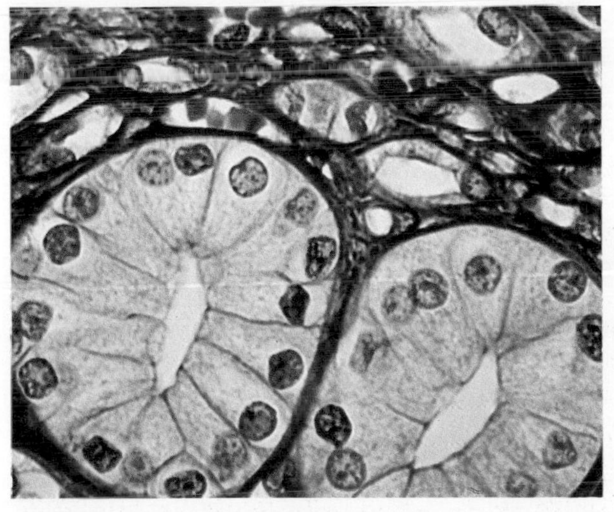

Figure 6 Cells in the kidney (×800). The boundaries of animal cells are often hard to see but the cells forming the two tubules (page 129) seen here in transverse section are quite clear. The nuclei are stained pink.

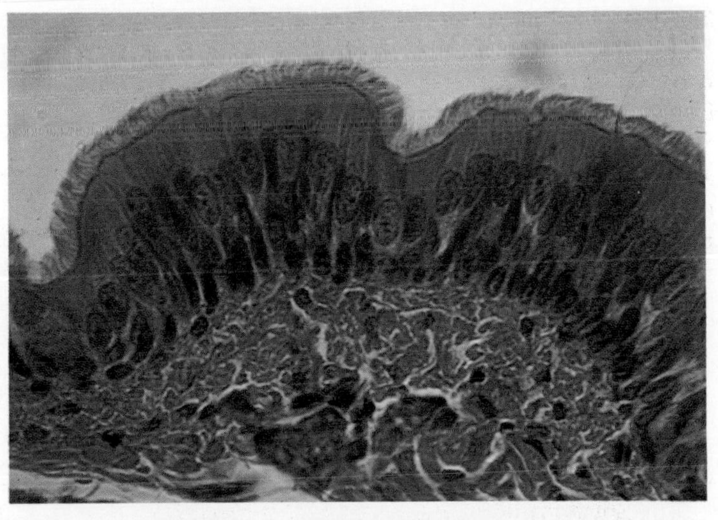

Figure 7 Cells forming the lining of the windpipe (×550). These tall cells form a layer only one cell thick. Their boundaries are not very clear but the cells can be made out because each one has a nucleus. The free border of each cell has a fringe of cilia (see Figure 13*a* on page 6).

outside the cell, or breaking down food material in order to produce energy. The cytoplasm may also contain food reserves such as starch grains or oil droplets.

Nucleus. Each cell has one nucleus and this is usually a rounded structure contained in the cytoplasm. The function of the nucleus is to control the chemical changes which take place in the cytoplasm. It will control, for example, the kinds of chemicals the cell will make and how much of them is produced.

The nucleus also controls cell division, as shown in Figure 10. A cell without a nucleus cannot reproduce. Once a cell has stopped dividing, chemicals from the nucleus will control what kind of cell it will become, e.g. a blood cell or a liver cell.

Plant cells

Figure 8 shows some typical plant cells. These are different from animal cells in several ways:

(a) They all have a **cell wall** surrounding the cell membrane. This is a non-living layer containing **cellulose**, which allows liquids and dissolved substances to pass freely through it. Under the microscope, the cell wall makes plant cells quite distinct and easy to see.

(b) Most plant cells have a large, fluid-filled space called a **vacuole**. The vacuole contains **cell sap**, a watery solution of sugars, salts and sometimes pigments (coloured material). This large, central vacuole pushes the cytoplasm aside so that most of it forms just a thin lining inside the cell wall. It is the outward pressure of the vacuole on the cytoplasm and cell wall which makes plant cells and their tissues firm (see page 20). Animal cells may sometimes have small vacuoles in their cytoplasm but they are usually produced to do a particular job, such as remove excess water.

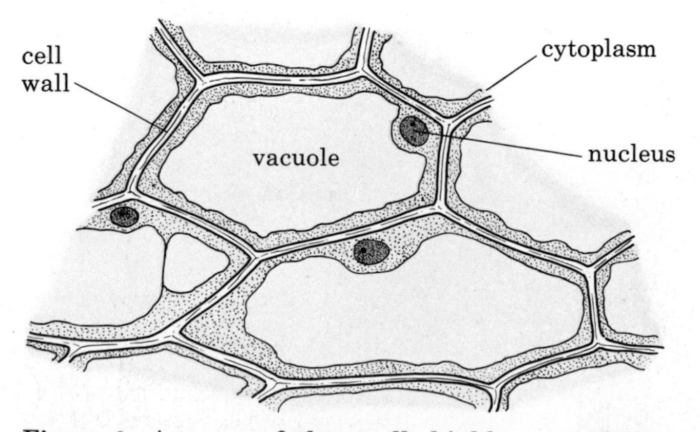

Figure 8 A group of plant cells highly magnified to show cell structures. This is how the plant cells in Figure 2 would appear in a microscope section.

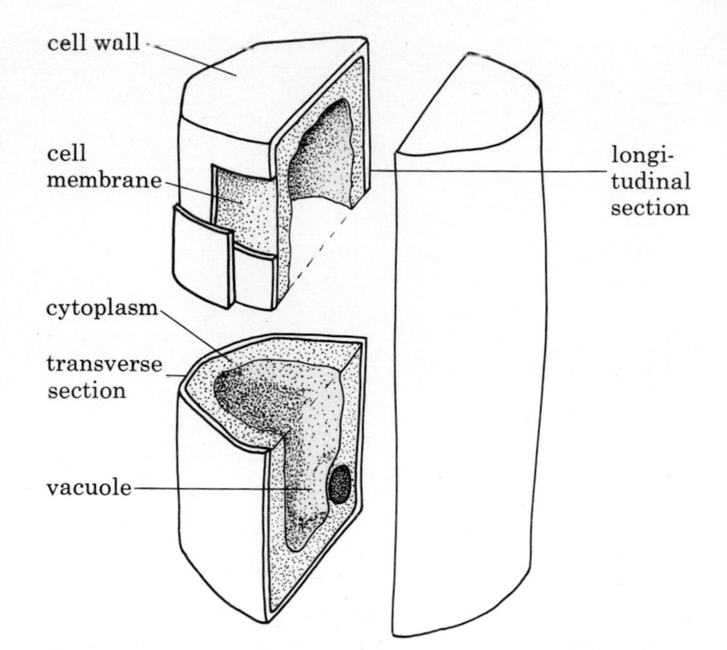

Figure 9 A plant cell in section. The shape of a cell seen in a section depends on which way it was cut. This diagram shows the same cell as it might appear in a transverse and a longitudinal section.

(c) In the cytoplasm of plant cells are many tiny structures called **plastids** which are not present in animal cells. If they contain the green substance **chlorophyll**, they are called **chloroplasts** (page 24). The colourless plastids usually contain starch.

QUESTIONS

1 (a) What structures are usually present in all cells, whether they are from an animal or from a plant?

(b) List the differences in structure between plant and animal cells by drawing up a table in two columns, one headed 'Plant cells' and one headed 'Animal cells'.

2 In what way does the red blood cell shown in Figure 1*a* on page 111 differ from most other animal cells?

CELL DIVISION AND CELL SPECIALIZATION

Cell division

When plants and animals grow, their cells increase in numbers by dividing in the growing regions of the organism. Typical growing

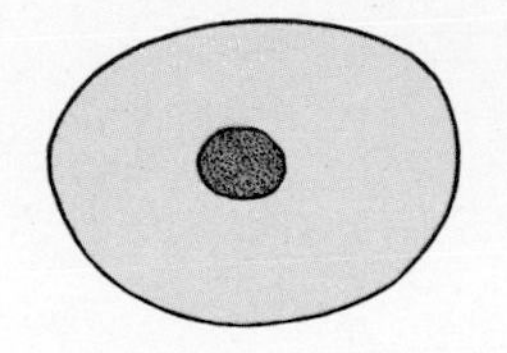

(*a*) Animal cell about to divide.

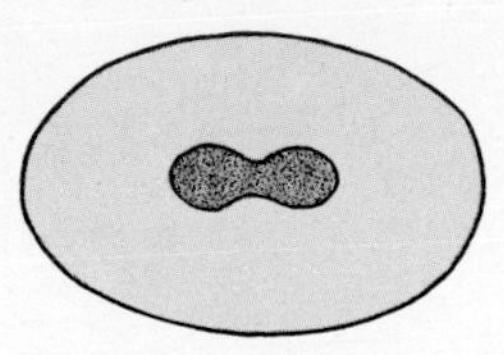

(*b*) The nucleus divides first.

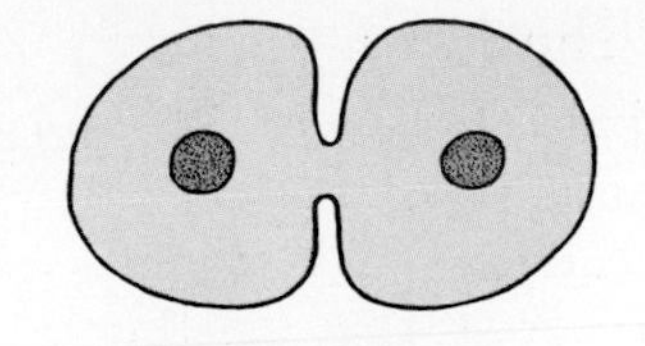

(*c*) The daughter nuclei separate and the cytoplasm pinches off between the nuclei.

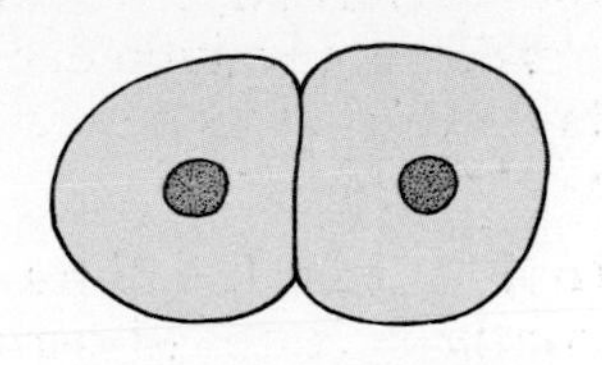

(*d*) Two cells are formed. One may keep the ability to divide, and the other may become specialized.

Figure 10 Cell division in an animal cell

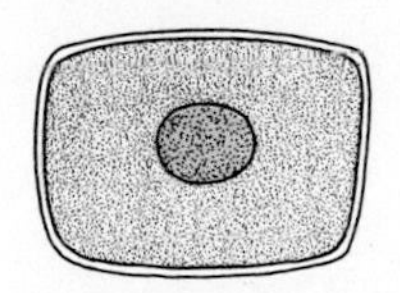

(*a*) A plant cell about to divide has a large nucleus and no vacuole.

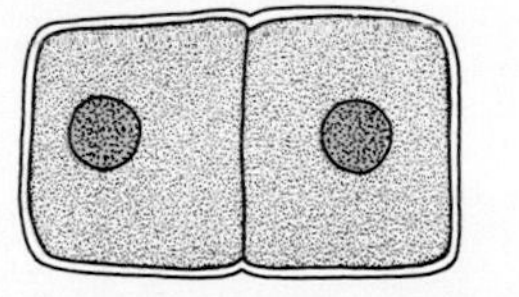

(*b*) The nucleus divides first. A new cell wall develops and separates the two cells.

(*c*) The cytoplasm adds layers of cellulose on each side of the new cell wall. Vacuoles form in the cytoplasm of one cell.

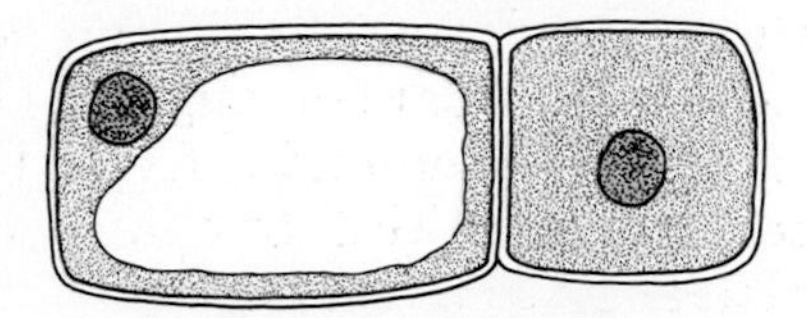

(*d*) The vacuoles join up to form one vacuole. This takes in water and makes the cell bigger. The other cell will divide again.

Figure 11 Cell division in a plant cell

regions are the ends of bones, layers of cells in the skin, root tips and buds. Each cell divides to produce two daughter cells. Both daughter cells may divide again, but usually one of the cells grows and changes its shape and structure to do one particular job—in other words, it becomes **specialized**. At the same time it loses its ability to divide any more. The other cell is still able to divide and so continue the growth of the tissue. **Growth** is therefore the result of cell division, followed by cell enlargement and, in many cases, cell specialization.

Figure 10 shows the process of cell division in an animal cell. The events in a plant cell are shown in Figure 11. Because of the cell wall, the cytoplasm cannot simply pinch off in the middle, and a new wall has to be laid down between the two daughter cells. Also, a new vacuole has to form.

Figure 12 shows the pattern of cell division that takes place at the growing point of a simple seaweed. The cells divide and expand to make the tip longer and wider. Once cells 4a–d have formed they cannot divide again but may become specialized, e.g. 4b and 4c may help carry food up to the growing point.

Figure 12 Growth by cell division and cell elongation at the tip of a seaweed

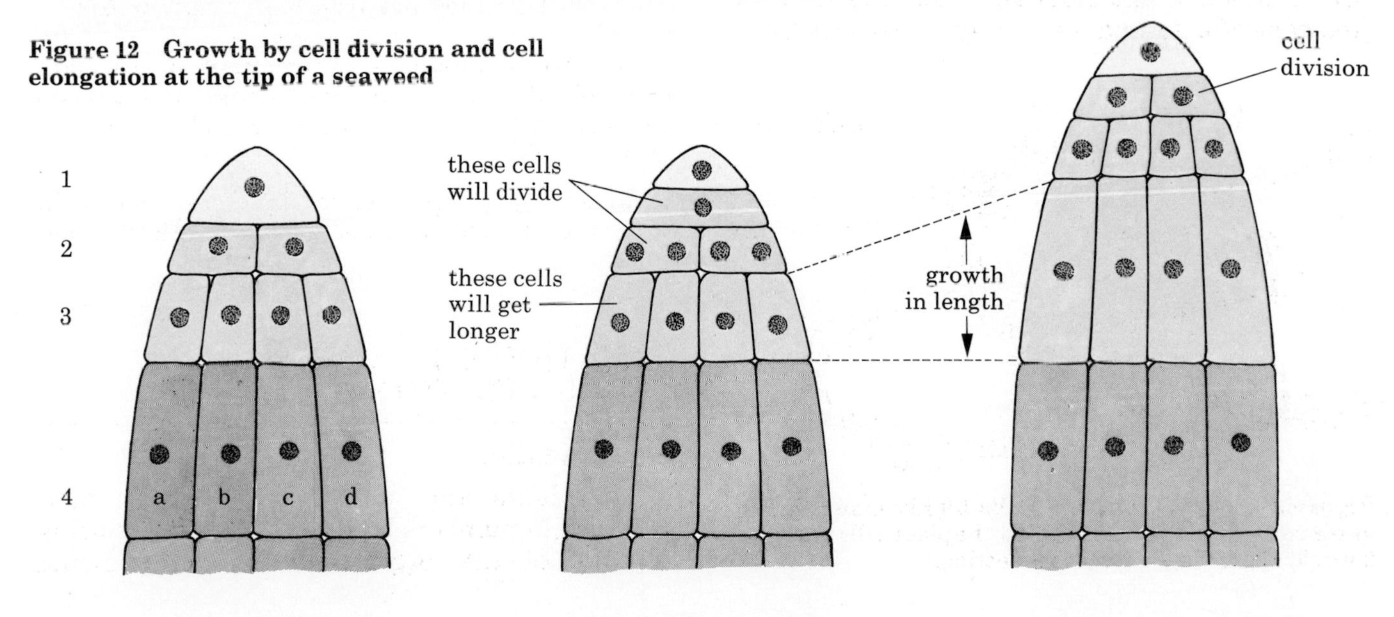

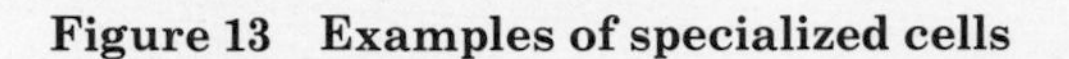

Figure 13 Examples of specialized cells

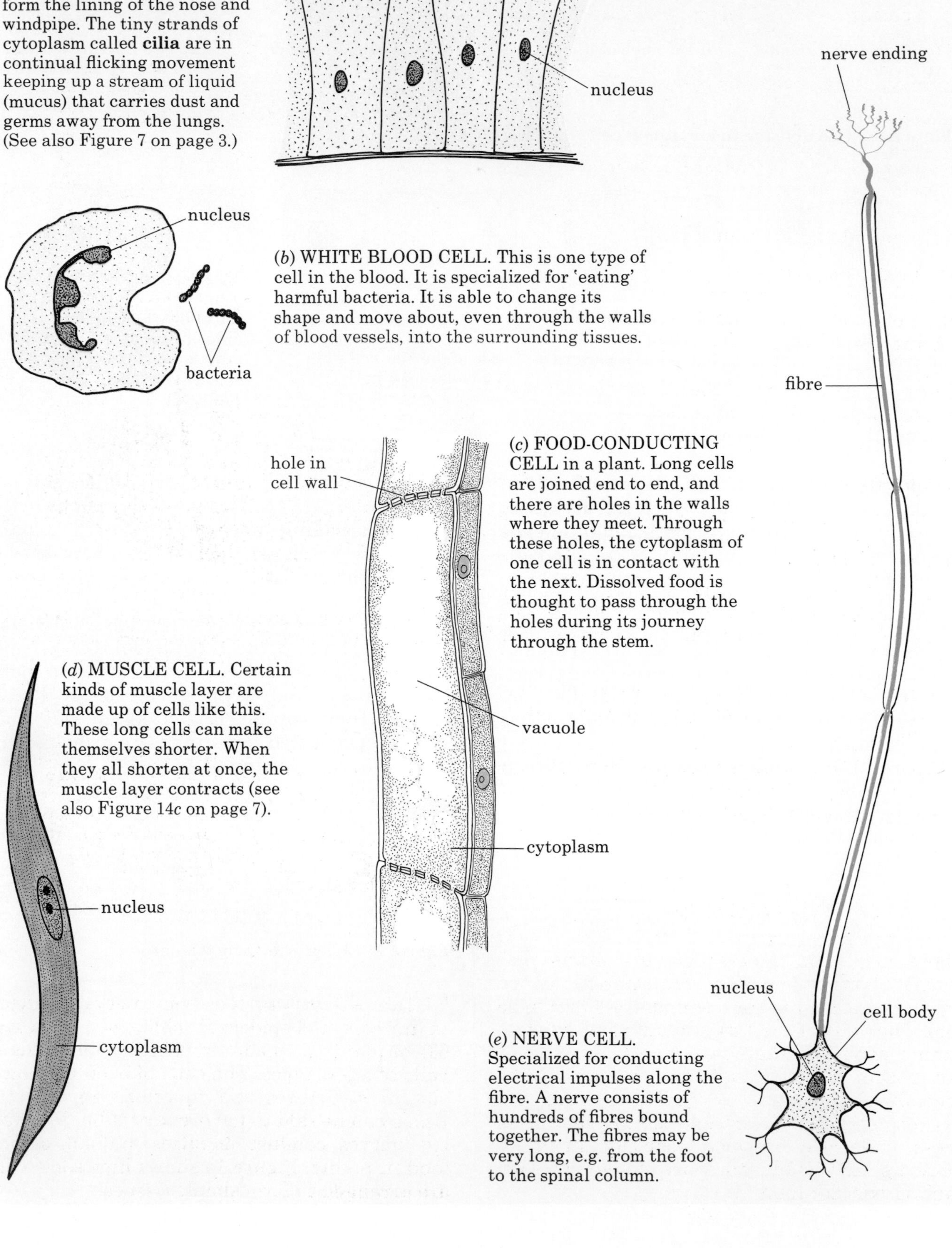

(*a*) CILIATED CELLS. These form the lining of the nose and windpipe. The tiny strands of cytoplasm called **cilia** are in continual flicking movement keeping up a stream of liquid (mucus) that carries dust and germs away from the lungs. (See also Figure 7 on page 3.)

(*b*) WHITE BLOOD CELL. This is one type of cell in the blood. It is specialized for 'eating' harmful bacteria. It is able to change its shape and move about, even through the walls of blood vessels, into the surrounding tissues.

(*c*) FOOD-CONDUCTING CELL in a plant. Long cells are joined end to end, and there are holes in the walls where they meet. Through these holes, the cytoplasm of one cell is in contact with the next. Dissolved food is thought to pass through the holes during its journey through the stem.

(*d*) MUSCLE CELL. Certain kinds of muscle layer are made up of cells like this. These long cells can make themselves shorter. When they all shorten at once, the muscle layer contracts (see also Figure 14*c* on page 7).

(*e*) NERVE CELL. Specialized for conducting electrical impulses along the fibre. A nerve consists of hundreds of fibres bound together. The fibres may be very long, e.g. from the foot to the spinal column.

Specialization of cells

Cells often become specialized. This means that

(a) they can carry out a particular task,
(b) they develop a distinct shape, and
(c) special kinds of chemical changes take place in their cytoplasm.

For example, **nerve cells** (Figure 13*e*)

(a) carry messages to and from the brain,
(b) are very long and thin, and
(c) their chemical reactions help them to carry electrical impulses,

while **muscle cells** (Figure 13*d*)

(a) produce movement,
(b) are also very long, and
(c) their chemical reactions enable them to contract (get shorter).

Figure 13 gives some other examples of specialized plant and animal cells.

QUESTIONS

3 Make a list of events, in the correct order, which take place (a) when an animal cell divides, (b) when a plant cell divides.

4 Look at Figure 4 on page 134. When a Malpighian cell of the skin divides, which daughter cell becomes specialized and which one keeps the ability to divide again?

5 How does the structure of the food-conducting cell of a plant (Figure 13*c*) help it to carry out its function?

6 Look at Figure 2*c* on page 59. (a) Whereabouts in a leaf are the food-carrying cells? (b) What other specialized cells are there in the leaf?

TISSUES AND ORGANS

There are some microscopic organisms that consist of one cell only and can carry out all the processes necessary for their survival (see page 198). The cells of the larger plants and animals cannot survive on their own. A muscle cell could not obtain its own food and oxygen. Other specialized cells have to provide the food and oxygen needed for the muscle cell to live. Unless these cells are grouped together in large numbers and made to work together, they cannot exist for long.

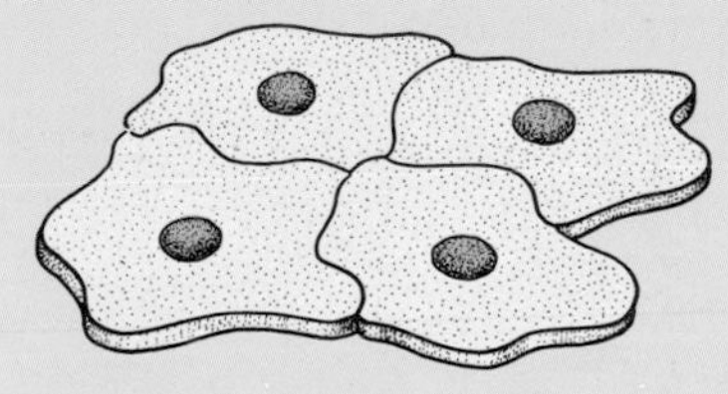

(*a*) Cells forming an epithelium, a thin layer of tissue, e.g. that lining the mouth cavity.

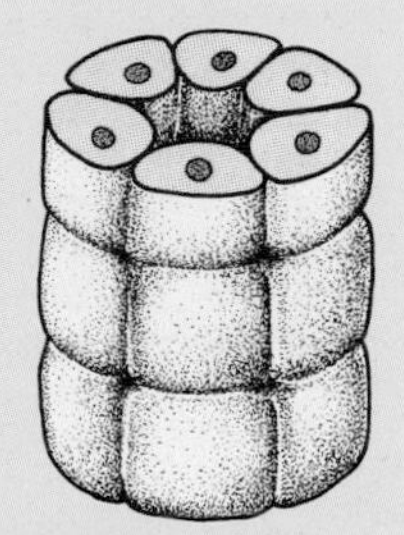

(*b*) Cells forming a fine tube, e.g. a kidney tubule (see Figure 6 on page 3).

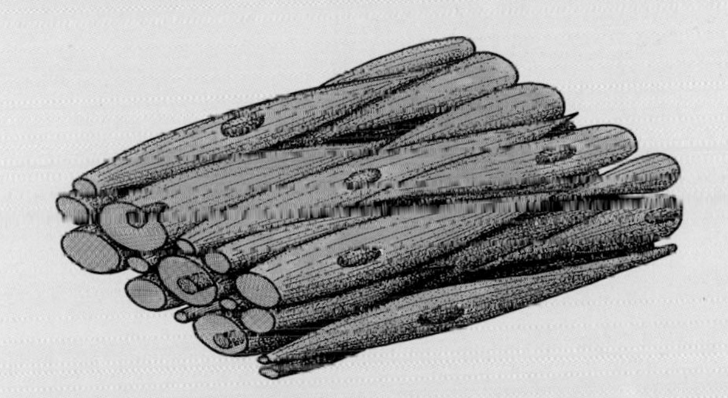

(*c*) Muscle cells forming a sheet of muscle tissue (see Figure 16 on page 9). Blood vessels, nerve fibres and connective tissue will also be present.

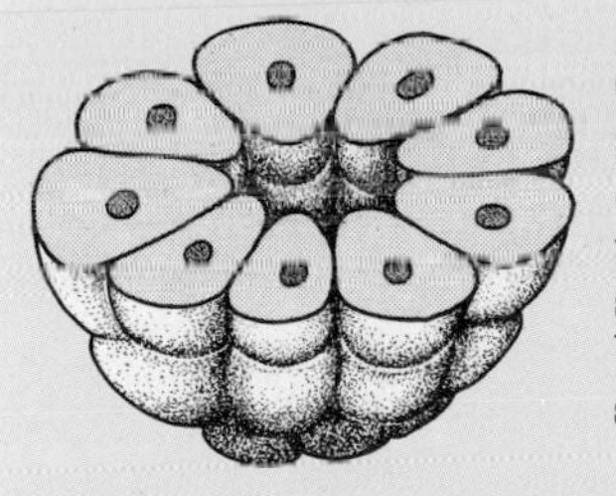

(*d*) Cells forming part of a gland. The cells make chemicals which are released into the central space and are carried away by a tubule like the one shown in (*b*).

Figure 14 How cells form tissues

Tissue. A tissue such as bone, nerve or muscle in animals, and epidermis, phloem or pith (page 63) in plants, is made up of many hundreds of cells of a few types. The cells of each type have similar structures and functions so that the tissue can be said to have a particular function, e.g. nerves conduct impulses, phloem carries food in plants. Figure 14 shows how some cells are arranged to form simple tissues.

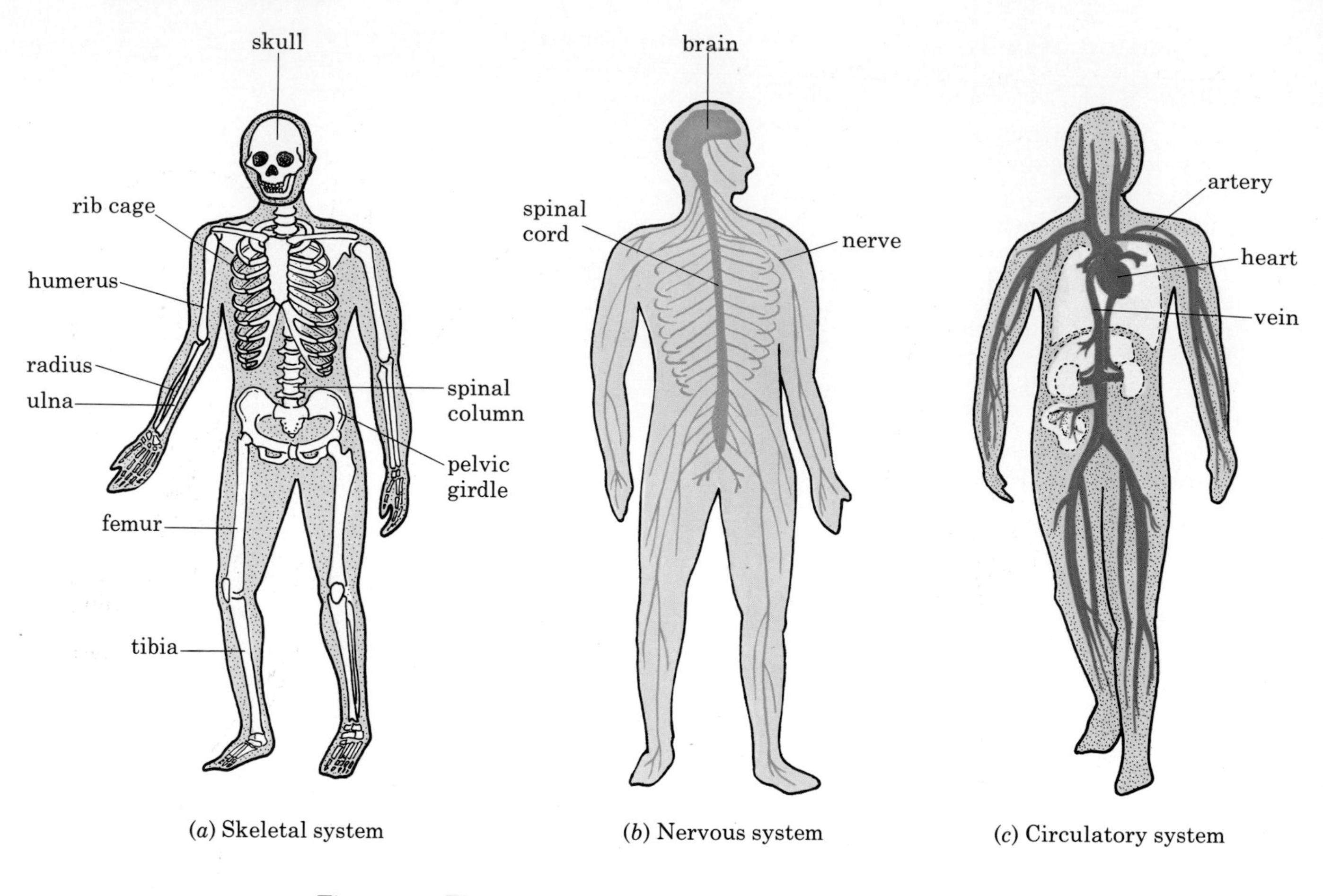

(*a*) Skeletal system (*b*) Nervous system (*c*) Circulatory system

Figure 15 Three examples of systems in the human body

Organs consist of several tissues grouped together to make a structure with a special function. For example, a muscle is an organ which contains long muscle cells held together with connective tissue and supplied with blood vessels and nerves. When a nerve impulse arrives in the muscle, it makes a large number of the muscle cells contract at the same time. To do this, the muscle cells need energy. This is provided by the food and oxygen brought by the blood vessels. The heart, lungs, stomach, intestines, brain and eyes are examples of organs in animals. In flowering plants, the root, stem and leaves are the organs. The tissues of the leaf are epidermis, palisade tissue, spongy tissue, xylem and phloem (see pages 25 and 59–61).

A system usually refers to a group of organs whose functions are closely related. For example, the heart and blood vessels make up the **circulatory system**; the brain, spinal cord and nerves make up the **nervous system** (Figure 15). In a flowering plant, the stem, leaves and buds make up a system called the shoot (page 58).

An organism is formed by the organs and systems working together to produce an independent plant or animal.

Figure 16 shows the relationship between cell, tissue, organ, system and organism.

QUESTIONS

7 Say whether you think the following are cells, tissues, organs or organisms: lungs (page 122), skin (page 133), root hair (page 68), mesophyll (page 60), multi-polar neurone (page 165).

8 How many complete cells are shown in Figure 19*a* on page 23?

9 What tissues are shown in the following drawings: Figure 9 on page 153; Figure 8 on page 167?

10 Look at Figure 5 on page 104. Which of the structures shown do you think are *organs*? What *system* is represented by the drawing? (The first heading on page 104 is a 'give-away'.)

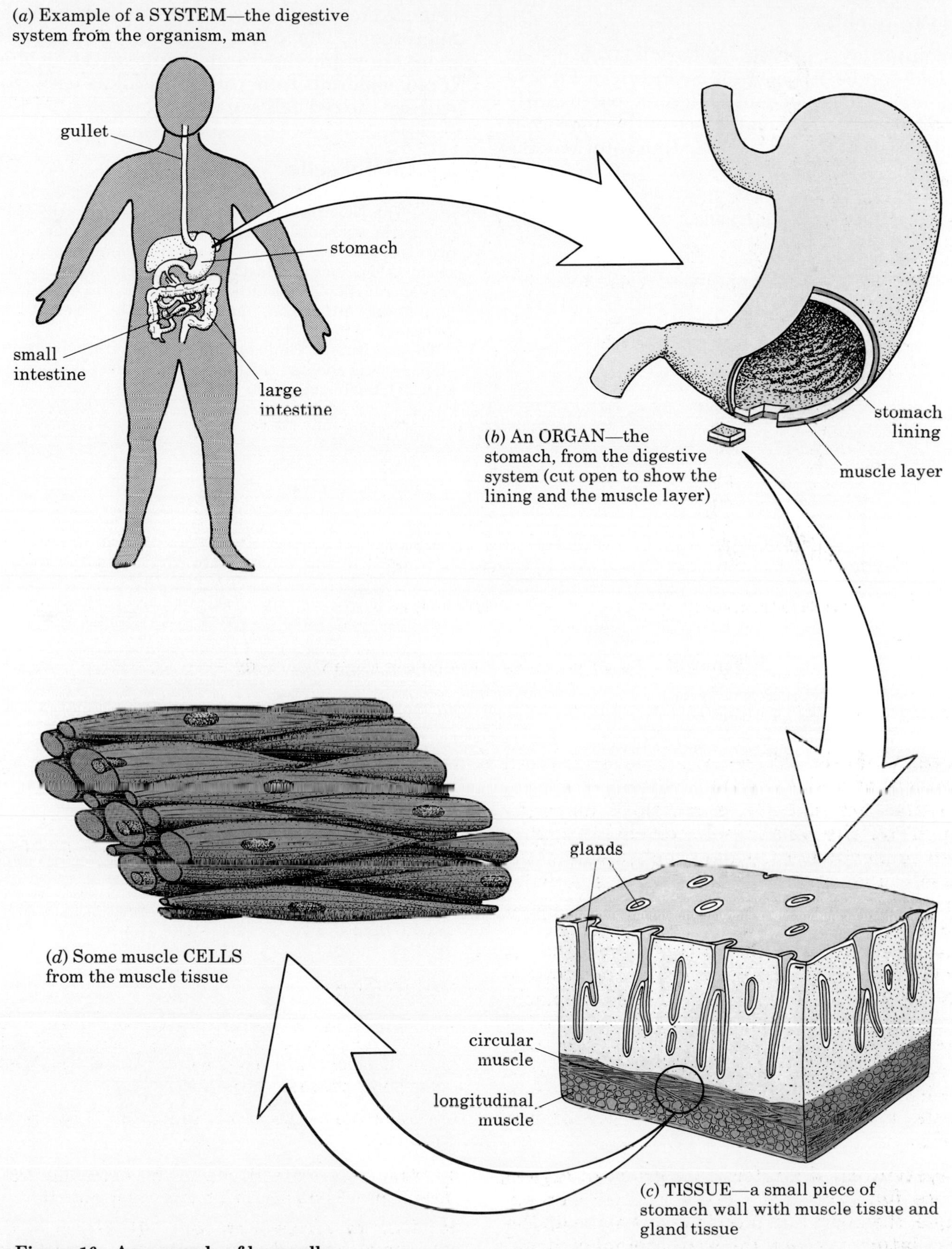

Figure 16 An example of how cells, tissues, organs and systems are related

PRACTICAL WORK

1 Plant cells

The outer layer of cells (epidermis) from a stem or leaf can be stripped off as shown in Figures 17*a* and *b*. A piece of onion is particularly suitable for this. A small piece of this epidermis is placed in a drop of weak iodine solution on a microscope slide and covered with a cover-slip (Figure 17*c*). The tissue is then studied under the microscope. The iodine will stain the cell nuclei pale yellow and the starch grains will stain blue. If red epidermis from rhubarb stalk is used, you will see the red cell sap in the vacuoles.

Figure 17 Studying plant cells

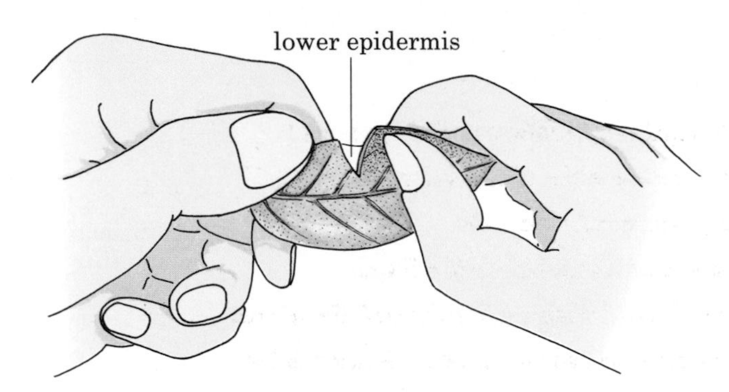

(*a*) Tear a leaf or piece of onion to obtain a small piece of epidermis, or . . .

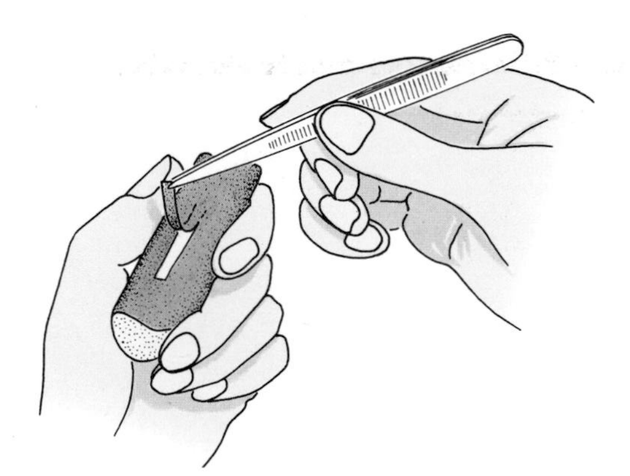

(*b*) Peel a strip of red epidermis from a piece of rhubarb stalk.

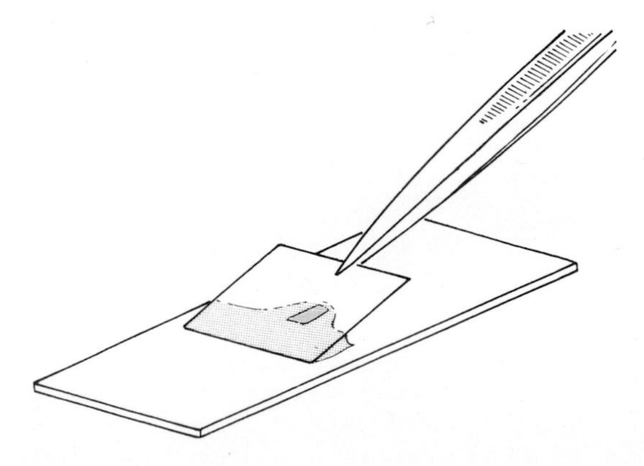

(*c*) Place the epidermis in a drop of weak iodine solution on a slide, and carefully lower a cover slip over it.

2 Animal cells

NOTE. The Department of Education and Science recommends that schools no longer use the technique which involves studying the epithelial cells which appear in a smear taken from the inside of the cheek, because of the very small risk of transmitting the AIDS virus. Some Local Education Authorities may forbid the use of this technique in their schools.

The Institute of Biology suggests that if the procedure on page 11 is adopted the risk is negligible. ('Biologist' **35** (4) p. 211, September 1988).

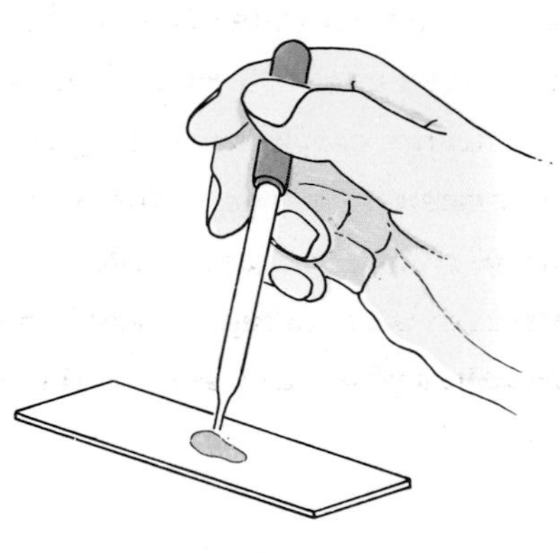

Figure 18 Add a drop of methylene blue

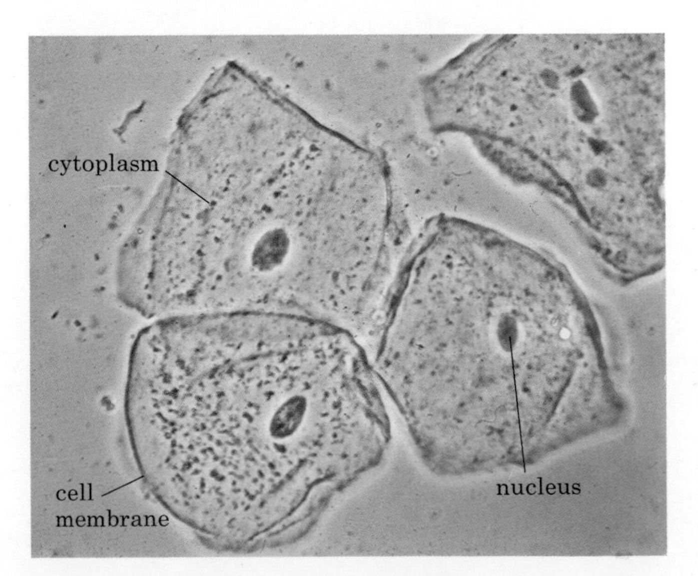

Figure 19 Human cells from the cheek lining (×1000). These cells are not stained but seen by a special kind of lighting.

Take a cotton bud from a freshly opened pack and rub it lightly on the inside of the cheek and gums. Then rub the bud on to a clean slide and drop the bud into a container of absolute alcohol. Cover the smear on the slide with a few drops of methylene blue solution (Figure 18) and then examine it under the microscope (Figure 19). Place used slides in laboratory disinfectant before washing them.

Various alternatives to cheek smears have been put forward, for example:

(a) Press some Sellotape on to a 'well-washed' wrist. When you remove the tape and study it under the microscope, you can see cells with nuclei. A few drops of methylene blue solution will stain the cells and make the nuclei more distinct (Figure 19).

(b) Press a microscope slide against the cornea of a fresh or refrigerated bullock's eye. Conjunctival cells stick to the slide and can be stained with methylene blue.

BASIC FACTS

- **Nearly all plants and animals are made up of thousands or millions of tiny cells.**
- **All cells contain cytoplasm and a nucleus, and are enclosed in a cell membrane.**
- **Plant cells also have a cellulose cell wall and a large central vacuole.**
- **Many chemical reactions take place in the cytoplasm to keep the cell alive.**
- **The nucleus directs the chemical reactions in the cell and also controls cell division.**
- **Cells are often specialized in their shapes and activities to carry out particular jobs.**
- **Large numbers of similar cells packed together form a tissue.**
- **Different tissues arranged together form organs.**
- **A group of related organs make up a system.**
- **An organism is an independent living creature with many different systems in its body.**
- **Growth takes place as a result of cell division and cell enlargement.**

2 Activities in cells

Cells need to make new substances, such as new cytoplasm or new cell wall material, in order to grow and divide. For this they must take in simple chemical substances through the cell membrane. The simple substances are then built up by a series of chemical reactions, into the more complicated substances that the cell needs. To make these chemical reactions work, the cytoplasm must provide both **energy** and **enzymes**. Enzymes are chemicals made in the cell. The energy is obtained by breaking down large food molecules into smaller molecules.

ENZYMES

Enzymes are proteins (page 260) made in the cells. Protein molecules are quite big; they have a definite shape and contain nitrogen atoms as well as atoms of carbon, hydrogen and oxygen.

An enzyme acts as a **catalyst**, that is, a chemical substance which speeds up a reaction but does not get used up in the reaction. One enzyme can be used many times over (see Figure 10 on page 30).

Figure 1*a* is a diagram of how an enzyme molecule might work to join two other molecules together and so form a more complicated substance.

An example of an enzyme-controlled reaction such as this is the joining up of two glucose molecules to form a molecule of maltose (see page 259.

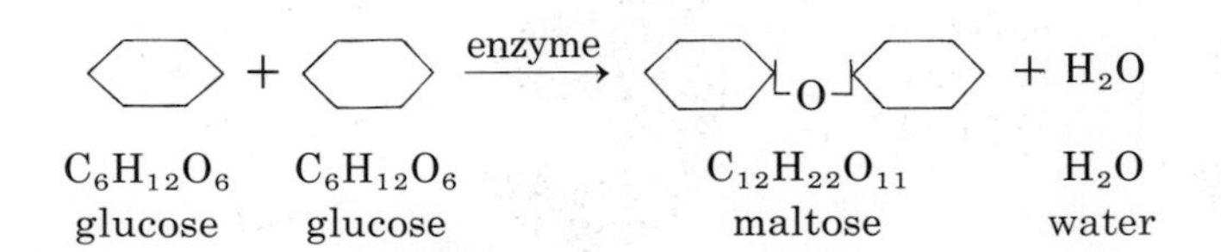

In a similar way, hundreds of glucose molecules might be joined together, end to end, to form a long molecule of starch to be stored in the plastid of a plant cell. The glucose molecules might also be built up into a molecule of cellulose to be added to the cell wall. Protein molecules are built up by enzymes which join together tens or hundreds of amino acid molecules (page 258). These proteins are added to the cell membrane, to the cytoplasm or to the nucleus of the cell. They may also be the proteins which act as enzymes.

After the new substance has been formed, the enzyme is set free to start another reaction. Molecules of the two substances might have combined without the enzyme being present but they would have done so very slowly. By bringing the substances close together, the enzyme molecule makes the reaction take place much more rapidly. A chemical reaction which would take hours or days to happen on its own takes only a few seconds when the right enzyme is present.

Figure 1*b* shows an enzyme speeding up a chemical change but this time it is a reaction in which the molecule of a substance is split into smaller molecules. If starch is mixed with water it will break down very slowly to sugar, taking several years. In your saliva there is an enzyme called **amylase** which can break down starch to sugar in minutes or seconds. Inside a cell, many of the 'breaking-down' enzymes are helping to break down glucose to carbon dioxide and water in order to produce energy (see below).

Enzymes and temperature

Most chemical reactions are speeded up by an increase in temperature. In many cases a rise of 10°C will double the rate of reaction in a cell. This is equally true for enzyme-controlled reactions, but above 45°C the enzymes break down. Enzymes are proteins, and heat alters the shape of protein molecules. Egg-white is a protein. When it is heated, its molecules change shape and the egg-white goes from a clear, runny liquid to a white solid and cannot be changed back again. Figure 1 shows how the shape of an enzyme molecule could be very important if it has to fit the substances on which it acts. Above 45°C the shapes of enzyme molecules are altered and they no longer work. The enzymes are said to have been **denatured** by heating them. This is one reason why temperatures above 45°C kill off most organisms. One way to test whether a substance is an enzyme is to heat it to boiling point. If it can still carry out its reactions after this, it can't be an enzyme.

Enzymes and pH

Acid or alkaline conditions alter the chemical properties of proteins, including enzymes. Most enzymes work best at a particular level of acidity or alkalinity (pH). The protein-digesting enzyme in your stomach, for example, works well at an acidity of pH 2. At this pH, the enzyme, amylase, from your saliva cannot work at all. Inside the cells, most enzymes will work best in neutral conditions (pH 7.0).

Enzymes are specific

This means simply that an enzyme which normally acts on one substance will not act on a different one. Figure 1*a* shows how the shape of an enzyme could decide what substances it combines with. The enzyme in Figure 1*a* has a shape which exactly fits the substances on which it acts, but would not fit the substance in *b*. Thus, an enzyme which joins amino acids up to make proteins will not also join up glucose molecules to make starch. If a reaction takes place in stages, e.g.

$$A + B \longrightarrow AB \quad \text{(stage 1)}$$

$$AB + C \longrightarrow ABC \quad \text{(stage 2)}$$

a different enzyme may be needed for each stage.

The names of enzymes usually end with **-ase** and they are often named according to the substance on which they act. For example, an enzyme which acts on proteins may be called a **protease**.

Intra- and extra-cellular enzymes

All enzymes are made inside cells. Most of them remain inside the cell to speed up reactions in the cytoplasm and nucleus. These are called **intra-cellular enzymes** (*intra-* means 'inside'). In a few cases, the enzymes made in the cells are let out of the cell to do their work outside. These are **extra-cellular enzymes** (*extra-* means 'outside'). Fungi (page 198) and bacteria (page 197) release extra-cellular enzymes in order to digest their food. A mould growing on a piece of bread releases starch-digesting enzymes into the bread and absorbs the soluble sugars which the enzyme produces from the bread. In the digestive systems of animals (page 102), extra-cellular enzymes are released into the stomach and intestines in order to digest the food.

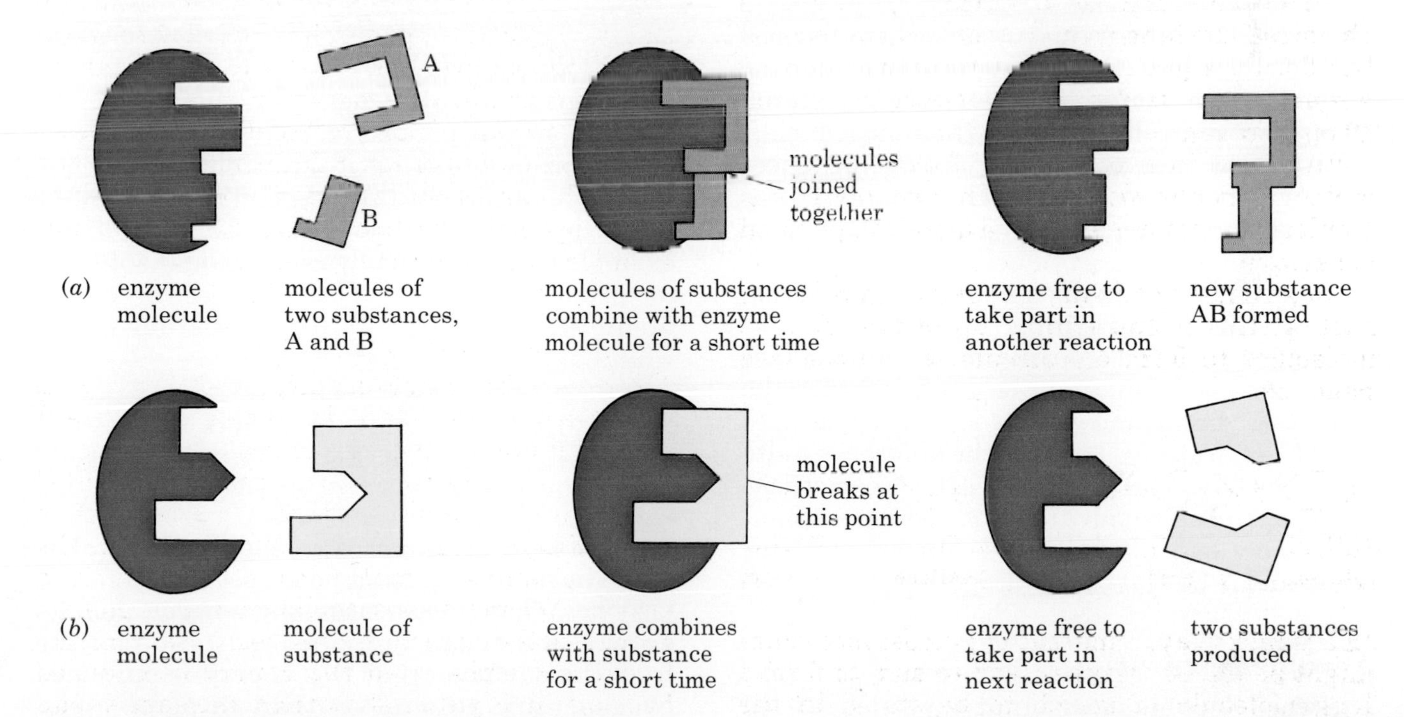

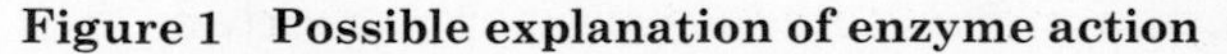
Figure 1 Possible explanation of enzyme action

QUESTIONS

1 What substances must a cell take in through the cell membrane in order to make (a) starch, (b) protein?

2 There are cells in your salivary glands which make an extra-cellular enzyme, amylase. Would you expect these cells to make intra-cellular enzymes as well? Explain your answer.

3 Apple cells contain an enzyme which turns the tissues brown when an apple is peeled and left for a time. Boiled apple does not go brown (Figure 2). Explain why the boiled apple behaves differently.

Figure 2 Enzyme activity in apple. Slice A has been freshly cut. B and C were cut two days earlier but C was dipped in boiling water for one minute.

PRACTICAL WORK

For experiments with enzymes see page 109.

ENERGY FROM RESPIRATION

Most of the processes taking place in cells need energy to make them happen. Building up proteins from amino acids or making starch from glucose needs energy. When muscle cells contract or nerve cells conduct electrical impulses, they use energy. This energy comes from the food which cells take in. The food mainly used for energy in cells is glucose.

The process by which energy is produced from food is called **respiration**.

Respiration is a chemical process which takes place in cells. It must not be confused with the process of breathing which is also sometimes called 'respiration'. To make the difference quite clear, the chemical process in cells is sometimes called **cellular respiration** or **tissue respiration**. The use of the word 'respiration' for breathing is best avoided altogether.

Aerobic respiration

The word **aerobic** means that oxygen is needed for this chemical reaction. The food molecules are combined with oxygen. The process is called **oxidation** and the food is said to be **oxidized**. Since all food molecules contain carbon, hydrogen and oxygen atoms, the process of oxidation converts the food to carbon dioxide (CO_2) and water (H_2O) and, at the same time, sets free energy which the cell can use to drive other reactions.

Aerobic respiration can be summed up by the equation

$$\underset{\text{glucose}}{C_6H_{12}O_6} + \underset{\text{oxygen}}{6O_2} \xrightarrow{\text{enzymes}} \underset{\text{carbon dioxide}}{6CO_2} + \underset{\text{water}}{6H_2O} + \underset{\text{energy}}{2830\text{ kJ}}$$

The 2830 kilojoules is the amount of energy you would get by completely oxidizing 180 grams of glucose to carbon dioxide and water. In the cell, the energy is not released all at once. The oxidation takes place in a series of small steps and not in one jump as the equation suggests. Each small step needs its own enzyme and at each stage a little energy is released (Figure 3).

Although the energy is used for the processes mentioned above, some of it always appears as heat. In 'warm-blooded' animals some of this heat is retained to keep the body temperature above that of the surroundings. In 'cold-blooded' animals the heat may build up for a time and allow the animal to move about faster. In plants the heat is lost to the surroundings as fast as it is produced.

Anaerobic respiration

The word **anaerobic** means 'in the absence of oxygen'. In this process, energy is still released from food by breaking it down chemically but the reactions do not use oxygen though they do often produce carbon dioxide. A common example is the action of yeast on sugar solution to produce alcohol. This process is called **fermentation**. The following equation shows what happens:

$$\underset{\text{glucose}}{C_6H_{12}O_6} \xrightarrow{\text{enzymes}} \underset{\text{alcohol}}{2C_2H_5OH} + \underset{\text{carbon dioxide}}{2CO_2} + \underset{\text{energy}}{118\text{ kJ}}$$

As with aerobic respiration, the reaction takes place in small steps and needs several different enzymes. The yeast uses the energy for its growth and living activities, but you can see from the equation that less energy is produced by anaerobic respiration than in aerobic respiration. This is because the alcohol still

Figure 3 Aerobic respiration

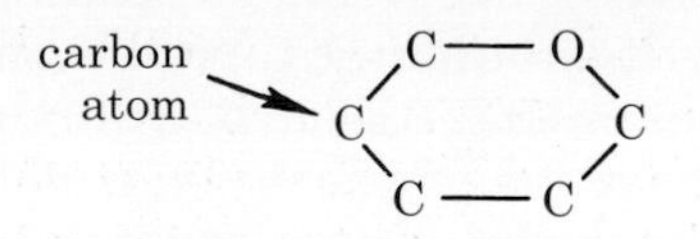

(*a*) Molecule of glucose (H and O atoms not all shown)

(*b*) The enzyme attacks and breaks the glucose molecule into two 3-carbon molecules

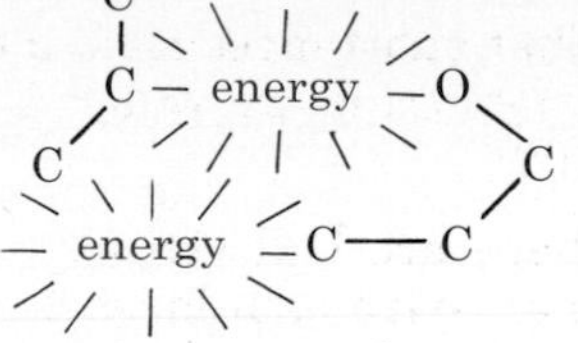

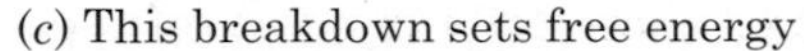

(*c*) This breakdown sets free energy

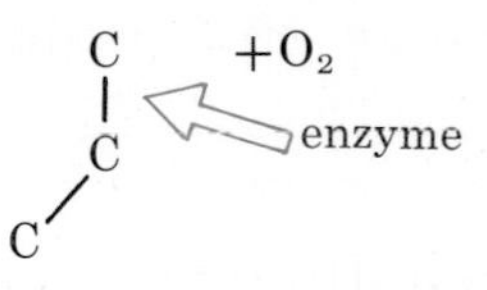

(*d*) Each 3-carbon molecule is broken down to carbon dioxide

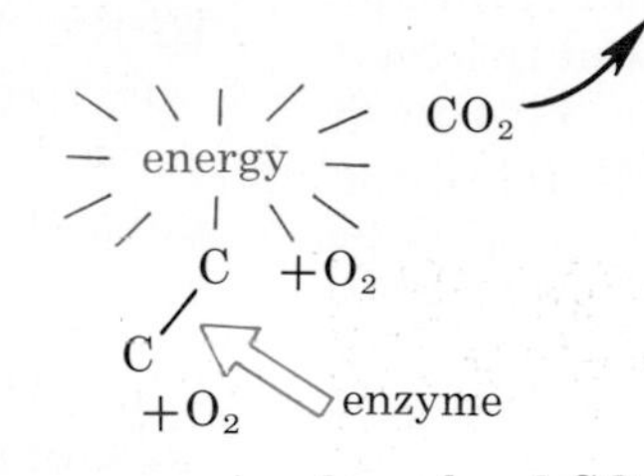

(*e*) More energy is released and CO_2 is produced

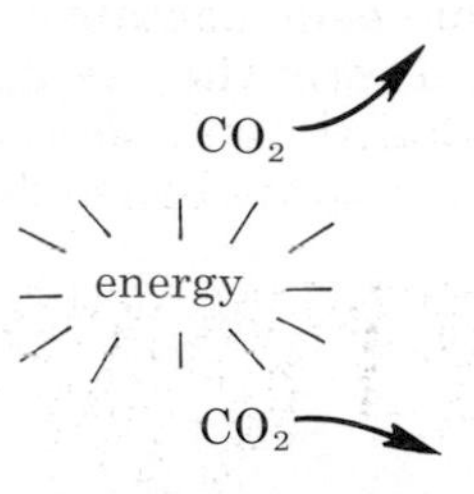

(*f*) The glucose has been completely oxidized to carbon dioxide (and water), and all the energy released

contains a great deal of energy which the yeast is unable to use.

In animals, the first stages of respiration in contracting muscle cells are anaerobic and produce **pyruvic acid** (the equivalent of the yeast's alcohol). Only later on is the pyruvic acid completely oxidized to carbon dioxide and water.

$$\text{glucose} \xrightarrow{\text{enzymes}} \text{pyruvic acid} \xrightarrow[\text{}]{\text{enzymes and oxygen}} CO_2 + H_2O$$

ANAEROBIC STAGE AEROBIC STAGE

QUESTIONS

4 (a) In which parts of a living organism does respiration take place?

(b) If, in one word, you had to say what respiration was about, which word would you choose from this list: breathing, energy, oxygen, cells, food?

5 What chemical substances (a) from outside the cell, (b) from inside the cell, must be provided for aerobic respiration to take place? (c) What are the products of aerobic respiration?

6 Victims of drowning who have stopped breathing are sometimes revived by a process called 'artificial respiration'. Why would a biologist object to the use of this expression? ('Resuscitation' is a better word to use.)

EXPERIMENTS ON RESPIRATION

If you look below at the chemical equation which represents aerobic respiration you will see that a tissue or an organism which is respiring should be (*a*) using up food, (*b*) using up oxygen, (*c*) giving off carbon dioxide, (*d*) giving off water and (*e*) producing energy:

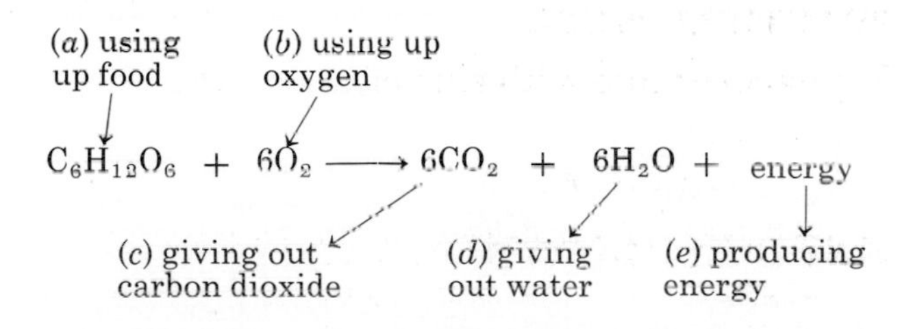

If we wish to test whether aerobic respiration is taking place:

(*d*) Giving off water vapour is not a good test because non-living material will give off water vapour if it is wet to start with.

(*a*) Using up food can be tested by seeing if an organism loses weight. This is not so easy as it seems because most organisms lose weight as a result of evaporation of water and this may have nothing to do with respiration.

(*b*), (*c*) and (*e*) are fairly easy to demonstrate either with whole organisms or pieces of living tissue, and three simple experiments will now be described.

Seeds are used as the living organisms because when they start to grow (germinate) there is a high level of chemical activity in the cells. The seeds are easy to obtain and to handle and they fit into small-scale apparatus. Blowfly maggots can be used as animal material.

(c) Giving off carbon dioxide in respiration

Put some germinating wheat grains in a large test-tube. Cover the mouth of the tube with aluminium foil. After 15–20 minutes take a sample of the air from the test-tube. Do this by pushing a glass tube attached to a 10 cm^3 plastic syringe through the foil and into the test-tube (Figure 4*a*). Withdraw the syringe plunger enough to fill the syringe with air from the test-tube. Now slowly bubble this air sample through a little clear lime water in a small test-tube (Figure 4*b*). Cover the mouth of the small test-tube and shake the lime water up.

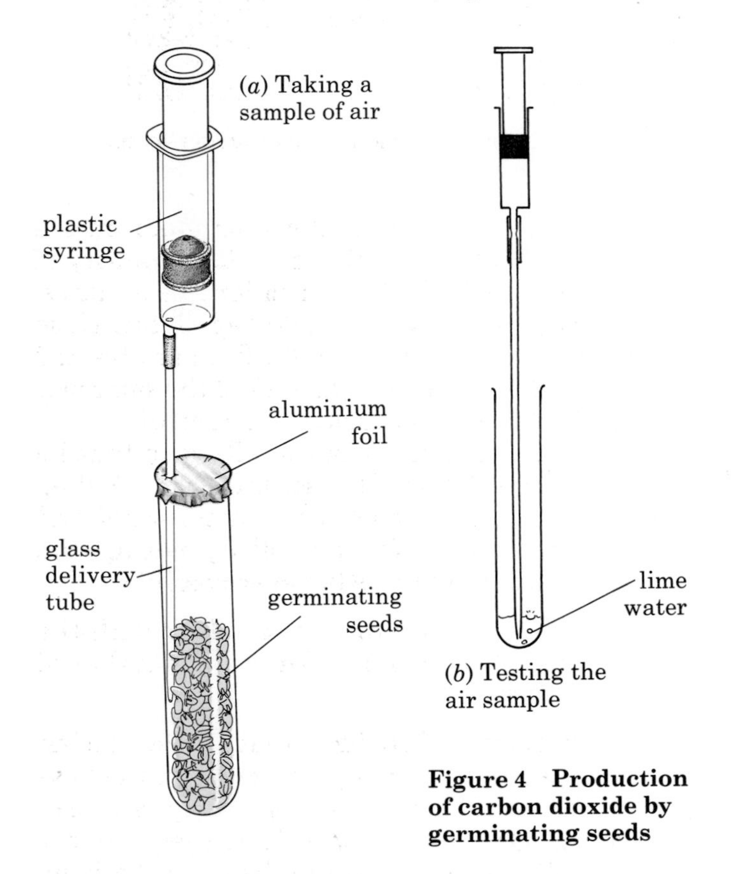

Figure 4 Production of carbon dioxide by germinating seeds

Result. The lime water will go milky.

Interpretation. Lime water turning milky is evidence of carbon dioxide but it could be argued that the carbon dioxide came from the air or that the seeds give off carbon dioxide whether or not they are respiring. The only way to disprove these arguments is to do a **control experiment** (see page 17).

Control. Boil some of the germinating wheat grains before starting the experiment. When you set up the experiment, put an equal amount of boiled wheat grains in a large test-tube and cover the mouth of the tube with aluminium foil exactly as you did for the living seeds. When you test the air from the living seeds, also test the air from the dead seeds. It should not turn the lime water milky. This means that the carbon dioxide did not come from the air, nor was it given off by the dead seeds. It must be a *living* process in the seeds which produces carbon dioxide and this process is likely to be respiration. Since you stopped *all* living processes by boiling the seeds in the control experiment, you have not been able to prove that it was respiration rather than some other chemical change which produced the carbon dioxide.

(b) Using up oxygen during respiration

The apparatus in Figure 5 is a **respirometer**, which can measure the rate of respiration (it is a 'respire meter') by seeing how quickly oxygen is taken up. Germinating seeds are placed in the test-tube and, as they use up the oxygen for respiration, the level of liquid in the delivery tubing will go up.

There are two snags to this. One is that the organisms usually give out as much carbon dioxide as they take in oxygen. So there may be no change in the total amount of air in the test-tube and the liquid level will not move. This snag is overcome by placing **soda-lime** in the test-tube. Soda-lime will absorb carbon dioxide as fast as the organisms give it out. So only the uptake of oxygen will affect the amount of air in the tube. The second snag is that quite small changes in temperature will make the air in the test-tube expand or contract and so cause the liquid to rise or fall whether or not respiration is taking place. To overcome this, the test-tube is

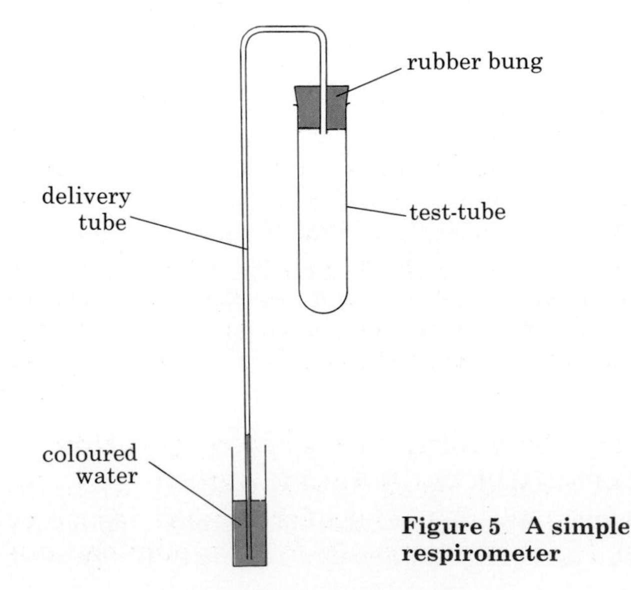

Figure 5 A simple respirometer

kept in a beaker of water (a water bath). The temperature of water changes far more slowly than that of air, so there will not be much change during a 30-minute experiment.

Control. To show that it is a *living* process which uses up oxygen a similar respirometer is prepared but containing an equal quantity of germinating seeds which have been killed by boiling. The apparatus is finally set up as shown in Figure 6 and left for 30 minutes.

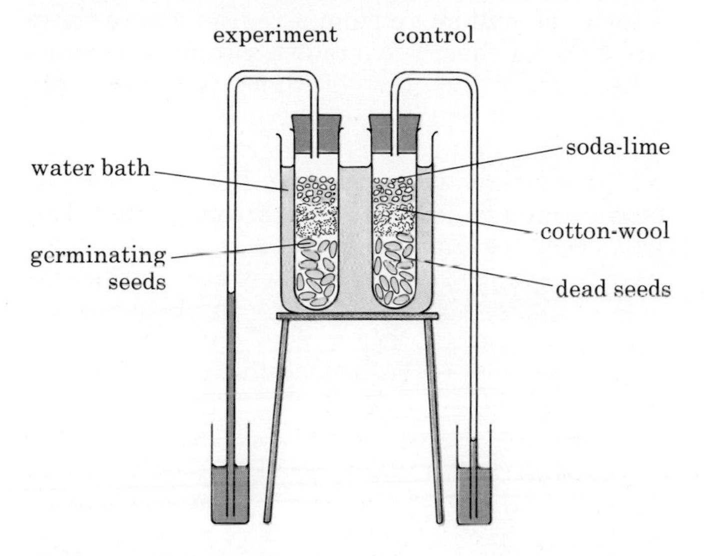

Figure 6 Uptake of oxygen by germinating seeds

Result. The level of liquid in the experiment goes up more than in the control. The level in the control may not move at all.

Interpretation. The rise of liquid in the delivery tubing shows that the living seedlings have taken up part of the air. It does not prove that it is *oxygen* which has been taken up. Oxygen seems the most likely gas, however, because (a) there is only 0.03 per cent carbon dioxide in the air to start with and (b) the other gas, nitrogen, is known to be less active than oxygen.

If the experiment is allowed to run for a long time, the uptake of oxygen could be checked at the end by placing a lighted splint in each test-tube in turn. If some of the oxygen has been removed by the living seedlings, the flame should go out more quickly that it does in the tube with dead seedlings.

(e) Producing energy in respiration

Fill a small vacuum flask with wheat grains which have been soaked for 24 hours and rinsed in 1 per cent formalin for five minutes (this

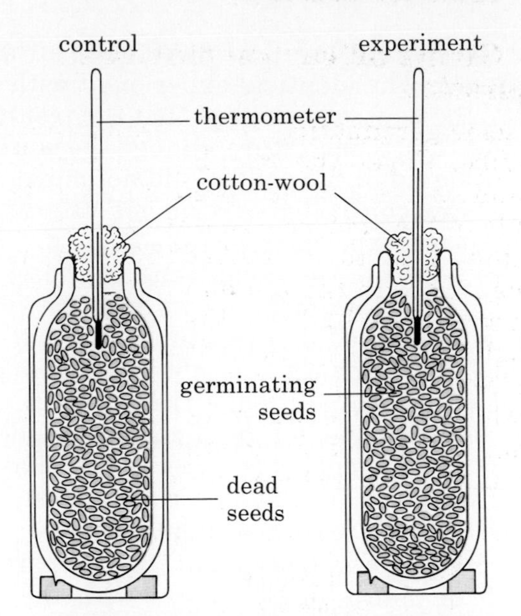

Figure 7 Energy release in germinating seeds

solution will kill any bacteria or fungi on the surface of the grains). Kill an equal quantity of soaked grains by boiling them for five minutes. Cool the boiled seeds in cold tap water, rinse them in 1 per cent formalin for five minutes and then put them in a vacuum flask of the same size as the first one. This flask is the control.

Place a thermometer in each flask so that its bulb is in the middle of the seeds (Figure 7). Plug the mouth of each flask with cotton-wool and leave both flasks for two days, noting the thermometer readings whenever possible.

Result. The temperature in the flask with the living seeds will be 5–10 °C higher than that of the dead seeds.

Interpretation. Provided that there are no signs of the living seeds going mouldy, the heat produced must have come from *living processes* in the seeds, because the dead seeds in the control did not give out any heat. There is no evidence that this process is respiration rather than any other chemical change but the result is what you would expect *if* respiration does produce energy.

CONTROLLED EXPERIMENTS

In most biological experiments, a second experiment called a **control** is set up. This is to make sure that the results of the first experiment are due to the conditions being studied and not to some other cause which has been overlooked.

In the experiment in Figure 6, the liquid rising up the tube could have been the result of the test-tube cooling down, so making the air inside it contract. The identical experiment with dead seeds—the control—showed that the result was not due to a temperature change, because the level of liquid in the control did not move.

The term 'controlled experiment' refers to the fact that the experimenter (a) sets up a control and (b) controls the conditions in the experiment. In the experiment shown in Figure 6, the seeds are enclosed in a test-tube, and soda-lime is added. This makes sure that any uptake or output of oxygen will make the liquid go up or down, and that the output of carbon dioxide will not affect the results. The experimenter has controlled both the amount and the composition of the air available to the germinating seeds.

If you did an experiment to compare the growth of plants in the house with the same plants growing in a greenhouse, you could not be sure whether it was the extra light or the high temperature of the greenhouse which caused better growth. This would not, therefore, be a properly controlled experiment. You must alter only one condition at a time, either the light or the temperature, and then you can compare the results with the control experiment.

A properly controlled experiment, therefore, alters only one condition at a time and includes a control which shows that it is this condition and nothing else which gave the result.

QUESTIONS

7 In an experiment like Figure 6 on page 17, the growing seeds took in 5 cm^3 oxygen and gave out 7 cm^3 carbon dioxide. What change in volume will take place (a) if no soda-lime is present, (b) if soda-lime is present?

8 The germinating seeds in Figure 7 on page 17 will release the same amount of heat whether they are in a beaker or a vacuum flask. Why then is it necessary to use a vacuum flask for this experiment?

9 What is the purpose of the control in the experiment to show carbon dioxide production (Figure 4 on page 16)?

HOW SUBSTANCES GET IN AND OUT OF CELLS

Diffusion

The molecules of a gas like oxygen are moving about all the time. So are the molecules of a liquid, or a substance like sugar dissolved in a liquid. As a result of this movement, the

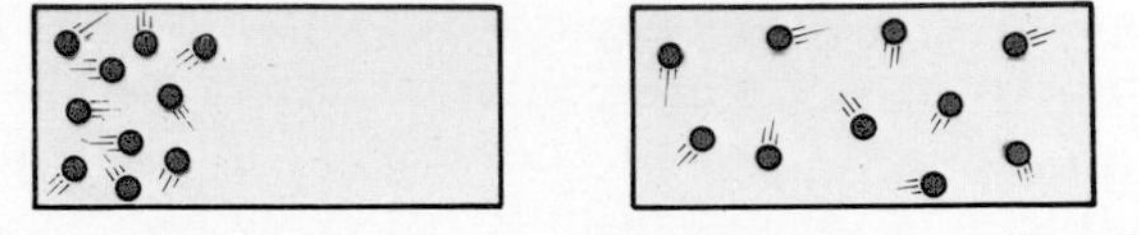

Figure 8 Diffusion

molecules spread themselves out evenly to fill all the available space (Figure 8). This process is called **diffusion**. One effect of diffusion is that the molecules of a gas, a liquid or a dissolved substance will move from a region where there are a lot of them (i.e. concentrated) to regions where there are few of them (i.e. less concentrated) until the concentration everywhere is the same. Figure 9*a* is a diagram of a cell with a high concentration of molecules (e.g. oxygen) outside and a low concentration inside. The

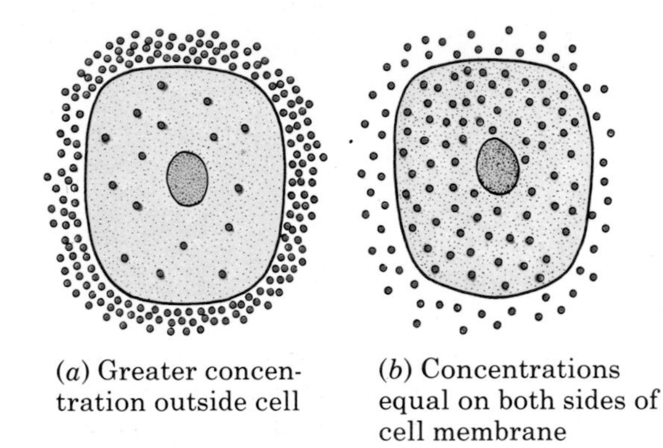

Figure 9 Molecules entering a cell by diffusion

effect of this difference in concentration is to make the molecules diffuse into the cell until the concentration inside and outside is the same (Figure 9*b*).

Whether this will happen or not depends on whether the cell membrane will let the molecules through. Small molecules such as water (H_2O), carbon dioxide (CO_2) and oxygen (O_2) can pass through the cell membrane fairly easily. So diffusion tends to equalize the concentration of these molecules inside and outside the cell all the time.

When a cell uses up oxygen for its aerobic respiration, the concentration of oxygen inside the cell falls and so oxygen molecules diffuse into the cell until the concentration is raised again. During tissue respiration, carbon dioxide is produced and so its concentration inside the cell goes up. Once again diffusion takes place, but this time the molecules move out of the cell. Thus, diffusion can explain how a cell takes in its oxygen and gets rid of its carbon dioxide.

Active transport

If diffusion were the only method by which a cell could take in substances, it would have no control over what went in or out. Anything that was more concentrated outside would diffuse into the cell whether it was harmful or not. Substances which the cell needed would diffuse out as soon as their concentration inside rose above that outside the cell. In fact, the cell membrane has almost complete control over the substances which enter and leave the cell. Many substances which the cell needs, such as glucose, are probably taken up by special chemical processes. These processes are not properly understood and may be quite different for different substances, but they are all described as **active transport**. Two facts are quite clear; for active transport of substances across the cell membrane, the cell needs to provide energy from respiration and, of course, enzymes.

QUESTIONS

10 Look at Figure 10. Try to explain why more water molecules diffuse from left to right than from right to left.

11 The text says that most protein molecules are kept inside the cell membrane. Name one example of a protein that has to get out of certain types of gland cell (see page 12 and Figure 14*d* on page 7).

Osmosis

Osmosis is the special name used to describe the diffusion of water across a membrane, from a weak solution to a stronger solution. In biology this usually means the diffusion of water into or out of cells. Osmosis is just one special kind of diffusion because it is only water molecules and their movement we are considering.

Figure 8 showed that molecules will diffuse from a region where there are a lot of them to a region where they are fewer in number; that is, from a region of highly concentrated molecules to a region of lower concentration.

Pure water has the highest possible concentration of water molecules; it is 100 per cent water molecules (Figure 10*a*). However, when a substance such as sugar dissolves in water, the sugar molecules attract some of the water molecules and stop them moving freely. So, in effect, the concentration of free water molecules is reduced by the presence of a dissolved substance. In Figure 10*b* the sugar molecules have 'captured' half of the water molecules and so the water concentration is reduced by 50%.

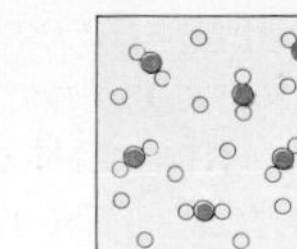
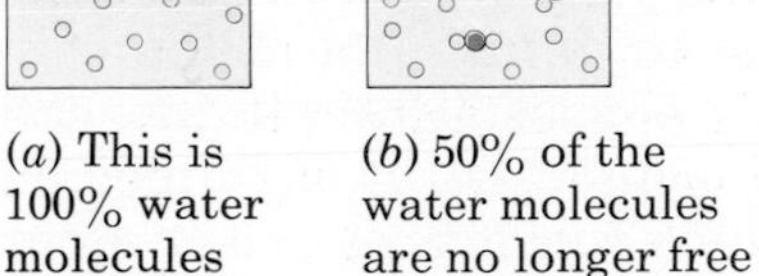

(*a*) This is 100% water molecules

(*b*) 50% of the water molecules are no longer free

(*c*)

Figure 10

If the pure water and the sugar solution are placed in contact (Figure 10*c*), the sugar molecules will diffuse to the left where there are fewer sugar molecules. (There is plenty of space between the water molecules for the sugar molecules to move in.) Water molecules will diffuse to the right because there are fewer free water molecules there (100 per cent on the left; only 50 per cent on the right).

So water will always diffuse from a weak solution to a strong solution because there are more free water molecules in the weak solution than there are in the strong solution.

Animal cells

In Figure 11 an animal cell is shown very simply. The red circles represent molecules dissolved in water in the cytoplasm. They may be sugar, salt or protein molecules. The blue circles represent the water molecules.

The cell is shown surrounded by pure water. Nothing is dissolved in the water; it has 100 per cent concentration of water molecules. So the concentration of water molecules outside the cell is greater than the concentration of free water molecules inside the cell. The water outside the

Figure 11 Osmosis in an animal cell

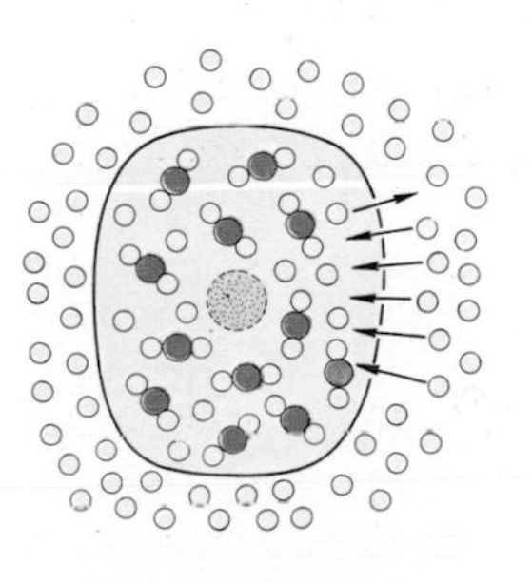

(*a*) There is a higher concentration of free water molecules outside the cell than inside, so water diffuses into the cell.

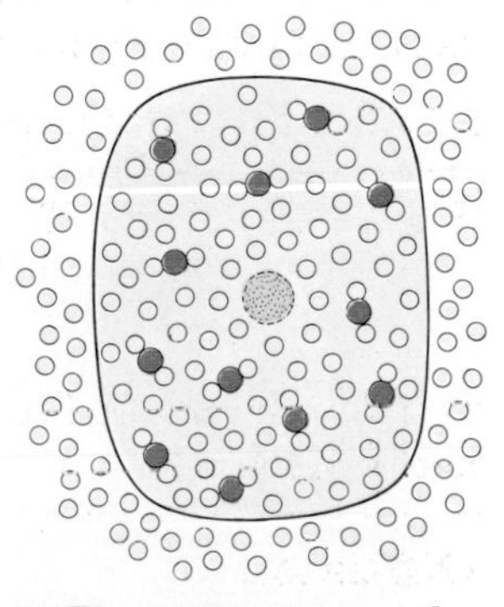

(*b*) The extra water makes the cell swell up. (Note that molecules are really far too small to be seen at this magnification.)

cell will diffuse into the cell. In a similar way, the dissolved substances inside the cell will tend to move out to a place of lower concentration.

For the water to diffuse into the cell or the dissolved substances to diffuse out of the cell, they must be able to pass through the cell membrane. The cell membrane, as explained on page 19, controls which substances can enter or leave the cell.

The membrane allows water to go through either way. It is **freely permeable** to water. So in our example, water can move freely in or out of the cell.

The cell membrane is **partially permeable** to most of the substances dissolved in the cytoplasm. So although the concentration of substances inside is high, the partially permeable membrane will not allow them to diffuse out.

The water molecules move into and out of the cell, but because there are more of them on the outside, they will move in faster than they move out. The water outside the cell does not have to be 100 per cent pure water. As long as the concentration of water outside is higher than that inside, water will diffuse in.

Water entering the cell will make it swell up, and unless the extra water is forced out in some way the cell will burst.

For this reason, it is very important that the cells in your body are surrounded by a liquid which has the same concentration as the liquid inside the cells. The outside liquid is called tissue fluid (see page 116) and its concentration depends on the concentration of the blood. The blood's concentration is controlled by the brain and kidneys, as described on page 131. The outer layer of your skin is not very permeable to water so you are in no danger of swelling up and exploding in your bath. In any case, the kidneys would soon remove any water you absorbed.

Plant cells

Both the cytoplasm and the cell sap in the vacuole of a plant cell contain salts and sugars. If a plant cell is placed in pure water, the water will diffuse through the cell wall and cell membrane into the vacuole, which will tend to swell up as a result. The cell wall of a mature plant cell cannot be stretched, so the extra water taken in can only increase the pressure of the vacuole, which pushes out against the cell wall (Figure 12).

This has a similar effect to inflating a soft bicycle tyre. The tyre represents the firm cell wall, the floppy inner tube is like the cytoplasm and the air inside corresponds to the vacuole. If enough air is pumped in, it pushes the inner tube against the tyre and makes the tyre hard. A plant cell with the vacuole pushing out on the cell wall is said to be **turgid** and the vacuole is exerting **turgor pressure** on the cell wall.

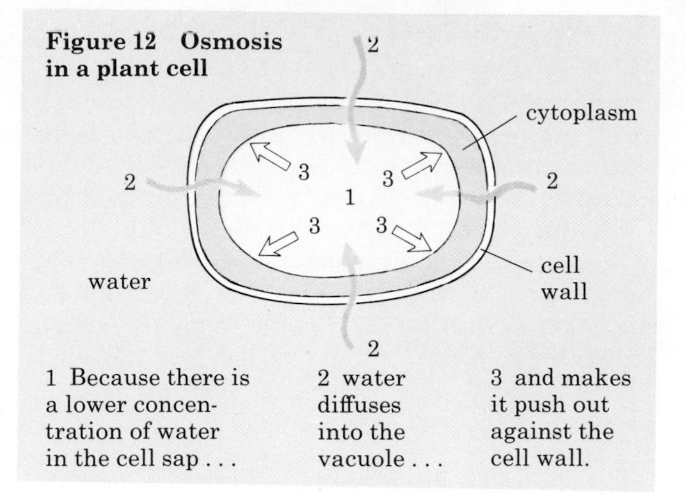

Figure 12 Osmosis in a plant cell

If all the cells in a leaf and stem are turgid, the stem will be firm and upright and the leaves held out straight. If the vacuoles lose water for any reason, the cells will lose their turgor and become **flaccid**. A leaf with flaccid cells will be limp and the stem will droop. A plant which loses water to this extent is said to be **wilting** (Figure 13).

Figure 13 Shoot of Busy Lizzie (left) wilting, and (right) recovered and turgid.

QUESTIONS

12 If an animal cell is surrounded by a salt solution with a concentration equal to that of the solution inside the cell, would you expect water to diffuse (a) in, (b) out, (c) both ways equally, or (d) in neither direction?

13 Describe and explain what you would expect to happen to an animal cell surrounded by a solution more concentrated than the solution inside the cell.

EXPERIMENTS ON DIFFUSION AND OSMOSIS

1 Diffusion of gases

Small squares of red litmus paper are moistened with water and pushed into a wide glass tube by means of a glass rod or piece of wire, so that they are equally spaced out. The squares will stick to the inside of the glass because they are wet. Both ends of the tube are closed with corks but one of the corks contains a pad of cotton-wool soaked in ammonia solution (Figure 14). Ammonia gas diffuses down the tube from the cotton-wool. As the gas reaches each square of litmus paper, the litmus turns blue (ammonia is an alkali). The time needed for each square to go completely blue is noted. The experiment is repeated with a fresh tube but using a much stronger solution of ammonia. The rate of diffusion will be faster than before because the ammonia molecules at the cotton-wool end are more concentrated.

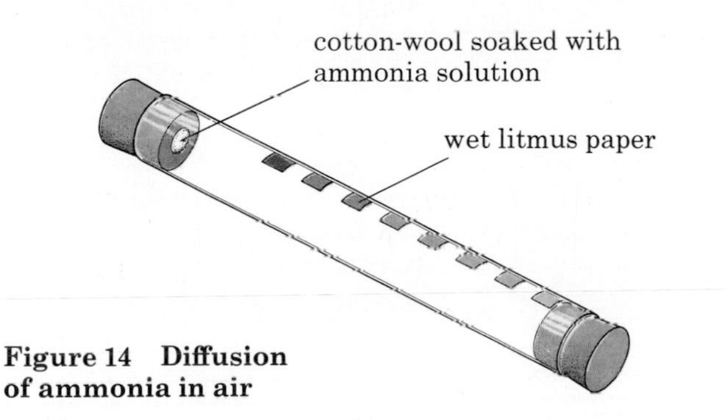

Figure 14 Diffusion of ammonia in air

2 Osmosis and turgor

The next two experiments use narrow tubing made of cellulose which is partially permeable. It allows water to diffuse freely but slows down the diffusion of sugar molecules. It is used in kidney machines because it lets molecules of harmful waste products out of the blood but retains the cells and proteins.

A 20 cm length of cellulose tubing is soaked in water and knotted securely at one end, then 3 cm^3 of a strong sugar solution is placed in the tubing using a graduated pipette (Figure 15*a*), and the open end of the tube is also knotted (Figure 15*b*). The filled tube should be quite floppy (Figure 15*c*). The cellulose tubing is now placed in a test-tube of water for 30–45 minutes (Figure 15*d*). Water diffuses into the sugar solution faster than it diffuses out so that the water pressure builds up inside the cellulose tubing and makes it turgid. This is similar to what happens when a plant cell becomes turgid.

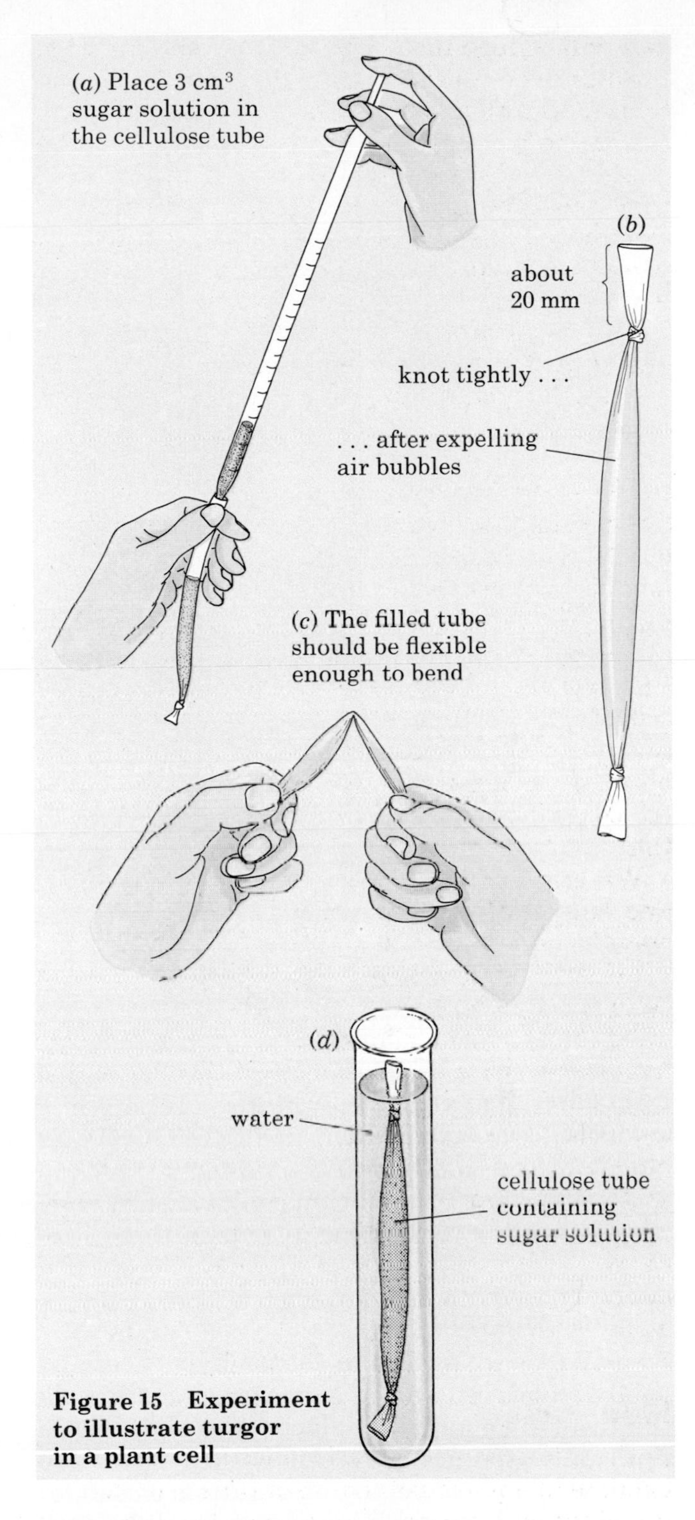

Figure 15 Experiment to illustrate turgor in a plant cell

3 Osmosis and water flow

A piece of cellulose tubing is soaked, knotted at one end and filled with sugar solution as in the previous experiment but it is then fitted over the end of a piece of glass capillary tubing and held in place with an elastic band. The capillary tube is pushed into the cellulose tubing until the solution enters the capillary. Then the capillary

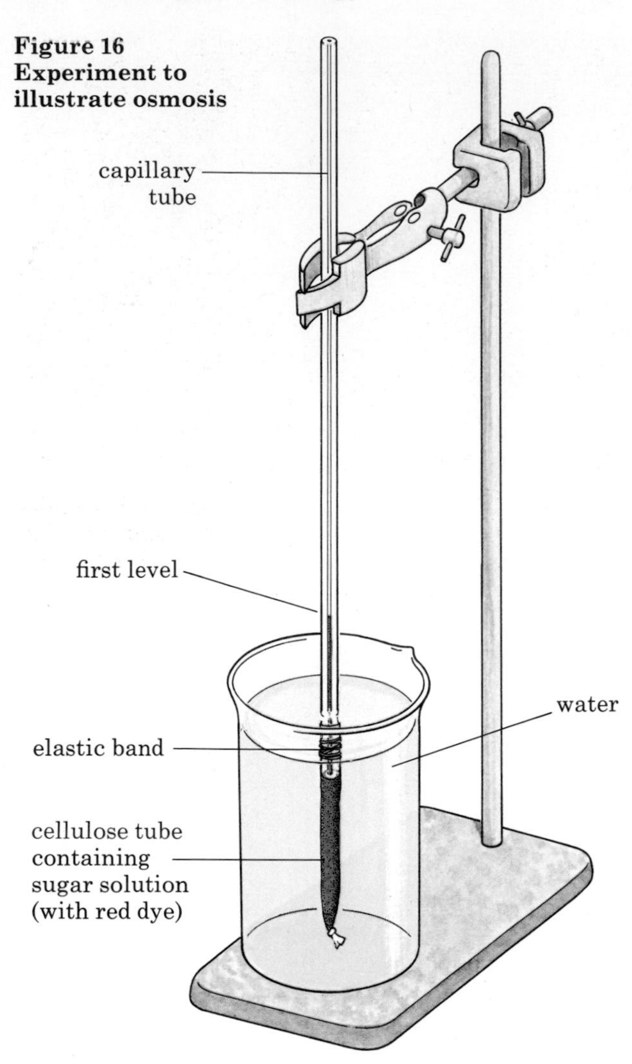

Figure 16 Experiment to illustrate osmosis

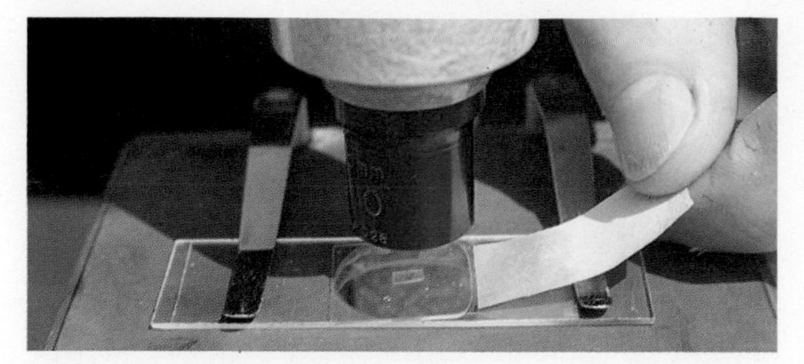

Figure 17 Changing the water for salt solution.

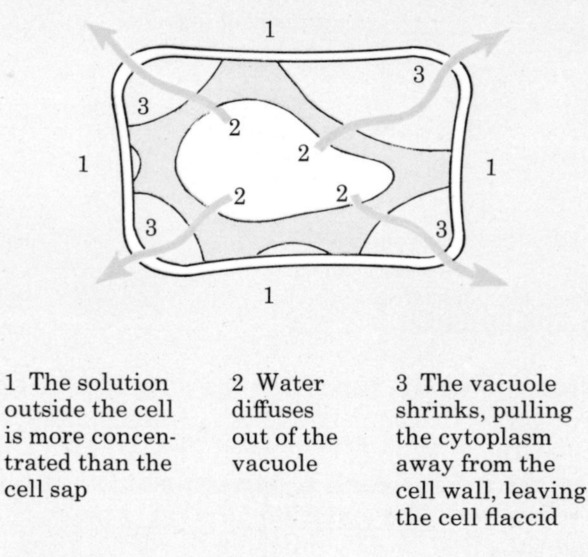

Figure 18 Plasmolysis

tube is clamped, with the cellulose tubing in a beaker of water as shown in Figure 16. Water molecules enter the cellulose tubing faster than they escape and the solution can be seen moving up the capillary tube after a few minutes.

4 Plasmolysis

A small piece of epidermis (outer layer of cells) is peeled from a red area of a piece of rhubarb stalk (Figure 17*b* on page 10), placed on a slide with a drop of water and covered with a cover-slip (Figure 17*c* on page 10). A solution of salt or sugar is placed at one edge of the cover-slip with a pipette and then drawn under the cover-slip by placing a piece of blotting paper on the opposite side (Figure 17). If a group of cells is watched closely, the red cell sap in their vacuoles will be seen to shrink and pull the cytoplasm away from the cell wall. The cells are now said to be **plasmolysed** (Figure 19*b*). The explanation for this is given in Figure 18. If water is now drawn under the cover-slip, the vacuoles will take in water by osmosis and return to their original shape. Rhubarb epidermis is chosen for this experiment because the red cell sap shows up well, but onion epidermis can be used as in Figure 19.

Plant cells rarely become plasmolysed in nature, but too much chemical fertilizer on the soil could plasmolyse the cells in plant roots.

QUESTIONS

14 In Experiment 3 (Figure 16), what do you think would happen (a) if a much stronger sugar solution was placed in the cellulose tube, (b) if the beaker contained a weak sugar solution instead of water, (c) if the sugar solution was in the beaker and the water was in the cellulose tube?

15 In Experiment 2 (Figure 15), what might happen if the cellulose tube filled with sugar solution was left in the water for several hours?

16 Figure 18 explains why the vacuole shrinks in Experiment 4. Give a brief explanation of why it swells up again when the cell is surrounded by water.

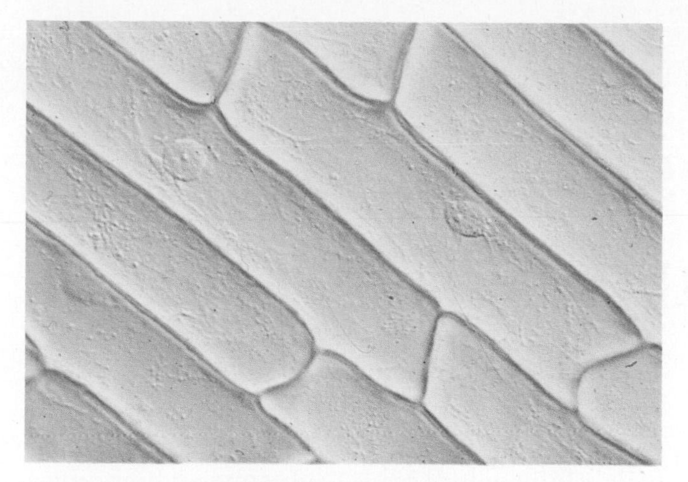

Figure 19(*a*) Turgid cells (×170). The cells are in a strip of epidermis from an onion scale (see Figure 8, page 4). The cytoplasm is pressed against the inside of the cell wall by the vacuole. Nuclei can be seen in three of the cells.

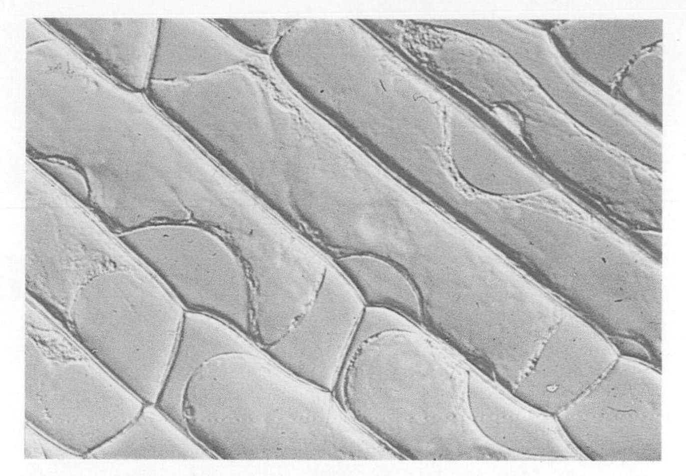

Figure 19(*b*) Plasmolysed cells (×170). The same cells as they appear after treatment with salt solution. The vacuole has lost water by osmosis, shrunk and pulled the cytoplasm away from the cell wall.

BASIC FACTS

ENZYMES

- **All cells make enzymes in their cytoplasm.**
- **Enzymes are needed to speed up all the chemical reactions in living cells.**
- **Each kind of reaction needs its own enzyme.**
- **Enzymes are destroyed by temperatures above 45 °C.**

RESPIRATION

- **Respiration is the way cells get energy from food.**
- **Aerobic respiration needs oxygen; anaerobic respiration does not.**
- **Both kinds of respiration produce carbon dioxide as a waste product.**

EXPERIMENTS

- **Experiments which test whether respiration is going on may show the uptake of oxygen, the output of carbon dioxide and the production of energy.**
- **In a controlled experiment, the scientist (a) controls the conditions so that only one condition is altered at a time and (b) sets up a control to show that the result is due only to this condition.**

DIFFUSION AND OSMOSIS

- **Diffusion is the result of molecules of liquid, gas or dissolved solid moving about.**
- **The molecules of a substance diffuse from a region where they are very concentrated to a region where they are less concentrated.**
- **Osmosis is the diffusion of water through a partially permeable membrane.**
- **Water diffuses from a dilute solution of salt or sugar to a stronger solution because the stronger solution contains less water.**
- **Cells take up water from weak solutions but lose water to strong solutions because of osmosis.**

3 Photosynthesis and nutrition in plants

All living organisms need food. They need it as a source of raw materials from which to build new cells and tissues for growth. They also need food as a source of energy, as a kind of 'fuel' with which to drive essential living processes and bring about chemical changes (see pages 14 and 92). Animals take in food, digest it, and use the digested products to build their tissues or to produce energy. Plants, on the other hand, actually *make* the food they need and then use it for energy and growth. The process by which plants make their food is called **photosynthesis** (*photos* means 'light'; *synthesis* means 'building-up').

PHOTOSYNTHESIS

From carbon dioxide in the air, and water from the soil, a plant can build sugars. A sugar molecule contains the elements carbon, hydrogen and oxygen (e.g. glucose, $C_6H_{12}O_6$). The carbon dioxide molecules, CO_2, provide the carbon and oxygen; the water molecules, H_2O, provide the hydrogen. From these simple compounds, CO_2 and H_2O, the plant is able to build sugar molecules, $C_6H_{12}O_6$. For this process, it needs enzymes, which are present in its cells, and energy, which it obtains from sunlight.

The process takes place mainly in the cells of the leaves and is summarized in Figure 1. Water is absorbed from the soil by the roots and carried in the water vessels of the veins, up the stem to the leaf. Carbon dioxide is absorbed from the air through the stomata (pores in the leaf, see page 60). In the leaf cells, the carbon dioxide and water are combined to make sugar; the energy for this reaction comes from sunlight which has been absorbed by the green pigment, **chlorophyll**. The chlorophyll is present in the **chloroplasts** of the leaf cells and it is inside the chloroplasts that the reaction takes place. Chloroplasts (Figure 1*d*) are small, green structures present in the cytoplasm of the leaf cells. Chlorophyll is the substance which gives leaves and stems their green colour. It is able to absorb energy from light and use it to split water molecules into hydrogen and oxygen. The oxygen escapes from the plant and the hydrogen molecules are added to carbon dioxide molecules to form sugar.

A working definition of photosynthesis is

> **the build-up of sugars from carbon dioxide and water by green plants using energy from sunlight which is absorbed by chlorophyll.**

A chemical equation for photosynthesis is

$$\underset{\text{carbon dioxide}}{6CO_2} + \underset{\text{water}}{6H_2O} \xrightarrow{\text{light energy}} \underset{\text{glucose}}{C_6H_{12}O_6} + \underset{\text{oxygen}}{6O_2}$$

though this really represents only the beginning and end of the process and does not show the many steps in between.

You can see from the equation that one product of photosynthesis is oxygen. Therefore, in daylight, when photosynthesis is going on in green plants, they will be taking in carbon dioxide and giving out oxygen. This exchange of gases is the opposite of that resulting from respiration (page 14) but it must not be thought that green plants do not respire. The energy they need for all their living processes—apart from photosynthesis—comes from respiration and this is going on all the time, using up oxygen and producing carbon dioxide. During the daylight hours, plants are photosynthesizing as well as respiring, so that all the carbon dioxide produced by respiration is used up by photosynthesis. At the same time, all the oxygen needed by respiration is provided by photosynthesis. Only when the rate of photosynthesis is faster than the rate of respiration will carbon

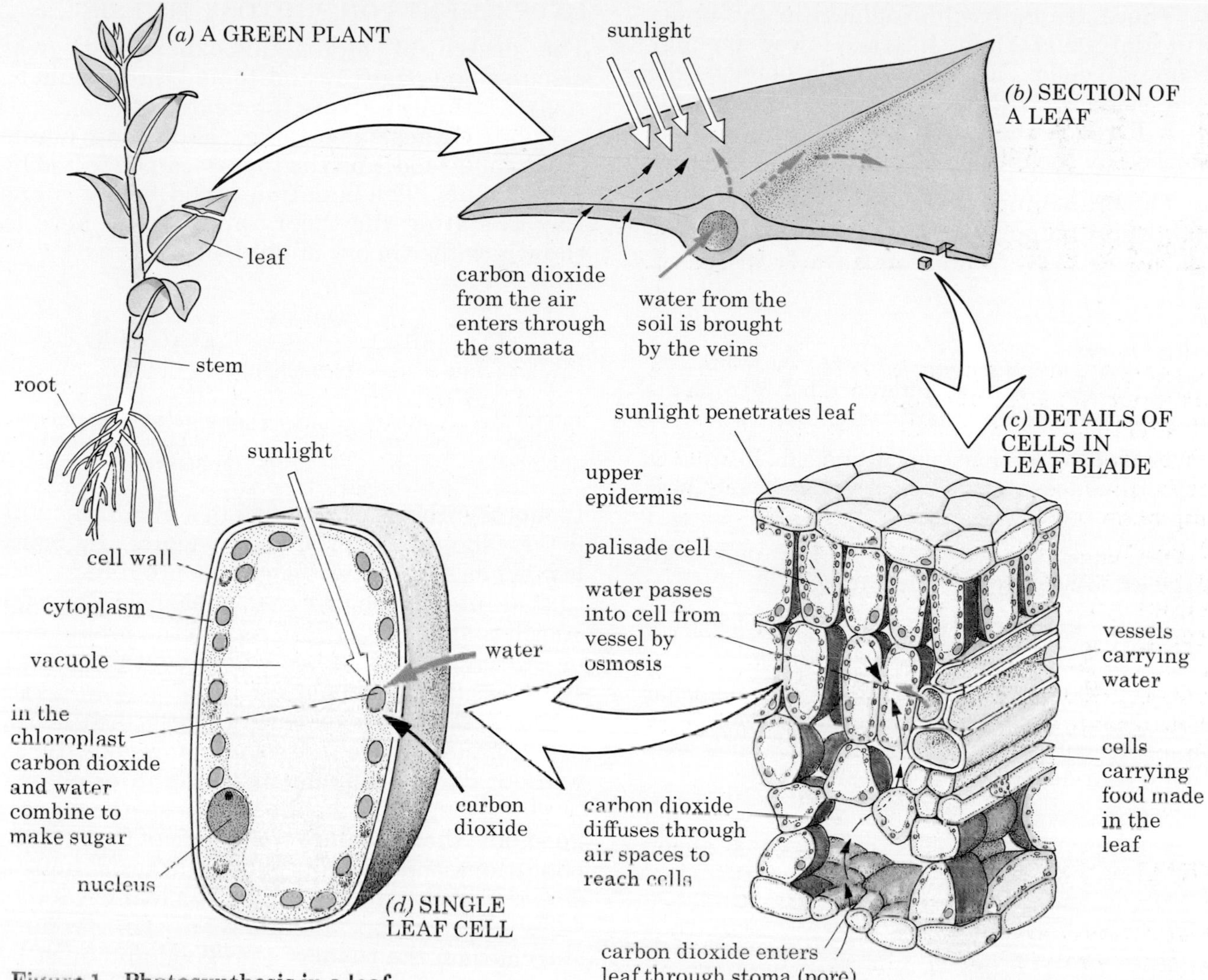

Figure 1 Photosynthesis in a leaf

dioxide be taken in and the excess oxygen be given out (Figure 2).

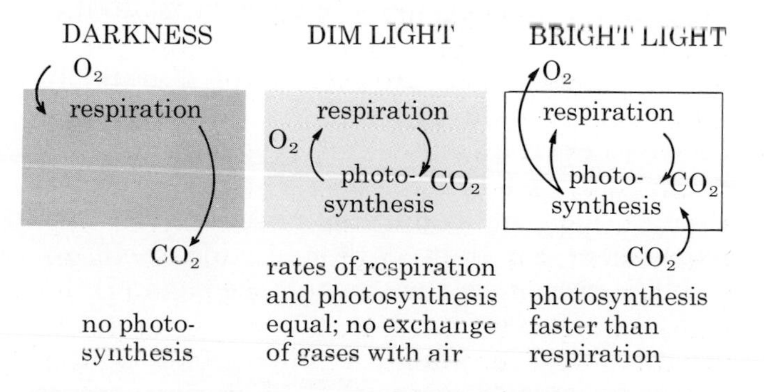

Figure 2 Respiration and photosynthesis

Adaptation of leaves for photosynthesis

When biologists say that something is **adapted**, they mean that its structure is well suited to its function. The detailed structure of the leaf is described on pages 58 to 61, and although there are wide variations in leaf shape the following general statements apply to a great many leaves, and are illustrated in Figures 1*b* and *c*.

(a) Their broad, flat shape offers a large surface area for absorption of sunlight and carbon dioxide.

(b) Most leaves are thin and the carbon dioxide has to diffuse across only short distances to reach the inner cells.

(c) The large spaces between cells inside the leaf provide an easy passage through which carbon dioxide can diffuse.

(d) There are many stomata (pores) in the lower surface of the leaf. These allow the exchange of carbon dioxide and oxygen with the air outside.

(e) There are more chloroplasts in the upper (palisade) cells than in the lower (spongy mesophyll) cells. The palisade cells, being on the upper surface, will receive most sunlight and this will reach the chloroplasts without being absorbed by too many cell walls.

(f) The branching network of veins provides a good water supply to the photosynthesizing cells. No cell is very far from a water vessel.

QUESTIONS

1 What is the essential difference between plants and animals in the way they obtain food?

2 What substances must a plant take in, in order to carry on photosynthesis? Where does it get these substances from?

3 What chemical process provides a plant with energy to carry on all its living activities?

4 What might be the advantage to most leaves of being thin?

5 Measurements on a leaf show that it is giving out carbon dioxide and taking in oxygen. Does this prove that photosynthesis is *not* going on in the leaf? Explain your answer.

6 Using the equation on page 24, say what you would accept as experimental evidence that photosynthesis is taking place in a plant.

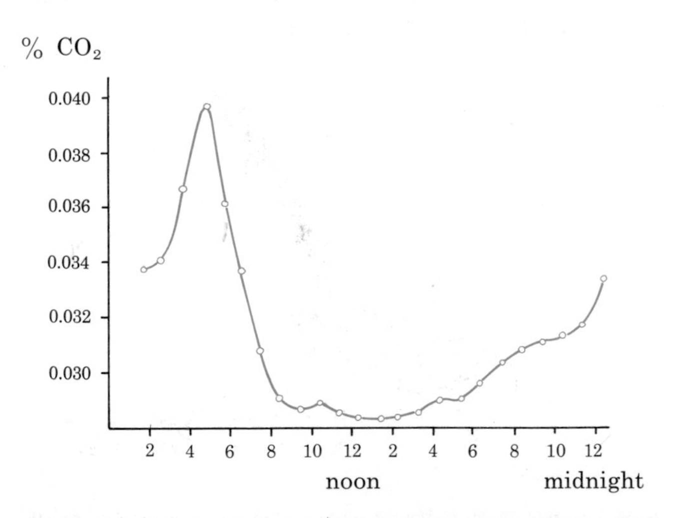

Figure 3 Daily changes in concentration of carbon dioxide one metre above a plant crop (From Verma and Rosenberg, *Span*, 1979)

7 Figure 3 is a graph showing the average daily change in the carbon dioxide concentration, one metre above an agricultural crop in July. From what you have learned about photosynthesis and respiration, try to explain the changes in the carbon dioxide concentration.

EXPERIMENTS ON PHOTOSYNTHESIS

The design of biological experiments was discussed on pages 17 and 18 and this should be revised before studying the next section.

Photosynthesis is a theory about how plants make their food, and the theory can be tested by experiments. The equation given below is one way of stating the theory and is used here to show how the theory might be tested.

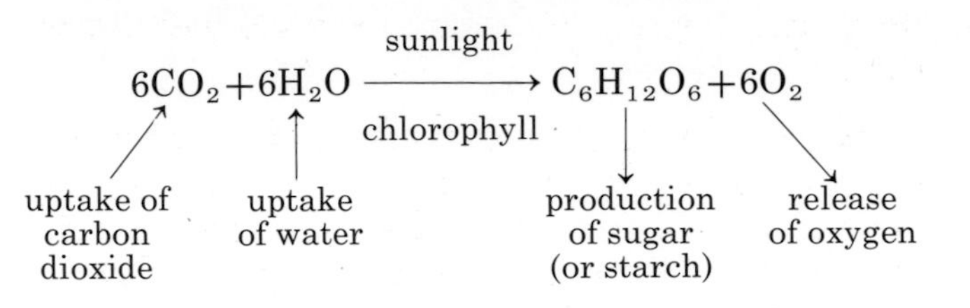

If photosynthesis is going on in a plant, then the leaves should be producing sugars. In many leaves, as fast as the sugar is produced, it is turned into starch. Since it is easier to test for starch than for sugar, we regard the production of starch in a leaf as evidence that photosynthesis has taken place.

The first three experiments described below are designed to see if the leaf can make starch without carbon dioxide, or sunlight, or chlorophyll, in turn. If the photosynthesis theory is a good one, then the lack of any one of these three conditions should stop photosynthesis, and so stop the production of starch. If a leaf without a supply of carbon dioxide can still produce starch, then the theory is no good and must be altered or rejected.

In designing the experiments, it is very important to make sure that only *one* condition is altered. If, for example, the method of keeping light from a leaf also cuts off its carbon dioxide supply, it would be impossible to decide whether it was the lack of light or lack of carbon dioxide which stopped the production of starch. To make sure that the experimental design has not altered more than one condition, a **control** is set up in each case. This is an identical situation, except that the condition missing from the experiment, e.g. light, carbon dioxide or chlorophyll, is present in the control (see page 17).

Destarching a plant. If the production of starch is your evidence that photosynthesis is taking place, then you must make sure that the leaf does not contain any starch at the beginning of the experiment. This is done by **destarching** the leaves. It is not possible to remove the starch chemically, without damaging the leaves, so a plant is destarched simply by leaving it in darkness for two or three days. Potted plants are destarched by leaving them in a dark cupboard for a few days. In the darkness, any starch in the leaves will be changed to sugar and carried away from the leaves to other parts of the plant. For plants in the open, the experiment is set up on the day before the test. During the night, most of the starch will be removed from the leaves. It is better still to wrap the leaves to be tested in aluminium foil for two days and then test one of the leaves to see that no starch is present.

Testing a leaf for starch. Iodine solution (yellow) and starch (white) form a deep blue colour when they mix. The test for starch, therefore, is to add iodine solution to a leaf to see if it goes blue. First, however, the leaf has to be treated as follows:

1 The leaf is detached and dipped in boiling water for half a minute. This kills the cytoplasm by destroying the enzymes in it, and so prevents any further chemical changes. It also makes the cell more permeable to iodine solution.

2 The leaf is boiled in alcohol (ethanol), using a water bath (Figure 4), until all the chlorophyll is dissolved out. This turns the leaf whitish and makes any colour changes caused by iodine easier to see.

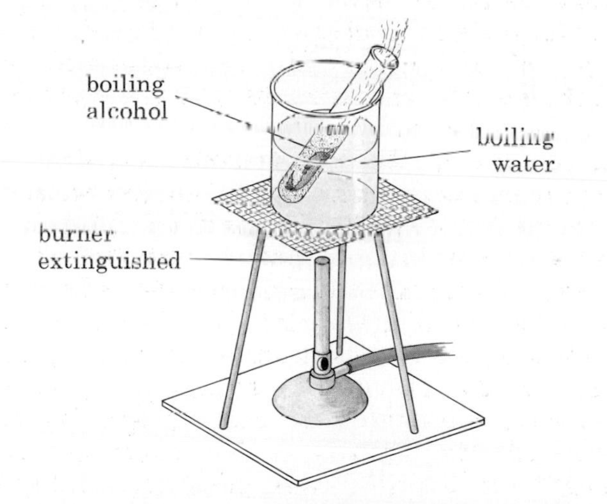

Figure 4 To remove chlorophyll from a leaf

3 Alcohol makes the leaf brittle and hard, but it can be softened by dipping it once more into the hot water. Then it is spread flat on a white surface such as a glazed tile.

4 Iodine solution is placed on the leaf. Any parts which turn blue have starch in them. If no starch is present, the leaf is merely stained yellow or brown by the iodine.

Experiment 1 Is chlorophyll necessary for photosynthesis?

It is not possible to remove chlorophyll from a leaf without killing it, and so a **variegated** leaf, which has chlorophyll only in patches, is used. A leaf of this kind is shown in Figure 5*a*. The white

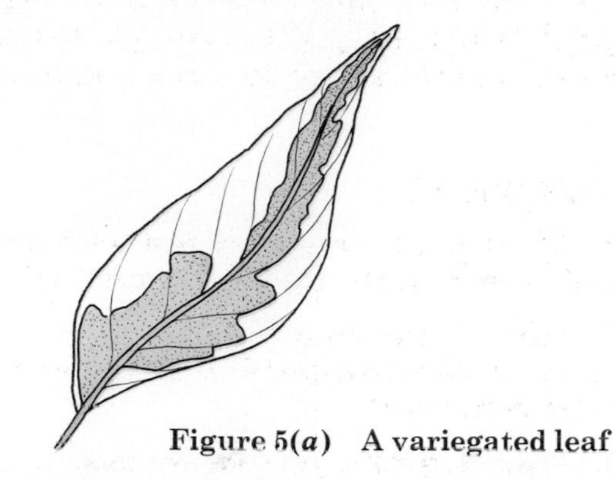

Figure 5(*a*) A variegated leaf

part of the leaf serves as the experiment, because it lacks chlorophyll, while the green part with chlorophyll is the control. After being de-starched, the leaf—still on the plant—is exposed to daylight for a few hours. It is then removed from the plant, drawn carefully to show where the chlorophyll is (i.e. the green parts), and tested for starch as described above.

Result. Only the parts that were previously green turn blue with iodine. The parts that were white stain brown (Figure 5*b*).

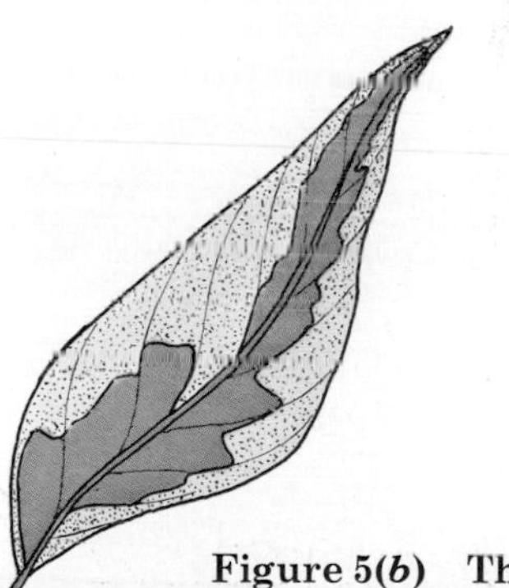

Figure 5(*b*) The same leaf after testing with iodine

Interpretation. Since starch is present only in the parts which originally contained chlorophyll, it seems reasonable to suppose that chlorophyll is needed for photosynthesis. (It must be remembered, however, that there are other possible interpretations which this experiment has not ruled out; for example, starch could be made in the green parts and sugar in the white parts. Such alternative explanations could be tested by further experiments.)

Experiment 2 Is light necessary for photosynthesis?

A simple shape is cut out from a piece of aluminium foil to make a stencil which is attached to a destarched leaf (Figure 6*a*). After four to six hours of daylight, the leaf is removed from the plant and tested for starch.

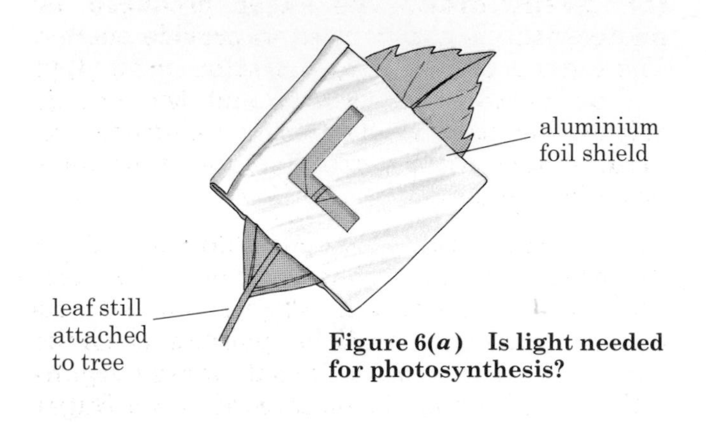

Figure 6(*a*) Is light needed for photosynthesis?

Result. Only the areas which had received light go blue with iodine.

Interpretation. As starch has not formed in the areas which received no light, it seems that light is needed for starch formation and thus for photosynthesis. (You could argue that the aluminium foil had stopped carbon dioxide from entering the leaf and that it was shortage of carbon dioxide rather than absence of light which prevented photosynthesis taking place. A

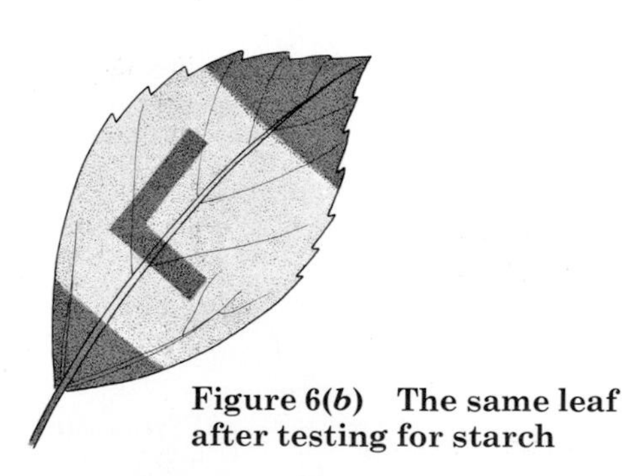

Figure 6(*b*) The same leaf after testing for starch

further control could be designed, using transparent material instead of aluminium foil for the stencil.)

Experiment 3 Is carbon dioxide needed for photosynthesis?

Two destarched potted plants are watered and the shoots enclosed in polythene bags, one of which contains soda-lime to absorb carbon dioxide from the air (the experiment), while the other has sodium hydrogencarbonate solution to produce carbon dioxide (the control), as shown in Figure 7. Both plants are placed in light for several hours and a leaf from each is then removed and tested for starch.

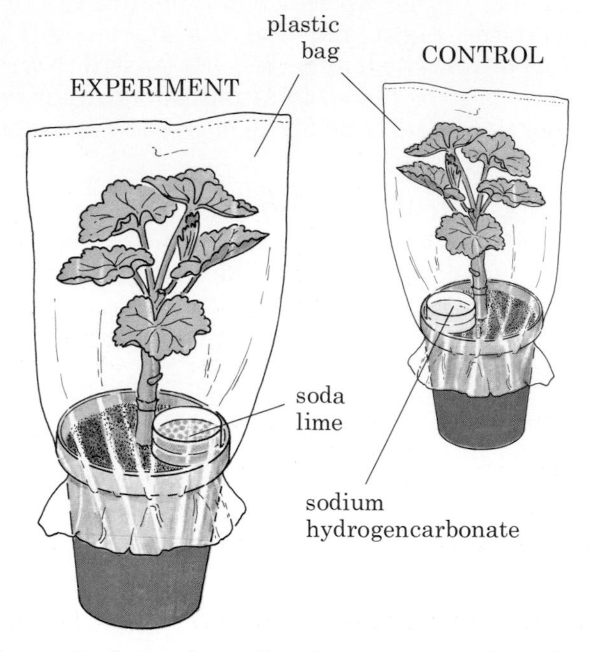

Figure 7 Is carbon dioxide necessary for photosynthesis?

Result. The leaf which had no carbon dioxide does not turn blue while the one from the polythene bag containing carbon dioxide does turn blue.

Interpretation. The fact that starch was made in the leaves which had carbon dioxide, but not in the leaves which had no carbon dioxide, suggests that this gas must be necessary for photosynthesis. (The control rules out the possibility that high humidity or high temperature in the plastic bag prevents normal photosynthesis.)

Experiment 4 Is oxygen produced during photosynthesis?

A short-stemmed funnel is placed over some Canadian pond weed in a beaker of water. A test-tube filled with water is placed upside-down over the funnel stem (Figure 8). The funnel is raised above the bottom of the beaker to allow the water to circulate. The apparatus is placed in sunlight, and bubbles of gas soon appear from the cut stems, rise and collect in the test-tube. When sufficient gas has collected, the test-tube is removed and a glowing splint inserted. A control experiment should be set up in a similar way but placed in a dark cupboard. Little or no gas will be collected.

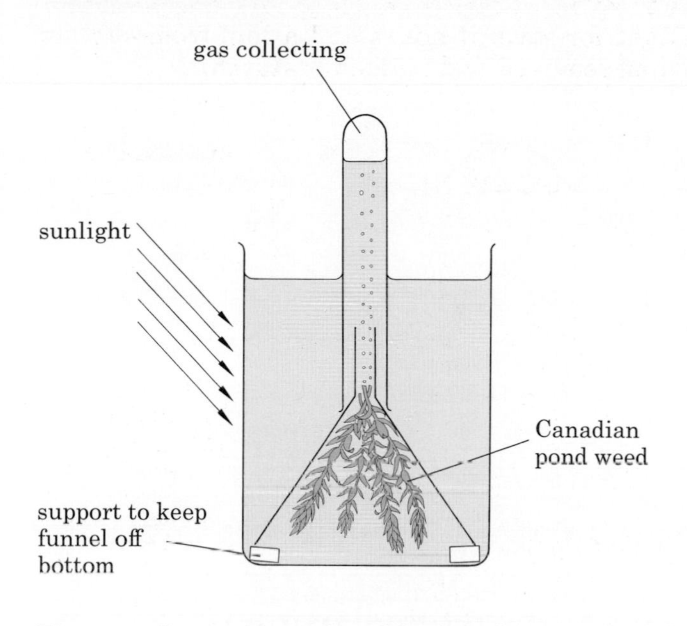

Figure 8 Is oxygen produced during photosynthesis?

Result. The glowing splint bursts into flames.

Interpretation. The relighting of a glowing splint does not prove that the gas collected is *pure* oxygen, but it does show that, in the light, this particular plant has given off a gas which is much richer in oxygen than ordinary air.

QUESTIONS

8 Iris leaves produce sugar but not starch. How would you have to change the design of Experiments 1, 2 and 3 if you wanted to use iris leaves?

9 A green plant makes sugar from carbon dioxide and water. Why do we not try the experiment of depriving a plant of water to see if that stops photosynthesis?

10 Does the method of destarching a plant take for granted the results of Experiment 2? Explain your answer.

11 In Experiment 2, an extra control was suggested to see whether the aluminium foil stencil had prevented carbon dioxide as well as light from getting into a leaf. If the stencil was made of clear plastic (a) how would its effect differ from that of the aluminium foil stencil, and (b) what result would you expect (i) if the stencil *had* interfered with the supply of carbon dioxide and (ii) if it *had not*?

12 Why do you think a pond weed, rather than a land plant is used for Experiment 4? In what way might this choice make the results less useful?

Use of the products of photosynthesis

The starch built up in the leaves is changed back into sugar, carried out of the leaves in the veins and distributed to other parts of the plant which might use it in any of the following ways (Figure 9).

(a) Respiration. The sugar produced by photosynthesis can be used to provide energy. The sugar is oxidized by respiration (page 14) to carbon dioxide and water, and the energy released is used to drive other chemical reactions such as the building up of proteins described below.

(b) Storage. Sugar which is not needed for respiration is turned into starch and stored. Some plants store it as starch grains in the cells of their stems or roots. Other plants such as the potato or parsnip have special storage organs (tubers) for holding the reserves of starch. Sugar may be stored in the fruits of some plants; grapes, for example, contain a large amount of glucose.

(c) Synthesis of other substances. As well as sugars for energy and starch for storage, the plant needs cellulose for its cell walls, fatty substances for its cell membranes, proteins for its cytoplasm and pigments for its flower petals, etc. All these substances are built up (synthesized) from the sugar molecules produced in photosynthesis.

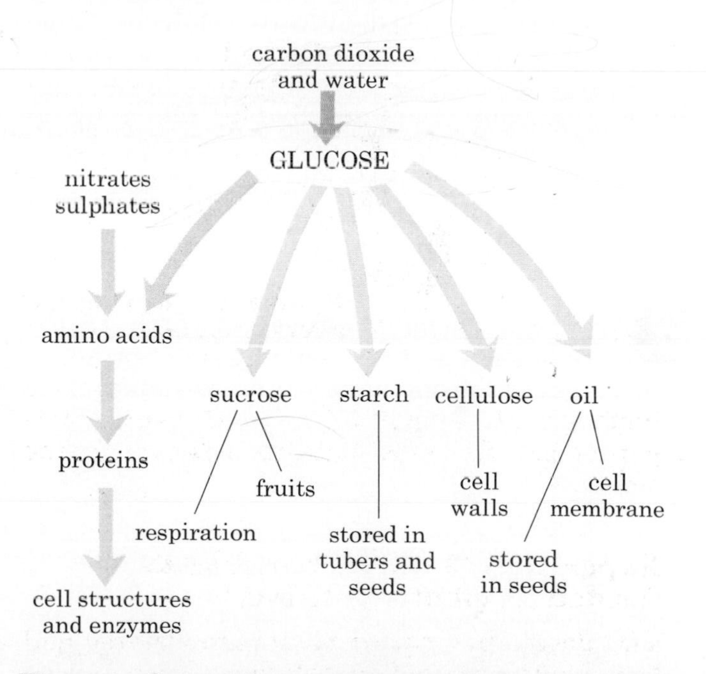

Figure 9 Green plants can make all the materials they need from carbon dioxide, water and salts

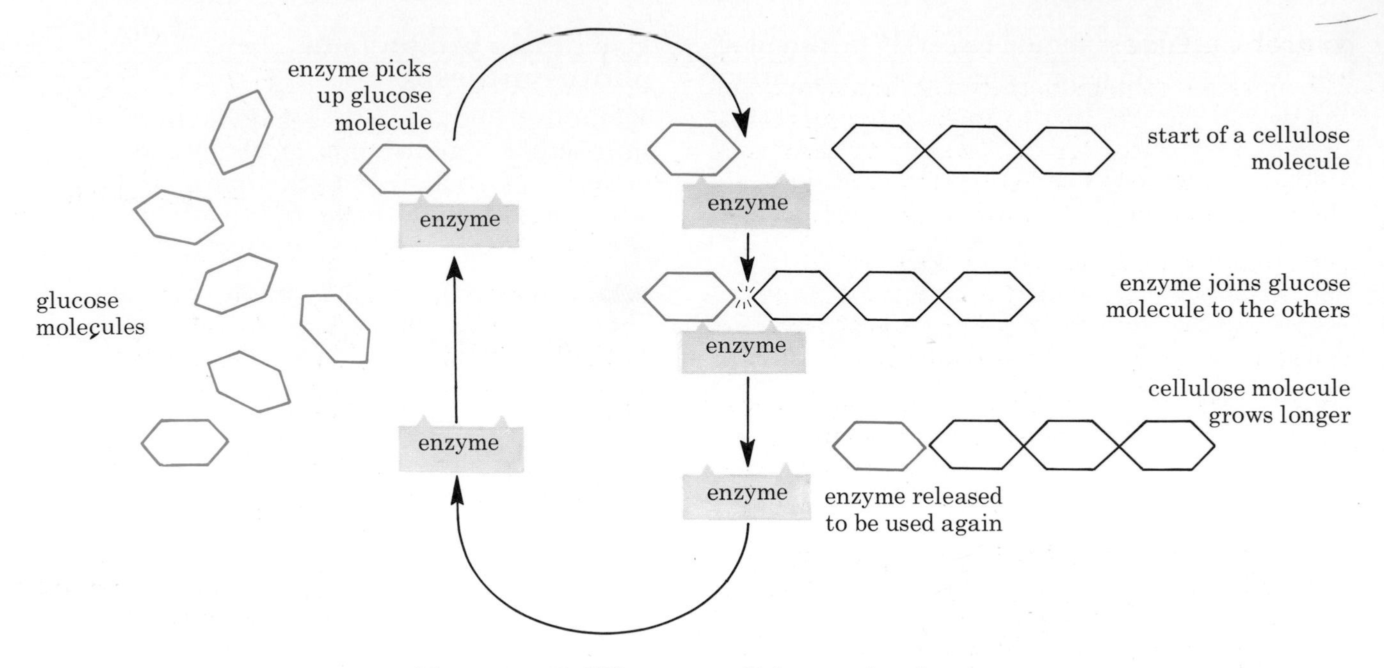

Figure 10 Building up a cellulose molecule

By joining hundreds of glucose molecules together, the long chain molecules of cellulose (Figure 10) are built up and added to the cell walls.

Amino acids (see page 93) are made by combining **nitrogen** with sugar molecules. These amino acids are then joined together to make the proteins which form the enzymes and the cytoplasm of the cell. The nitrogen for this synthesis comes from **nitrates** which are absorbed from the soil by the roots.

Proteins also need **sulphur** molecules and these are absorbed from the soil in the form of **sulphates** (SO_4). **Phosphorus** is needed for many enzyme reactions and is taken up as phosphates (PO_4).

The chlorophyll molecule needs **magnesium** (Mg). This metallic element is also obtained in salts from the soil (see the list of salts on page 31).

All these chemical processes, such as the uptake of salts and the building up of proteins, need energy from respiration to make them happen.

QUESTION

13 What substances does a green plant need to take in, to make (a) sugar, (b) proteins? What must be present in the cells to make reactions (a) and (b) work?

THE SOURCES OF MINERAL ELEMENTS

The mineral elements needed by plants are absorbed from the soil in the form of salts. For example, a plant's needs for potassium (K) and nitrogen (N) might be met by absorbing the salt **potassium nitrate** (KNO_3). Salts like this come originally from rocks which have been broken down to form the soil (page 40). They are continually being taken up by plants or washed out by rain. They are replaced partly from the dead remains of plants and animals. When these organisms die and their bodies decay, the salts they contain are released back into the soil. This process is explained in some detail, for nitrates, on page 36.

In arable farming, the ground is ploughed and whatever is grown is removed. There are no dead plants left to decay and replace the mineral salts. The farmer must replace them by using animal manure or artificial fertilizers in measured quantities over the land. Below is a list of three artificial fertilizers used in agriculture.

(a) Ammonium nitrate (NH_4NO_3). The formula shows that this is a rich source of nitrogen.

(b) Superphosphate. A mixture containing calcium, phosphate and sometimes sulphate.

(c) Compound NPK (N = nitrogen, P = phosphorus, K = potassium). A mixture of ammonium sulphate, ammonium phosphate and potassium chloride.

Water cultures

It is possible to demonstrate the importance of the various mineral elements by growing plants in water cultures. A full water culture is a solution in water of the salts which provide all the necessary elements for healthy growth, e.g.

potassium nitrate for potassium and nitrogen
magnesium sulphate for magnesium and sulphur
potassium phosphate for potassium and phosphorus
calcium nitrate for calcium and nitrogen

From these elements, plus the carbon dioxide, water and sunlight needed for photosynthesis, a green plant can make all the substances it needs for a healthy existence (Figure 11).

Figure 11 Soil-less culture. The tomatoes are growing in a solution of mineral salts circulated in the plastic troughs. This method increases production and avoids the need for soil sterilization.

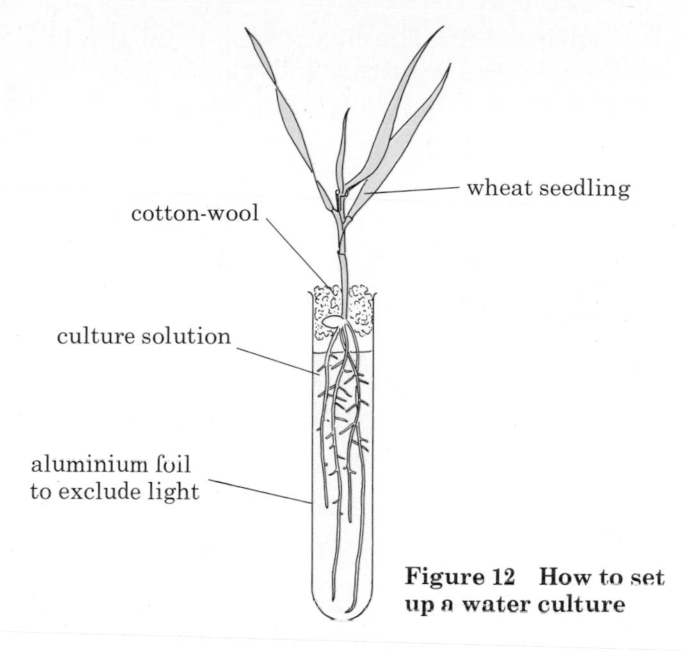

Figure 12 How to set up a water culture

Experiment 5 The importance of different mineral elements

Wheat seedlings are placed in test-tubes containing water cultures as shown in Figure 12. The tubes are covered with aluminium foil to keep out light and so stop the growth of green algae (microscopic green plants; see page 199). Some of the solutions have one of the elements missing. For example, in the list of chemicals above, magnesium chloride is used instead of magnesium sulphate and so the solution will lack sulphur. In a similar way, solutions lacking nitrogen, potassium and phosphorus can be prepared.

The seedlings are left to grow in these solutions for a few weeks, keeping the tubes topped up with distilled water.

Result. Figure 13 shows the kind of results which might be expected from wheat seedlings. Generally, the plants in a complete culture will be tall and sturdy, with large, dark green leaves. The plants lacking nitrogen will usually be stunted and have small, pale leaves. In the absence of magnesium, chlorophyll cannot be made, and these plants will be small with yellow leaves.

Figure 13 Result of water culture experiment

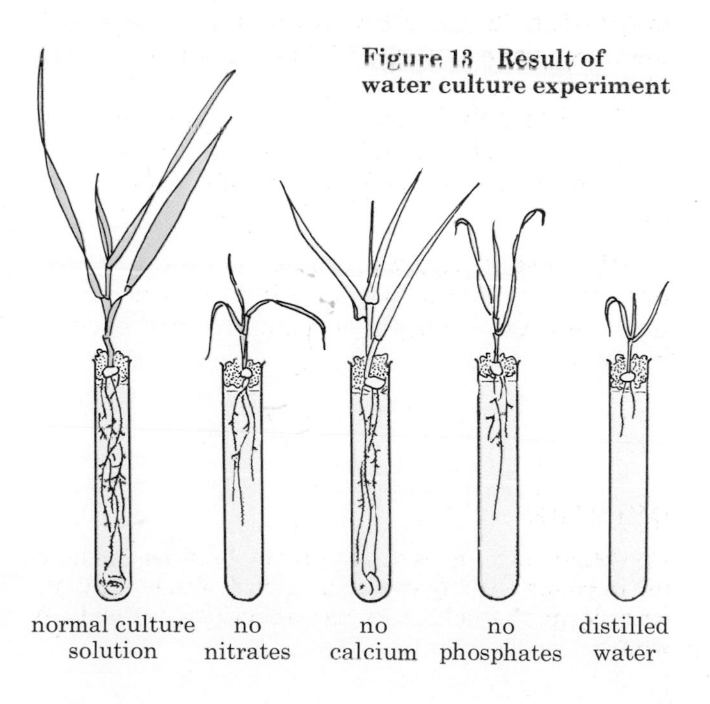

Interpretation. The healthy plant in the full culture is the control and shows that this method of raising plants does not affect them. The other, less healthy plants show that a full range of mineral elements is necessary for normal growth.

Figure 14 shows the result of a glass-house experiment in which a group of plants was deprived of nitrate and phosphate.

Figure 14 Effect of mineral salts on growth. The chrysanthemums on the right were planted at the same time as those on the left but were not given so much nitrate and phosphate.

QUESTIONS

14 What salts would you put in a water culture which is to contain *no* nitrogen?

15 What mineral elements do you think are provided by (a) bone meal (page 152), (b) dried blood (page 111)?

16 How can a floating pond plant, like duckweed, survive without having its roots in soil?

BASIC FACTS

- **Photosynthesis is the way plants make their food.**
- **They combine carbon dioxide and water to make sugar.**
- **To do this, they need energy from sunlight, which is absorbed by chlorophyll.**
- **Plant leaves are adapted for the process of photosynthesis by being broad and thin, with many chloroplasts in their cells.**
- **From the sugar made by photosynthesis, a plant can make all the other substances it needs, provided it has a supply of mineral salts like nitrate, phosphate and potassium.**
- **Experiments to test the photosynthesis theory are designed to exclude light, or carbon dioxide, or chlorophyll, to see if the plant can still produce starch.**

4 The interdependence of living organisms

'Interdependence' means the way in which living organisms depend on each other in order to remain alive, grow and reproduce. For example, bees depend for food on pollen and nectar from flowers. Flowers depend on bees for pollination (page 75). Bees and flowers are therefore interdependent.

FOOD CHAINS AND FOOD WEBS

One important way in which organisms depend on each other is for their food. Many animals, such as rabbits, feed on plants. Such animals are called **herbivores**. Animals called **carnivores** eat other animals. A **predator** is a carnivore which kills and eats other animals. A fox is a predator which preys on rabbits. **Scavengers** are carnivores which eat the dead remains of animals killed by predators. These are not hard and fast definitions. Predators will sometimes scavenge for their food and scavengers may occasionally kill living animals.

Basically, all animals depend on plants for their food. Foxes may eat rabbits, but rabbits feed on grass. A hawk eats a lizard, the lizard has just eaten a grasshopper but the grasshopper was feeding on a grass blade. This relationship is called a **food chain**. Another example of a food chain is:

lettuce→snail→thrush→sparrow hawk

The organisms at the beginning of a food chain are usually very numerous while the animals at the end of the chain are often large and few in number. The **food pyramid** in Figure 1 shows this relationship. There will

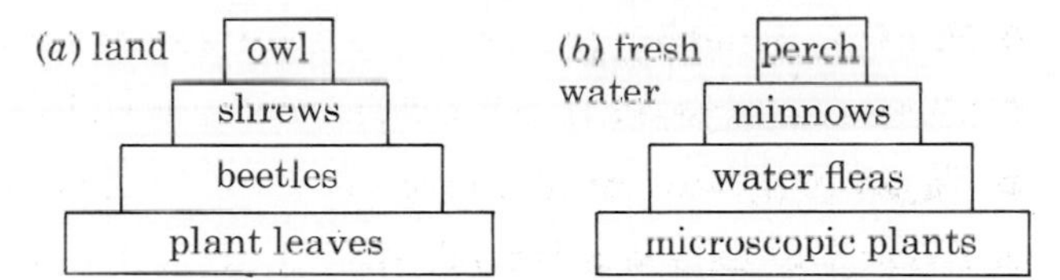

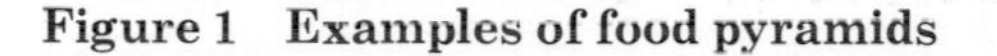

Figure 1 Examples of food pyramids

be millions of microscopic, single-celled green plants (page 198) in a pond. These will be eaten by the larger but less numerous water-fleas, which in turn will become the food of small fish like minnow and stickleback (Figure 2). The hundreds of small fish may be able to provide enough food for only four or five large carnivores, like pike or perch (Figure 3).

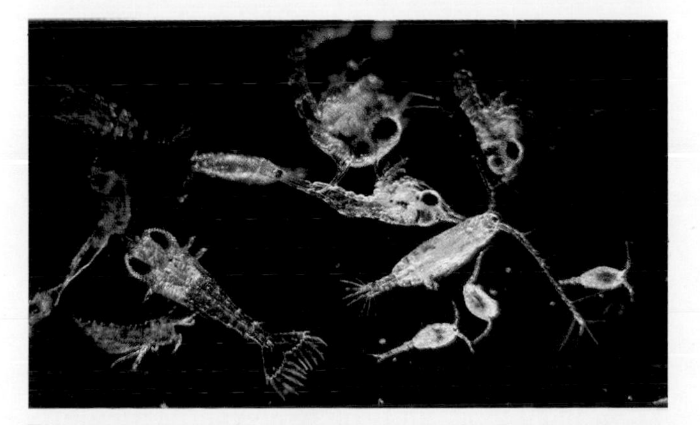

Figure 2 Animals at the bottom of a food pyramid. These are some of the microscopic animals (see Crustacea, page 206) which live near the surface of the sea and are eaten by fish.

Figure 3 A food chain in action. The pike is eating a stickleback. In this case, the stickleback may escape because its spines (see Figure 5*a* on page 233) will stick in the pike's mouth.

Food chains are not really as simple as this, because most animals eat more than one type of food. A fox, for example, does not feed entirely on rabbits but takes beetles, rats and blackberries in its diet. To show these relationships more accurately, a **food web** can be drawn up (Figure 4).

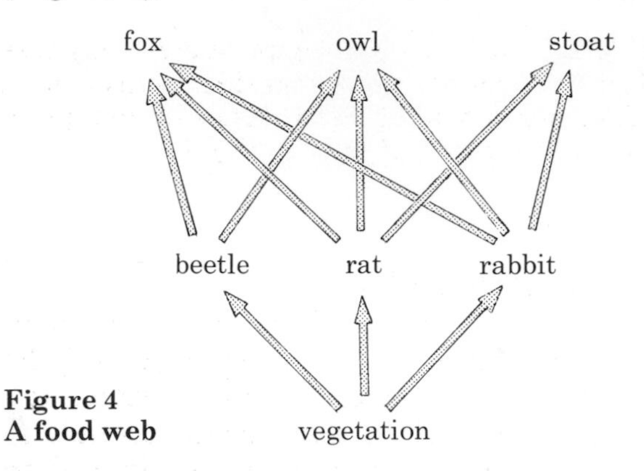

Figure 4
A food web

The food webs for land, sea and fresh water, or for ponds, rivers and streams will all be different. Food webs will also change with the seasons when the food supply changes.

If some event interferes with a food web, all the organisms in it are affected in some way. For example, if the rabbits in Figure 4 were to die out, the foxes, owls and stoats would eat more beetles and rats. Something like this happened in 1954 when the disease myxomatosis wiped out nearly all the rabbits in England. Foxes ate more voles, beetles and blackberries, and attacks on lambs and chickens increased. Even the vegetation was affected because the tree seedlings which the rabbits used to nibble off were allowed to grow. As a result, woody scrubland started to develop on what had been grassy downs. Figure 5 shows a similar effect.

Figure 5 The effect of grazing. Ten years ago the area in the background was fenced off. Notice how the young trees have grown since the sheep were prevented from nibbling off the seedlings.

If you take the idea of food chains one step further you will see that all living organisms depend on sunlight and photosynthesis (page 24). Green plants make their food by photosynthesis, which needs sunlight. Since all animals depend, in the end, on plants for their food, they therefore depend indirectly on sunlight. A few examples of our own dependence on photosynthesis are given below.

bread	cheese	honey
↑	↑	↑
flour	milk	bees
↑	↑	↑
wheat grains	cow	nectar
↑	↑	↑
wheat grows by photosynthesis	grass	flowers
	↑	↑
	grass grows by photosynthesis	flowering plants grow by photosynthesis

Nearly all the energy released on the Earth can be traced back to the sun. Coal comes from tree-like plants, buried millions of years ago. These plants absorbed sunlight for their photosynthesis when they were alive. Petroleum was formed, also millions of years ago, from the partly decayed bodies of microscopic plants which lived in the sea. These too had absorbed sunlight for photosynthesis.

Today it is possible to use mirrors and solar panels to collect energy from the sun directly, but the best way, so far, of trapping and storing energy from sunlight is to grow plants and make use of their products, such as starch, sugar, oil, alcohol and wood, for food or as energy sources for other purposes. For example, sugar from sugar cane can be fermented (page 14) to alcohol and used as motor fuel instead of petrol.

Saprophytes

There are a number of organisms which have not been fitted into the food webs or food chains described so far. Among these are the **saprophytes**. Saprophytes do not obtain their food by photosynthesis, nor do they kill and eat living animals or plants. Instead they feed on dead and decaying matter such as dead leaves in the soil or rotting tree trunks (see Figure 6). The most numerous examples are the fungi, such as mushrooms, toadstools or moulds (see Figure 11 on page 221), and the bacteria, particularly those which live in the soil. They produce extracellular enzymes (page 13) which digest the decaying matter and then they absorb the

Figure 6 Saprophytes. These toadstools are getting their food from the rotting tree stump.

soluble products back into their cells. In so doing, they remove the dead remains of plants and animals which would otherwise collect on the Earth's surface. They also break these remains down into substances which can be used by other organisms. Some bacteria, for example, break down the protein of dead plants and animals and release nitrates which are taken up by plant roots and there built into new amino acids and proteins (page 29). This use and re-use of materials in the living world is called **recycling**.

Figure 7 shows the general idea of recycling. The green plants are the organisms which actually build up chemical substances for food, by photosynthesis. They are therefore called the **producers**. The animals which eat the plants and each other are the **consumers**. The bacteria and fungi, especially those in the soil, are called the **decomposers** because they break down the dead remains and release the chemicals for the plants to use again. Two examples of recycling, one for carbon and one for nitrogen, are described below.

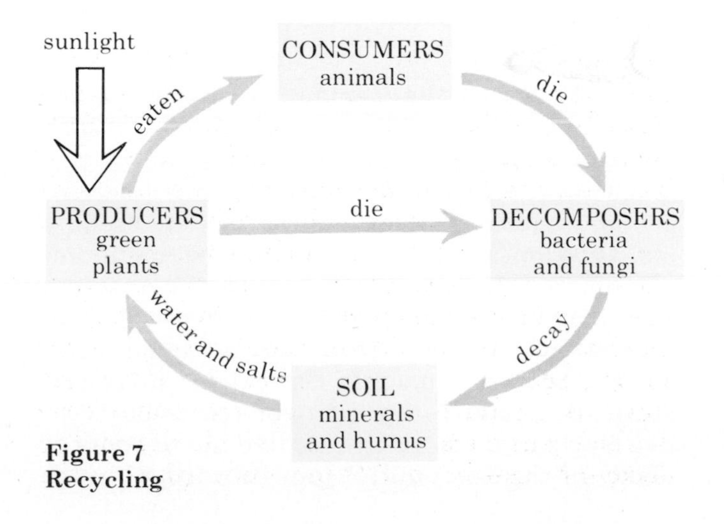

Figure 7 Recycling

QUESTIONS

1 Describe briefly all the possible ways in which the following might depend on each other: grass, earthworm, blackbird, oak tree, soil.

2 Explain how the following foodstuffs are produced as a result of photosynthesis: wine, butter, eggs, beans.

3 An electric motor, a car engine and a race-horse can all produce energy. Show how this energy comes, originally, from sunlight. What forms of energy on the Earth are *not* derived from sunlight?

THE CARBON CYCLE

Carbon is an element which occurs in all the compounds which make up living organisms. Plants get their carbon from carbon dioxide in the atmosphere and animals get their carbon from plants. The carbon cycle, therefore, is mainly concerned with what happens to carbon dioxide (Figure 8).

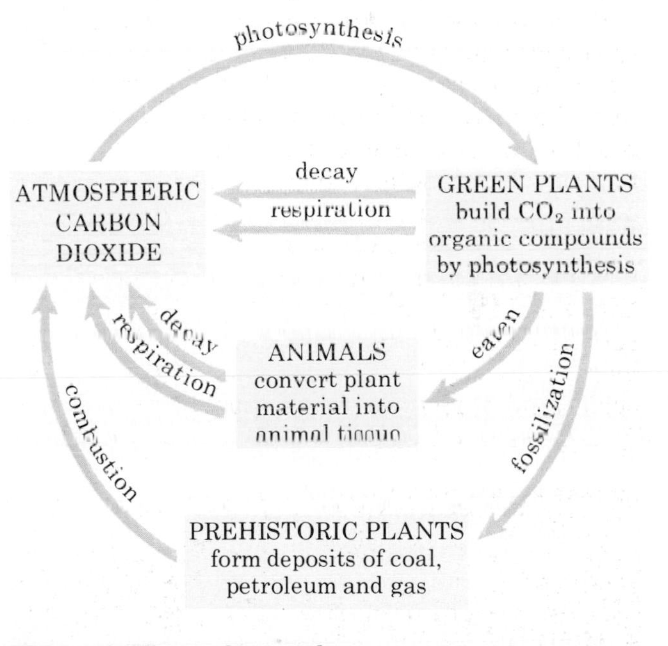

Figure 8 The carbon cycle

Removal of carbon dioxide from the atmosphere

Green plants remove carbon dioxide from the atmosphere as a result of their photosynthesis. The carbon of the carbon dioxide is built first into a carbohydrate such as sugar. Some of this is changed into starch or the cellulose of cell walls, and the proteins, pigments and other compounds of a plant. When the plants are eaten by animals, the organic plant material is digested, absorbed and built into the compounds making up the animals' tissues. Thus the carbon atoms from the plant become part of the animal.

Addition of carbon dioxide to the atmosphere

(a) Respiration. Plants and animals obtain energy by oxidizing carbohydrates in their cells to carbon dioxide and water (page 14). The carbon dioxide and water are excreted and so the carbon dioxide returns once again to the atmosphere.

(b) Decay. The organic matter of dead animals and plants is used by saprophytes, especially bacteria and fungi, as a source of energy. These micro-organisms decompose the plant and animal remains and turn the carbon compounds into carbon dioxide.

(c) Combustion (burning). When carbon-containing fuels such as wood, coal, petroleum and natural gas are burned, the carbon is oxidized to carbon dioxide ($C+O_2 \rightarrow CO_2$). The hydrocarbon fuels, such as coal and petroleum, come from ancient plants which have only partly decomposed over the millions of years since they were buried.

So, an atom of carbon which today is in a molecule of carbon dioxide in the air may tomorrow be in a molecule of cellulose in the cell wall of a blade of grass. When the grass is eaten by a cow, the carbon atom may become part of a glucose molecule in the cow's blood stream. When the glucose molecule is used for respiration, the carbon atom will be breathed out into the air once again as carbon dioxide. The same kind of cycling applies to nearly all the elements of the Earth. No new matter is created, but it is repeatedly rearranged. A great proportion of the atoms of which you are composed will, at one time, have been part of other organisms.

THE OXYGEN CYCLE

Animals and plants use up oxygen from the air for their respiration. Oxygen is also used up when substances burn. We depend on the photosynthesis of green plants to replace the oxygen at the same rate as it is used up.

Today, man's activities affect these cycles. For example, the nitrogen present in sewage from man's body waste may not be recycled to the land which produces his food but discharged into the sea. The carbon fuels are being burned in great quantities, adding more and more carbon dioxide to the atmosphere.

QUESTIONS

4 Outline the events that might happen to a carbon atom in a molecule of carbon dioxide which entered the stoma in the leaf of a potato plant, and became part of a starch molecule in a potato tuber which was then eaten by a man. Finally the carbon atom is breathed out again in a molecule of carbon dioxide.

5 Large areas of tropical forest, particularly in South America, are being cut down to make way for roads, cities and agriculture. What effect might this have on the carbon dioxide level in the Earth's atmosphere?

THE NITROGEN CYCLE

When a plant or animal dies, its tissues decompose, mainly as a result of the action of saprophytic bacteria. One of the important products of this decay is **ammonia** (a compound of nitrogen, NH_3), which is washed into the soil.

Processes which add nitrates to the soil

Nitrifying bacteria. These are bacteria living in the soil which use the ammonia from decaying organisms as a source of energy (like we use glucose in respiration). In the process of getting energy from ammonia, the bacteria produce **nitrates**. Although plant roots can take up ammonia in the form of its compounds, they take up nitrates more readily, so the nitrifying bacteria increase the fertility of the soil by making nitrates available to the plants.

Nitrogen-fixing bacteria. This is a special group of nitrifying bacteria which can absorb nitrogen as a gas from the air spaces in the soil (page 41), and build it into compounds of ammonia. Nitrogen gas cannot itself be used by plants. When it has been made into a compound of ammonia, however, it can easily be changed to nitrates by other nitrifying bacteria. The process of building the gas, nitrogen, into compounds of ammonia is called **nitrogen fixation**. Some of the nitrogen-fixing bacteria live freely in the soil. Others live in the roots of plants of the pea family (**leguminous plants**), where they cause swellings called **root nodules** (Figure 9). These leguminous plants are able to thrive in soils where nitrates are scarce, because

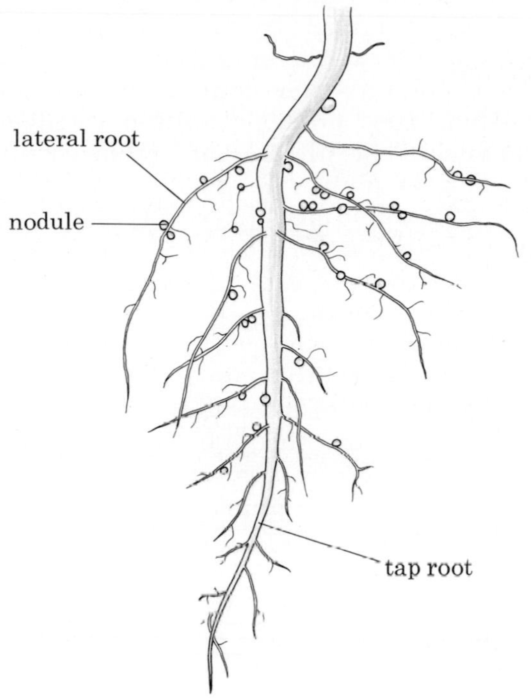

Figure 9 Root nodules of a leguminous plant

the nitrogen-fixing bacteria in their nodules make compounds of nitrogen available for them. Leguminous plants are also included in crop rotations (see page 38) to increase the nitrate content of the soil.

Processes which remove nitrates from the soil

Uptake by plants. Plant roots absorb nitrates from the soil and use them for making proteins (page 30).

Leaching. Nitrates are very soluble (i.e. dissolve easily in water), and as rain water passes through the soil it dissolves the nitrates and carries them away in the run-off or to deeper layers of the soil. This is called **leaching**.

Denitrifying bacteria. These are bacteria which obtain their energy by breaking down nitrates to nitrogen gas which then escapes from the soil into the atmosphere.

These processes are summed up in Figure 10. Although the diagram refers only to nitrogen, a similar cycle could be drawn for sulphur, phosphate, potassium and other minerals.

QUESTION

6 Very briefly explain the difference between nitrifying, nitrogen-fixing and denitrifying bacteria.

MANURING AND CROP ROTATION

Manuring

In a natural community of plants and animals, the processes which remove nitrates from and add nitrates to the soil are in balance. In agriculture, most of the crop is usually removed so that there is little or no organic matter left on

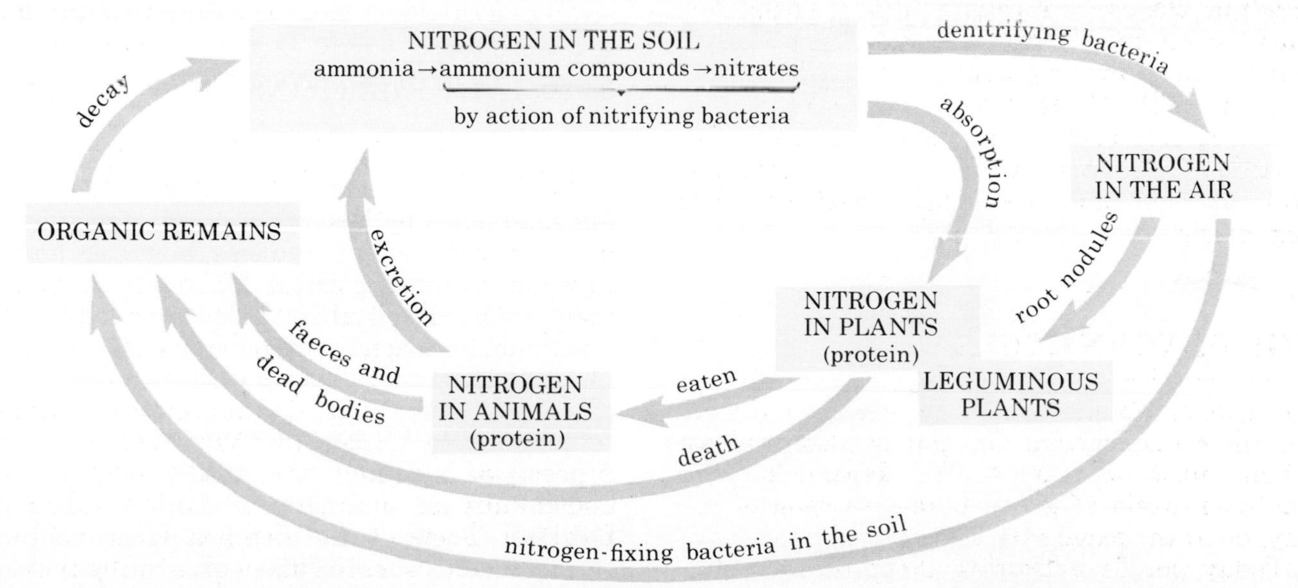

Figure 10 The nitrogen cycle

the soil for the nitrifying bacteria to act on. In a mixed farm (one with animals as well as plant crops) the animal manure, mixed with straw, is ploughed back into the soil and thus replaces the nitrates and other minerals removed by the crop. It also gives the soil a good structure and improves its water-holding properties (page 45).

When animal manure is not available in large enough quantities, artificial fertilizers are used. These are mineral salts made on an industrial scale. Examples are ammonium nitrate (for nitrogen), superphosphates (for phosphates) and compound NPK fertilizer for nitrogen, phosphorus and potassium (see page 30). These are spread on the soil in carefully calculated amounts to provide the minerals, particularly nitrogen, phosphorus and potassium that the plants need. Figures 11 and 12 show how these artificial fertilizers increase the yield of crops from agricultural land, but they do little to maintain a good soil structure because they contain no humus (see pages 40 and 54).

Figure 11 Wheat grown on experimental plots. The strips of wheat have received different amounts and types of fertilizer. The growth on each strip is measured to determine which fertilizer is most effective.

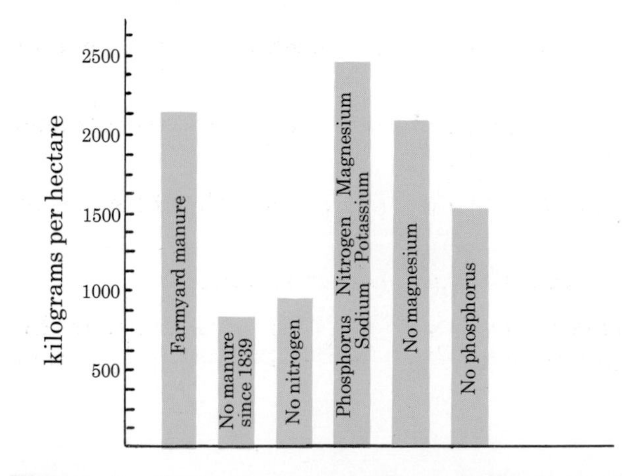

Figure 12 Average yearly wheat yields from 1852 to 1925; Rothamsted experimental station, Broadbalk Field.

Crop rotation

Different crops make differing demands on the soil; potatoes and tomatoes use much potassium, for example. By changing the crop grown from year to year, no single group of minerals is continuously removed from the soil. Leguminous crops such as clover and beans may help to replace the nitrogen content of the soil because their root nodules contain nitrogen-fixing bacteria. The nitrates are released into the soil when the plant dies and decays.

The use of artificial fertilizers has made crop rotation, at least for the reasons above, largely unnecessary. However, turning arable land over to grass for a year or two does improve the crumb structure of the soil (page 40), its drainage and other properties. Rotation also reduces the chances of infectious diseases that can enter the crop through the soil. For example, repeated crops of potatoes in the same field will increase the population of the fungus causing the disease 'potato blight'. If potatoes are not planted for a few years, the crop will suffer less from this disease when potatoes are grown again.

QUESTIONS

7 To judge from Figure 12 which mineral element seems to have the most pronounced effect on the yield of wheat? Explain your answer.

8 Draw up two columns headed A and B. In A list the processes which add nitrates to the soil and in B list those which remove nitrates from the soil. How might the activities of man alter the normal balance between these two processes?

SPECIAL RELATIONSHIPS

In this chapter the inter-relationships between organisms have been described in general terms and they apply to most of the plants and animals on the Earth. Some organisms, however, have a special relationship which does not fit any of the patterns described so far. Two examples of such relationships are parasitism and symbiosis.

Parasitism and symbiosis

A **parasite** is a plant or animal which feeds on another living organism without necessarily killing it (Figure 13). A flea is a parasite which gets its food by sucking small quantities of blood from a mammal such as a dog. The dog is called

Figure 13 A parasitic fungus. These are the spore-producing structures of a bracket fungus. The main part of the fungus is inside the tree trunk. Sooner or later the tree will be killed.

the **host** of the parasite, and the relationship between the two animals is called **parasitism**. Parasitic relationships are one-sided because the parasite gets all the benefit and the host gets none, though in many cases the host is not harmed by the parasite. Sometimes, however, the parasite does damage the health of its host. The bacteria which cause diseases such as typhoid and tuberculosis are parasites which live in the human body. Their activities and their waste products cause the symptoms of disease.

In the stomachs of cattle and sheep there live large numbers of bacteria. They cause no symptoms of illness and, in fact, they are thought to be of value to the animals because they help to digest the cellulose in its food (see page 211). The relationship in this case is called **symbiosis** and not parasitism. The cow benefits from the relationship because it is better able to digest grass. The bacteria are thought to benefit by having an abundant supply of food, though they are themselves digested when the grass moves along the cow's digestive tract.

The nitrogen-fixing bacteria in root nodules (page 37) provide a further example of symbiosis. The plant benefits from the extra nitrates that the bacteria provide, while the bacteria are protected in the plant's cells and can also use the sugars made by the plant's photosynthesis.

It is often difficult to produce good evidence that both organisms are getting some benefit from so-called symbiotic relationships and it is not always easy to make a distinction between parasitism, symbiosis and a number of other similar relationships.

BASIC FACTS

- **Green plants are the starting-point for food chains and food webs.**
- **Green plants are the producers; animals are consumers; bacteria and fungi are decomposers.**
- **All food production and most of the energy production on the Earth depend on sunlight.**
- **Even a small change in a food web is likely to affect all the animals and plants which are involved in the food web.**
- **The carbon cycle involves (a) the uptake of carbon dioxide by green plants for their photosynthesis and (b) the release of carbon dioxide from respiration and burning.**
- **The nitrogen cycle keeps a steady supply of nitrates in the soil mainly as a result of the activities of nitrifying and nitrogen-fixing bacteria.**
- **Soil fertility is improved by adding artificial fertilizers or manure, and by crop rotation.**
- **Parasitism and symbiosis involve a close individual relationship between two organisms. If the association benefits only one organism it is called parasitism. If it benefits both, it is called symbiosis.**

5 Soil

COMPONENTS OF SOIL

Soil is a mixture of (a) sand and clay particles, (b) humus, (c) water, (d) air, (e) dissolved salts and (f) bacteria and other micro-organisms (Figures 1 and 2).

(a) Sand and clay particles. These are formed from rocks which have been weathered and broken down by the action of wind, rain, rivers and glaciers. Sand particles are larger than clay particles and have a different chemical composition. The sand is mainly **silicon oxide** while clay is a mixture of **aluminium oxide** and silicon oxide. **Crumbs** of soil consist of groups of clay and sand particles stuck together by humus. The crumb structure of a soil affects its drainage, air content and general fertility.

(b) Humus. This is formed from decaying organic matter, e.g. dead remains of plants and animals, and animal droppings. Gardeners often use the term 'humus' to describe coarse organic material such as leaf mould, peat or compost.

Figure 1 A soil profile. Thin, chalky soil. The soil layer varies from a few centimetres to 30 or 40 cm. In this case, there is a layer of soil about 25 cm thick on top of limestone rock.

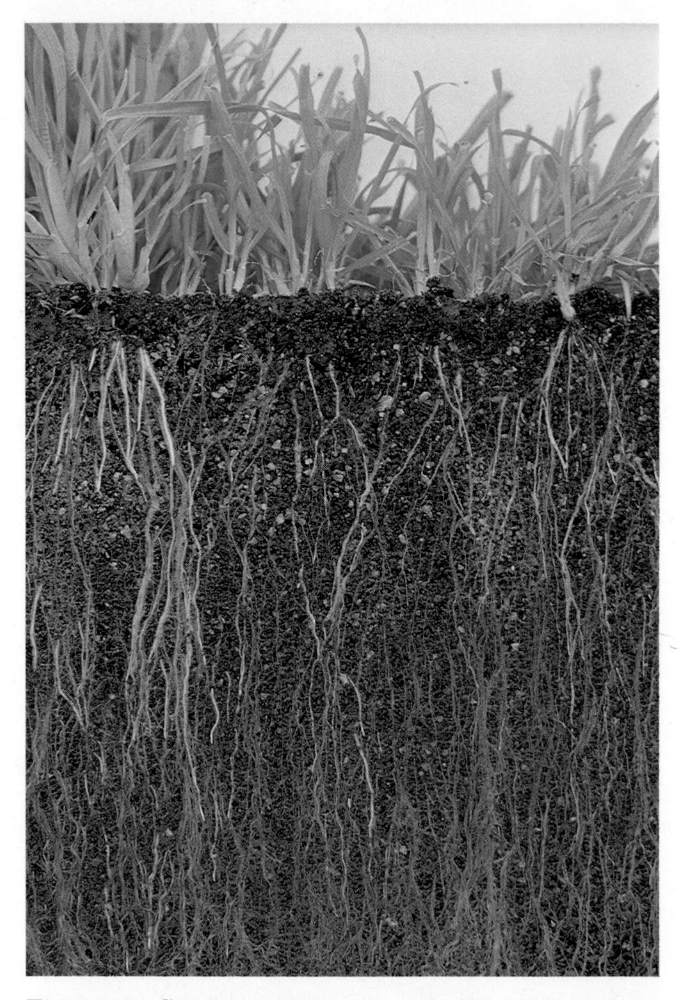

Figure 2 Good agricultural soil. The soil is much deeper than that in Figure 1, and the plant roots can be clearly seen.

When soil scientists refer to humus, however, they usually mean the organic material which actually forms part of the soil crumbs. This humus comes from the same decaying plant and animal material as the gardener's 'humus' but has decayed a bit further. It has become a black, jelly-like material which helps to 'glue' the sand and clay particles together in the soil crumbs.

Humus therefore makes an important contribution to the soil's crumb structure. Coarse humus makes the soil lighter—that is, easier to

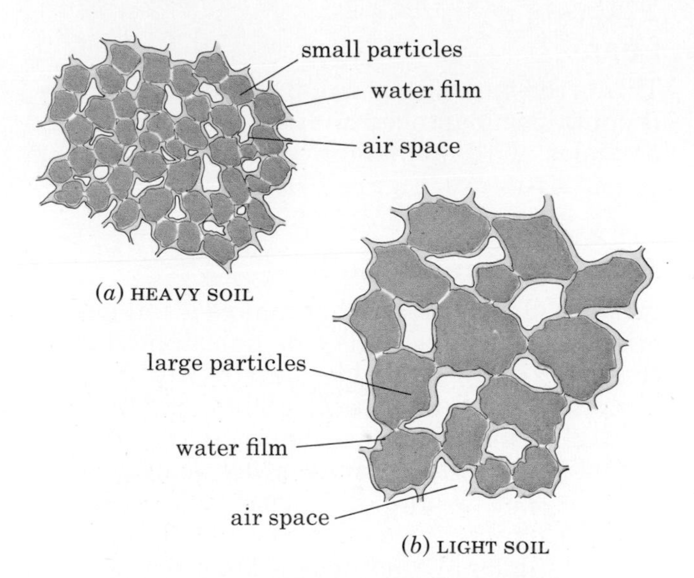

Figure 3 Structure of light and heavy soils

dig or plough—and its spongy texture helps to hold water in the soil. Both grades of humus are broken down by bacteria and release mineral salts such as nitrates and phosphates, as described on page 36, so increasing the fertility of the soil.

(c) Water. In a moist soil, water gets into the crumbs and forms a thin film over the particles (Figure 3). The water film is held to the particles by the force of **capillary attraction** (page 258). The root hairs (page 64) of plants come into close contact with the soil particles and absorb water from this film. In order to do this, however, they have to develop a 'suction' force big enough to overcome the capillary attraction.

(d) Air. Air fills the spaces between the soil crumbs. The larger the crumbs, the bigger will be the air spaces. The air in these spaces supplies the oxygen needed by the roots of plants and by the other organisms living in the soil. In a waterlogged soil, the air spaces become filled with water and so reduce this oxygen supply.

(e) Mineral salts. Salts of potassium, iron, magnesium, phosphates, sulphates and nitrates are present in solution in the soil water. They come from rock particles and from the action of bacteria on the organic matter, humus, in the soil (see 'The nitrogen cycle' on page 36). These salts are taken up by the roots of plants and used to build up the substances needed for their cells (see page 30).

(f) Soil organisms. The soil contains a wide variety of small animals such as mites, insects and worms (Figure 4). There is also a large population of micro-organisms such as fungi and bacteria. All these organisms affect the

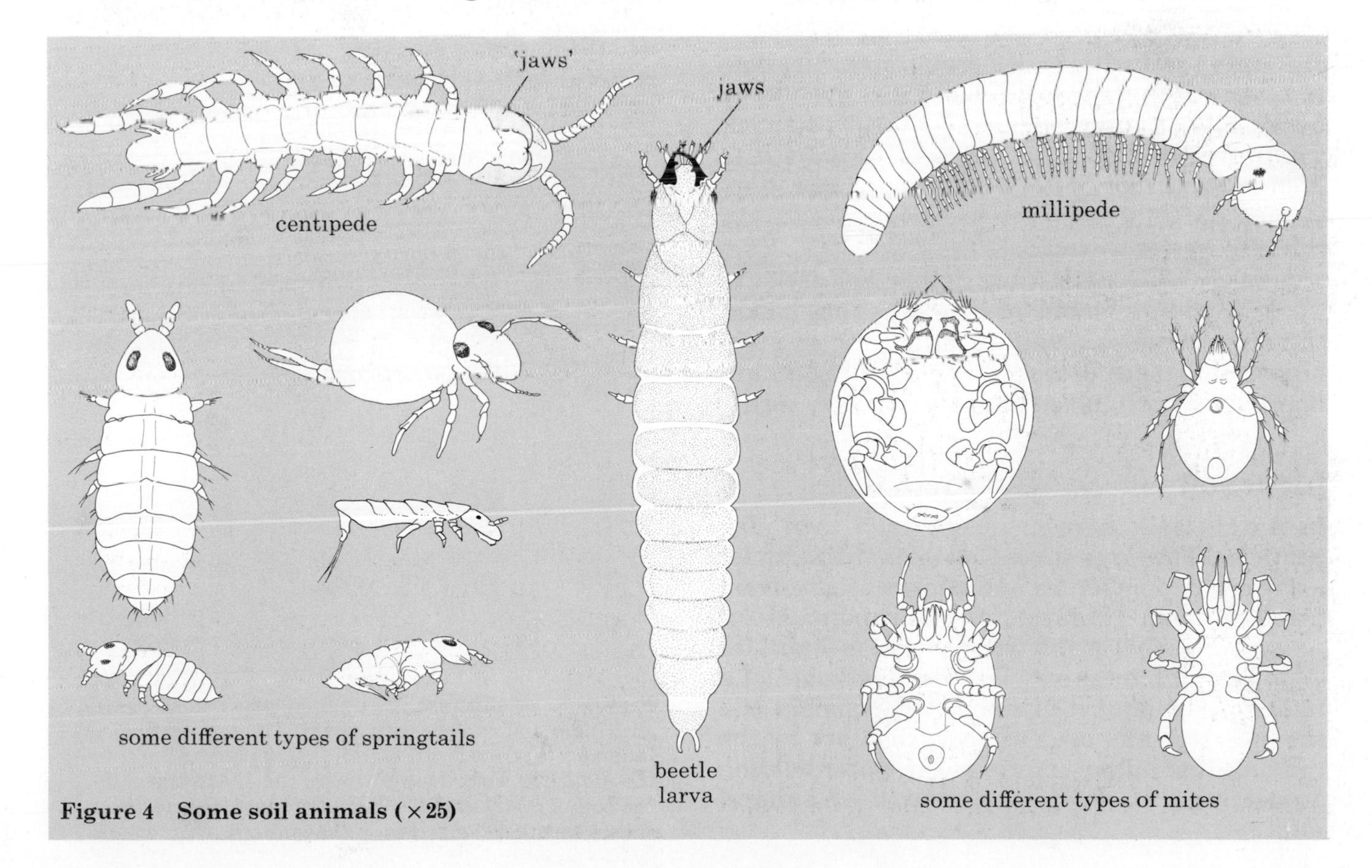

Figure 4 Some soil animals (×25)

structure and properties of the soil, but the fungi and bacteria are very important. They act on the organic remains of plants and animals, break them down to a fine humus and release the salts that are needed by plants.

QUESTIONS

1 List all the things you would expect to find if you carefully analysed a sample of soil.

2 What do plant roots require from the soil?

TYPES OF SOIL

Heavy soil

A heavy soil has a high proportion of clay particles. Partly because of their small size, the water film round the clay particles makes them stick together more strongly than sand particles. This makes the soil 'heavy' to dig or plough because it is sticky when wet. When it is dry it forms hard clods, which are difficult to break up. Because there are only small air spaces between clay particles, the amount of air in the soil is small. The small spaces also slow down the drainage of water through the soil (Figure 3*a*).

A heavy soil with a good crumb structure does not necessarily show all these disadvantages, because the crumbs behave like large particles. The slow drainage of a heavy soil allows it to hold more water in times of drought and stops the soluble salts being leached away (page 37) by the rain in wet weather.

A heavy soil can be made lighter and easier to work by adding humus or lime. The lime makes the clay particles clump together and so gives bigger air spaces between them. Adding humus improves the crumb structure of a heavy soil.

Light soil

In a light soil there are more sand than clay particles. The large spaces between the particles leave more room for air and allow water to drain through easily (Figure 3*b*). The sand particles are easy to separate so that the soil is 'light' for digging and ploughing. The disadvantages of a light soil are that it dries out very quickly and the mineral salts are easily washed out by the rain. Adding humus improves the water-holding properties and provides a source of mineral salts.

Loam

This is the best type of soil for most agriculture. It contains a balanced mixture of clay and sand particles with abundant humus and a good crumb structure. Figure 2 shows a loam soil.

Acid and alkaline soils

It is not always clear what makes a soil acid or alkaline. Soil in a chalky or limestone district (Figure 1) will probably be slightly alkaline because of the particles of calcium carbonate in the soil. A light, sandy soil may be slightly acid because the rain washes away soluble compounds that would otherwise neutralize the acids. A chalky soil could have a pH (see page 260 and Figure 5) as high as 8. Most British soils are slightly acid with pH values of 6.4–6.9. The

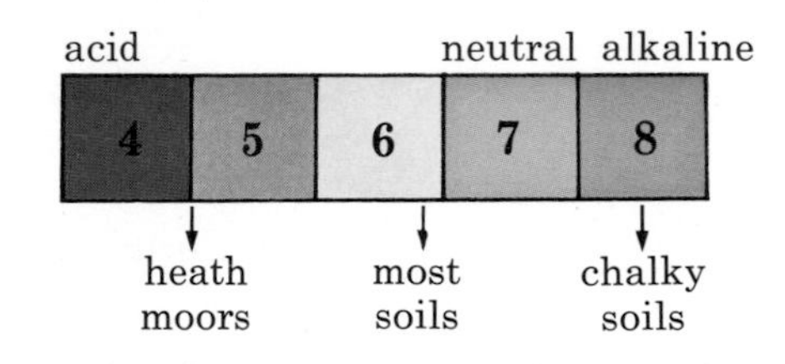

Figure 5 The pH scale of acidity

very acid soil of a heath or moorland (Figure 7) could have a pH as low as 4.5. Acid soils are usually infertile because the acidity makes the mineral salts very soluble and so they are easily washed away by the rain. In very alkaline soils

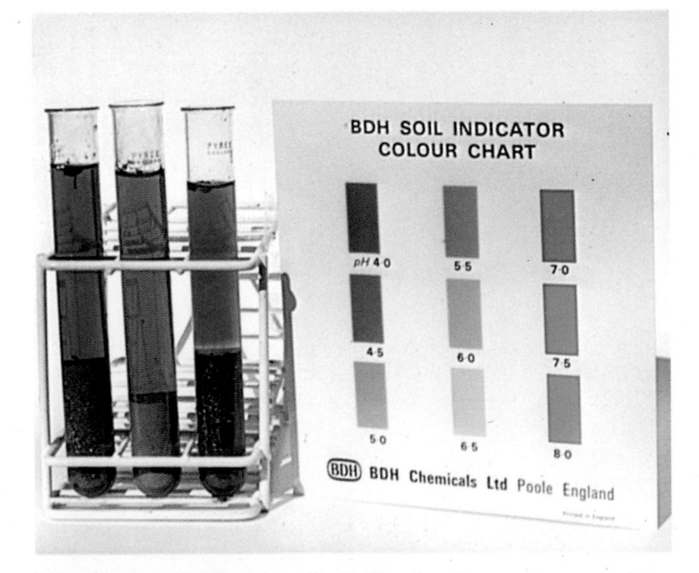

Figure 6 A pH test. The pH of soil can be tested by shaking a soil sample with water. The soil settles and an indicator is added to the water. By comparing the colour of the indicator with the colours on the chart, the acidity of the soil can be judged. What pH values do you think are shown by the soil samples in the photograph?

Figure 7 Heathland. The bracken and heather are typical of a light, sandy and probably acid soil.

there may be plenty of minerals, but they are so insoluble that the plants cannot absorb them.

Some plants are better able to tolerate acid conditions than others. Lettuce and onions, for example, will not grow well if the soil is more acid than pH 6, whereas rhododendrons and heather will only grow well in acid soils with a pH as low as 5.5. In agriculture, acid soils are treated with lime to raise the pH to about 6.5.

Soil fertility may be increased by adding manure and artificial fertilizers or by crop rotation. These processes are discussed more fully on pages 30 and 38.

QUESTIONS

3 In each case state one advantage and one disadvantage of (a) a light soil, (b) a heavy soil

4 What agricultural practices and climatic conditions might affect the amount of air in a soil?

5 In what two ways might lime improve a soil's fertility?

6 Revise pages 38 and 40 and then make a list of all the things you could do to improve the fertility of agricultural land.

LIFE IN THE SOIL

Except to the keen gardener, soil may look dull and lifeless, but in fact it contains millions of organisms moving, growing, reproducing and competing with each other for food. At the bottom of the food pyramid (page 33) are the bacteria which get their food from the humus by digesting it and absorbing the solutions into their cells. Soil fungi are tiny, thread-like structures which feed in a similar way to bacteria.

The bacteria and fungi are important in the soil because they break down the humus and re-cycle the minerals (see page 40). Some bacteria and single-celled plants also 'fix' nitrogen (see page 36) and so increase the nitrates in the soil. Soil bacteria and fungi are mostly classed as saprophytes (page 34) because they digest dead matter and absorb the soluble products.

Other soil organisms are threadworms (nematodes) and a large number of tiny insects such as the springtails shown in Figure 4. The spider-like mites may feed on decaying vegetation, on the bacteria, or on the single-celled animals which eat the bacteria.

Several different types of earthworm live in burrows which they make in the soil. They swallow the soil and pass it through their digestive tract. As it passes through, some of the humus is dissolved by enzymes and absorbed into the worm's body. The undigested soil is passed out in the burrows or on to the surface of the ground as worm casts. The worm casts dry out, crumble and so continually add a layer of fine soil to the surface. The worm's burrows (Figure 8) improve the soil's drainage

Figure 8 Earthworm in its burrow

and some types of worm pull leaves into their burrows to feed on them. This increases the humus content of the deeper layers of the soil. In general, it is thought that earthworms improve soil fertility by helping to mix up the humus-rich and the humus-poor layers. It must be admitted, however, that there is not much experimental evidence to show that the productivity of a soil is increased by earthworm activity.

Some of the predators at the top of the food chains in the soil are the centipedes and beetle larvae. Figure 4 shows the large pointed 'jaws' of these animals. Centipedes spear small insects, such as springtails, with their 'jaws' and inject a paralysing poison before eating them.

Some of the soil organisms may be harmful to crops. The 'leather-jacket' larva of the daddy-long-legs eats the roots of wheat plants; some nematode worms damage potato plants (potato root 'eelworm'). These harmful organisms can be killed by putting **pesticides** (poisons which kill pests) in the soil. There is always a danger that these pesticides will kill a great many of the harmless organisms as well and so upset the balance of life in the soil (Figure 16 on page 54). Any change in this balance is likely to affect the structure and fertility of the soil.

QUESTION

7 (a) Construct a simple food chain which involves soil organisms.

(b) Draw a cycle similar to Figure 7 on page 35 but using only organisms which live in the soil. Are there any 'producers' in the soil?

EXPERIMENTS WITH SOIL

1 Observation of mineral particles in soil

Some dry soil is sieved to remove stones and particles larger than about 3mm. The soil is crushed lightly to break up the crumbs and 50 grams of it is placed in a small, flat-sided bottle. The bottle is filled with water almost to the top, the cap screwed on and the bottle shaken for at least 30 seconds to disperse the soil throughout the water. The bottle is then allowed to stand for 10 to 15 minutes so that the soil settles down (Figure 9). The large particles will fall most rapidly and form the bottom layer. Smaller particles will remain suspended in the water. The larger particles of humus will float to the top. As the soil particles settle, they may form distinct layers. The boundaries between the

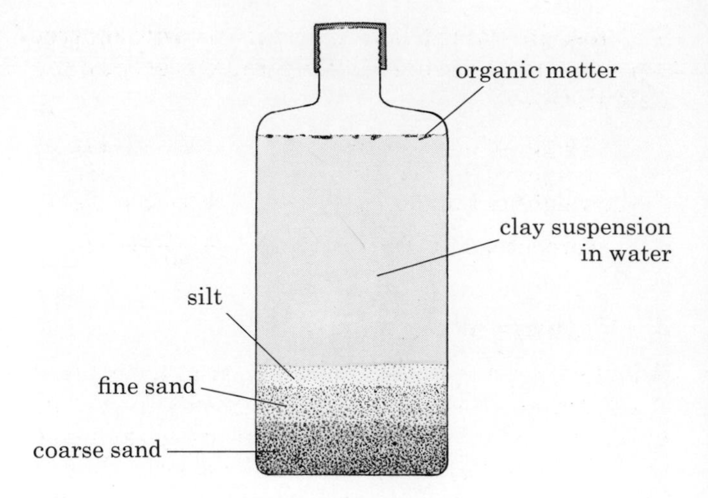

Figure 9 Mineral particles in the soil

layers can be marked on the outside of the bottle with a marker pen and the depths measured. The composition of different soil samples can then be compared.

2 Weight of water in the soil

A sample of soil is placed in a weighed evaporating basin which is then weighed again. The basin is heated in an oven at 100°C for two days to drive off the soil water. The basin and soil are then weighed again. Heating and reweighing are repeated until two weighings give the same result, showing that all the soil water has been evaporated.

The difference between the second and final weighings will give the weight of water that was present at first. Temperatures higher than 100°C must not be used because they will cause the humus to burn away and so give an extra loss of weight. Here is a sample calculation:

weight of basin		200g
weight of basin + moist soil		250g
∴ weight of soil	250 − 200 =	50g
final weight of basin + dry soil		240g
∴ loss in weight	250 − 240 =	10g
percentage water in moist soil	$\frac{10 \times 100}{50}$	= 20%

3 Weight of humus in soil

The dry soil from the previous experiment is placed in a metal tray and heated strongly over a bunsen flame to burn off all the humus. Heating is continued until smoke stops coming off and the charred organic matter has disappeared. This leaves only the grey or reddish mineral particles. When cool, the soil is weighed again.

The loss in weight is due to the organic matter being burnt to carbon dioxide and water. Sample calculation:

weight of dry soil	40g
weight of 'burnt' soil	38g
∴ weight of humus	40−38= 2g
percentage humus in dry soil	$\frac{2 \times 100}{40} = 5\%$

4 Volume of air in soil

A large glass jar is filled with water and the level A is marked. A small tin can is placed in the jar and removed full of water so that the level drops to B (Figure 10*a*). The volume of water in the tin can is found by pouring it into a measuring cylinder. The can is now stamped, open end down, into the soil until its base is level with the soil (Figure 10*b*). The can is dug out and the soil cut level to the top of the can. The volume of soil in the can is the same as the volume of water removed from the glass jar. The soil is now loosened from the can and tipped into the glass jar, breaking up all the lumps at the same time (Figure 10*c*). Adding the soil to the jar will raise the water level again, but because air has escaped from the soil, the level will not return to A. Water is now added to the jar from a measuring cylinder to bring the level back to A (Figure 10*d*). The volume of water added, is the same as the volume of air which escaped from the soil. Sample calculation:

volume of the tin can	200 cm³
volume of water added to jar	28 cm³
∴ volume of air in 200 cm³ soil	28 cm³
percentage air in soil	$\frac{28 \times 100}{200} = 14\%$

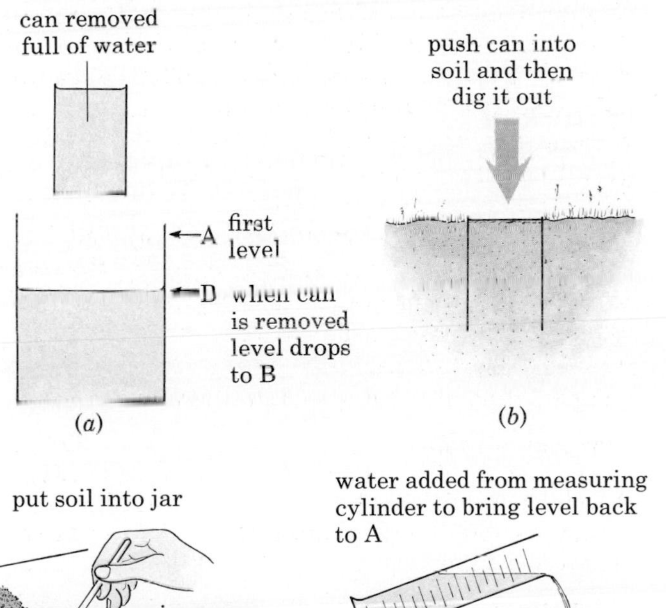

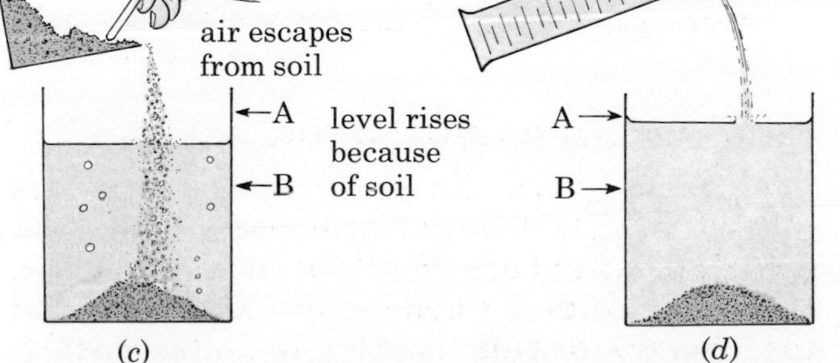

Figure 10 To measure the volume of air in a soil sample

5 The permeability (porosity) of soil

The permeability of a soil means how easily water will pass through it. Two glass funnels are plugged with glass wool and half-filled with equal volumes of sandy and clay soils respectively. Both are then covered with water and the levels kept the same by topping up during the experiment. Thus, there will be no difference in the water pressure in the two funnels (Figure 11). The water that runs through

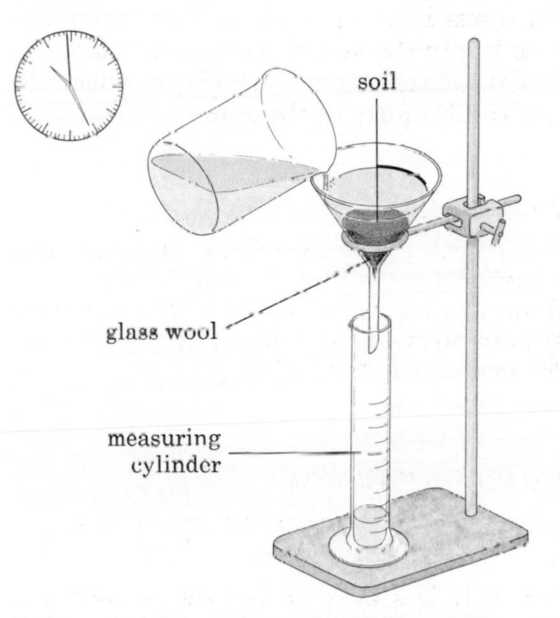

Figure 11 Comparing the permeability of soils to water. (Only one of the two funnels is shown.)

in a given time is collected in a measuring cylinder. The sandy soil is more permeable to water than a clay soil and will allow far more water to run through in a fixed time.

6 A comparison of the water-holding properties of soils

Two equal sized plastic cups have their bases perforated and lined with glass fibre. One cup is partly packed with dry sand and the other is

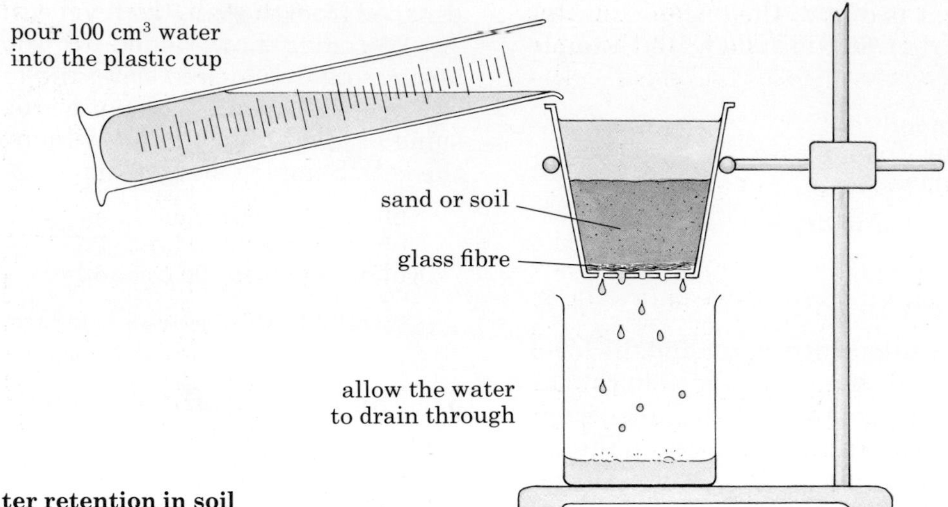

Figure 12 Water retention in soil

packed to the same depth with dry soil, rich in organic matter. The cups are supported, in turn, over a beaker and a known volume of water is poured through the soil from a measuring cylinder (Figure 12). When all the water has drained through, it is returned to the measuring cylinder to see how much has been retained by the soil. The experiment is repeated with the other sample using the same volume of water as before.

It is expected that the soil rich in organic matter will hold a greater proportion of water than the sand, thus showing one of the advantages of organic matter in the soil.

7 Capillary attraction in soil

Two glass tubes 1cm or more wide, and about 50cm long are packed with dry sand. One tube contains fine sand and the other contains coarse sand. The ends are plugged with glass wool and the tubes clamped upright in a beaker of water (Figure 13). The water travels up through the sand by capillary attraction (page 258), and its level can be seen by the darker colour of the sand. The levels are measured and compared after one day. The results show that water travels further in the fine sand. The smaller the particles in a soil, the greater is the capillary attraction.

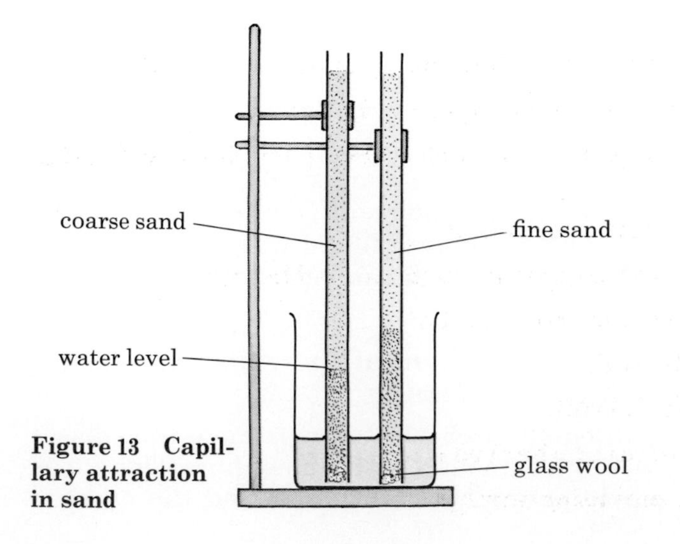

Figure 13 Capillary attraction in sand

8 The presence of micro-organisms in the soil

Two Petri dishes of sterile, vegetable agar are prepared as described on page 47. In one of them are sprinkled some particles of soil (Figure 14)

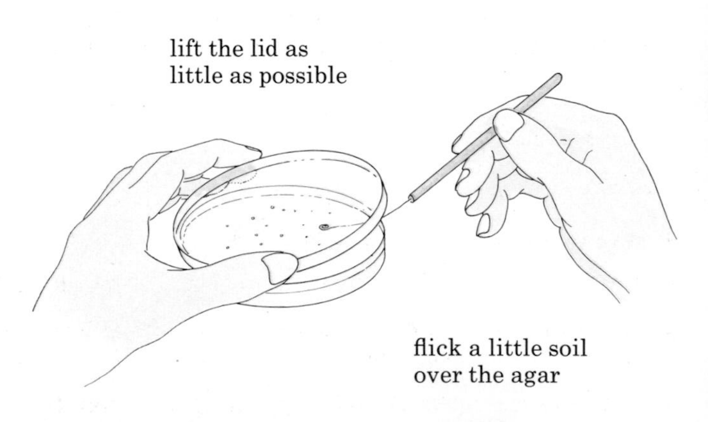

Figure 14 Culturing micro-organisms from soil

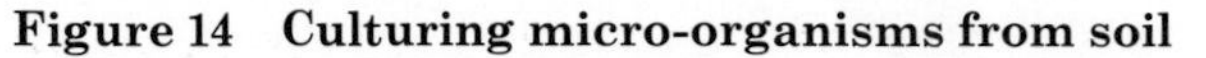

and in the other, the control, are scattered some particles of sand that have been sterilized by heating. If bacteria or fungi are present in the soil, they will appear as colonies on the surface of the agar in a few days. The absence of any colonies from the control will prove that the micro-organisms were in the soil and did not come from the air, from the dish, from the agar or from the instruments used to scatter the particles.

9 Extraction of some of the larger soil animals

An apparatus such as the one in Figure 15 is set up with a soil sample. The heat from the lamp gradually dries out the soil from the top and so drives the insects and mites down into the Petri dish where they are trapped in a preserving fluid. After a few days, the 'catch' is examined under the low power of the microscope.

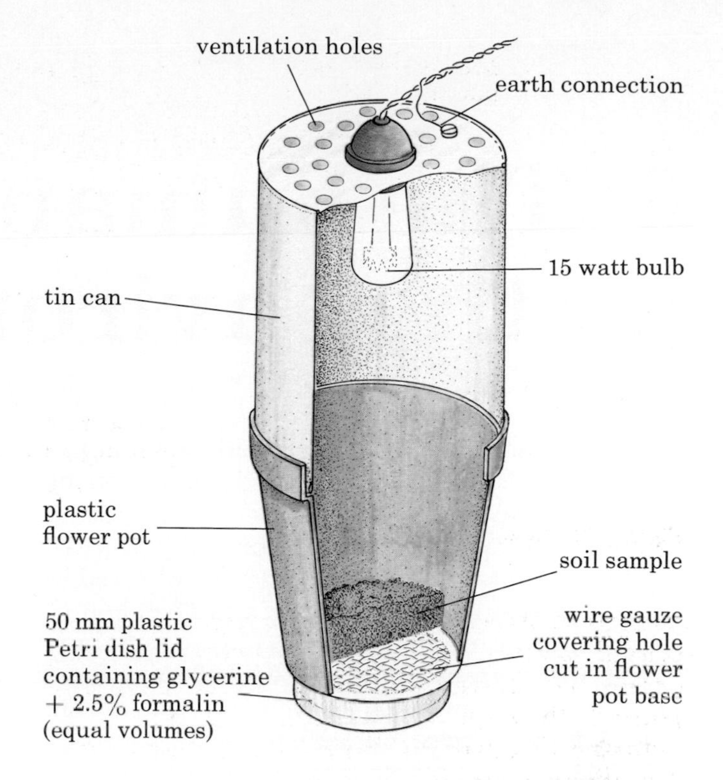

Figure 15 Extraction of small animals from the soil

QUESTIONS

8 Why is it necessary to start with dry soil in order to measure the weight of humus in the soil?

9 Would you expect a soil which is very permeable to water (Experiment 5) to also retain a lot of water (Experiment 6)? Explain your answer.

Preparation of culture plates for soil micro-organisms

This should be done by the teacher or technician.

The Petri dishes are sterilized by super-heated steam in an autoclave or pressure cooker for 15 minutes at a pressure of 1 kg per cm^2. This kills any bacteria that are already on the glassware. The culture medium is made by adding 2 g agar (page 258) and 25 cm^3 vegetable juice (e.g. tomato juice) to 100 cm^3 hot distilled water. This mixture is stirred until all the substances are dissolved, and then sterilized in an autoclave. When it is fairly cool but still liquid, the agar is poured into the Petri dishes which are covered at once. The agar sets to a jelly when cool. After the soil particles have been added to the dishes, the lids should be sealed on with adhesive tape and not opened again. When the experiment is finished, the dishes must be sterilized before opening them up and disposing of the cultures.

BASIC FACTS

- **Soil crumbs are made up of clay and sand particles stuck together with humus.**
- **The amount of air and water in a soil depends partly on the size of the soil crumbs.**
- **A heavy soil is one with a lot of clay particles; it is sticky, hard to work and may not drain well if the crumb structure is poor.**
- **A light soil has more sand particles, is easy to work and drains well.**
- **Humus is formed from decaying organisms or parts of organisms such as dead leaves.**
- **Humus makes the soil better drained, easier to work and more fertile.**
- **Plant roots absorb water and mineral salts from the soil.**
- **Large numbers of bacteria and fungi are present in the soil.**
- **They play an important part in breaking down plant and animal remains.**
- **There are many other organisms living in the soil, including earthworms.**

6 The human impact on the environment

A few thousand years ago, most of the humans on the Earth probably obtained their food by gathering leaves, fruits or roots and by hunting animals. The population was probably limited by the amount of food that could be collected in this way. Human faeces, urine and dead bodies were left on or in the soil and so played their part in the nitrogen cycle (page 36). Life may have been short and many babies may have died from starvation or illness, but man fitted into the food web and nitrogen cycle like any other animal.

Once agriculture had been developed, it was possible to support much larger populations and the balance between man and his environment was upset. The effects of the human population on the environment are complicated and difficult to study. They are even more difficult to forecast. In their ignorance, humans have destroyed many plants and animals, and great areas of natural vegetation. Unless humans limit their own numbers and treat their environment with more care and understanding, they could make the Earth's surface impossible to live on and so cause their own extinction.

The account which follows mentions just some of the ways in which our activities damage our own environment.

THE EFFECT ON FOOD WEBS

The hunting of animals

One obvious way to upset a food web is to remove some of the plants or animals which form part of it. If the tawny owls were removed from the food web in Figure 1 we would expect the numbers of shrews to increase because fewer were being eaten by the owls. The numbers of woodlice and earthworms might then go down because there were more shrews to eat them. The effect of the rabbit disease, myxomatosis, on a food web has been described on page 34.

In 1910, in the Grand Canyon (USA) national game reserve, an attempt was made to protect the deer population by shooting the animals which ate them. These were cougars, wolves, bobcats and coyotes. After fourteen years, the deer population had increased from about 4000 to 100000, and the environment could not support them. The grass was overgrazed, the trees and young shrubs were destroyed by browsing and the deer were dying from starvation in large numbers. Human interference with the food web had not only destroyed hundreds of cougars, wolves and coyotes but threatened to lead to the destruction of the environment and the deer which lived there (Figure 2).

More commonly, humans kill animals for food, profit or 'sport'. Over-fishing has reduced

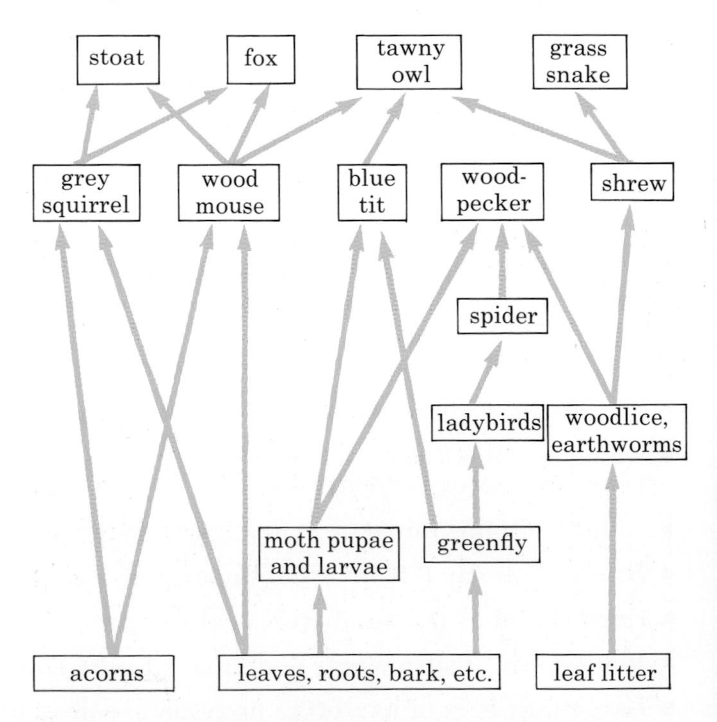

Figure 1 The food web of an oak tree (only a small sample of animals is shown) (From P. W. Freeland, *School Science Review*, 1973)

some fish stocks to the point where they cannot reproduce fast enough to keep up their numbers.

Animals like the leopard and tiger have been reduced to dangerously low levels by hunting, in order to sell their skins (Figure 3). The blue whale's numbers have been reduced from about 2000000 to 6000 as a result of intensive hunting.

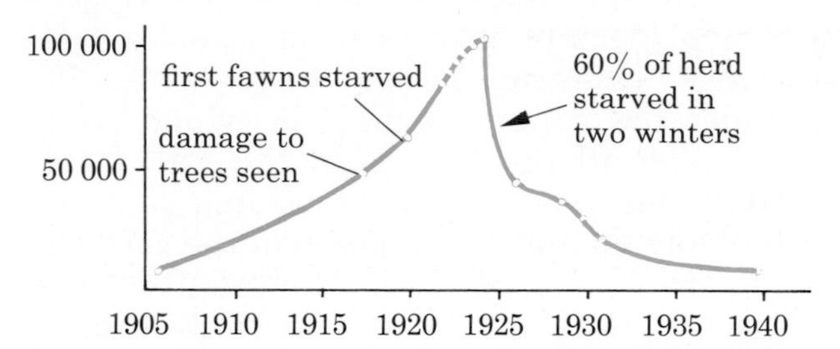

Figure 2 The result of human interference with a food web: changes in deer population after killing predators (From J. A. Barker, *In the Balance*, Evans, 1975)

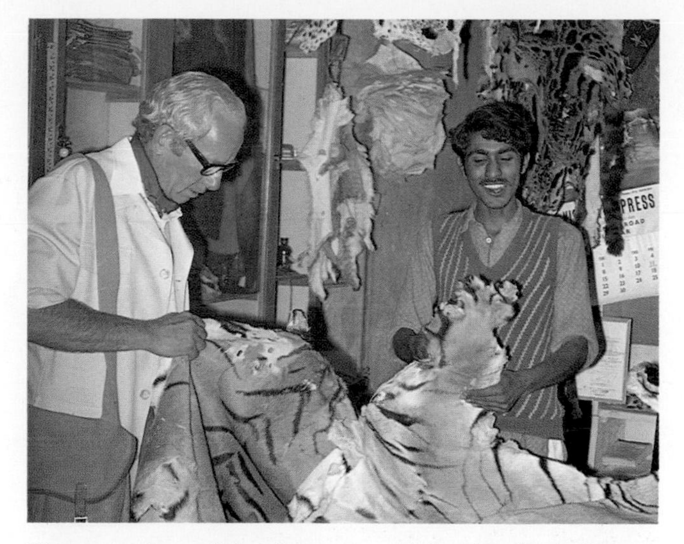

Figure 3 One of the threats to wild animals. As long as the sale and use of skins is permitted, there will be people ready to kill these animals.

Agriculture

(a) Monoculture. The whole point of crop farming is to remove a mixed population of trees, shrubs, wild flowers and grasses and replace it with a dense population of only one species such as wheat or beans (Figures 4 and 5). This is called a **monoculture**. Figure 1 is a simplified diagram of a food web which can be supported by a single oak tree. Similar food webs could be constructed for the grasses and wild flowers and shrubs. Clearly, a field of barley could not support such a mixed population of creatures. Indeed, every attempt is made to destroy any organisms such as rabbits, insects or pigeons, which try to feed on the crop plant.

So, the balanced life of a natural plant and animal community is pushed out from the land taken over for agriculture. It is left to survive only in small areas of woodland, heath or hedgerow. We have to decide on a balance between the amount of land to be used for agriculture, roads or building and the amount of

Figure 4 Natural vegetation. A mixed population of wild plants growing naturally on open ground.

Figure 5 A monoculture. Only barley is allowed to grow. All competition is removed.

Figure 6 Bulldozing a road through a rain forest. The road not only destroys the natural vegetation, it also opens up the forest to further exploitation.

land left alone in order to keep a rich variety of wild-life on the Earth's surface (Figure 6).

(b) Pesticides. This is a general name for any chemicals which destroy agricultural pests. In order to maintain a monoculture, plants which compete with the crop plant for root space, soil minerals and sunlight are killed by chemicals called **herbicides** (Figures 7 and 8). The crop plants are protected against fungus diseases by spraying them with chemicals called **fungicides** (Figure 9). To destroy insects which eat and damage the plants, the crops are sprayed with **insecticides**.

Pesticide	*Kills*
insecticide	insects
fungicide	parasitic fungi
herbicide	'weed' plants

The trouble with nearly all these pesticides is that although they kill the harmful organisms, they also kill harmless or beneficial ones.

In about 1960, a group of chemicals, including one called **dieldrin**, were used as insecticides to kill wireworms and other insect pests in the soil. It was also used as a 'seed dressing'. If seeds were dipped in the chemical before planting, it prevented certain insects from attacking the seedlings. This was thought to be better than spraying the soil with dieldrin which would have killed all the insects in the soil. Unfortunately pigeons, rooks, pheasants and partridges dug up and ate so much of the seed that the dieldrin poisoned them. Thousands of these birds were poisoned and, because they were part of a food web, birds of prey and foxes, which fed on them, were also killed.

Figure 8 Weed control in kale. The area in the foreground has been treated with a herbicide which has prevented the growth of the yellow-flowered charlock.

Figure 7 Weed control by spraying. The field has been planted with sugar beet seed. Before the seed germinates, the field is sprayed to stop weeds growing.

Figure 9 Control of fungus diseases. The barley on the right is grown from seed treated with a fungicide. The untreated barley (left) is affected by a mildew fungus.

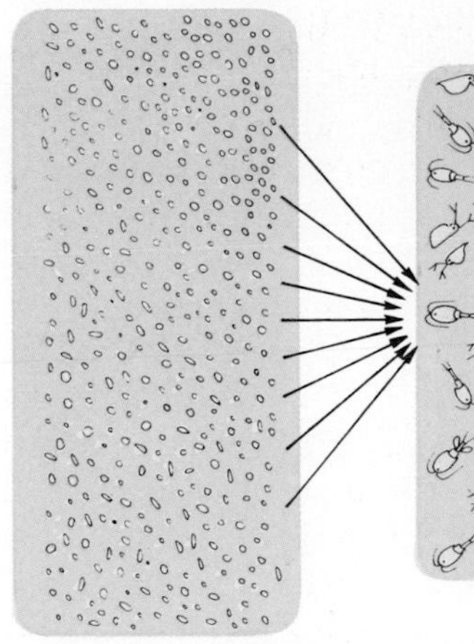

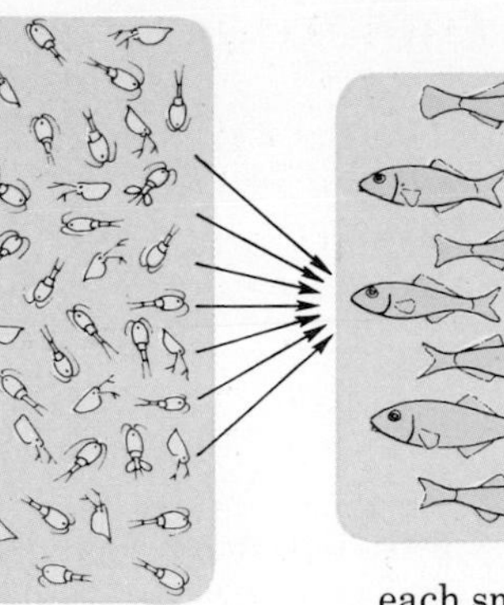

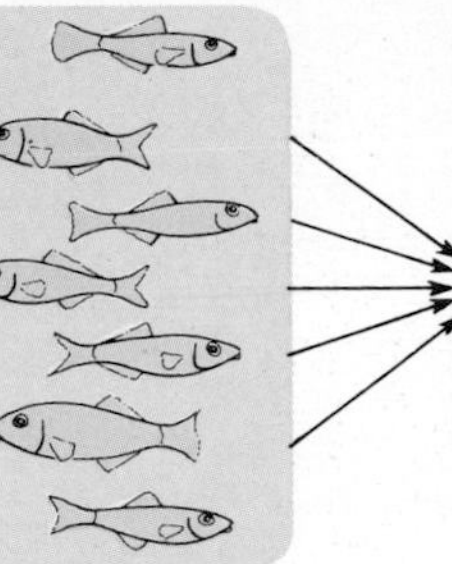

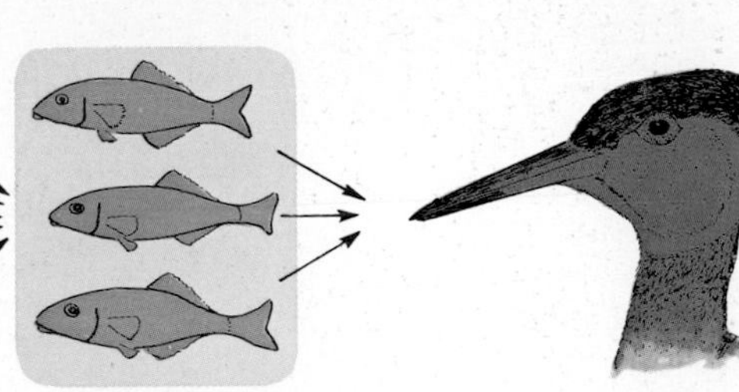

Figure 10 How a pesticide becomes more concentrated as it moves along a food chain. The intensity of colour represents the concentration of DDT.

The concentration of insecticide increases as it passes along a food chain (Figure 10). Clear Lake in California was sprayed with DDT to kill gnat larvae. The insecticide made only a weak solution of 0.015 parts per million (ppm) in the lake water. The microscopic plants and animals which fed in the lake water built up concentrations of about 5 ppm in their bodies. The small fish which fed on the microscopic animals had 10 ppm. The small fish were eaten by larger fish, which in turn were eaten by birds called grebes. The grebes were found to have 1600 ppm of DDT in their body fat and this high concentration killed large numbers of them.

These new insecticides had been thoroughly tested in the laboratory to show that they were harmless to man and other animals when used in low concentrations. It had not been foreseen that the insecticides would become more and more concentrated as they passed along the food chain.

Insecticides like this are called 'persistent' because they last a long time without breaking down. This makes them good insecticides but they also 'persist' for a long time in the soil, in rivers, lakes and the bodies of animals, including man. This is a serious disadvantage.

Eutrophication

On page 30 it was explained that plants need a supply of nitrates for making their proteins, and a source of phosphates for many chemical reactions in their cells. The rate at which plants grow is often decided by how much nitrate and phosphate they can obtain. In recent years, the amount of nitrate and phosphate in our rivers and lakes has been greatly increased. This is called **eutrophication** and the following processes are the main causes.

(a) Discharge of treated sewage. In a sewage treatment plant, human waste is broken down by bacteria and made harmless, but the breakdown products include phosphates and nitrates. When the water from the sewage treatment is discharged into rivers, the large quantity of phosphate and nitrate it contains allows the microscopic plant-life to grow very rapidly (Figure 11).

Figure 11 Growth of algae in a canal. Abundant nitrate and phosphate from treated sewage and from farmland make this growth possible.

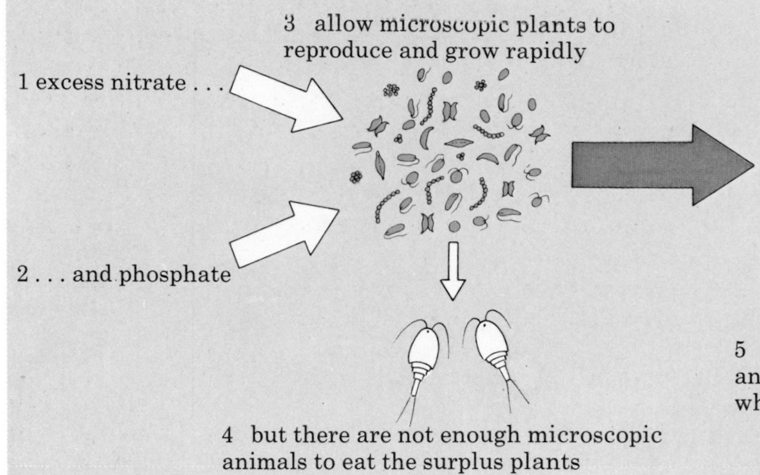

Figure 12 Effects of eutrophication

(b) Use of detergents. Detergents contain a lot of phosphate. This is not removed by sewage treatment and is discharged into rivers. This large amount of phosphate also encourages growth of microscopic plants.

(c) Agriculture. Intensive agriculture has led to nitrates being washed out of the soil by rain. The nitrates find their way into streams and rivers where they allow the overgrowth of the microscopic green plants.

(d) 'Factory farming'. Chickens, calves and pigs are often reared in large sheds instead of in open fields. Their urine and faeces are washed out of the sheds with water. If this mixture gets into streams and rivers it supplies an excess of nitrates and phosphates for the microscopic plants.

The microscopic plants are at the bottom of a food chain but with the extra nitrates and phosphates from the processes listed above, they increase far too rapidly to be kept in check by the microscopic animals which normally eat them. So they die and fall to the bottom of the river or lake. Here, their bodies are broken down by bacteria. The bacteria need oxygen to carry out this breakdown and the oxygen is taken from the water (Figure 12). So much oxygen is taken that the water becomes deoxygenated and can no longer support animal life. Fish and other organisms die from suffocation (Figure 13).

It is possible to reduce eutrophication by (1) using detergents with less phosphates, (2) using agricultural fertilizers that do not dissolve so easily and (3) using animal wastes on the land instead of letting them reach rivers.

QUESTIONS

1 What might be the effect of the removal of earthworms from the food web in Figure 1?

2 Give five examples of a monoculture.

3 What might be the effect on the food web of Figure 1 of spraying the tree with an insecticide?

4 At one time, elm trees were sprayed with DDT to kill the beetles which carried Dutch elm disease. In the autumn, the sprayed leaves fell to the ground and were eaten by earthworms. From the food web in Figure 1, suggest what effects this might have had on other organisms.

5 Explain briefly why too much nitrate could lead to too little oxygen in river water.

Figure 13 Fish killed by pollution. The water may look clear but is so short of oxygen that the fish have died from suffocation.

THE EFFECT ON THE SOIL

Soil erosion

Bad methods of agriculture lead to soil erosion. This means that the soil is blown away by the wind, or washed away by rainwater. Erosion may occur for a number of reasons.

(a) Cutting down trees. Trees can grow on hillsides even when the soil layer is quite thin. When the trees are cut down and the soil is ploughed, there is less protection from the wind and rain. Heavy rainfall tends to wash the soil off the hillsides into the rivers. The hillsides are left bare and the rivers become choked up with mud which can cause floods (Figures 14 and 15).

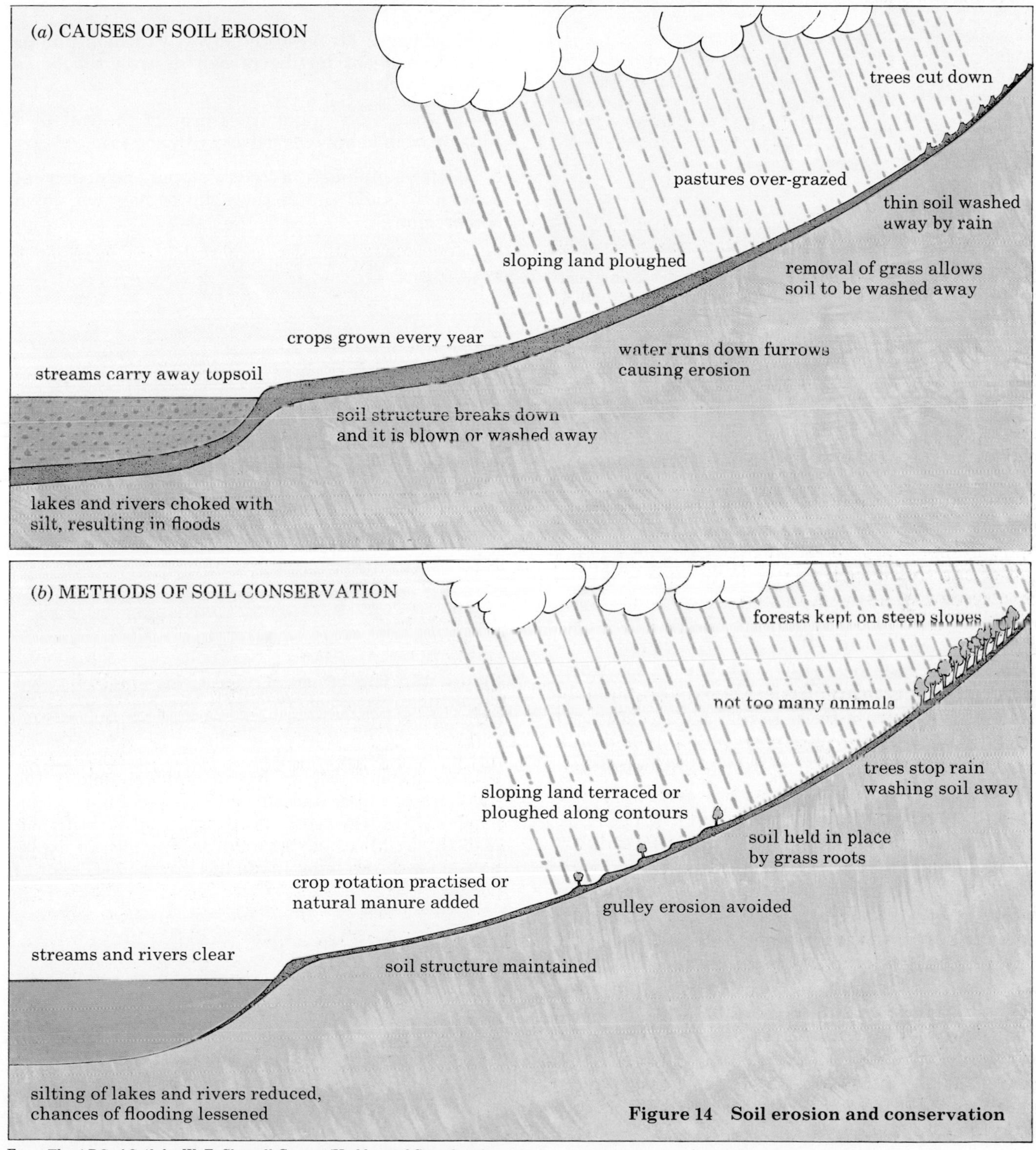

Figure 14 Soil erosion and conservation

From *The ABC of Soils* by W. E. Shewell-Cooper (Hodder and Stoughton)

Figure 15 Soil erosion. If trees are removed from steeply sloping ground, there is a danger that the topsoil will be eroded. The tussock grass in this area has not prevented the soil being washed away by the rain.

(b) Bad farming methods. If land is ploughed year after year and treated only with chemical fertilizers, the soil's crumb structure (see page 40) may be destroyed and it becomes dry and sandy. In strong winds it can be blown away as dust, leading to the formation of 'dust bowls', as in central USA in the 1930s, and even to deserts.

(c) Over-grazing. If too many animals are kept on a pasture, they eat the grass down almost to the roots, and their hooves trample the surface soil into a hard layer. As a result, the rain water will not penetrate the soil and so it runs off the surface, carrying the soil with it.

Use of pesticides

The effect of insecticides on food webs was described on page 50. When insecticides get into the soil, they kill the insect pests but they also kill other organisms. The effects of this on the soil's fertility are not very clear. An insecticide called **aldrin** was found to reduce the number of species of soil animals in a pasture to half the original number. Ploughing up a pasture also reduces the number of species to the same extent, so the harm done by the insecticide is not obvious.

Earthworms are not much affected by insecticides like DDT but they have been killed by insecticides containing copper. When earthworms are killed, the soil quality gets poorer because humus is no longer brought into the deeper layers of soil by the earthworm's feeding activities.

QUESTIONS

6 Read pages 38 and 40 and say why the continuous use of chemical fertilizers can destroy the soil's crumb structure.

7 In what ways might trees protect the soil in a hillside from being washed away by the rain?

8 If a farmer ploughs a steeply sloping field, in what direction should the furrows run to help cut down soil erosion?

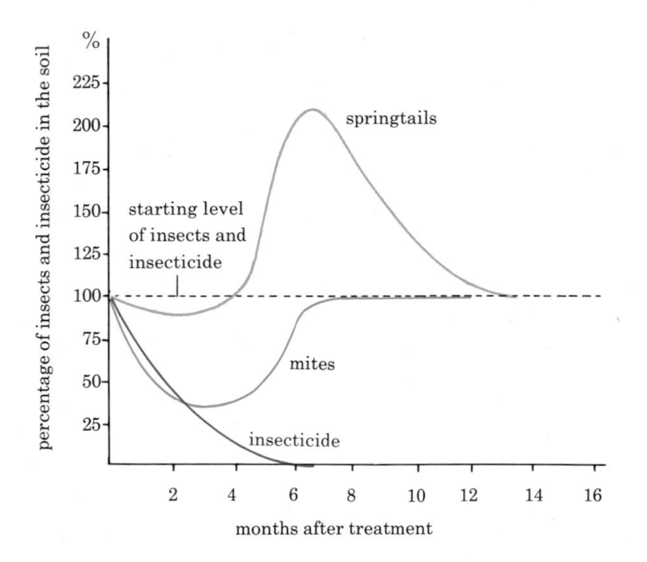

Figure 16 The effect of insecticide on some soil organisms (From *Soil Pollutants and Soil Animals* by Clive A. Edwards, © 1969 Scientific American Inc.)

9 The graph in Figure 16 shows the change in the numbers of mites and springtails (see Figure 4 on page 41) in the soil after treating it with an insecticide. Mites eat springtails. Suggest an explanation for the changes in numbers over the 16-month period.

THE EFFECT ON WATER

Human activity sometimes pollutes streams, rivers (Figure 17), lakes and even coastal waters. This affects the living organisms in the water and sometimes poisons humans or infects them with disease.

Sewage. Diseases like typhoid and cholera are caused by certain bacteria when they get into the human intestine. The faeces passed by people suffering from these diseases will contain these harmful bacteria. If the bacteria get into drinking water they may spread the disease to hundreds of other people. For this reason, among others, untreated sewage must not be emptied into rivers. It is treated at the sewage works so that all the solids are removed and the water discharged into rivers is free from all the harmful bacteria and poisonous chemicals (but see 'Eutrophication' on page 51).

Eutrophication. When nitrates and phosphates from farmland and sewage escape into water they cause excessive growth of microscopic green plants. This may result in a serious oxygen shortage in the water as explained on page 51.

Chemical pollution. Many industrial processes produce poisonous waste products. Electro-plating, for example, produces waste containing copper and cyanide. If these chemicals are released into rivers they poison the animals and plants and could also poison humans who drink the water. It is estimated that the River Trent receives 850 tonnes of zinc, 4000 tonnes of nickel and 300 tonnes of copper each year from industrial processes.

In 1971, 45 people in Minamata Bay in Japan died and 120 were seriously ill as a result of mercury poisoning. It was found that a factory had been discharging a compound of mercury into the bay as part of its waste. Although the mercury concentration in the sea was very low, its concentration was increased as it passed through the food chain (see page 51). By the time it reached the people of Minamata Bay, in the fish and other sea food which formed a large part of their diet, it was concentrated enough to cause brain damage, deformity and death.

Figure 17 River pollution. The river is badly polluted by the effluent from a paper mill.

Figure 18 Oil pollution. Oiled sea birds like this guillemot cannot fly to reach their feeding grounds. They also poison themselves by trying to clean the oil from their feathers.

High levels of mercury have also been detected in the Baltic Sea and in the Great Lakes of North America.

Oil pollution of the sea is becoming a familiar event. In 1967, a tanker called the *Torrey Canyon* ran on to rocks near Land's End and 100000 tonnes of crude oil spilled into the sea. Thousands of sea birds were killed by the oil (Figure 18), and the detergents which were used to try and disperse the oil killed many more birds and sea creatures living on the coast. Since 1967, there have been even greater spillages of crude oil from tankers and off-shore oil wells.

QUESTIONS

10 What are the possible dangers of dumping and burying poisonous chemicals on the land?

11 Before most water leaves the water works, it is exposed for some time to the poisonous gas, chlorine. What do you think is the point of this?

THE EFFECT ON THE AIR

Some factories and all motor vehicles release poisonous substances into the air. Factories produce smoke and sulphur dioxide; cars produce lead compounds, carbon monoxide and the oxides of nitrogen which lead to 'smog'.

Figure 19 Air pollution. Much of this discharge may be steam, but invisible chemicals like sulphur dioxide will also be present.

Smoke. This consists mainly of tiny particles of carbon and tar which come from burning coal either in power stations or in the home. The tarry drops contain chemicals which may cause cancer. When the carbon particles settle, they blacken buildings and damage the leaves of trees. Smoke in the atmosphere cuts down the amount of sunlight reaching the ground. For example, since the Clean Air Act of 1956, London has received 70 per cent more sunshine in December.

Smoke also caused the dense 'pea-soup' fogs of industrial districts. When the water droplets in these fogs were inhaled, they contributed to illness and death from bronchitis. The Clean Air Acts of 1956 and 1968 have effectively stopped these lethal fogs in Britain.

Sulphur dioxide. Coal and oil contain sulphur. When these fuels are burned, they release sulphur dioxide (SO_2) into the air. Although the tall chimneys of factories (Figure 19) send smoke and sulphur dioxide high into the air, the sulphur dioxide dissolves in rain water and forms an acid. When this acid falls on buildings, it slowly dissolves the limestone and mortar. When it falls on plants, it reduces their growth and damages their leaves.

Lead. Compounds of lead are mixed with petrol to improve the performance of motor cars. The lead is expelled with the exhaust gases into the air. In some areas of heavy traffic it may reach levels which are dangerous, because lead can cause damage to the brain in children.

Most of the forms of pollution described in this chapter could be prevented provided we were prepared to pay the cost of the necessary measures.

QUESTIONS

12 If compounds of lead and mercury get into the body, they are excreted only very slowly. Why do you think this makes them dangerous poisons even when they are in low concentrations in the air or the water?

13 It costs money to prevent harmful chemicals escaping into the air from factories and cars. The effects of pollution also cost a great deal of money. List some of the ways in which the effects of pollution (a) affect our health and (b) cost us money.

BASIC FACTS

- **The plants and animals in a food web are so interdependent that even a small change in the numbers of one group has a far-reaching effect on all the others.**
- **Human hunting activities and farming upset the natural balance between other living organisms.**
- **Pesticides kill insects, weeds and fungi that could destroy our crops.**
- **Pesticides help to increase agricultural production but they kill other organisms as well as pests.**
- **A pesticide or pollutant which starts off at a low, safe level can become dangerously concentrated as it passes along a food chain.**
- **Eutrophication of lakes and rivers results in the excessive growth of algae followed by an oxygen shortage when the algae die and decay.**
- **Soil erosion results from removal of trees from sloping land, use of only chemical fertilizers on ploughed land and putting too many animals on pasture land.**
- **Humans pollute lakes and rivers with industrial waste and sewage effluent.**
- **Humans pollute the sea with crude oil and factory wastes.**
- **Humans pollute the air with smoke and sulphur dioxide from factories, and lead and carbon dioxide from motor vehicles.**

SECTION TWO

Flowering Plants

7 Plant structure and function

Figure 1 shows a young sycamore plant. It is typical of many flowering plants in having a **root system** below the ground and a **shoot** above ground. The shoot consists of an upright stem, with leaves and buds. The buds on the side of the stem are called **lateral buds**. When they grow, they will produce branches. The bud at the tip of the shoot is the **terminal bud** and when it grows, it will continue the upward growth of the stem. The lateral buds and the terminal buds may also produce flowers.

The leaves make food by photosynthesis (page 24) and pass it back to the stem. The stem carries this food to all parts of the plant which need it and also carries water and dissolved salts from the roots to the leaves and flowers. In addition, the stem supports and spaces out the leaves so that they can receive sunlight and absorb carbon dioxide which they need for photosynthesis. An upright stem also holds the flowers above the ground, helping the pollination by insects or the wind (page 74). A tall stem may help in seed dispersal later on (page 80).

The roots anchor the plant in the soil and prevent it falling over or being blown over by the wind. They also absorb the water and salts the plant needs for making food in the leaves.

The structure and functions of the plant organs will be considered in more detail in this chapter.

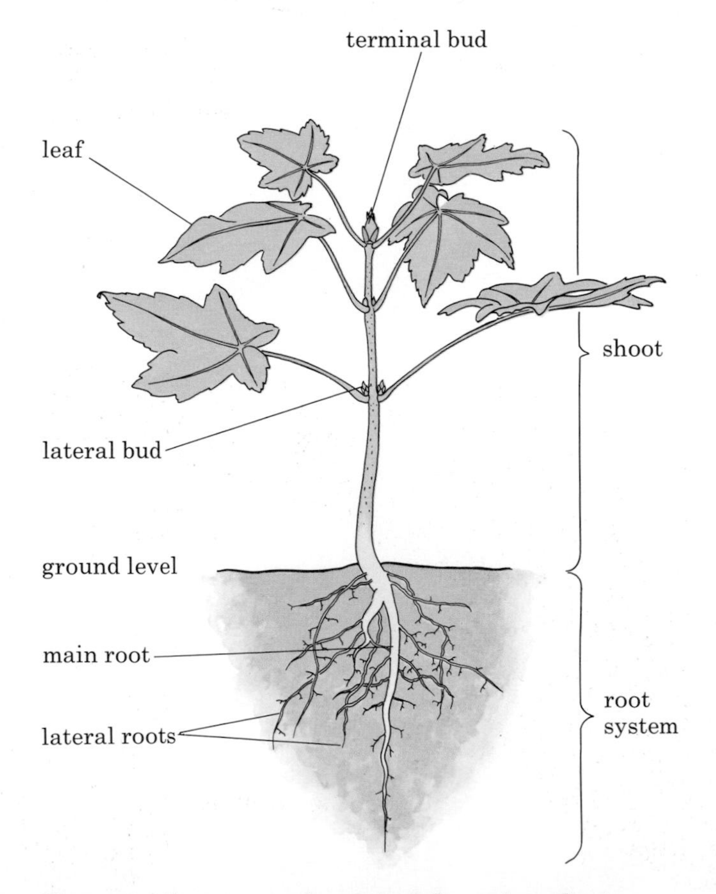

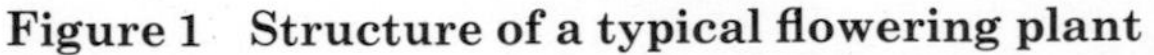

Figure 1 Structure of a typical flowering plant

LEAF

Figure 2*a* shows a typical leaf of a broad-leaved plant. It is attached to the stem by a **leaf stalk** which continues into the leaf as a **midrib**. Branching from the midrib is a network of veins which deliver water and salts to the leaf cells and carry away the food made by them. As well as carrying food and water, the network of veins forms a kind of skeleton which supports the softer tissues of the leaf blade. The **leaf blade** is broad and thin. Figure 2*c* shows a vertical section through a small part of a leaf blade, and Figure 3 is a photograph of a leaf section under the microscope.

Epidermis

This is a single layer of cells on the upper and lower surface of the leaf. The epidermis helps to keep the leaf's shape. The closely fitting cells (Figure 2*c*) reduce evaporation from the leaf and prevent bacteria and fungi from getting in.

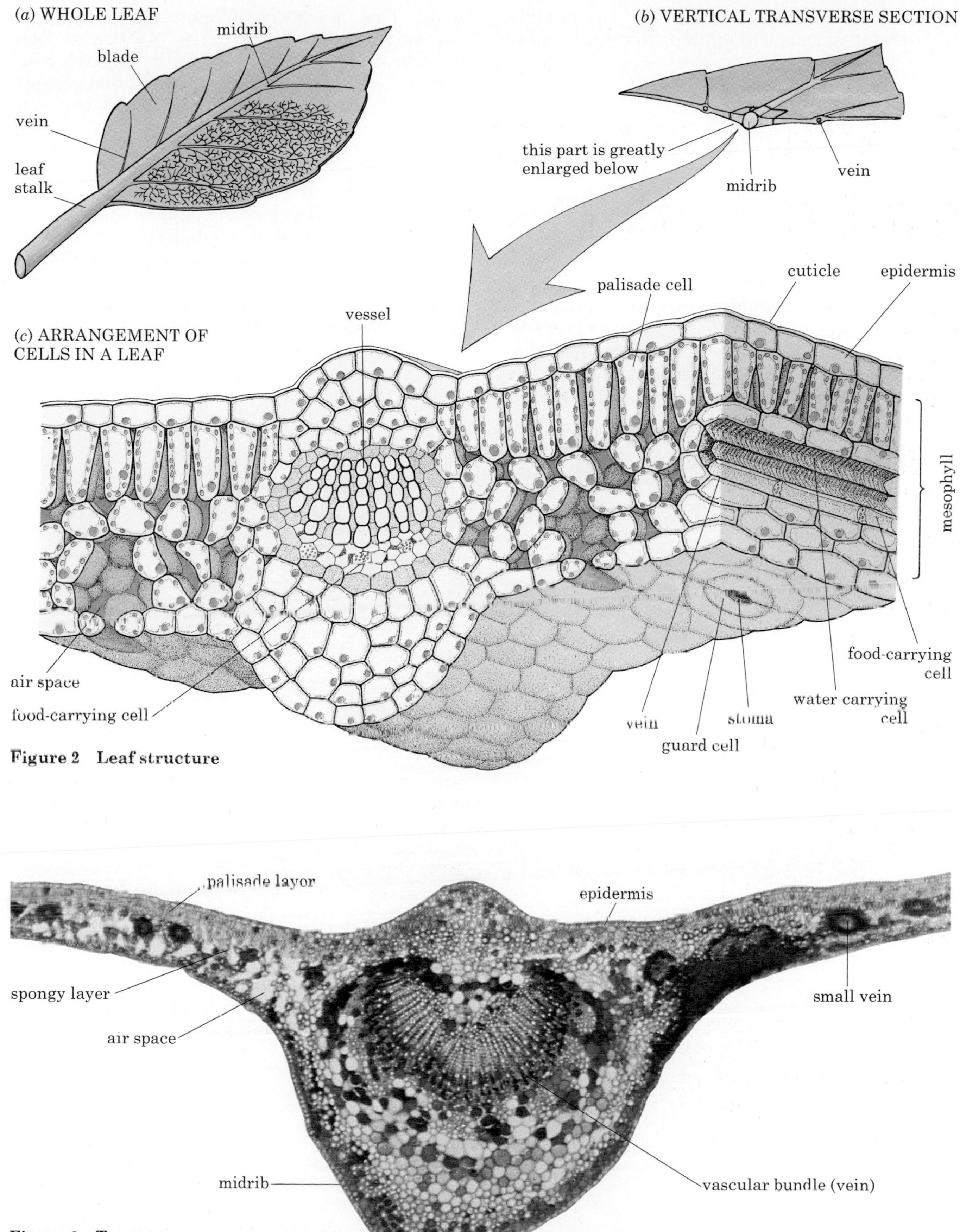

Figure 2 Leaf structure

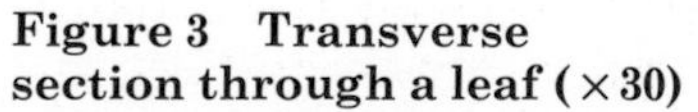

Figure 3 Transverse section through a leaf (×30)

There is a thin waxy layer called the **cuticle** over the epidermis which helps to reduce water loss.

Stomata

In the epidermis (the lower epidermis in most leaves), there are openings called **stomata** (singular=stoma). Each stoma is formed by a pair of **guard cells** which have a curved shape (Figure 4). The guard cells can take in water by osmosis from neighbouring cells and so increase the turgor pressure of their vacuoles (page 20). When their turgor pressure increases, the guard cells curve apart and open the stoma (Figure 5). When the guard cells lose water, they straighten up and close the stoma. Generally speaking, the stomata are open in the daytime and closed at night. This means that when photosynthesis is occurring in the light (page 24), the carbon dioxide needed can diffuse in through the stomata. In darkness, when photosynthesis stops, the stomata close and prevent the leaf losing too much water by evaporation.

Mesophyll

The tissue between the upper and lower epidermis is called mesophyll (Figure 2*c*). It consists of two zones, the upper, **palisade mesophyll** and the lower, **spongy mesophyll** (Figure 6). The palisade cells are usually long and contain many chloroplasts. The spongy mesophyll cells vary in shape and fit loosely together, leaving many air spaces between them. The function of the palisade cells and—to

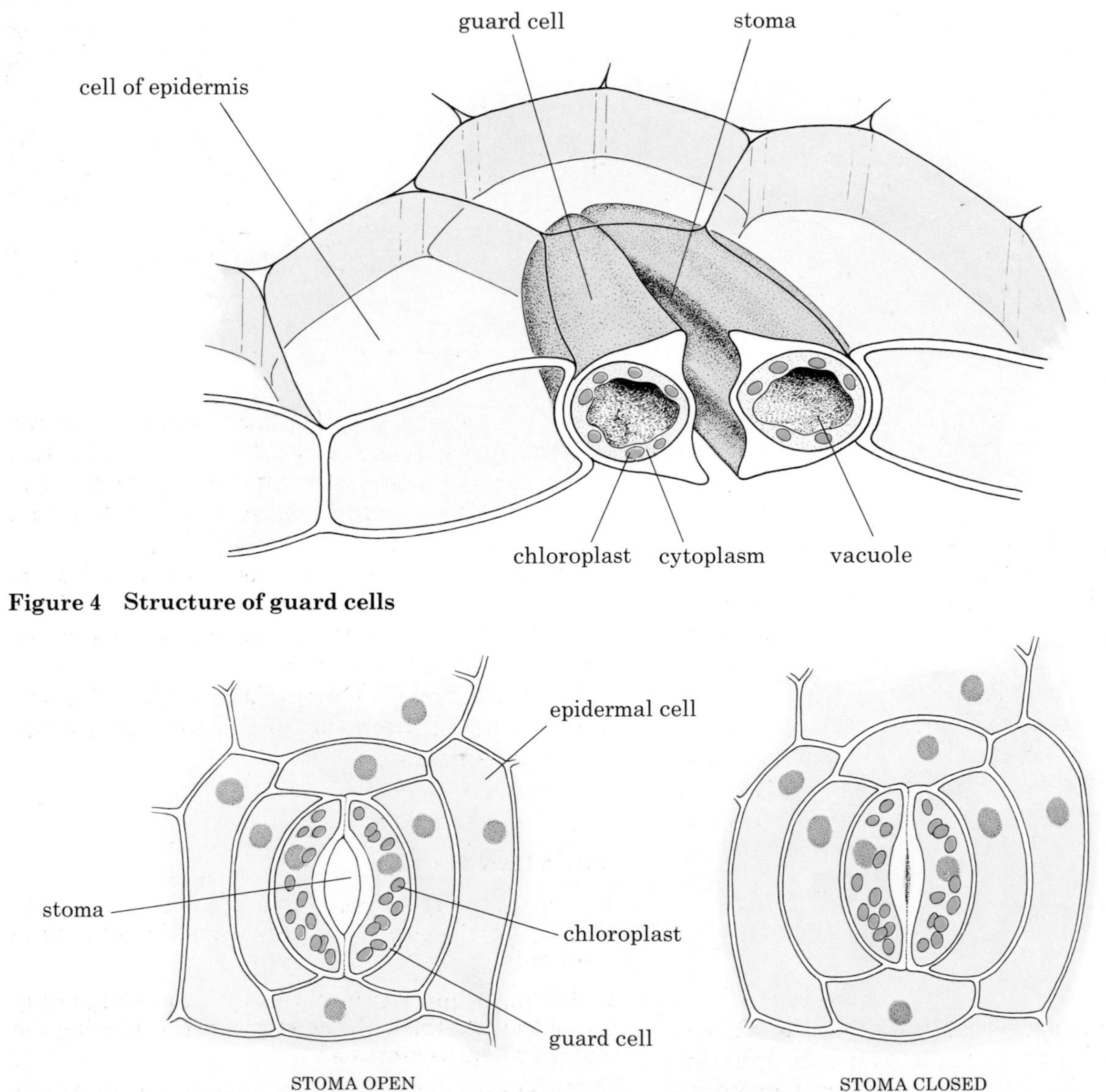

Figure 4 Structure of guard cells

Figure 5 Appearance of a stoma in leaf epidermis

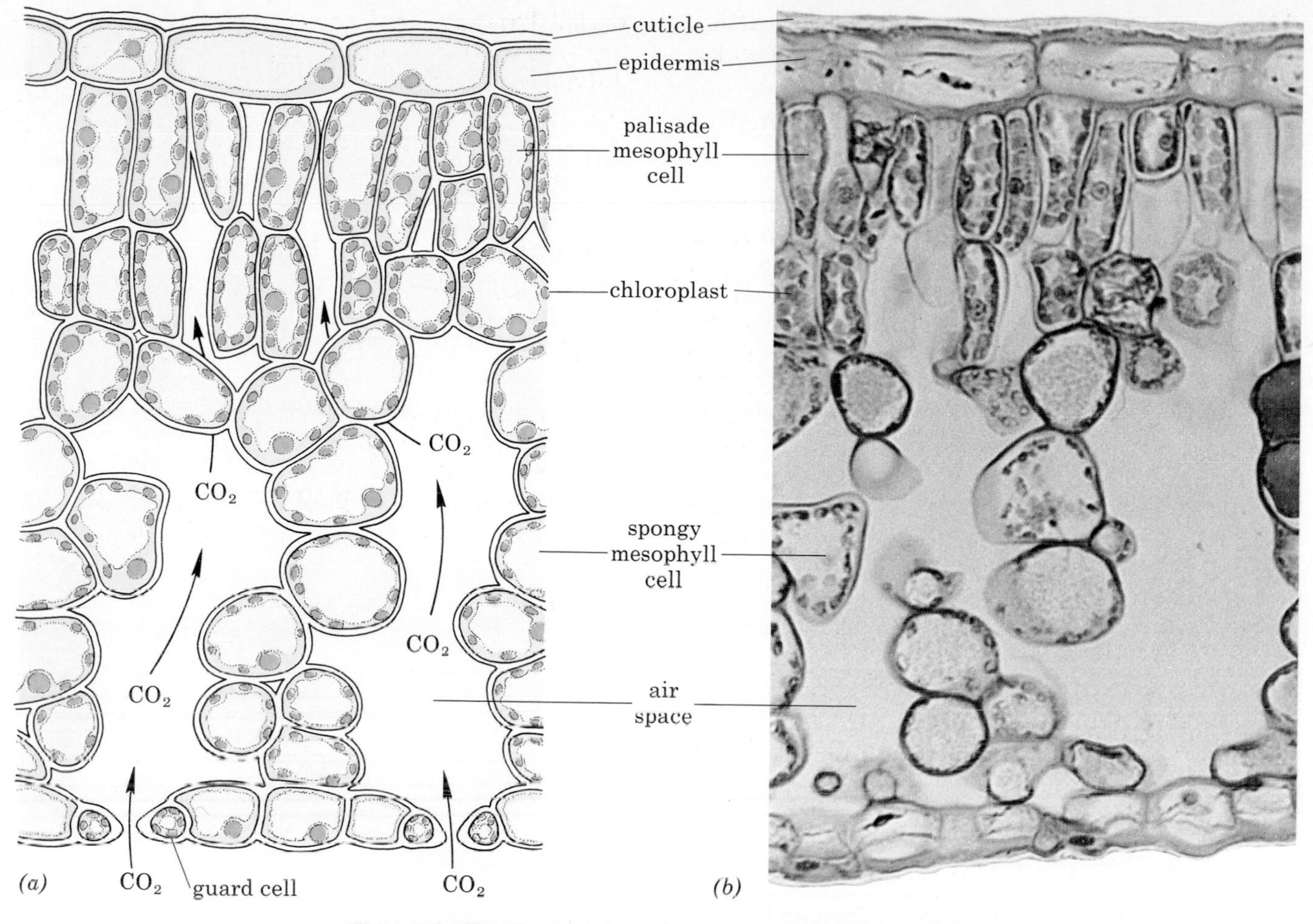

Figure 6 Vertical section through a leaf blade (×300)

a lesser extent—the spongy mesophyll cells is to make food by photosynthesis. Their chloroplasts absorb sunlight and use its energy to join carbon dioxide and water molecules to make sugar molecules as described on page 24. In daylight, when photosynthesis is rapid, the mesophyll cells are using up carbon dioxide. As a result, the concentration of carbon dioxide in the air spaces falls to a low level and more carbon dioxide diffuses in (page 18) from the outside air, through the stomata (Figure 6). This diffusion continues on through the air spaces, up to the cells which are using carbon dioxide. These cells are also producing oxygen as a by-product of photosynthesis. When the concentration of oxygen in the air spaces rises, it diffuses out through the stomata.

Veins

The water needed for making sugar by photosynthesis is brought to the mesophyll cells by the veins. The cells take in the water by osmosis (page 19) because the concentration of water molecules in a leaf cell, which contains sugars, will be less than the concentration of water in a vein, which does not contain sugars. The branching network of leaf veins means that no cell is very far from a water supply.

The sugars made in the mesophyll cells are passed to the phloem cells (see page 62) of the veins, and these cells carry the sugars away from the leaf into the stem.

The ways in which a leaf is thought to be well adapted to its function of photosynthesis are listed on page 25.

QUESTIONS

1 Look at Figure 6*a*. Why do you think that photosynthesis does not take place in the cells of the epidermis?

2 During bright sunlight, what gases are (a) passing out of the leaf through the stomata, (b) entering the leaf through the stomata?

3 What types of leaves do you know which do not have any midrib?

STEM

Figure 7 is a diagram of a stem cut across (transversely) and down its length (longitudinally) to show its internal structure.

Epidermis

Like the leaf epidermis, this is a single layer of cells which helps to keep the shape of the stem and cuts down the loss of water vapour. There are stomata in the epidermis which allow the tissues inside to take up oxygen and get rid of carbon dioxide. In woody stems, the epidermis is replaced by bark which consists of many layers of dead cells.

Vascular bundles

Running the length of the stem are veins or vascular bundles (Figures 7*b* and 7*c*). The inner part of each vascular bundle consists of a tissue called **xylem** (pronounced 'zylem'). Xylem contains long, empty, dead cells which form microscopic tubes called **vessels**. These vessels carry water and salts from the root system, through the stem and into the leaf veins. The outer part of the vascular bundle is formed by a tissue called **phloem** (pronounced 'flow-em'). This contains living cells joined end to end (Figure 7*c*). They carry food made by the leaf mesophyll cells to all parts of the plant.

Water and salts will nearly always be travelling up the stem from the roots to the

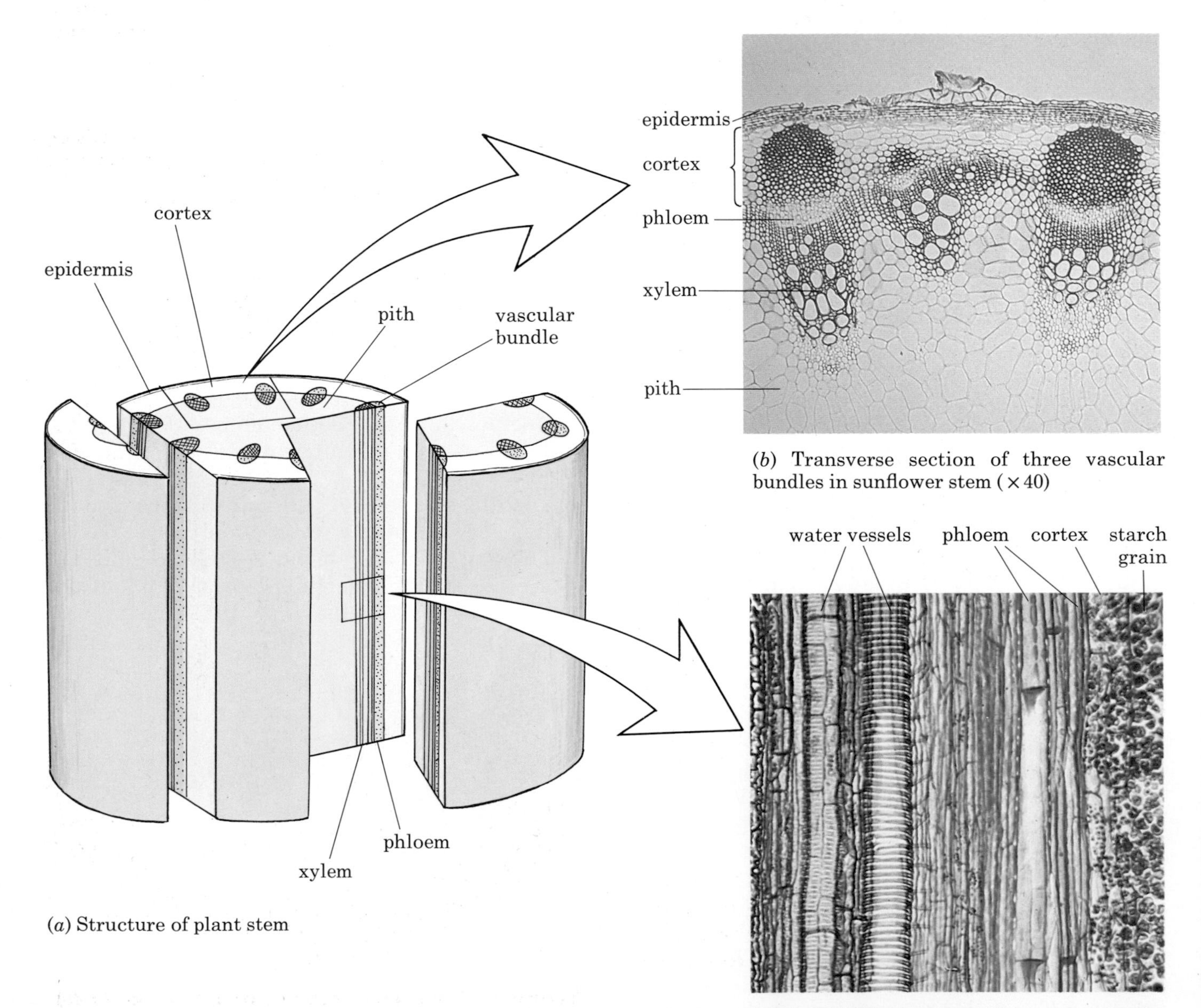

(*a*) Structure of plant stem

(*b*) Transverse section of three vascular bundles in sunflower stem (×40)

(*c*) Longitudinal section of vascular bundle (×160)

Figure 7 Stem structure

leaves through the xylem. The phloem may be carrying the food made in the leaves down the stem to the roots and storage organs like potatoes, or up the stem to growing buds, flowers or developing fruits.

The vascular bundles contain many strong fibres, and the way that the bundles are arranged in a circular pattern (Figure 7*a*) just beneath the epidermis makes the stem strong. This arrangement of strengthening tissues is well suited to resist the sideways bending caused by strong winds.

The tissue between the vascular bundles and the epidermis is called the **cortex**. Its cells often store starch. In green stems, the outer cortex cells contain chloroplasts and make food by photosynthesis. The central tissue of the stem is called **pith**. The cells of the pith and cortex act as packing tissues and help to support the stem in the same way as a lot of blown-up balloons packed tightly into a plastic bag would form quite a rigid structure.

QUESTIONS

4 What structures help to keep the stem's shape and upright position?

5 What are the differences between xylem and phloem?

ROOT

Figure 8 shows the internal structure of a typical root. The vascular bundle is in the centre of the root (Figure 9), unlike the stem where the bundles form a cylinder in the cortex. This central position is well suited to stand up to the pulling force on the roots when the shoot is being blown about by the wind.

The xylem carries water and salts from the root to the stem. The phloem will bring food from the stem to the root, to provide the root cells with substances for their energy and growth.

Outer layer and root hairs

There is no distinct epidermis in a root. At the root tip are several layers of cells forming the **root cap**. These cells are continually replaced as fast as they are worn away as the root tip is pushed through the soil.

In a region above the root tip, where the root has just stopped growing, the cells of the outer

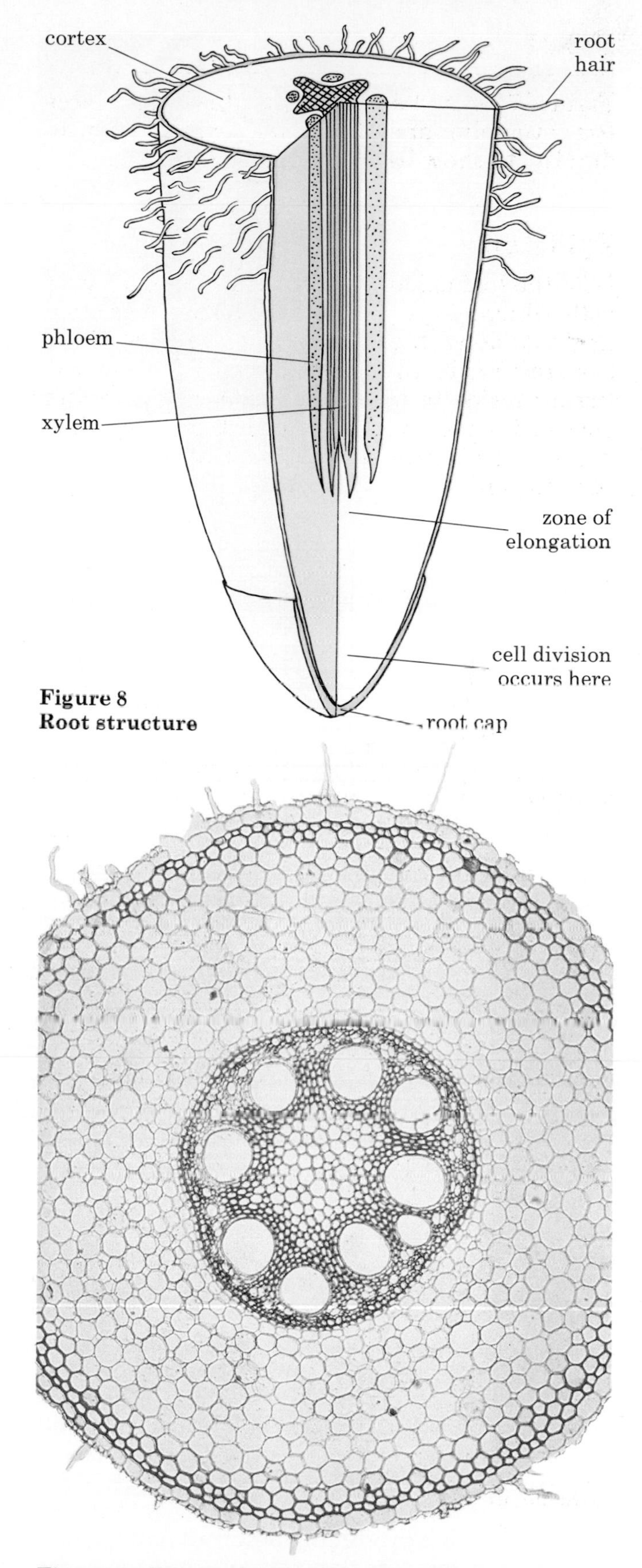

Figure 8
Root structure

Figure 9 Transverse section through a root (×40). Notice that the vascular bundle is in the centre. Some root hairs can be seen in the outer layer of cells.

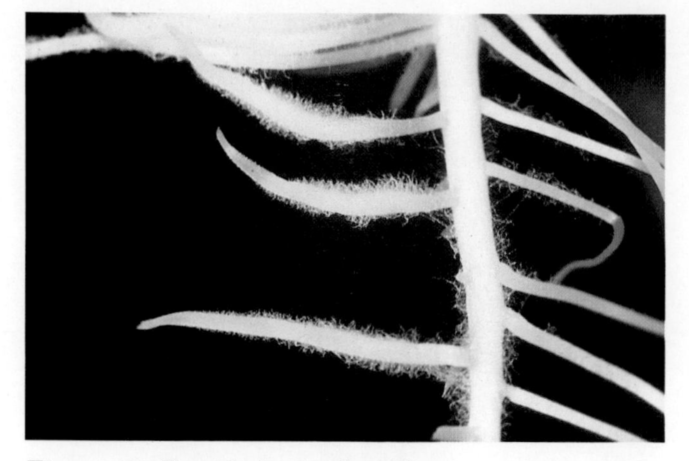

Figure 10 Root hairs (×5) as they appear on a root grown in moist air

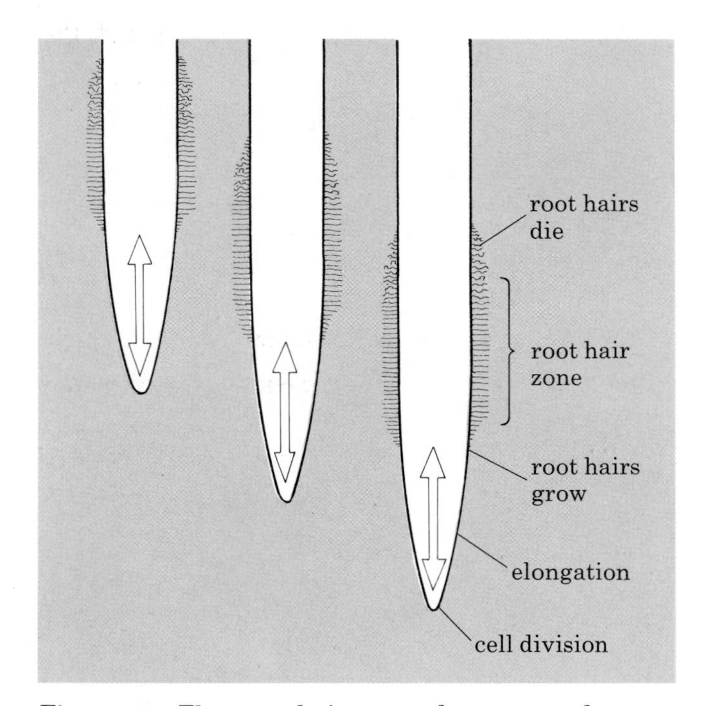

Figure 11 The root hair zone changes as the root grows

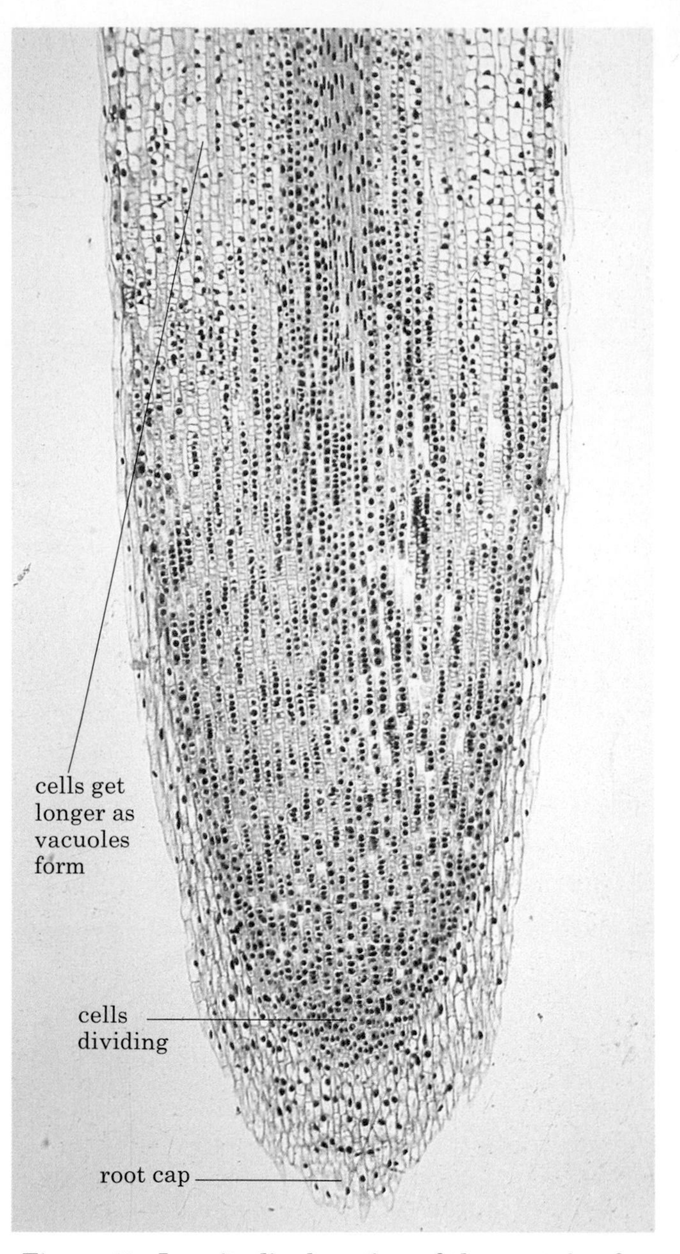

Figure 12 Longitudinal section of the root tip of an onion plant (×60)

layer produce tiny, tube-like outgrowths called root hairs (Figure 6 on page 68). These can just be seen as a downy layer on the roots of seedlings grown in moist air (Figure 10). In the soil, the root hairs grow between the soil particles and stick closely to them. The root hairs take up water from the soil by osmosis, and absorb mineral salts by diffusion or active transport (page 19). Root hairs remain alive for only a short time. The region of root just below a root hair zone is producing new root hairs, while the root hairs at the top of the zone are shrivelling (Figure 11). Above the root hair zone, the cell walls of the outer layer become less permeable. This means that water cannot get in so easily.

Growing point

At the tip of the root, the cells are dividing rapidly and producing a large number of new cells (Figure 12). Just above this region, the cells start to absorb water by osmosis. At this stage the cell walls are quite soft and as the vacuole expands it will make the cell longer (see Figure 11 on page 5). Hundreds of cells getting longer at the same time will push the root tip down through the soil.

QUESTIONS

6 State briefly the functions of the following: xylem, palisade cell, root hair, root cap, stoma, epidermis.

7 If you were shown a short, cylindrical structure cut from part of a plant, how could you tell whether it was a piece of stem or a piece of root?

8 Describe the path taken by (a) a carbon dioxide molecule from the air and (b) a water molecule from the soil, until they reach a mesophyll cell of a leaf to be made into sugar.

9 Why do you think that root hairs are produced only on the parts of the root system that have stopped growing?

10 Discuss whether you would expect to find a vascular bundle in a flower petal.

BASIC FACTS

- **The shoot of a plant consists of the stem, leaves, buds and flowers.**
- **The leaf makes food by photosynthesis in its mesophyll cells.**
- **The water for photosynthesis is carried in the leaf's veins.**
- **The carbon dioxide for photosynthesis enters the leaf through the stomata and diffuses through the air spaces in the leaf.**
- **Sunlight is absorbed by the chloroplasts in the mesophyll cells.**
- **The food made in the leaf is carried away in the phloem cells.**
- **The stem supports the leaves and flowers.**
- **It contains vascular bundles (veins).**
- **The water vessels in the veins carry water up the stem to the leaves.**
- **The phloem in the veins carries food up or down the stem to wherever it is needed.**
- **The roots hold the plant in the soil and absorb the water and mineral salts needed by the plant for making sugars and proteins.**
- **The root hairs make very close contact with soil particles and are the main route by which water and mineral salts enter the plant.**

8 Transport of substances in plants

MOVEMENT OF WATER BY TRANSPIRATION

The main force which draws water from the soil and through the plant is caused by a process called **transpiration**. Water evaporates from the leaves and causes a kind of 'suction' which pulls water up the stem (Figure 1). The water travels up the vessels in the vascular bundles (Figure 2) and this flow of water is called the **transpiration stream**. Figure 3 shows the cells in part of a leaf blade. As explained on page 20, the cell sap in each cell is exerting a turgor pressure outwards on the cell wall. This pressure forces some water out of the cell wall and into the air space between the cells. Here the water evaporates and the water vapour passes by diffusion through the air spaces in the mesophyll and out of the stomata. It is this loss of water vapour from the leaves which is called transpiration.

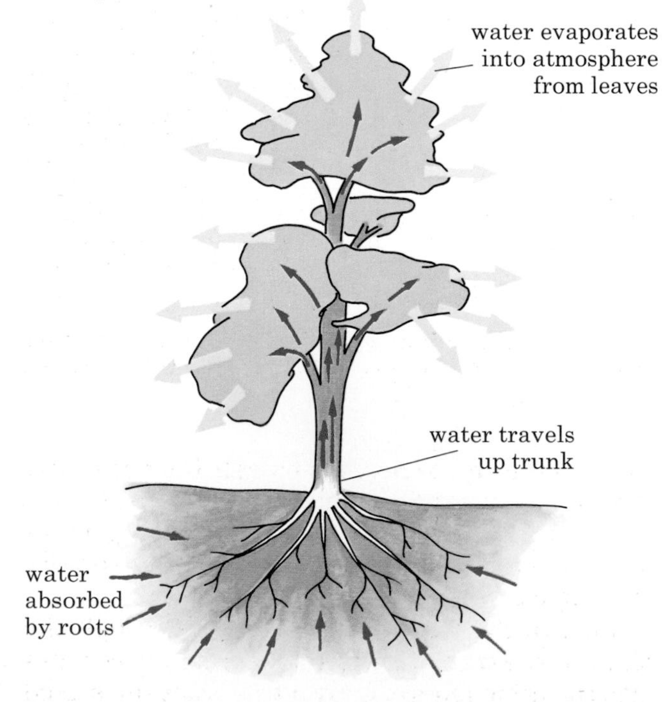

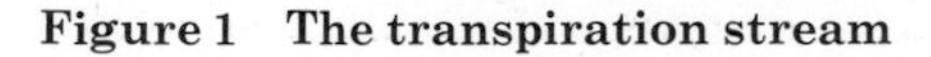
Figure 1 The transpiration stream

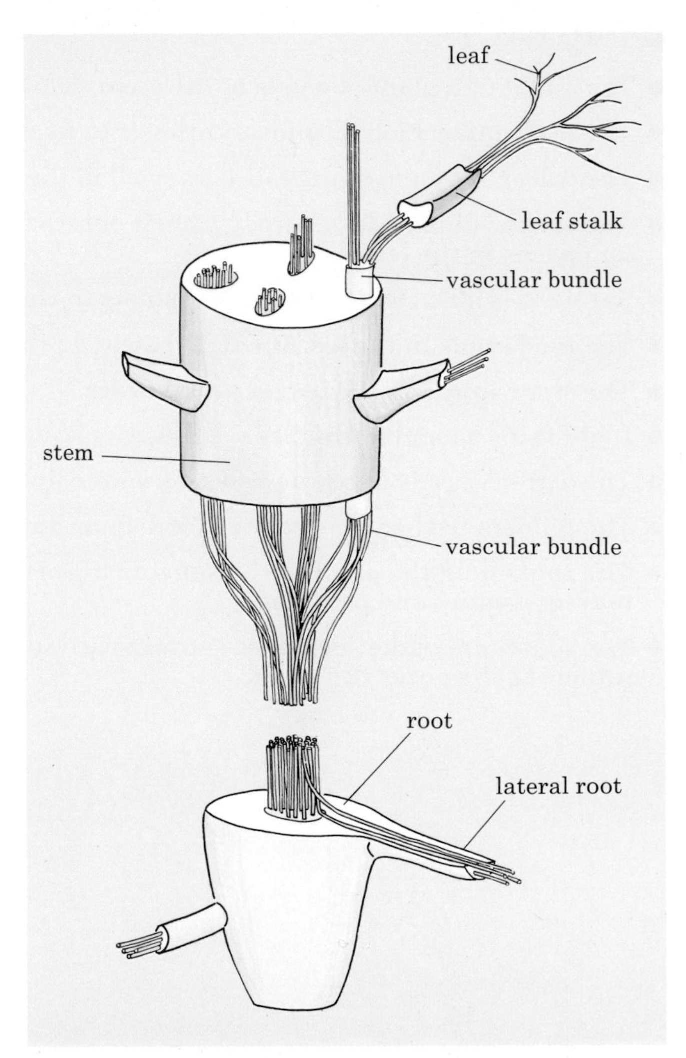

Figure 2 The distribution of vascular bundles in root, stem and leaf

The cell walls which are losing water in this way replace it by drawing water from the nearest vein. Most of this water travels along the cell walls without actually going inside the cells (Figure 4). Thousands of leaf cells are evaporating water like this and drawing water to replace it from the xylem vessels in the veins. As a result, water is pulled through the xylem vessels and up the stem from the roots. This

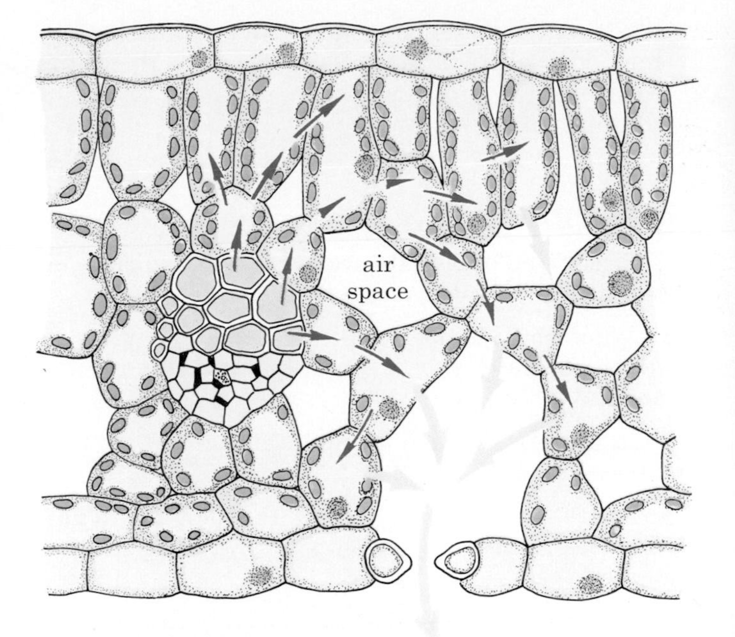

Figure 3 Movement of water through a leaf. The dark blue arrows show water leaving the vessels in a vein and passing from one cell to another. The light blue arrows show water evaporating from the cell walls.

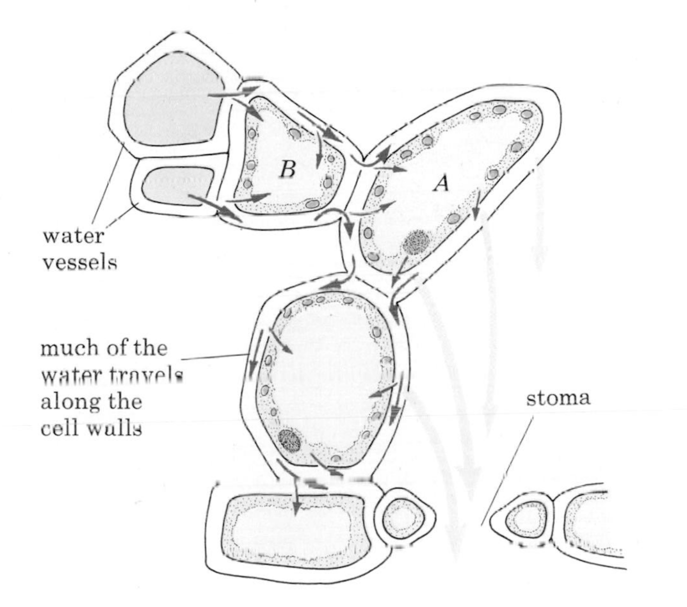

Figure 4 The probable pathway of water in leaf cells. (The cell walls are drawn abnormally thick to show water travelling in them.)

Figure 5 Californian redwoods. Some of these trees are over 100 metres tall. Transpiration from their leaves pulls water up the trunk.

transpiration pull is strong enough to draw up water 50 metres or more in trees (Figure 5).

In addition to the water passing along the cell walls, a small amount will pass right through the cells. When leaf cell *A* (Figure 4) loses water, its turgor pressure will fall. This fall in pressure allows the water in the cell wall to enter the vacuole and so restore the turgor pressure. In conditions of water shortage, cell *A* may be able to get water by osmosis from cell *B* more easily than *B* can get it from the xylem vessels. In this case, all the mesophyll cells will be losing water faster than they can absorb it from the vessels, and the leaf will wilt (see page 20).

Most of the water which travels through the plant in the transpiration stream simply escapes into the atmosphere. Only a tiny fraction is used up by photosynthesis (page 24) in the leaf cells.

Rate of transpiration

Transpiration is the evaporation of water from the leaves, so any change which increases or reduces evaporation will have the same effect on transpiration.

(a) Light intensity. Light itself does not affect evaporation but in daylight, the stomata (page 60) of the leaves are open. This allows the water vapour in the leaves to diffuse out into the atmosphere. At night, when the stomata close, transpiration is greatly reduced.

(b) Humidity. If the air is very humid, i.e. contains a great deal of water vapour, it can accept very little more from the plants and so transpiration slows down. In dry air, the diffusion of water vapour from the leaf to the atmosphere will be rapid.

(c) Air movements. In still air, the region round a transpiring leaf will become saturated with water vapour so that no more can escape from the leaf. In these conditions, transpiration would slow down. In moving air, the water vapour will be swept away from the leaf as fast as it diffuses out. This will speed up transpiration.

(d) Temperature. (1) Warm air can hold more water vapour than cold air. Thus evaporation or transpiration will take place more rapidly into warm air. (2) When the sun shines on the leaves, they will absorb heat as well as light. This warms them up and increases the rate of evaporation of water.

QUESTIONS

1 What kind of climate and weather conditions do you think will cause a high rate of transpiration?

2 What would happen to the leaves of a plant which was losing water by transpiration faster than it was taking it up from the roots?

3 In what two ways does sunlight increase the rate of transpiration?

4 Apart from drawing water through the plant, what else may be drawn up by the transpiration stream?

5 Transpiration has been described in this chapter as if it took place only in leaves. What other parts of a plant might transpire?

UPTAKE OF WATER AND SALTS

Uptake of water by the roots

Figure 6 shows a root hair in the soil. The cytoplasm of the root hair is selectively permeable to water. The soil water is a weak

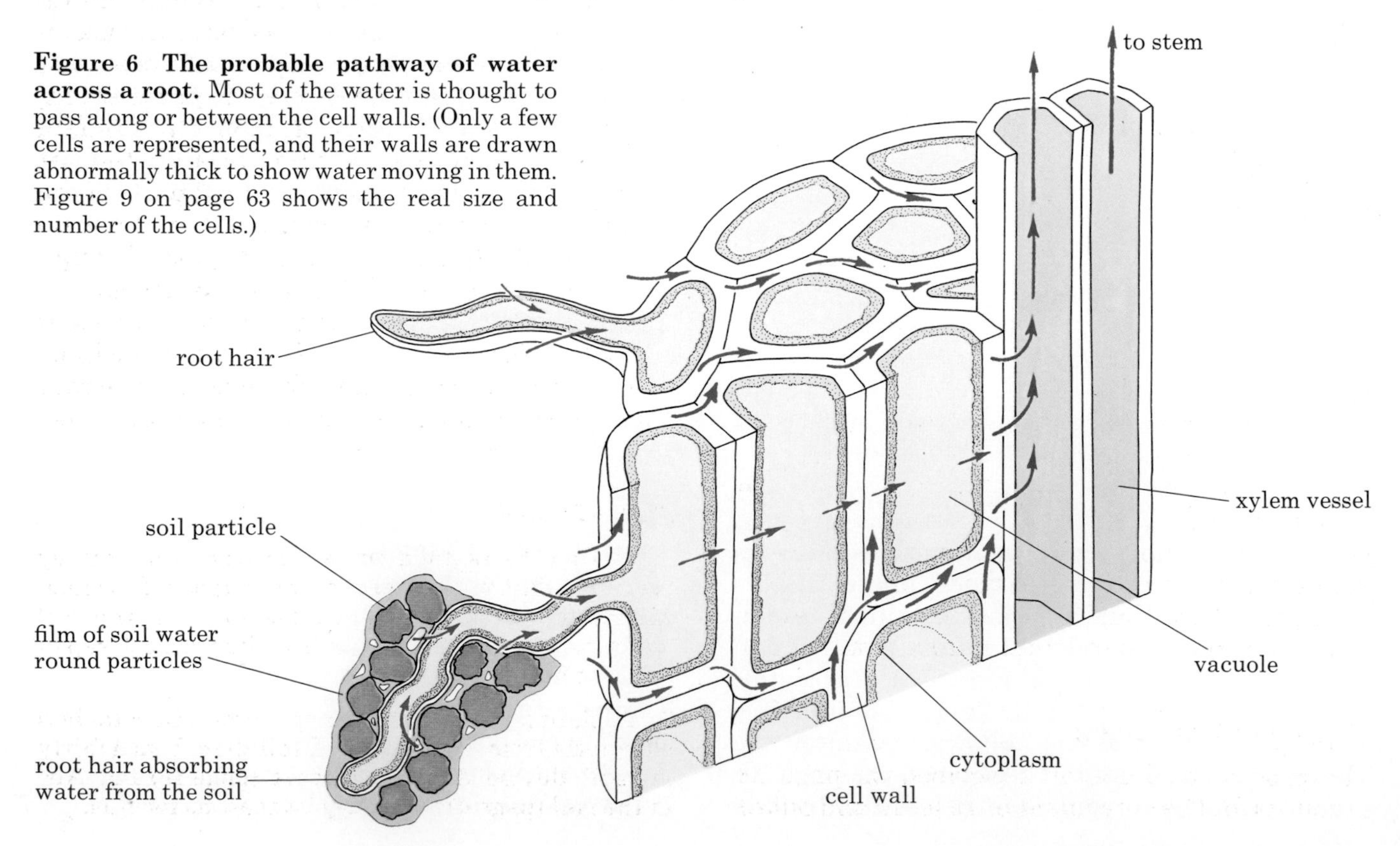

Figure 6 The probable pathway of water across a root. Most of the water is thought to pass along or between the cell walls. (Only a few cells are represented, and their walls are drawn abnormally thick to show water moving in them. Figure 9 on page 63 shows the real size and number of the cells.)

solution (i.e. more watery) than the cell sap and so water passes by osmosis (page 19) from the soil into the cell sap of the root hair cell. This flow of water into the root hair cell raises the cell's turgor pressure (page 20). So water is forced out through the cell wall into the next cell and so on, right through the cortex of the root to the xylem vessels. Here it is drawn up by the transpiration stream caused by evaporation from the leaves (Figure 7).

Most of the water travels across the root in, or between, the cell walls without entering the vacuoles. From this flow of water, the cells can absorb water by osmosis into the vacuoles, if their turgor pressure falls for any reason.

Uptake of salts by the roots

The method by which roots take up salts from the soil is not fully understood. Some salts may be carried in with the water drawn up by transpiration. Some may diffuse into the root hair cells and some may be taken up by active transport (page 19). The growing region of the root and the root hair zone seem to be most active in taking up salts. Most of the salts appear to be carried at first in the xylem vessels, though they soon appear in the phloem as well.

The salts are used by the plant's cells to build up essential molecules. Nitrates, for example, are built into amino acids by the roots and these are used later to make proteins.

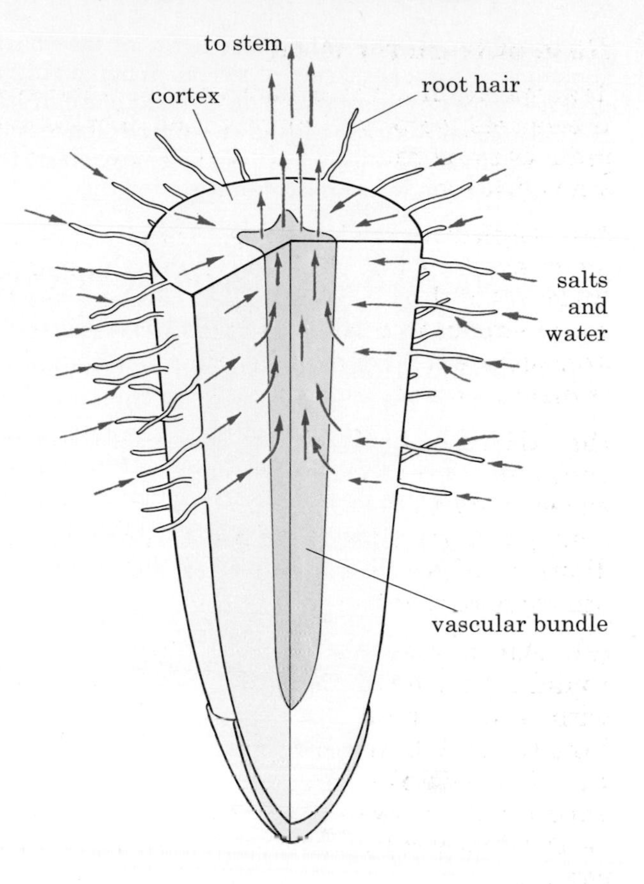

Figure 7 Intake of water and salts by a root

TRANSPORT OF FOOD AND GASES

Transport of food

On page 62 it was explained that the food made in the leaves by photosynthesis is carried away in the cells of the phloem. The food may travel up or down the stem according to which parts of the plant need it most. The carbohydrate is carried mainly as sucrose but the processes which move the dissolved substances are not understood.

We know that the movement depends on living processes in the phloem cells. Anything which harms the living cytoplasm of the cells also stops the transport of food materials. The transport of food and mineral salts in plants is sometimes referred to as **translocation**.

Transport of gases

The process of diffusion described on page 18 accounts for the movement of gases in and out of a plant. During respiration, oxygen is taken in and carbon dioxide given out. When photosynthesis is faster than respiration, carbon dioxide diffuses in and oxygen diffuses out (Figure 2 on page 25). In leaves and green stems, the gases enter and leave through the stomata (page 60). Then they diffuse through the air spaces between the cells to reach all parts of the plant organ. In woody plants, the stems have no stomata and the gases have to pass through small openings in the bark called **lenticels**.

Roots obtain their oxygen from the air spaces in the soil (page 41). Most of this oxygen will be dissolved in the soil water which enters the root through the growing region and the root hairs.

QUESTIONS

6 If root hairs take up water from the soil by osmosis, what would you expect to happen if so much nitrate fertilizer was put on the soil that the soil water became a stronger solution than the cell sap of the root hairs?

7 A plant's roots may take up water and salts less efficiently from a waterlogged soil than from a fairly dry soil. Revise 'Active transport' (page 19) and 'Air' in the soil (page 41) and suggest reasons for this.

8 Phloem forms the innermost layer of tree bark. The wood consists entirely of xylem. When a ring of bark is cut away from a tree trunk, the xylem can still carry water and salts to the leaves, and the leaves can still make food by photosynthesis. Nevertheless, the tree will die. Suggest an explanation for this.

EXPERIMENTS ON TRANSPORT IN PLANTS

1 To measure the rate of transpiration

Since transpiration is the *loss* of water vapour, the rate of transpiration can be measured by weighing a plant at intervals to see how much weight it loses. For example, a potted plant is watered, and the pot and soil are enclosed in a plastic bag tied round the plant's stem (Figure 8). This makes sure that any losses are due to

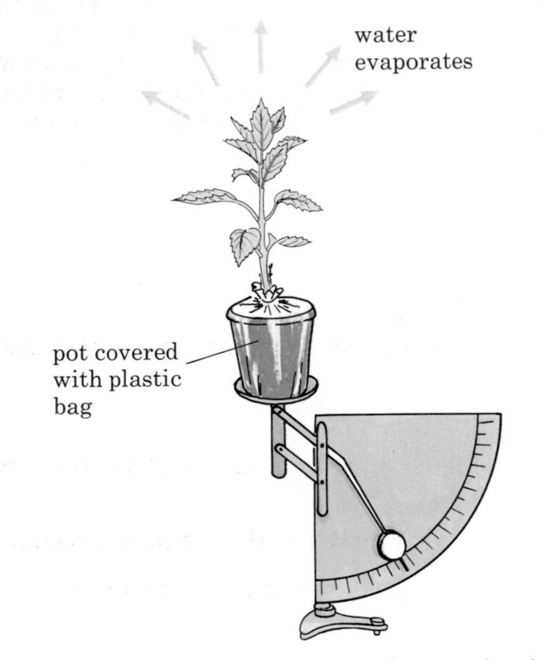

Figure 8 Measuring the rate of transpiration in a potted plant

evaporation from the shoot and not from the soil. The plant is weighed at intervals. If it loses, say, 56 grams in four hours, the rate of transpiration is 14 grams of water transpired per hour. This result assumes that any change in weight is entirely due to transpiration. In fact, there may be small gains in weight due to absorption of carbon dioxide in photosynthesis, or small losses due to the escape of carbon dioxide from respiration. In practice, over a few hours, these changes are very small compared with the losses due to transpiration.

A control experiment should be carried out. A pot of soil is watered and wrapped in a plastic bag exactly as before but without having a plant. If this is weighed over the same period as the experimental plant, there should be little or no change in weight. This shows that the plant is responsible for the loss in weight.

2 To find which surface of a leaf loses more water vapour

Filter paper is soaked in a five per cent solution of cobalt chloride, dried and cut into 5-mm squares. When dry the paper is blue, but it changes to pink when damp. Two squares of this cobalt chloride paper, held in forceps, are dried

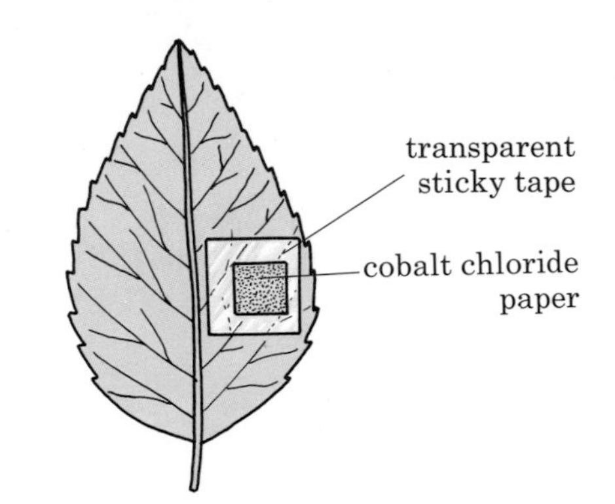

Figure 9 Which surface of a leaf loses more water vapour?

over a bench lamp or a small Bunsen flame until they are blue. They are then stuck to a leaf using transparent sticky tape (Figure 9). One square is stuck to the upper surface and the other to the lower surface. The cobalt chloride paper will turn pink as water vapour reaches it from the leaf. The side of the leaf with more stomata will release water vapour more quickly. On this side, therefore, the cobalt chloride will turn pink first. In the leaves of most trees and shrubs, the stomata are on the underneath surface only. The leaves of grasses and related plants, have stomata on both sides.

3 Transport in the vascular bundles

The shoots of several leafy plants are placed in a solution of one per cent methylene blue. The shoots are left in the light for 30 minutes or more. In some cases, after this time, the blue dye will appear in the leaf veins. If some of the stems are cut across, the dye will be seen in the vascular bundles (see Figure 7 on page 62). These results show that the dye, and therefore probably also the water, travels up the stem in the vascular bundles. Closer study would show that it travels in the xylem vessels.

4 To show that water vapour is given off during transpiration

The shoot of a recently watered potted plant, or a plant in the garden, is completely enclosed in a transparent, polythene bag which is tied round the base of the stem (Figure 10). The plant is allowed to remain for an hour or two in direct sunlight. The water vapour transpired by the plant will soon saturate the atmosphere inside the bag and drops of water will condense on the inside. The bag is removed and all the condensed water shaken into a corner so that it can be tested with cobalt chloride paper. A control experiment is set up using a shoot in a similar situation but from which all the leaves and flowers have been detached.

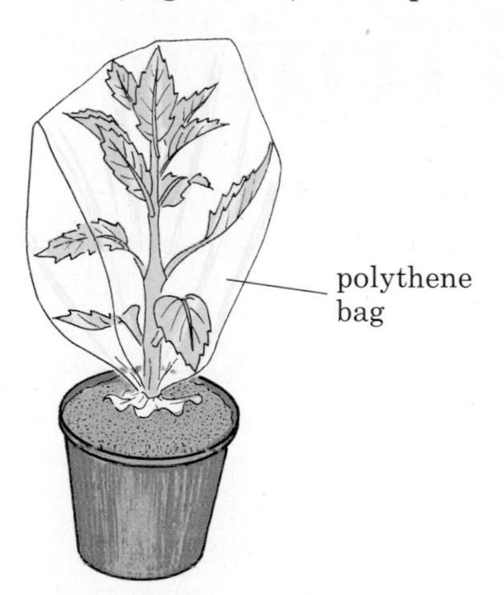

Figure 10 Method of collecting liquid from transpiration

QUESTIONS

9 A leafy shoot, plus the beaker of water in which it is placed, weighs 275 grams. Two hours later, it weighs 260 grams. An identical beaker of water, with no plant, loses 3 grams over the same period of time. What is the rate of transpiration of the shoot?

10 In Experiment 2, suggest (a) why you think forceps are used to handle the cobalt chloride paper squares, (b) why you think it was water vapour from the leaf and not water vapour from the air which made the cobalt chloride paper change from blue to pink.

BASIC FACTS

- **Transpiration is the evaporation of water from the leaves of a plant.**
- **The rate of transpiration is increased when light falls on the leaves, because this makes the stomata open.**
- **Transpiration is also increased by warm, dry and moving air.**
- **Transpiration from the leaves causes water to be pulled up the stem in the xylem vessels; this is the transpiration stream.**
- **Water brought by the xylem vessels to the leaves passes from cell to cell mainly in the cell walls.**
- **Water passes from cell to cell across the root until it reaches the xylem vessels in the centre.**
- **Root hairs also take up mineral salts from the soil.**
- **Food is carried by the phloem in the veins of leaves, stems and roots.**
- **The movement of oxygen and carbon dioxide through a plant takes place by diffusion through the air spaces between the cells.**
- **The rate of transpiration in a small plant is found by determining the weight it loses in a fixed time.**

9 Flowers, fertilization and fruits

Many people enjoy flowers because of their colour and scent. Wild flowers form an attractive part of the countryside. Potted plants and cut flowers are used to decorate our houses. The biological importance of flowers, however, is that they are the reproductive structures of the plant. They contain the plant's sexual organs. The male organs are the stamens which produce pollen. The female organs are the ovaries in which the seeds develop. The scent and colour of the petals are not for our benefit, but attract the insects which bring about pollination.

FLOWER STRUCTURE

The basic structure of a flower is shown in Figures 1 and 2.

Petals. These are usually brightly coloured and sometimes scented. They are arranged in a circle (Figure 1) or a cylinder (Figure 3). Most flowers have from five to ten petals and sometimes they are joined together to form a tube (Figure 4) and the individual petals can no longer be distinguished. The colour and scent of the petals attract insects to the flower and they will bring about pollination (page 74).

The flowers of grasses and many trees do not have petals but small, leaf-like structures which enclose the reproductive organs.

Sepals. Outside the petals is a ring of sepals. They are often green and much smaller than the petals. They may protect the flower when it is in the bud.

Stamens. The stamens are the male reproductive organs of a flower. Each stamen has a stalk called the **filament**, with an **anther** on the end. Flowers such as the buttercup and wild rose have many stamens; others such as the tulip have a small number, often the same as, or double, the number of petals or sepals. Each anther consists of four **pollen sacs** in which the pollen grains are produced by cell division. When the anthers are ripe, the pollen sacs split open and release their pollen.

Ovary. This is the female reproductive organ. Flowers such as the buttercup and blackberry have a large number of ovaries while others, such as the lupin have a single ovary. Inside the ovary there are one or more **ovules**. Each buttercup ovary contains one ovule but the lupin ovary contains several. The ovule will become a **seed**, and the whole ovary will become a **fruit**. (In biology, a fruit is the fertilized ovary

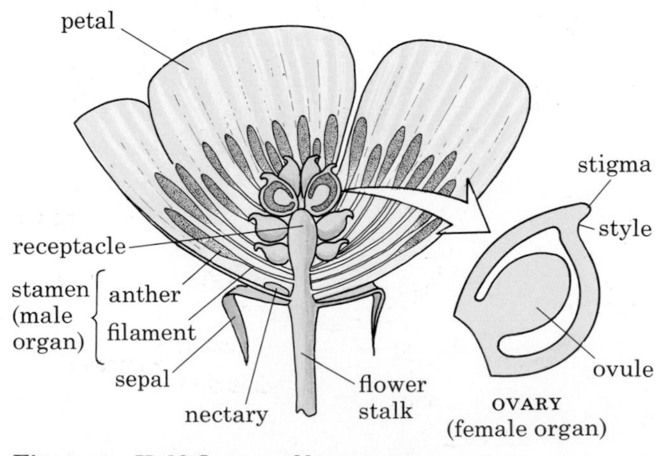

Figure 1 Half-flower of buttercup

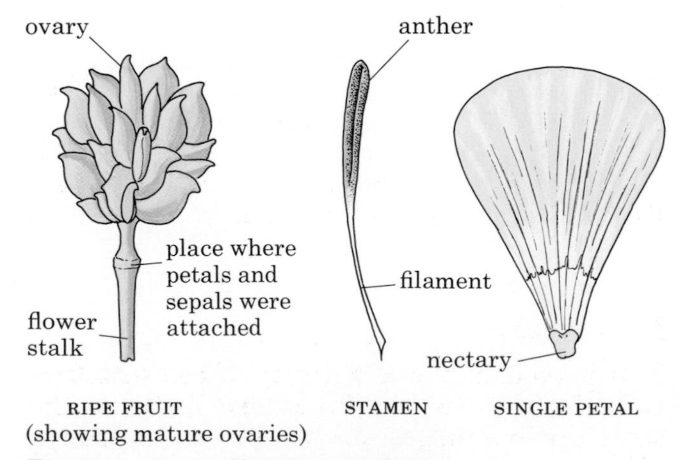

Figure 2 Parts of buttercup flower

Figure 3 Tulip flower. Some of the petals have been removed to show the structures inside. The stamens are black, and the green ovary has the stigma on top of it.

Figure 4 Daffodil flower cut in half. The inner petals form a tube. Three stamens are visible round the long style and the ovary contains many white ovules.

of a flower, not necessarily something to eat.)

Attached to the ovary are the **style** and the **stigma**. The stigma has a sticky surface and pollen grains will stick to it during pollination. The style may be quite short (buttercup, Figure 1) or very long (lupin, Figure 5).

Receptacle. The flower structures just described are all attached to the expanded end of a flower stalk. This is called the **receptacle** and, in a few cases after fertilization, it becomes fleshy and edible, e.g. strawberry (Figure 21), apple and pear.

Buttercup

This is an example of a flower whose structure is fairly easy to see (Figure 1). To show the structure in a diagram, a 'half-flower' is drawn. This shows the flower as it would appear if cut vertically down the middle. In the buttercup, with five petals, two and a half petals would appear in the drawing.

At the base of the petals are swellings called **nectaries**. These produce a sugary solution called **nectar**, which is collected by insects. There are five sepals, and neither the sepals nor the petals are joined together. There are about 60 stamens and 30–40 ovaries, each one containing a single ovule.

Lupin

The lupin flower is shown in Figures 5–7. There are five sepals but these are joined together forming a short tube. The five petals are of different shapes and sizes. The uppermost, called the **standard**, is held vertically. Two petals at the sides are called **wings** and are partly joined together. Inside the wings are two more petals joined together to form a boat-shaped **keel**.

The ovary is long, narrow and pod-shaped, with about ten ovules in it. The long style ends in a stigma just inside the pointed end of the keel. There are ten stamens, five long ones and five short ones. Their filaments are joined together at the base to form a sheath round the ovary.

The flowers of peas and beans are very similar to those of lupins.

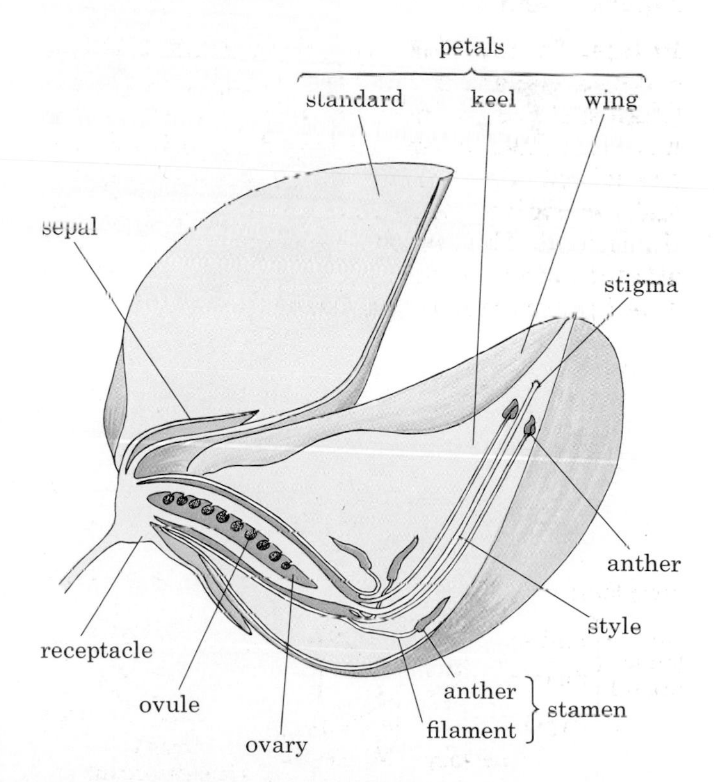

Figure 5 Half-flower of lupin

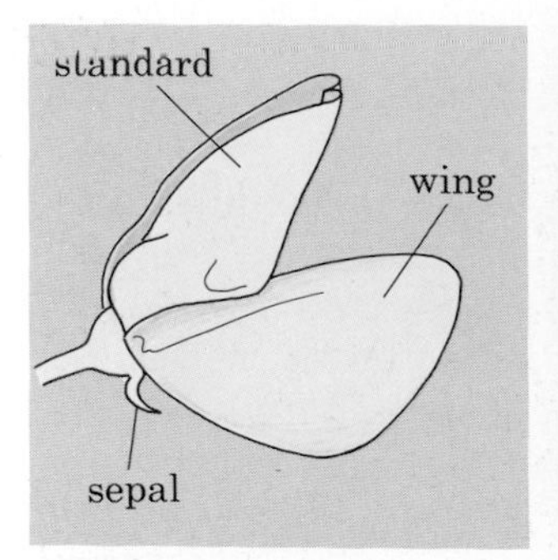

(*a*) Complete flower

The standard is yellow and pink. The wing is all pink

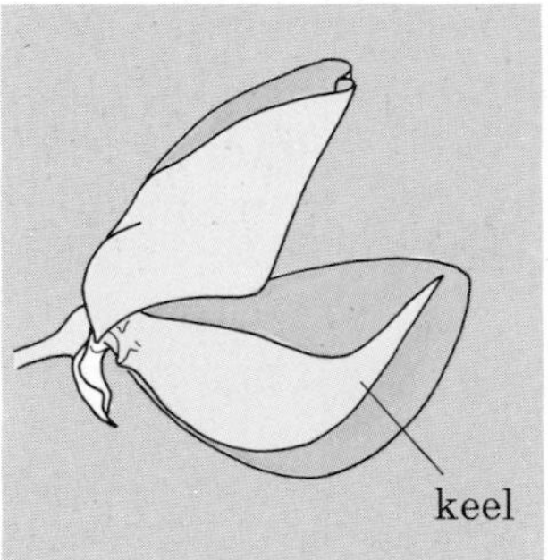

(*b*) One 'wing' has been removed

Notice the pollen being forced from the tip of the keel

stigma
stamen
filaments joined

(*c*) One side of the keel has been removed

There are five long stamens and five short ones

Figure 6 Taking a lupin flower apart

Figure 7 Lupin flowers. There are several hundred flowers on each stalk. The youngest flowers, at the top, have not yet opened. The oldest flowers are at the bottom and have been pollinated.

QUESTIONS

1 Working from outside to inside, list the parts of a flower.

2 What features of flowers attract insects?

3 Make a table to show how a lupin flower differs from a buttercup flower.

	Lupin flower	Buttercup flower
Sepals Petals Stamens Ovary		

POLLINATION

The transfer of pollen from the anthers to the stigma is called **pollination**. The anthers split open, exposing the microscopic pollen grains (Figure 8). The pollen grains are then carried away on the bodies of insects, or simply blown by the wind, and may land on the stigma of another flower. In **self-pollinating** flowers, the pollen which reaches the stigma comes from the same plant. In **cross-pollination**, the pollen is carried from the anthers of one plant to the stigma of another (of the same species).

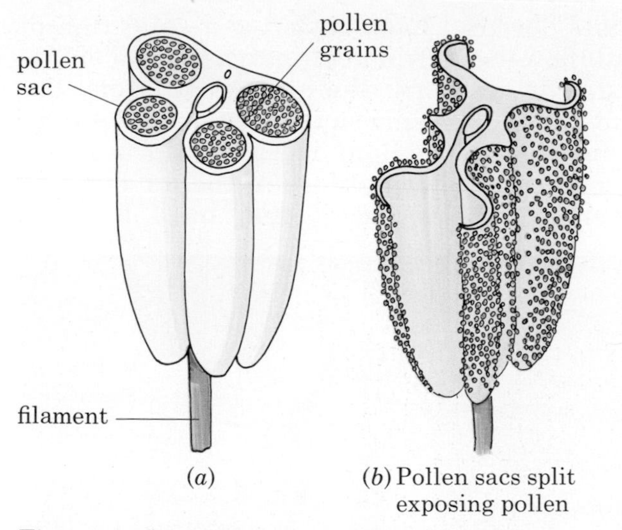

Figure 8 Structure of an anther (top cut off)

Pollination of the lupin

Lupin flowers have no nectar. The bees which visit them come to collect pollen which they take back to the hive for food. Other members of the lupin family—e.g. clover—do produce nectar.

The weight of the bee, when it lands on the flower's wings, pushes down these two petals and the petals of the keel. The pollen from the anthers has collected in the tip of the keel and as the petals are pressed down, the stigma and long stamens push the pollen out from the keel on to the underside of the bee (Figures 9 and 10*a*). The bee, with pollen grains sticking to its body, then flies to another flower. If this flower is older than the first one, it will already have lost its pollen. When the bee's weight pushes down the keel, only the stigma comes out and touches the insect's body, picking up pollen grains on its sticky surface (Figure 10*b*).

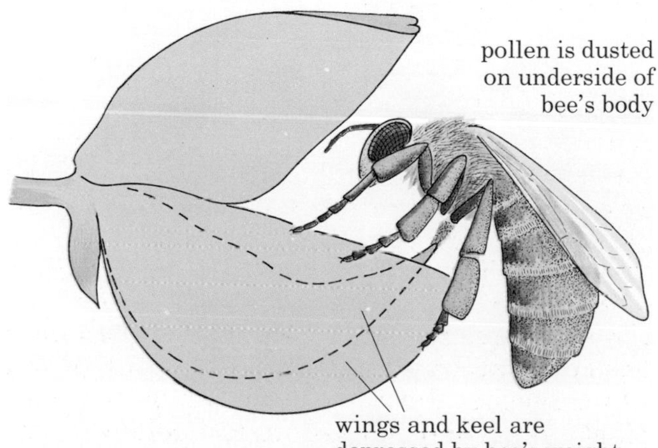

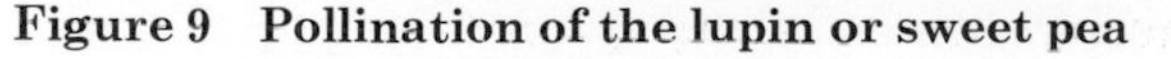

Figure 9 Pollination of the lupin or sweet pea

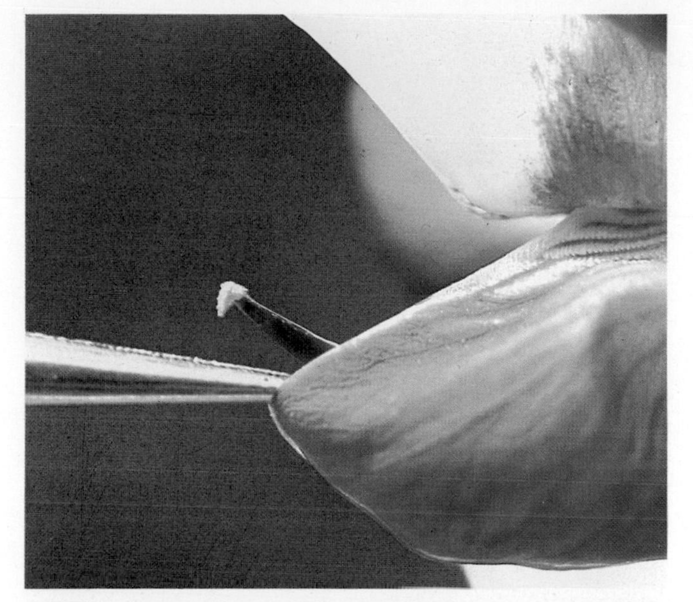

(*a*) The forceps are pushing the wings down as a bee's weight would. Pollen is being squeezed out of the tip of the keel.

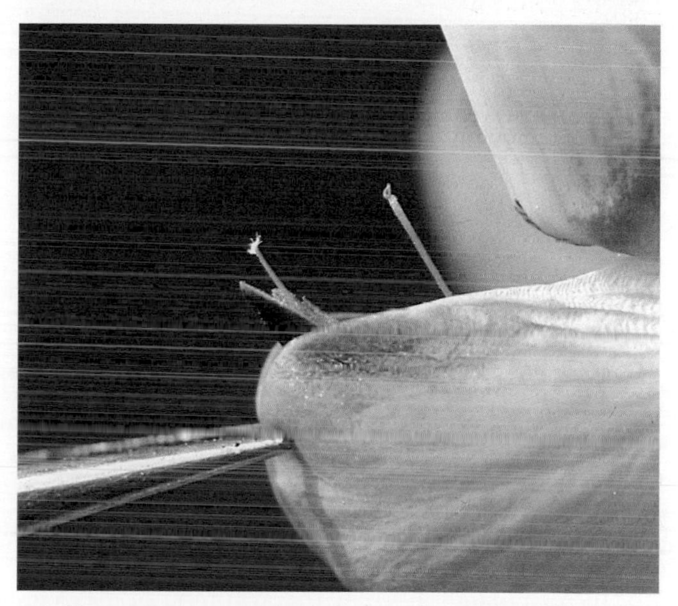

(*b*) The anthers have lost their pollen. The stigma now protrudes from the keel when the bee lands.

Figure 10 Pollination of the lupin

Pollination of the buttercup

Only a large, heavy insect such as a bee is able to pollinate the lupin. The buttercup flower may be pollinated by a variety of insects which walk about in the flower, looking for the nectar at the base of the petals. The anthers ripen before the ovaries so that an insect visiting a young flower will be dusted with pollen. When this insect visits an older flower, the ovaries will be ripe and their stigmas will pick up pollen from the insect's body.

Figure 11 Bumble bee visiting a sage flower (like the white deadnettle below). The stigma can be seen rubbing against the bee's back.

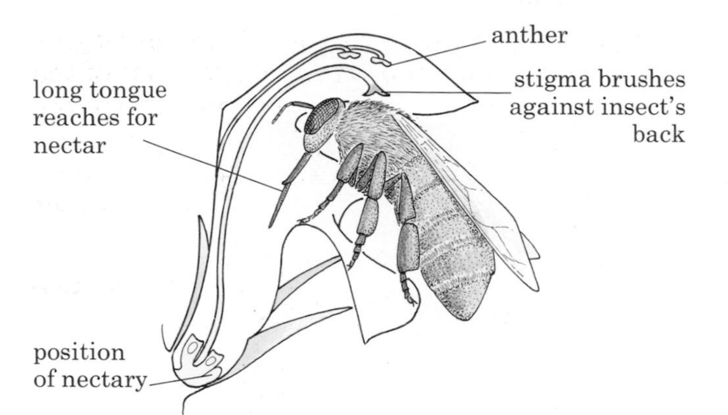

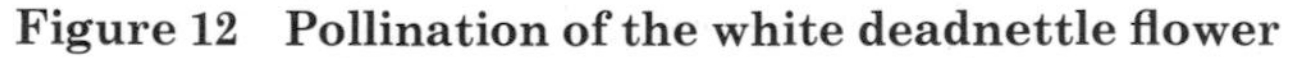

Figure 12 Pollination of the white deadnettle flower

Adaptation of flowers for insect pollination

Figure 11 shows a bee visiting a sage flower and Figure 12 shows what is happening inside a similar flower. Only an insect like a bee is large enough to brush against the stamens or stigma, so the flower is said to be adapted to pollination by bees. In general, insect-pollinated flowers are adapted to this method of pollination by having brightly coloured petals, nectar and scent. All these features attract insects to the flowers.

Wind pollination

Many grasses and trees do not depend on insects for pollination, but the pollen is released into the air and carried away by the wind. Most of the pollen will be wasted, but some will be trapped on the stigmas of the ripe flowers. Figure 13 shows the pollen falling from the male catkin of a birch tree. (Wind-pollinated plants often have the stamens and ovaries in separate, male or female flowers.) Each catkin is a collection of small flowers. In wind-pollinated flowers of this kind, the anthers often protrude from the flowers and are easily shaken about by the wind to dislodge the pollen. The pollen grains are smooth and light and float in the air for long distances. The stigmas often have feathery

Figure 13 Wind pollination. Clouds of pollen being shaken from birch catkins

Figure 14 Grass flowers. The anthers are a brown–purple colour. The stigmas are fluffy white structures. Notice that both these organs are protruding from the flowers, which are small and green.

branches and stick out from the flowers (Figure 14). This improves their chances of trapping pollen as it blows past on the wind.

The 'petals' are small and green, and the flowers have no scent and no nectar. This is what you would expect in flowers which have no need to attract insects.

QUESTIONS

4 If plants are growing in a greenhouse where insects cannot enter, how would you make sure that all the flowers were pollinated?

5 Why do you think a large insect such as a bee can pollinate a lupin flower, while a small insect like a fly cannot?

6 Draw up a table to compare the structures of flowers which are pollinated by insects and those pollinated by the wind. Compare each feature in turn, e.g. petals, stamens, stigmas, etc.

7 Which of the following trees would you expect to be pollinated by insects: apple, hazel, oak, cherry, horse-chestnut, sycamore?

FERTILIZATION AND FRUIT FORMATION

Pollination is complete when pollen from an anther has landed on a stigma. If the flower is to produce seeds, pollination has to be followed by a process called **fertilization**. In all living organisms, fertilization happens when a male sex cell and a female sex cell meet and join together (they are said to **fuse** together). The cell which is formed by this fusion develops into an embryo of an animal or a plant (Figure 15).

In animals, the male sex cell is the sperm and the female sex cell is the egg or ovum (page 138).

In flowering plants, the male sex cell is in the pollen grain; the female sex cell, called the **egg cell**, is in the ovule. For fertilization to occur, the nucleus of the male cell from the pollen grain

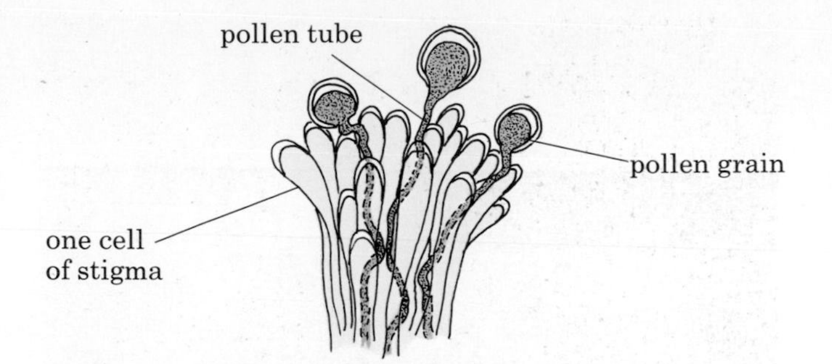

Figure 16 Pollen grains growing on the stigma of a crocus

has to reach the female nucleus of the egg cell in the ovule, and fuse with it. The following account explains how this happens.

Fertilization

The pollen grain absorbs liquid from the stigma and a microscopic **pollen tube** grows out of the grain (Figure 16). This tube grows down the

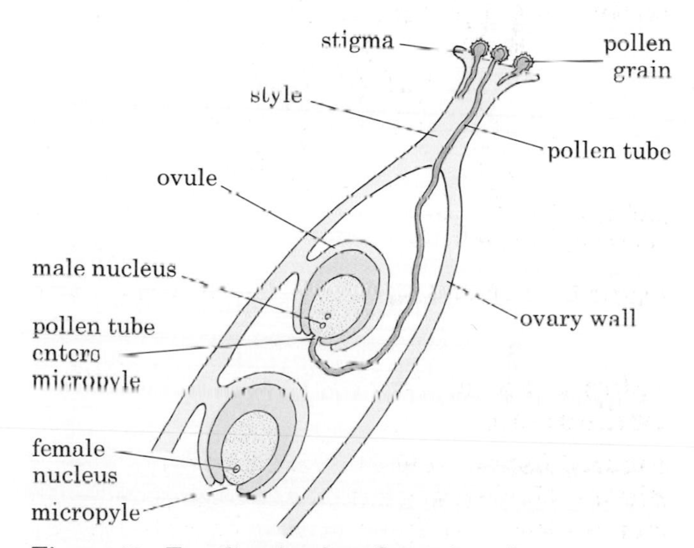

Figure 17 Fertilization in a flowering plant

style and into the ovary where it enters a small hole, the **micropyle**, in an ovule (Figure 17). The nucleus of the pollen grain travels down the pollen tube and enters the ovule. Here it

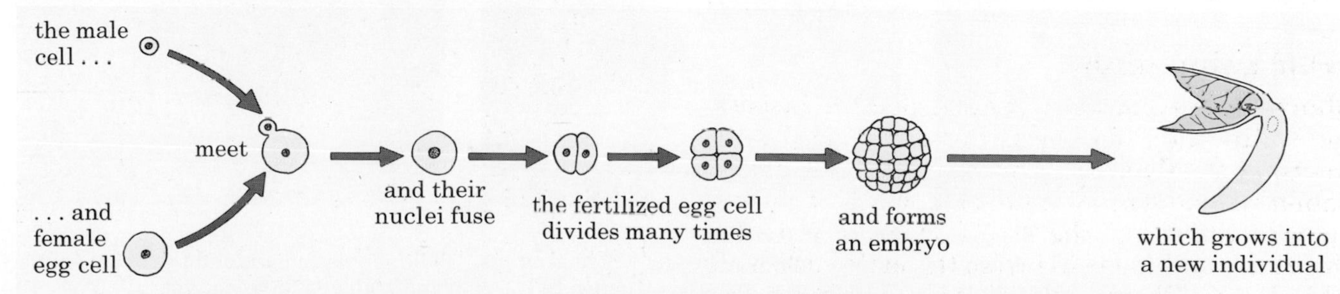

Figure 15 Fertilization. These events take place during fertilization in plants and animals

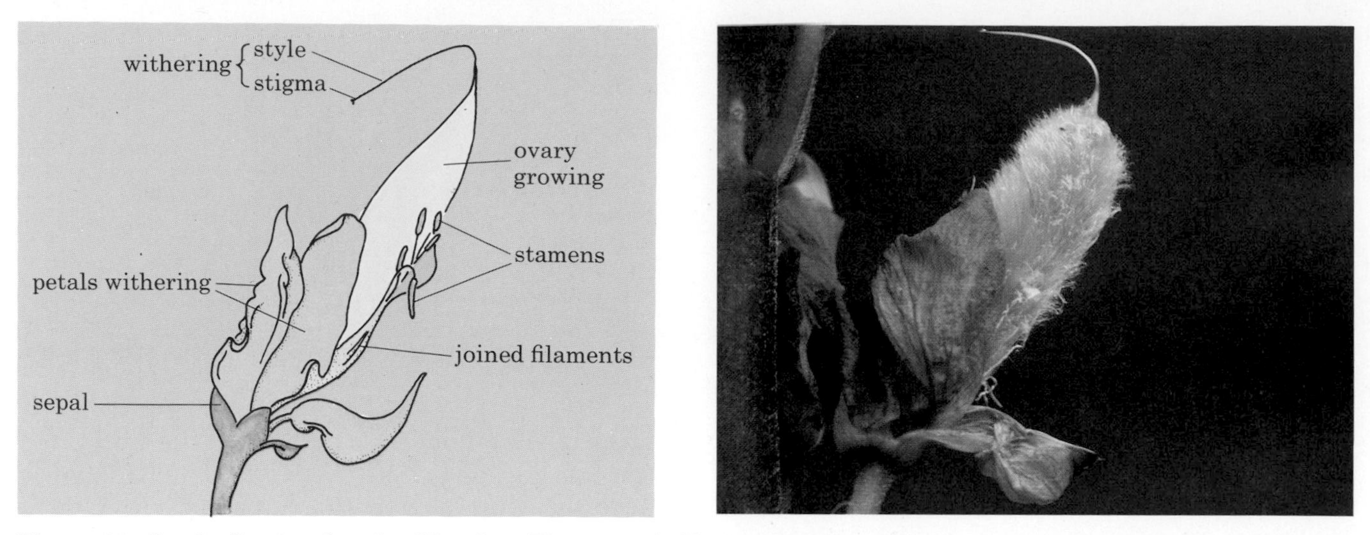

Figure 18 Lupin flower after fertilization. The ovary (still with the style and stigma attached) has grown much larger than the flower. The petals have turned brown and shrivelled.

combines with the nucleus of the egg cell. Each ovule in an ovary needs to be fertilized by a separate pollen grain.

Although pollination must occur before the ovule can be fertilized, pollination does not always result in fertilization. For example, if the pollen of a buttercup arrived on the stigma of a lupin, it would not produce a pollen tube. Self-pollination is prevented in some species because the pollen in a flower will not grow on its own stigma. The pollen and stigma are said to be **incompatible**.

From the description of pollination in the lupin, it will be seen that the pollen grains come into contact with the stigma of their own flower before they are shed, but they do not grow pollen tubes and therefore self-pollination of the flower does not occur.

Fruit and seed formation

After the pollen nucleus has fused with the egg nucleus, the egg cell divides many times and produces a miniature plant called an **embryo**. The embryo consists of a tiny root and shoot with two special leaves called **cotyledons**. Food made in the leaves of the parent plant is carried in the phloem to the cotyledons. The cotyledons grow so large with this stored food that they completely enclose the embryo (Figure 2*b* on page 82). The outer wall of the ovule becomes thicker and harder, and forms the seed coat or **testa** (see page 82).

As the seeds grow, the ovary also becomes much larger and the petals and stamens shrivel and fall off (Figure 18). The ovary is now called a **fruit**. The biological definition of a fruit is a

(*a*) Flowers of a tomato plant. There are six sepals and six petals. The stamens form a tube in the centre of the flower.

(*b*) After fertilization. The ovary has grown rapidly. The shrivelled remains of the flower are still attached to it.

(*c*) The ovary has formed a ripe fruit. The sepals remain and the dried petals and stamens have not yet fallen off.

Figure 19 Development of a tomato fruit

(*a*) In the lower flowers, the fertilized ovaries are small and green. The brown styles are still attached.

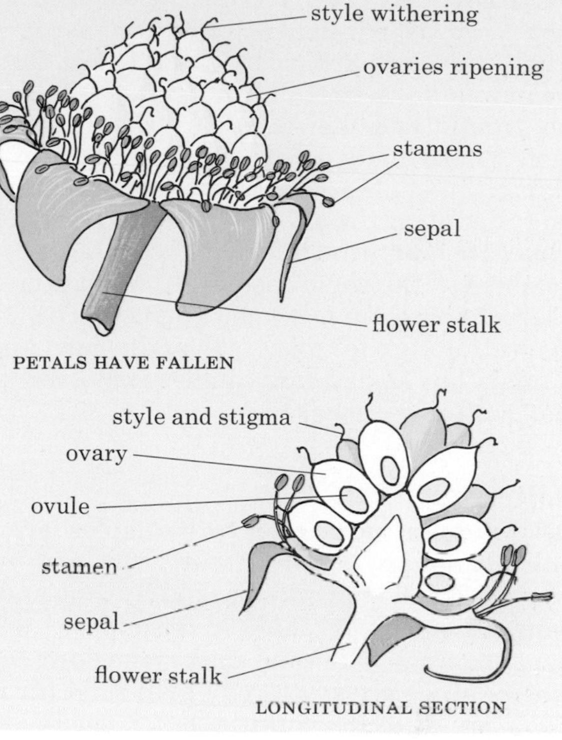

(*b*) Blackberry fruit

Figure 20 Fruit formation in a blackberry. After fertilization, the petals and stamens fall off and the ovaries swell.

fertilized ovary; it is not necessarily edible. In the lupin, the fertilized ovary forms a dry, hard pod, but in a related plant—the runner bean the ovary wall becomes fleshy and edible before drying out. A plum is a good example of a fleshy, edible fruit. Tomatoes (Figure 19) and cucumbers are also fruits although they are classed as vegetables in the shops. Blackberries (Figures 20*a* and *b*) and raspberries are formed by many small fruits clustered together. In strawberries (Figure 21), the fruits are the pips and the edible part is formed by the receptacle of the flower.

QUESTIONS

8 Which parts of a tomato flower (a) grow to form the fruit, (b) fall off after fertilization, (c) remain attached to the fruit?

9 Make a drawing to show what you think a ripe pea pod might look like if only four out of ten ovules in the ovary had been fertilized after pollination. (A pea flower is very similar to a lupin flower.)

10 Which of the following edible plant products do you think are, biologically, (a) fruits, (b) seeds, (c) neither: runner beans, peas, grapes, baked beans, marrow, rhubarb, tomatoes?

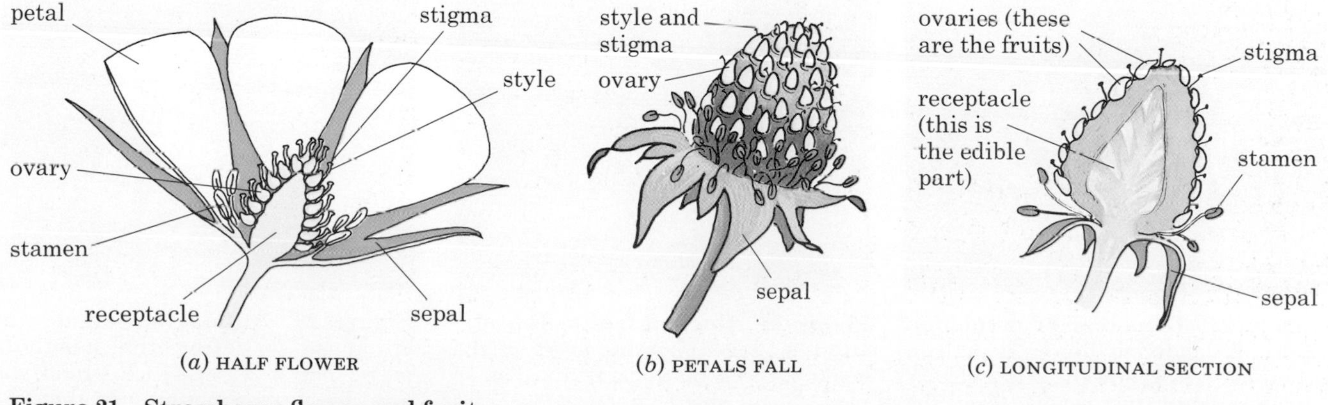

Figure 21 Strawberry flower and fruit

DISPERSAL OF FRUITS AND SEEDS

When the seeds are mature, the whole fruit or the individual seeds fall from the parent plant to the ground and the seeds may then germinate (page 82). In many plants, the fruits or seeds are adapted in such a way that they are carried a long distance from the parent plant. This reduces competition for light and water between members of the same species. It may also result in plants growing in new places. The main adaptations are for dispersal by the wind and by animals but some plants have 'explosive' pods that scatter the seeds.

Wind dispersal

(a) 'Pepper pot' effect. Examples are the white campion, poppy (Figure 22*a*) and antirrhinum. The flower stalk is usually long and the ovary becomes a dry, hollow capsule with one or more openings. The wind shakes the flower stalk and the seeds are scattered on all sides through the openings in the capsule.

(b) 'Parachute' fruits and seeds. Clematis, thistles, willow herb and dandelion (Figures 22*b* and 24) have seeds or fruits of this kind. Feathery hairs project from the fruit or seed and so increase its surface area. As a result, the seed 'floats' over long distances before sinking to the ground. It is therefore likely to be carried a long way from the parent plant by slight air currents.

(c) 'Winged' fruits. Fruits of the lime, sycamore (Figure 22*c*) and ash trees have wing-like outgrowths from the ovary wall, or leaf-like structures on the flower stalk. These 'wings' cause the fruit to spin as it falls from the tree and so slow down its fall. This delay increases the chances of the fruit being carried away in air currents.

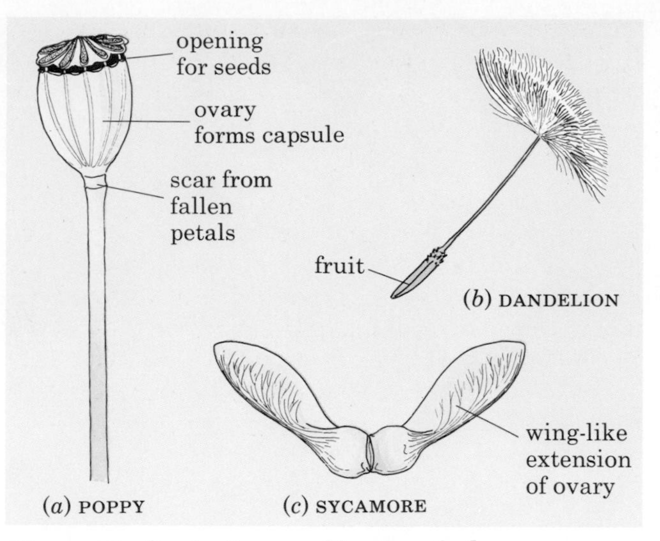

Figure 22 Seeds dispersed by the wind

Animal dispersal

(a) Mammals; hooked fruits. In herb bennet, agrimony (Figure 23) and goosegrass, hooks develop from the style, from the receptacle or

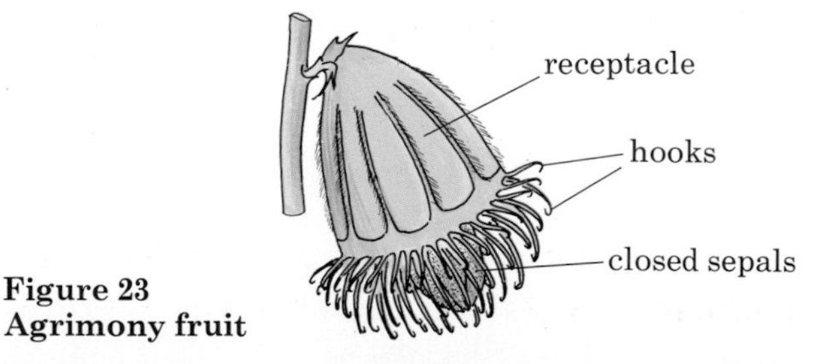

Figure 23 Agrimony fruit

Figure 24 Dispersal of dandelion fruits. A dandelion 'flower' is really a group of small flowers. Each ovary forms a fruit; the sepals form the parachute.

Figure 25 Hooked fruits. The situation is 'staged' but the effect of the hooks on the burdock fruits can be seen. Some small green goosegrass fruits are also present.

Figure 26 Animal dispersal? The dormouse is eating the hawthorn berry, but what evidence would you need to be reasonably sure that the seeds are dispersed in this way?

from the ovary wall. These hooks catch in the fur of passing mammals or in the clothing of people (Figure 25) and, later, some way from the parent, they fall off or are brushed or scratched off.

(b) Birds; succulent fruits. Fruits such as the blackberry and elderberry are eaten by birds and mammals (Figure 26). The hard pips containing the seed are not digested and so pass out with the droppings of the bird away from the parent plant. The soft texture and, in some cases, the bright colour of these fruits may be regarded as an adaptation to this method of dispersal.

'Explosive' fruits

The pods of flowers in the pea family, e.g. gorse, broom, lupin and vetches, dry in the sun and shrivel. The tough fibres in the fruit wall shrink and set up a tension. When the fruit splits in half down two lines of weakness, the two halves curl back suddenly and flick out the seeds (Figure 27).

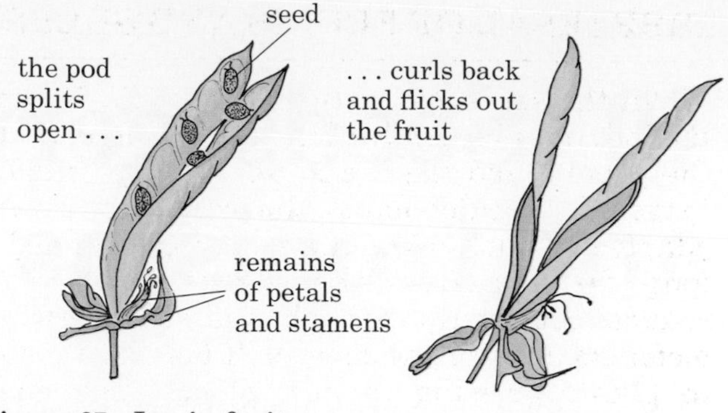

Figure 27 Lupin fruit

QUESTIONS

11 What are the advantages of a plant dispersing its seeds over a wide area? What disadvantages might there be?

12 Which methods of dispersal are likely to result in seeds travelling the greatest distances? Explain your answer.

13 Students sometimes confuse 'wind pollination' and 'wind dispersal'. Write a sentence or two about each to make the difference clear.

BASIC FACTS

- **Flowers contain the reproductive organs of plants.**
- **The stamen is the male organ. It produces pollen grains which contain the male gamete.**
- **The ovary is the female organ. It produces ovules which contain the female gamete and will form the seeds.**
- **Brightly coloured petals attract insects, which pollinate the flower.**
- **Pollination is the transfer of pollen from the anthers of one flower to the stigma of another.**
- **Pollination may be done by insects or by the wind.**
- **Flowers which are pollinated by insects are usually brightly coloured and have nectar. The shape of the petals may adapt the flower to pollination by one type of insect.**
- **Flowers which are pollinated by the wind are usually small and green. Their stigmas and anthers hang outside the flower where they are exposed to air movements.**
- **Fertilization occurs when a pollen tube grows from a pollen grain into the ovary and up to an ovule. The pollen nucleus passes down the tube and fuses with the ovule nucleus.**
- **After fertilization, the ovary grows rapidly to become a fruit and the ovules become seeds.**
- **Seeds and fruits may be dispersed by the wind, by animals, or by an 'explosive' method.**
- **Dispersal scatters the seeds so that the plants growing from them are less likely to compete with each other and with their parent plant.**

10 Seeds, germination and tropisms

The previous chapter described how a seed is formed from the ovule of a flower as a result of fertilization, and is then dispersed from the parent plant. If the seed lands in a suitable place it will **germinate**, i.e. grow into a mature plant. To understand the process of germination, the structure of the french bean will first be studied. The french bean plant is a dicot (see page 203).

SEED STRUCTURE AND GERMINATION OF THE FRENCH BEAN

Seed structure

The seed (Figures 1 and 2) contains a miniature plant, the **embryo**, which consists of a root or **radicle**, and a shoot or **plumule**. The embryo is attached to two leaves called the **cotyledons**, which are swollen with stored food. This stored food, mainly starch, is used by the embryo when it starts to grow. The embryo and cotyledons are enclosed in a tough seed coat or **testa**. The **micropyle**, through which the pollen tube entered (page 77), remains as a small hole in the testa and allows the seed to take up water before germinating.

Germination

The stages of the process are shown in Figure 3. The seed absorbs water and swells. After three or four days, the radicle bursts through the testa and grows down into the soil, pushing its way between soil particles and small stones. Its tip is protected by the root cap (see page 63). Branches, called **lateral roots**, grow out from

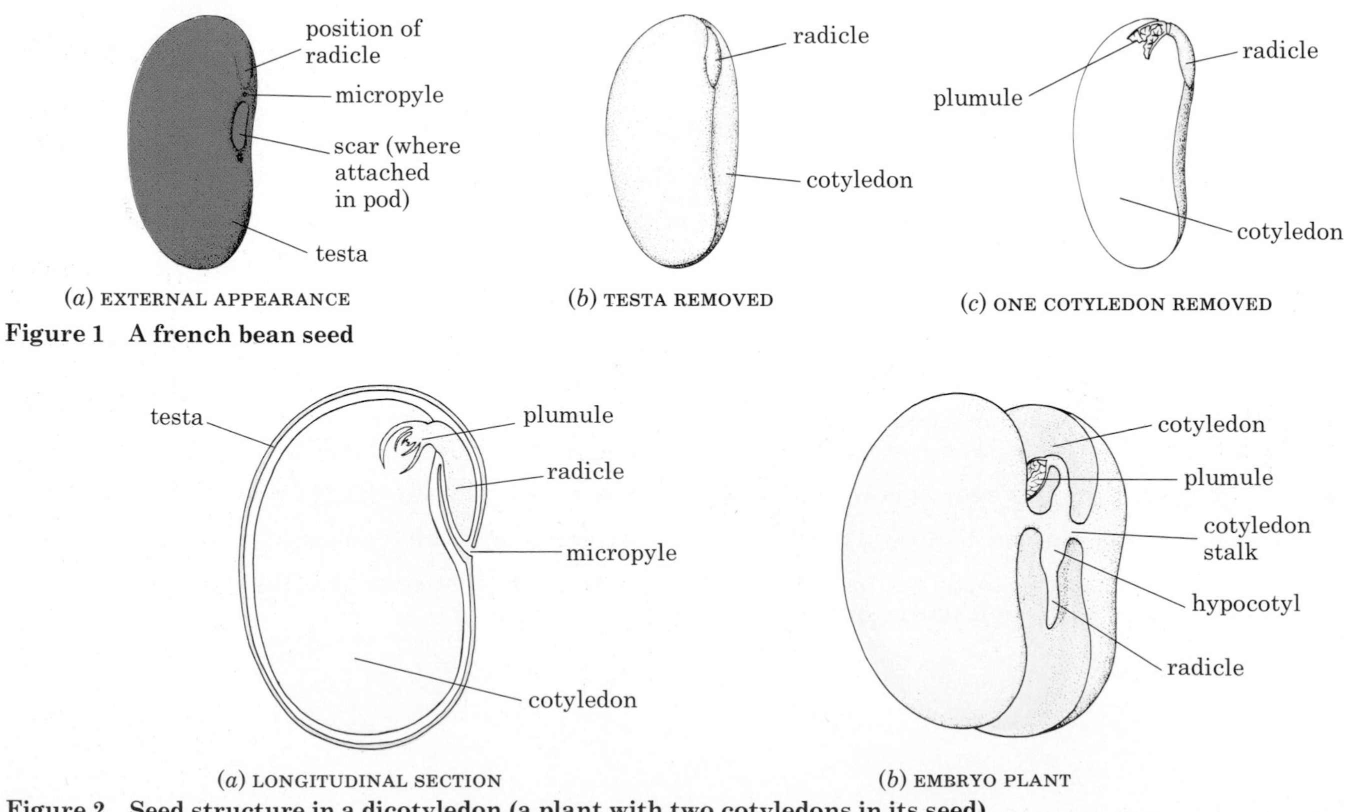

Figure 1 A french bean seed

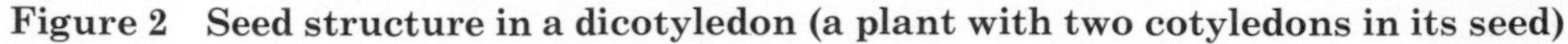

Figure 2 Seed structure in a dicotyledon (a plant with two cotyledons in its seed)

the side of the main root and help to anchor it firmly in the soil. On the main root and the lateral roots, microscopic **root hairs** grow out. These are fine outgrowths from some of the outer cells. They make close contact with the soil particles and absorb water from the spaces between them (see page 68).

A region of the embryo's stem, the **hypocotyl**, just above the radicle (Figure 2*b*), now starts to elongate. The radicle, by now, is firmly anchored in the soil so the rapidly growing hypocotyl arches upwards through the soil, pulling the cotyledons with it (Figure 4). Sometimes the cotyledons are pulled out of the testa, leaving it below the soil, and sometimes the cotyledons remain enclosed in the testa for a time. In either case, the plumule is well protected from damage while it is being pulled through the soil because it is enclosed between the cotyledons.

Once the cotyledons are above the soil, the hypocotyl straightens up and the leaves of the plumule open out. Up to this point, all the food needed for making new cells and producing energy has come from the cotyledons.

The main type of food stored in the cotyledons is starch. Before this can be used by the growing shoot and root, the starch has to be turned into a soluble sugar. In this form, it can be transported by the phloem cells. The change from starch to sugar in the cotyledons is brought about by enzymes which become active as soon as the seed starts to germinate. The cotyledons shrivel as their food reserve is used up, and they fall off altogether soon after they have been brought above the soil.

Figure 4 Germinating seedlings. Notice how the brown testa is shed and the white cotyledons turn green.

By now the plumule leaves have grown much larger, turned green and started to absorb sunlight and make their own food by photosynthesis (page 24). Between the plumule leaves

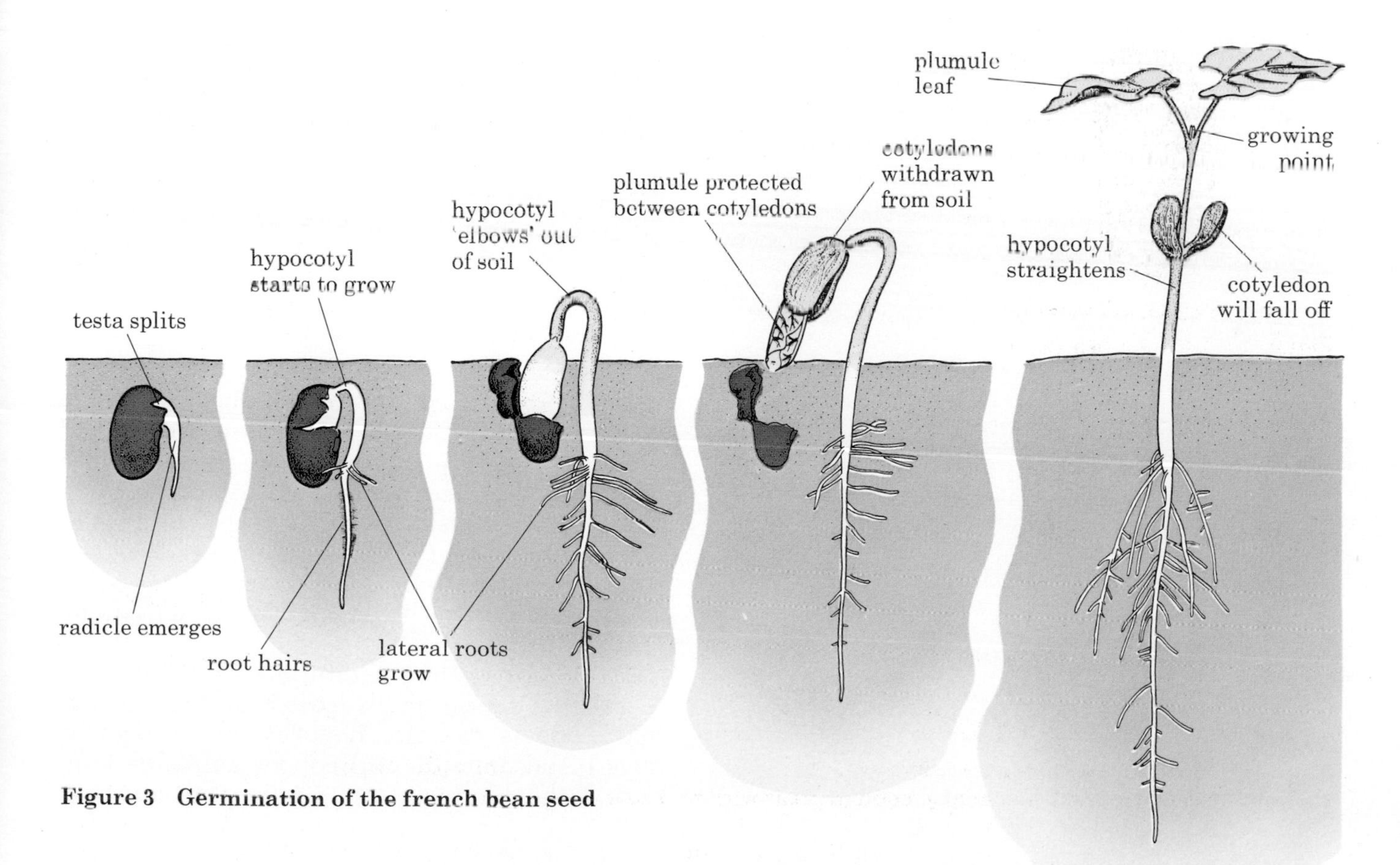

Figure 3 Germination of the french bean seed

is a growing point which continues the upward growth of the stem and the production of new leaves. The embryo has now become an independent plant, absorbing water and mineral salts from the soil, carbon dioxide from the air and making food in its leaves.

EXPERIMENTS ON GERMINATION

1 Region of growth in a root

Leave some peas in water for 24 hours and then wrap them in a roll of blotting paper as shown in Figure 5. After three days, the radicles will have grown about 10mm. Choose seedlings with

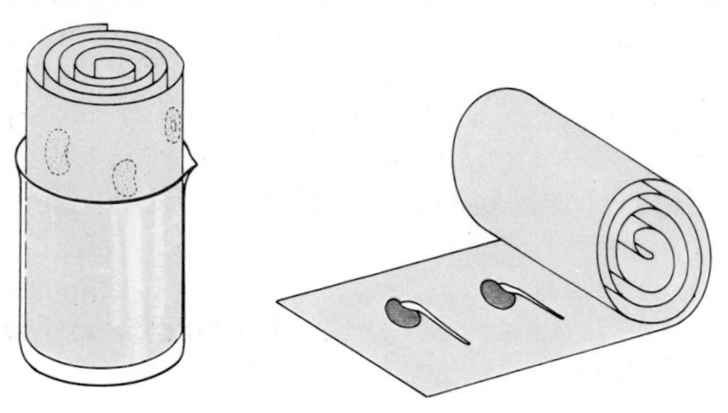

(*a*) Seeds held between moist blotting paper

(*b*) It can be unrolled to inspect the seedlings

Figure 5 One method of growing seedlings with straight radicles

straight radicles and mark the radicles with ink lines about 2mm apart. Figure 6 shows one way of doing this. Place three marked seedlings between two strips of moist cotton-wool in a Petri dish so that the seeds are held firmly but

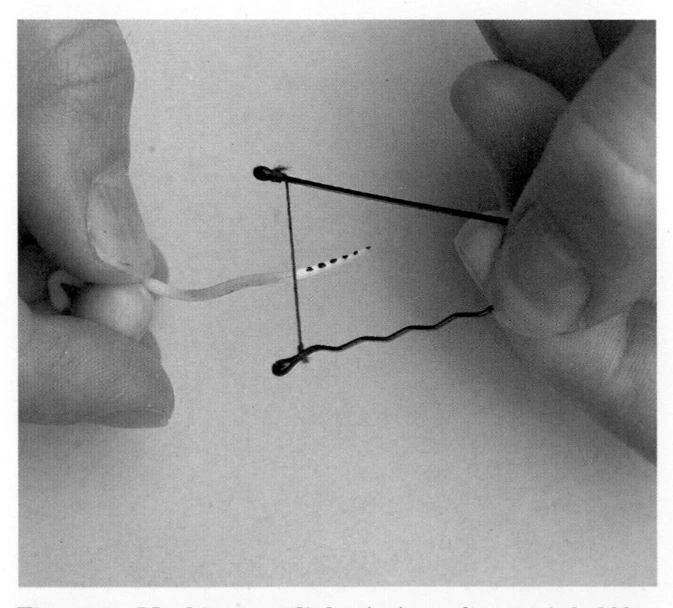

Figure 6 Marking a radicle. A piece of cotton is held by the hairpin and dipped into black ink.

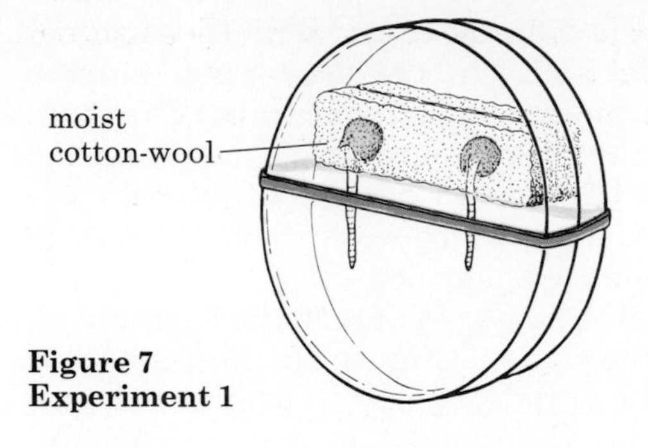

Figure 7 Experiment 1

the radicles are exposed and easily seen (Figure 7). Keep the lid of the Petri dish in place with an elastic band and leave the dish on its edge, with the radicles pointing downwards, for two days.

Result. The ink marks will have become most widely spaced in the region just behind the root tip (Figure 8).

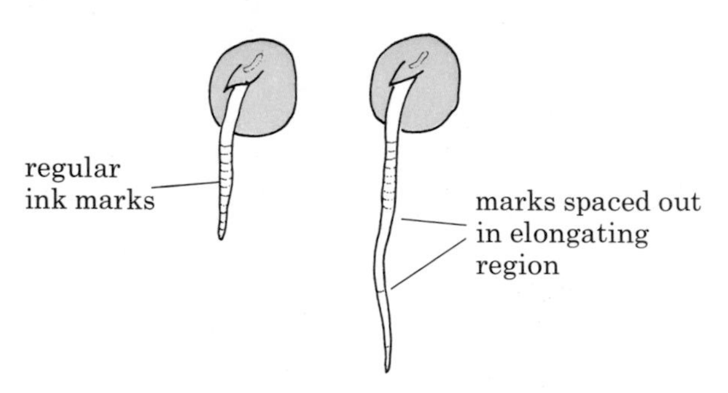

Figure 8 Region of growth in a radicle

Interpretation. The wide spacing of the marks shows that the region of most rapid growth is just behind the tip. In fact it is known that cell division takes place in the tip of the root but cell elongation occurs a short distance behind the tip (see Figure 12 on page 64).

2 Food reserves in seeds

Soak some peas in water for a day, then remove the testas and crush the cotyledons up in a little water with a mortar and pestle. Pour some of the crushed watery mixture into a test-tube to a depth of about 5cm and heat it to boiling point over a small Bunsen flame. Cool the tube under the tap and share the mixture between two test-tubes. To one of these add a few drops of iodine solution. If it turns blue, starch is present in the cotyledons.

To the other tube carefully add about 5mm (depth) dilute sodium hydroxide. (CARE: Sodium hydroxide is caustic. Report any spillage at once.) Add 5mm (depth) copper sulphate solution. If the mixture turns purple, it means that protein is present in the cotyledons. (See page 96 for other food tests.)

If wheat is germinated for a week, the grains can be shown to contain sugar when tested with Benedict's solution as described on page 96.

QUESTIONS

1 What are the functions of (a) the radicle, (b) the plumule and (c) the cotyledons of a seed?

2 During germination of the french bean, how are (a) the plumule, (b) the radicle protected from damage as they are forced through the soil?

3 List all the possible ways in which a growing seedling might use the food stored in its cotyledons.

4 At what stage of development is a seedling able to stop depending on the cotyledons for its food?

5 What do you think are the advantages to a germinating seed of having its radicle growing some time before the shoot starts to grow?

6 Figure 8 shows the result of an experiment to find the region of maximum growth in a root. Draw a diagram to show how the result would have appeared if (a) the root grew simply by adding new cells at the tip, and (b) if the root grew mainly at the point just below its attachment to the cotyledon.

7 Explain why sugar can be found in germinating wheat grains but not in ungerminated grains.

CONDITIONS FOR GERMINATION

For seeds to germinate they must be able to obtain water and oxygen. Also, most seeds will not start to germinate if the temperature is below a certain level.

Water

Water is absorbed from the soil by the radicle and root hairs (page 68). It is used (a) to build up new cytoplasm, (b) to enlarge the cells' vacuoles and make them expand during growth (page 5), (c) to carry sugar from the cotyledons to the growing regions and (d) to keep up the turgor pressure (page 20) in the cells of the seedling. Later on, water is needed for photosynthesis in the cotyledons.

Oxygen

In the early stages of germination, all the oxygen needed by the seed must come from the air spaces in the soil (page 41). The oxygen is needed for respiration (page 14) which provides the energy required for all the chemical processes in living cells. These processes are very rapid during germination and if the soil lacks oxygen because it is waterlogged or heavily compacted, germination may be prevented. (For experiments on the respiration of germinating seeds see pages 16 and 17.)

Temperature

On page 12 it was explained that a rise in temperature speeds up most chemical reactions, including those taking place in living organisms. Germination, therefore, occurs more rapidly at high temperatures, up to about 40°C. Above 45°C, the enzymes in the cells are destroyed and the seedlings would be killed. Below certain temperatures (e.g. 0–4°C) germination may not start at all in some seeds.

EXPERIMENTS ON THE CONDITIONS FOR GERMINATION

1 The need for water

Label three containers A, B and C and put dry cotton-wool in the bottom of each. Place an equal number of soaked seeds in each container. Leave A quite dry; add water to B to make the cotton-wool moist; add water to C until all the seeds are completely covered (Figure 9). Put lids on the containers and leave them all at room temperature for a week.

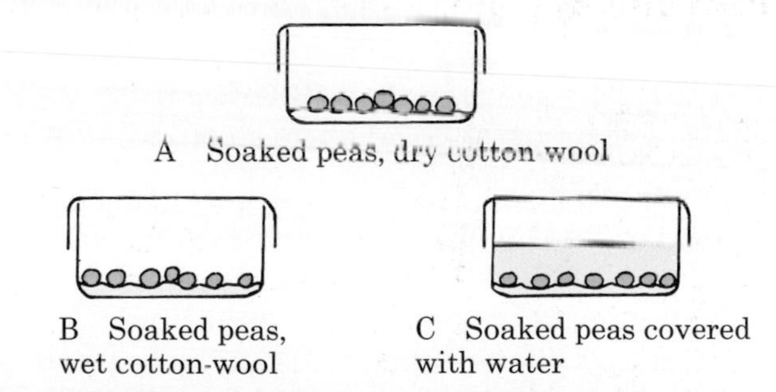

Figure 9 The need for water in germination

Result. The seeds in B will germinate normally. Those in A may have started to germinate but will have dried up at an early stage. The seeds in C may also have started to germinate but will probably not be as advanced as those in B and may have died and started to decay.

Interpretation. Although water is necessary for germination, too much of it may prevent germination by cutting down the oxygen supply to the seed.

2 The need for oxygen

Set up the experiment as shown in Figure 10. (CARE: Pyrogallic acid and sodium hydroxide are a caustic mixture. Use eye shields, handle the liquids with care and report any spillage at once.)

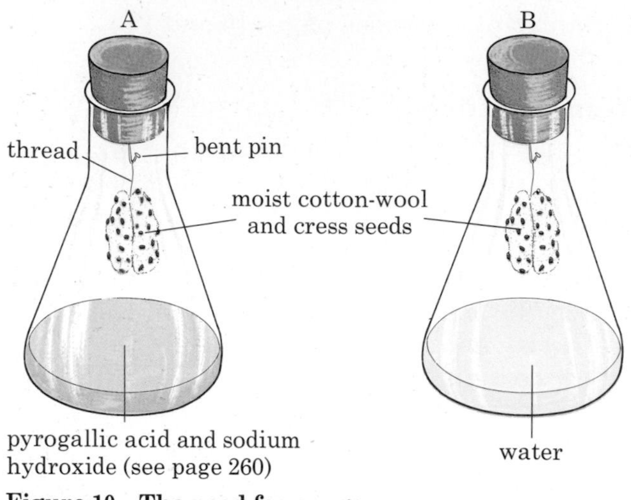

Figure 10 The need for oxygen

If the moist cotton-wool is rolled in some cress seeds, they will stick to it. The bungs must make an air-tight seal in the flask and the cotton-wool must not touch the solution. Pyrogallic acid and sodium hydroxide absorb oxygen from the air, so the cress seeds in flask A are deprived of oxygen. Flask B is the control (see page 17). This is to show that germination can take place in these experimental conditions provided oxygen is present. Leave the flasks for a week at room temperature.

Result. The seeds in flask B will germinate but there will be little or no germination in flask A.

Interpretation. The main difference between flasks A and B is that A lacks oxygen. Since the seeds in this flask have not germinated, it looks as if oxygen is needed for germination.

To show that the chemicals in flask A had not killed the seeds, the cotton-wool can be swapped from A to B. The seeds from A will now germinate.

3 Temperature and germination

Soak some maize grains for a day and then roll them up in three strips of moist blotting paper as shown in Figure 11. Put the rolls into plastic bags. Place one in a refrigerator (about 4°C), leave one upright in the room (about 20°C) and put the third one in a warm place such as over a radiator or—better—in an incubator set to

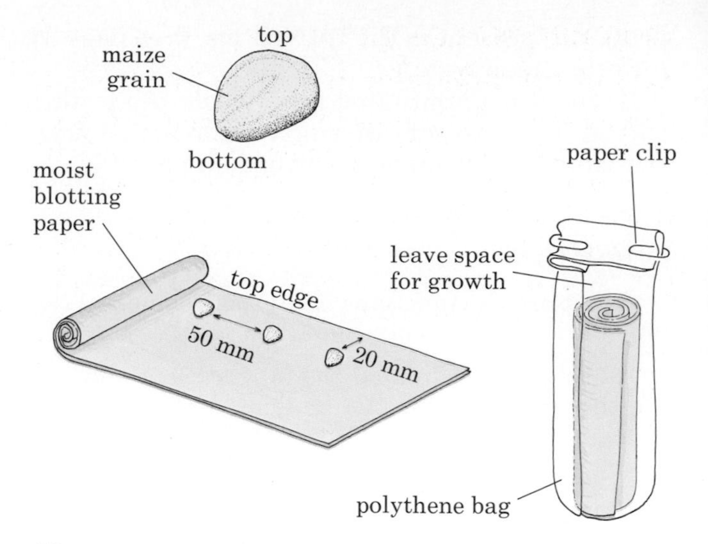

Figure 11 Temperature and germination. Roll the seeds in moist blotting paper, and stand the rolls upright in plastic bags.

30°C. After a week, examine the seedlings and measure the length of the roots and shoots.

Result. The seedlings kept at 30°C will be more advanced than those at room temperature. The grains in the refrigerator may not have started to germinate at all.

Interpretation. Seeds will not germinate below a certain temperature. The higher the temperature, the faster the germination, at least up to 35–40°C.

QUESTIONS

8 Why do you think that a control is described for Experiment 2 but not for Experiments 1 and 3?

9 Describe the natural conditions in the soil that would be most favourable for germination. How could a gardener try to create these conditions?

10 Suppose you planted some bean seeds in the garden but no plants appeared after three weeks. List all the possible causes for this failure. How could you investigate some of these possibilities?

TROPISMS

A **tropism** is a change in the direction of growth in a shoot or root as a result of the effect of light or gravity. For example, shoots will appear to grow towards a source of light if it comes from the side rather than from above the plant. This response to the direction of light is called **phototropism**. The response made to the direction of gravity is called **geotropism.**

EXPERIMENTS ON TROPISMS

1 Phototropism in cress seedlings

Spread some cress seeds thinly on moist cotton-wool in two Petri dishes and allow the seeds to germinate in total darkness for five or six days. (Keep the cotton-wool moist during this time.) Cut off the leafy parts (cotyledons plus plumule) from *half* the seedlings in each dish (Figure 12). Place one dish of seedlings directly under a bench lamp about 40cm away. Place the other dish the same distance away from a bench lamp shining from the side, and slightly below the dish (Figure 13). Screen the seedlings as far as possible from other sources of light, with sheets of card cut from the sides of boxes. Examine the seedlings from 2 to 24 hours later.

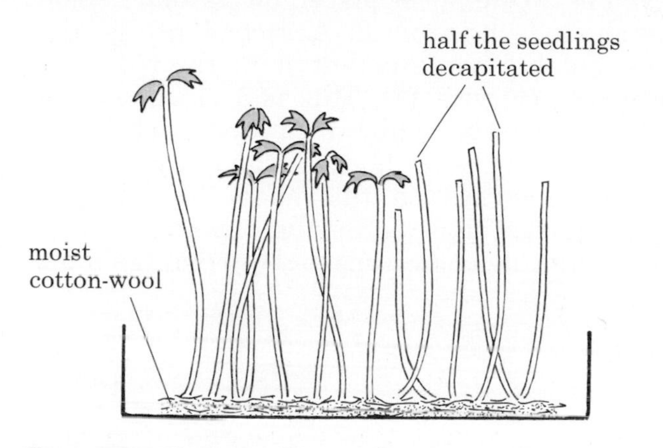

Figure 12 Cress seedlings after one week

Result. The stalks (hypocotyls) of the intact seedlings illuminated from the side will have curved towards the light. The hypocotyls with the cotyledons and plumules removed will not have curved at all. The hypocotyls illuminated from above will still be growing vertically.

Interpretation. Although the 'bending' takes place in the hypocotyl, it does not occur unless the cotyledons and plumule are present. A possible explanation is given under 'The auxin theory' (page 89). The simplest interpretation is that the one-sided lighting has stimulated the cress shoots to change their direction of growth. This response is called **positive phototropism**, i.e. the response is towards the source of light. Nearly all plant shoots are positively phototropic. By growing towards the source of light in this way, the shoot brings its leaves into the best position for photosynthesis. Figure 14 shows positive phototropism in sunflower seedlings.

Figure 14 Phototropism. The sunflower seedlings have received one-sided lighting for a day.

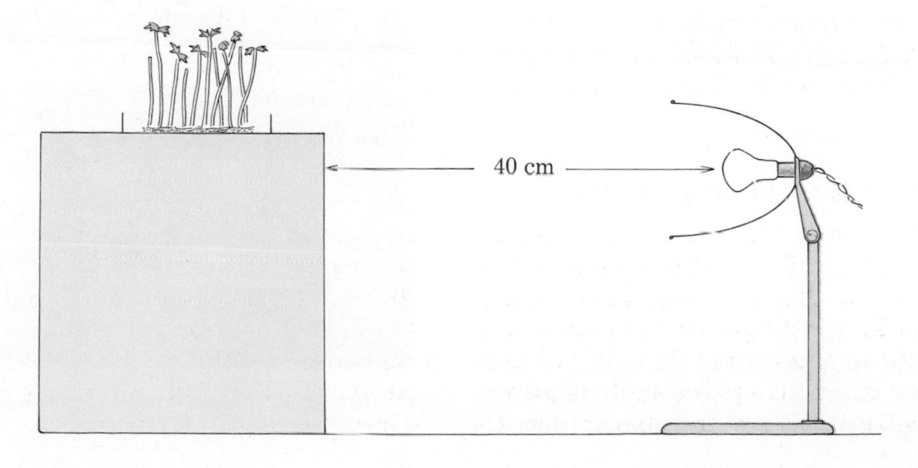

Figure 13 One-sided lighting for Experiment 1

2 Geotropism in pea radicles

Soak about ten peas in water for a day and then let them germinate in a roll of moist blotting paper (Figure 5 on page 84). After three days, choose six seedlings with straight radicles and set them up in two Petri dishes as described in Experiment 1, page 84, and shown in Figure 15. Keep the Petri dishes on their edges, but in one

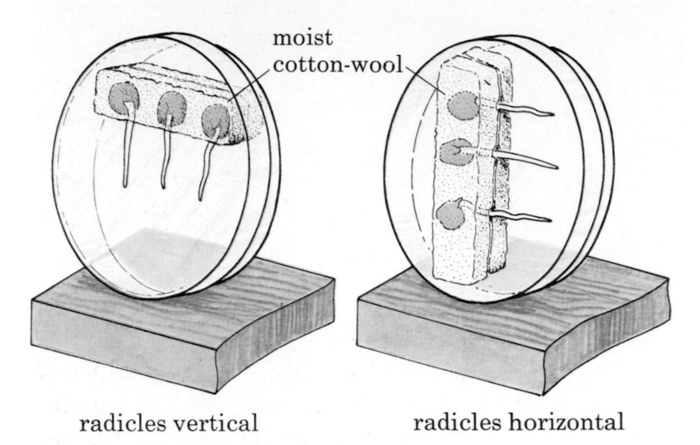

Figure 15 Geotropism in radicles

let the radicles point downwards and in the other make them point horizontally. Leave them in darkness for two days.

Result. The radicles which were horizontal will have changed their direction of growth and will be growing downwards (Figure 16). The radicles which were vertical will be straight.

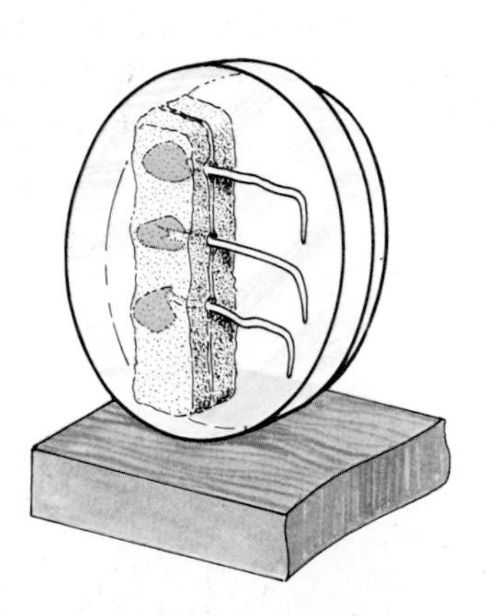

Figure 16 After two days

Interpretation. The only difference between the two dishes was that the horizontal radicles had the 'pull' of gravity acting on one side while the vertical radicles had gravity acting down their length. The 90° change in the direction of growth is therefore most likely to be a response to the one-sided stimulus of gravity. Roots are **positively geotropic**, i.e. they grow towards the 'pull' of gravity. This response will cause radicles to grow down into the soil, which is where they will obtain their water and mineral salts, and also anchor the plant firmly.

3 Geotropism in bean shoots

Some bean seedlings are grown in soil in small flower pots until the plumule leaves are well developed. One pot is placed on its side so that the shoot is horizontal. A second pot is also arranged horizontally but it is rotated on a **clinostat** (Figure 17). This is a clockwork or electric turntable which rotates the pot slowly about four times an hour. Although gravity is pulling sideways on this shoot, it will pull equally on all sides as the plant rotates.

The lighting conditions for both plants should

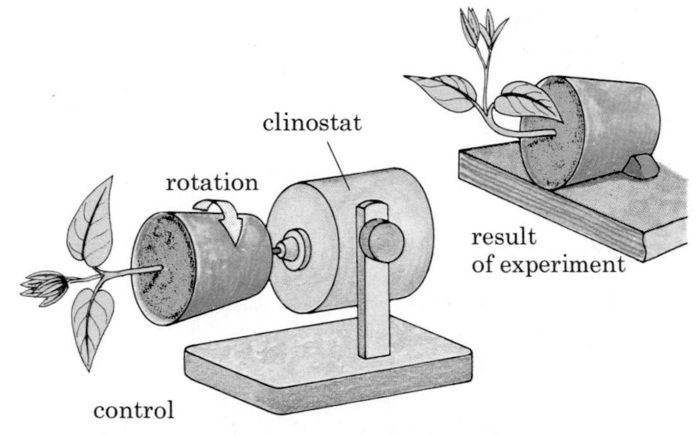

Figure 17 Geotropism in shoots

Figure 18 Negative geotropism. The tomato plant has been left on its side for 24 hours.

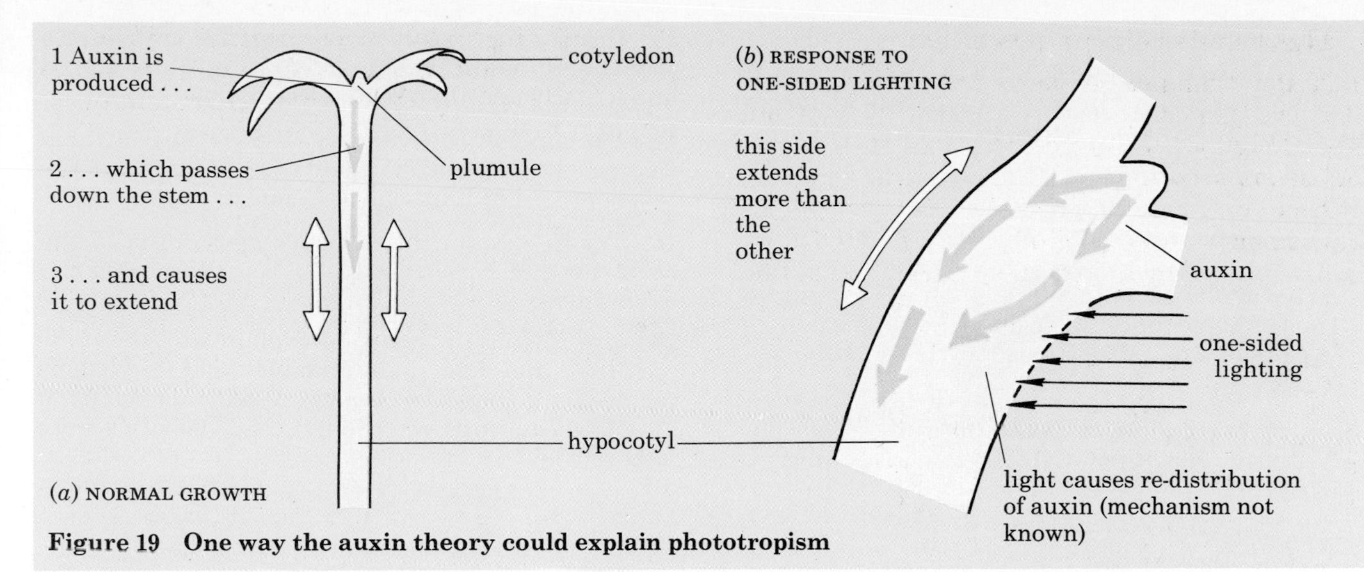

Figure 19 One way the auxin theory could explain phototropism

be the same or they should both be covered by cardboard boxes.

Result. After about 24 hours, the shoot in the clinostat is still growing horizontally. The shoot of the other plant has changed its direction of growth. By turning through 90° it is now growing vertically upwards.

Interpretation. The control with the clinostat shows that it is not just being horizontal that makes the shoot change direction but the fact that gravity is acting continuously on one side. The shoot is **negatively geotropic**, i.e. grows away from the direction of the gravitational pull. Most shoots are negatively geotropic and so produce upright stems. The advantages of an upright stem are mentioned on page 58.

The auxin theory of tropisms

This theory supposes that the growing parts of shoots produce a chemical called an **auxin**. From the tip of the shoot, the auxin passes down the stem and causes the cells just behind the tip to get longer. The elongation of cells in this region causes the stem to grow in length. If more auxin reached the left side of the stem than the right, the left side would grow faster than the right and so produce a curve to the right. Figure 19 suggests one way in which this theory could explain the positive phototropic response of a shoot.

In roots, it is thought that a growth substance, similar to auxin, reduces cell extension rather than increases it. Figure 20 shows how the theory could explain positive geotropism in roots.

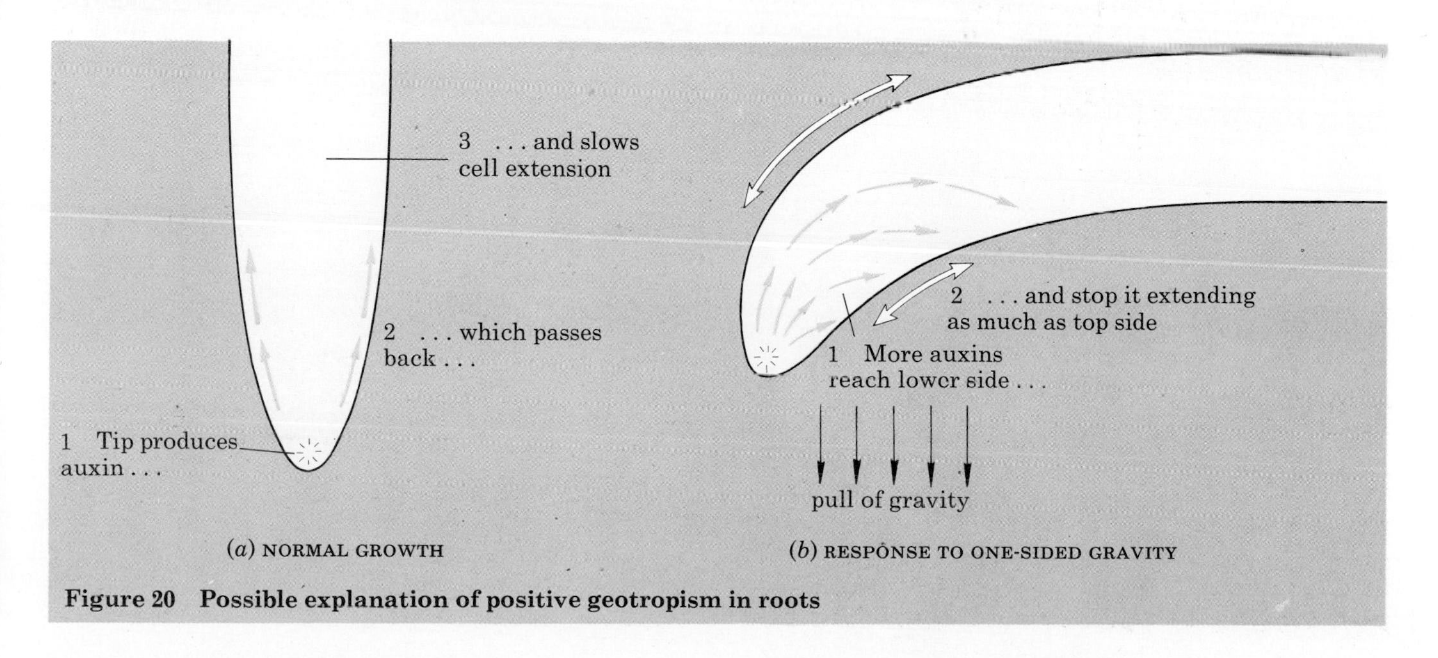

Figure 20 Possible explanation of positive geotropism in roots

There is a good deal of conflicting evidence about the part played by growth substances in causing tropisms. The account given above is only a simple theory, without very much supporting evidence.

QUESTIONS

11 What is the advantage to a plant of (a) positive geotropism in its roots, (b) negative geotropism in its shoots, (c) positive phototropism in its shoots?

12 Would you expect lateral roots (Figure 3 on page 83) to be (a) positively geotropic, (b) negatively geotropic, or (c) neither? Explain your reasoning.

13 Discuss the possible results on the growth of a shoot that might be observed if it were placed horizontally and illuminated from below.

14 Explain how the clinostat used for the control in Experiment 3 (page 88) could also be used for the same purpose in Experiments 1 and 2.

15 Try to explain the negative geotropism of the bean shoot in Experiment 3 (page 88) by using the auxin theory.

16 Use the auxin theory to explain why the cress seedlings that had their plumules and cotyledons removed in Experiment 1 (page 87) did not respond to one-sided lighting even though the hypocotyls were unharmed.

BASIC FACTS

- **A dicot seed consists of an embryo with two cotyledons enclosed in a seed coat (testa).**
- **The embryo consists of a small root (radicle) and shoot (plumule).**
- **The cotyledons contain the food store that the embryo will use when it starts to grow.**
- **When germination takes place, (a) the radicle of the embryo grows out of the testa and down into the soil and (b) the plumule is pulled backwards out of the soil.**
- **In many seeds, the cotyledons are also brought above the soil. They photosynthesize for a while before falling off.**
- **Germination is influenced by temperature and the amount of water and oxygen available.**
- **A tropism is a change in the direction of growth in a root or shoot, in response to a one-sided stimulus, usually light or gravity.**
- **Growing towards the stimulus is called a positive response. Growing away from the stimulus is called a negative response.**
- **The growth of roots and shoots is controlled by chemicals called auxins which are produced at the root or shoot tip and pass back to the rest of the structure.**

In the figure opposite, the veins are modelled in blue plastic, the arteries in red, the nerves in orange and the lymphatics in green. Not all these structures are represented. The lymphatics are shown only on the left and the nerves on the right.

SECTION THREE

Human Physiology

11 Food and diet

THE NEED FOR FOOD

All living organisms need food. An important difference between plants and animals is that green plants can make food in their leaves but animals have to take it in 'ready-made' by eating plants or the bodies of other animals. In all plants and animals food is used as follows:

(a) For growth; it provides the substances needed for making new cells and tissues.

(b) As a source of energy for the chemical reactions which take place in living organisms to keep them alive. When food is broken down during respiration (see page 14), the energy from the food can be used for activities such as movement, heart beat and nerve impulses.

(c) Replacement of worn and damaged tissues. The substances provided by food are needed to replace—for example—the millions of our red blood cells that break down each day, and to replace the skin which is worn away, and to repair wounds.

QUESTION

1 Suggest some actual examples of the uses which a tree might make of its food.

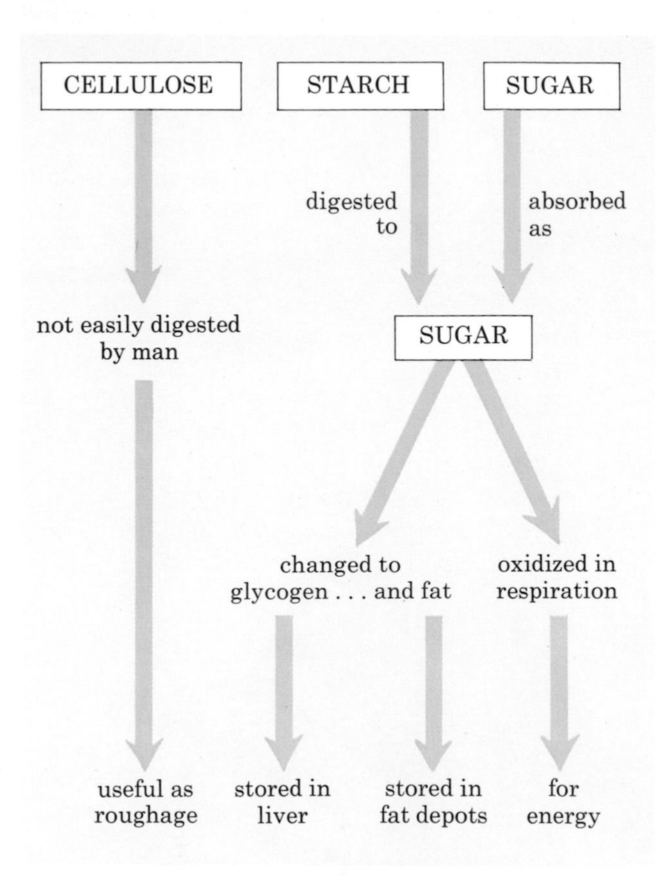

Figure 1 Digestion and use of carbohydrate

CLASSES OF FOOD

There are three classes of food: carbohydrates, proteins and fats.

Carbohydrates

Sugar and **starch** are important carbohydrates in our diet. Starch is abundant in potatoes, bread, rice and other cereals. Sugar appears in our diet mainly as **sucrose** (table sugar) which is added to drinks and many prepared foods such as biscuits and cakes.

Although *all* foods provide us with energy, carbohydrates are the cheapest and most readily available source of energy. They contain the elements carbon, hydrogen and oxygen (e.g. glucose is $C_6H_{12}O_6$). When carbohydrates are oxidized to provide energy by respiration they are broken down to carbon dioxide and water (see page 14). If we eat more carbohydrates than we need for our energy requirements, the excess is converted in the liver to either glycogen (see page 108) or fat. The glycogen is stored in the liver and muscles; the fat is stored in fat depots round the kidneys or under the skin (Figure 1).

The **cellulose** in the cell walls of all plant tissues is a carbohydrate. We probably do not derive very much nourishment from cellulose but it is important in the diet as **fibre** (see page 95) which helps to maintain a healthy digestive system.

Proteins

All plants contain some protein, but beans or cereals like wheat and maize are the best sources. Lean meat, fish, eggs, milk and cheese are important sources of animal protein.

Proteins, when digested, provide the chemical substances needed to build cells and tissues, e.g. skin, muscle, blood and bones. Neither carbohydrates nor fats can do this and so it is essential to include some proteins in the diet.

When proteins are digested, they are broken down to substances called **amino acids**. The amino acids are absorbed into the blood stream and used to build up different proteins. These proteins form part of the cytoplasm of cells and tissues. Figure 2 shows how a small protein is broken down by digestion to eleven amino acids (A–K). In the cells of the body the amino acids are joined up again in another pattern to form a different protein which the cells need.

The amino acids which are not used for making new tissues cannot be stored, but the body changes them to glycogen which can then be stored or oxidized to provide energy.

Chemically, proteins differ from both carbohydrates and fats because they contain nitrogen and sulphur as well as carbon, hydrogen and oxygen.

Fats

Animal fats are found in meat, milk, cheese, butter and egg yolk. Plant fats occur as oils in fruits (palm oil) and seeds (sunflower seeds), and are used for cooking and making margarine. Fats contain only carbon, hydrogen and oxygen but in different proportions to the carbohydrates.

Fats are used in the body to form parts of the cell structure such as the cell membrane. When oxidized in respiration, they give twice as much energy as either carbohydrates or proteins. Fats can be stored in the body, providing a means of long-term storage of energy in fat depots as mentioned above.

QUESTIONS

2 What sources of protein are available to a vegetarian who (a) will eat animal products but not meat itself, (b) will eat only plants and their products?

3 Why must all diets contain some protein?

4 In what senses can the fats in your diet be said to contribute to 'keeping you warm'?

DIET

A balanced diet must provide the requirements listed below.

(a) Sufficient energy

The amount of food taken in each day must be sufficient to provide the energy needed to stay alive and be active. Even during sleep, the body needs energy to keep the heart beating and the lungs filling and emptying. The energy needed during the waking hours depends on the age and activity of the person. A manual labourer needs more energy than a bus driver, for example. If the diet does not provide enough energy, reserves of glycogen and fat are used. When these run out, the person cannot keep up his normal rate of activity. Finally, he has to use up the protein of his own tissues to provide the

A-B-C-D-E-F-G-H-I-J-K

(*a*) This represents a small protein molecule. The letters are the amino acids

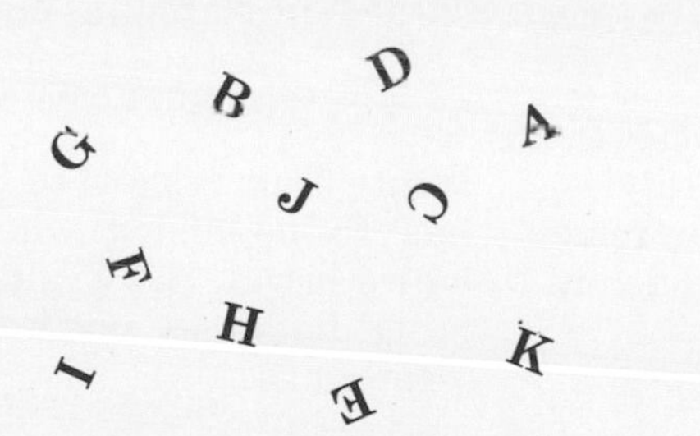

(*b*) The protein is digested and the amino acids are set free

B-J-K-D-A-F
|
E-C-G-H-I

(*c*) The same amino acids are built up into a different protein

Figure 2 Digestion and use of protein

necessary energy simply to stay alive, and he becomes thin and feeble.

(b) The correct proportions of proteins, fats and carbohydrates

The diet must include some protein for making and replacing the structures of the body. Pregnant women and growing children have special need of protein because of the new tissues they are producing. A shortage of protein in young children not only reduces growth, but leads to protein-deficiency disease and permanent mental retardation possibly because insufficient brain cells are made.

Insufficient carbohydrates and fats in the diet can result in a shortage of energy but an excess of these substances may lead to your putting on weight because the surplus is stored as fat in the fat depots.

(c) Mineral salts

The red cells of our blood need **iron** for their haemoglobin (page 111); our bones need **calcium phosphate**; the thyroid gland (page 171) requires **iodine** for making thyroxine and all cells need **sodium** and **potassium**. These mineral elements are normally obtained from the diet, though they can be supplied artificially (e.g. in tablets) if necessary. Lack of iron leads to **anaemia**. This is a shortage of haemoglobin, which means that the blood cannot carry enough oxygen. The anaemic person looks pale and lacks energy. Insufficient calcium and phosphorus, particularly in young children, leads to weak bones and teeth.

Most ordinary diets will contain adequate amounts of these mineral elements. Red meat, for example, contains iron, and milk contains calcium and phosphorus.

(d) Vitamins

These are chemical substances which play a part in essential chemical reactions in the body but cannot be used as a source of energy or be built into cell structures. They often work together with enzymes (page 12), but whereas we can build up our enzymes from amino acids, we cannot build up vitamins and have to obtain them ready-made in our diet. Like mineral salts,

Name and source of vitamin	Diseases and symptoms caused by lack of vitamin	Notes
Vitamin A		
Milk, butter, cheese, liver, cod-liver oil, fresh green vegetables.	Reduced resistance to diseases especially those affecting the skin. Poor night vision.	Green vegetables contain a yellow substance called *carotene* which is turned into vitamin A by the body.
Vitamin B_1		
Whole grains of cereals (e.g. wholemeal bread), lean meat, liver, yeast.	Wasting of tissues, paralysis, heart failure. These are extreme symptoms of a tropical deficiency disease called *beriberi*.	White flour is short of B_1 and this has to be added when making white bread. Badly planned slimming diets may not have enough vitamin B_1.
Vitamin C		
Oranges, lemons, black-currants, tomatoes, potatoes, green vegetables.	Bleeding under the skin at the joints. Swollen, bleeding gums; poor wound-healing. These are all symptoms of *scurvy*.	Scurvy is only likely to occur when fresh food is not available. Milk is deficient in vitamin C, so babies need other sources.
Vitamin D		
Cod-liver oil, butter, milk, cheese, egg-yolk, liver.	Children's bones do not harden properly, leading to deformities. This is called *rickets*.	Fats in the skin are changed into vitamin D by sunlight.

a mixed diet will usually contain all the vitamins we need, but a diet consisting of one food, such as rice, may be lacking in vitamins. A lack of any one vitamin leads to an illness which can be cured only by providing the missing vitamin.

The table on page 94 gives the sources and importance of four vitamins (fifteen or more are known).

Figure 4 Healthy diet. A diet made up from the correct proportions of these food samples would be balanced and healthy.

(e) Water

About 70 per cent of most tissue consists of water; it is an essential part of cytoplasm. Substances are carried round the body as a watery solution in the blood; the process of digestion takes place in water. Since we lose water by evaporation, sweating, urinating and breathing, we have to make good this loss by taking in water with the diet.

(f) Fibre

At one time called 'roughage', fibre consists mostly of plant cell walls, and occurs in all vegetables, fruit and bread. We do not have any enzymes for digesting cellulose and so it passes unchanged through the alimentary canal (page 102) as far as the large intestine. Here there are bacteria which can digest some of the substances in plant cell walls to form fatty acids (page 105) which are absorbed by the large intestine. Vegetable fibre, therefore, may supply some useful food materials but it has other important functions.

The fibre itself and the bacteria which multiply from feeding on it, add bulk to the contents of the large intestine and help it to retain water. This softens the faeces and reduces the time needed for the undigested residues to pass out of the body. Both of these effects help to prevent constipation and keep the colon in a healthy condition (see page 250).

Figure 3 shows the food value of four different kinds of food.

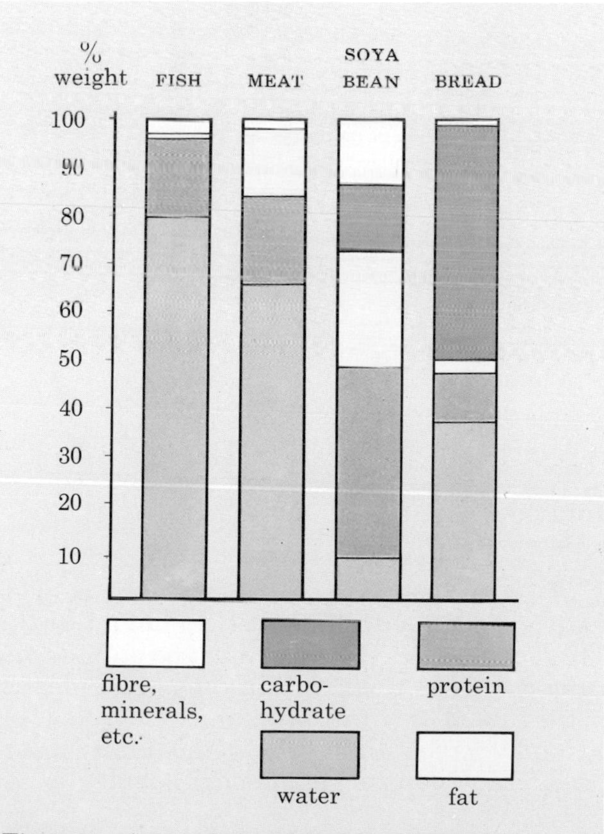

Figure 3 An analysis of four food samples

QUESTIONS

5 Figure 4 shows some examples of the food that would give a balanced diet. Consider each sample in turn and say what class of food or item of diet is mainly present. For example, the meat is mainly protein but will also contain some iron.

6 What is the value of leafy vegetables, such as cabbage and lettuce, in the diet?

7 From the information in Figure 3, suggest which item of food contains the best balance of nutrients.

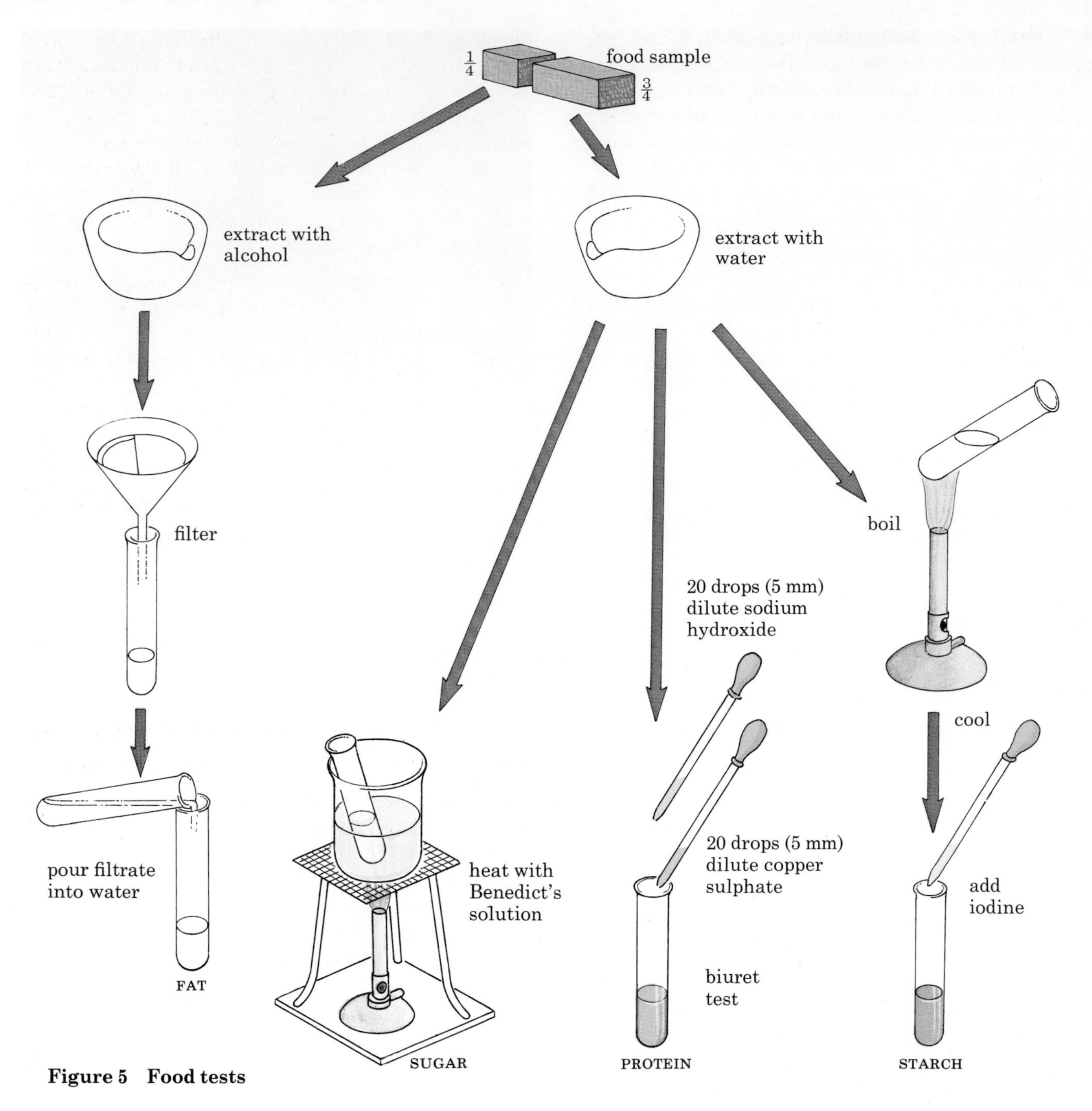

Figure 5 Food tests

7 Diets in Africa and South East Asia often consist largely of rice or maize. What is the disadvantage of diets like these which depend so heavily on one food source?

PRACTICAL WORK

Food tests

1 Test for starch. A little starch powder is shaken in a test-tube with some cold water and then boiled to make a clear solution. When the solution is cold, three or four drops of **iodine solution*** are added and a dark blue colour is produced.

2 Test for glucose. A little glucose is heated with some **Benedict's solution*** in a test-tube. The mouth of the test-tube must be pointed *away* from people as the solution tends to 'bump' out of the tube. The solution changes from clear blue to cloudy green, then yellow and finally to a red precipitate (deposit) of copper(I) oxide.

* See page 260 for the preparation of these solutions.

3 Test for protein (Biuret test). To a 1 per cent solution of albumen (the protein of egg-white) 5 cm^3 dilute sodium hydroxide is added (CARE: this solution is caustic), followed by 5 cm^3 1 per cent copper sulphate solution. A purple colour indicates protein.

4 Tests for fats. Two drops of cooking oil are thoroughly shaken with about 5 cm^3 ethanol in a dry test-tube until the fat dissolves. The alcoholic solution is poured into a test-tube containing a few cm^3 water. A cloudy white emulsion is formed. This shows that the solution contained some fat or oil.

Application of the food tests

The tests can be used on samples of food such as milk, raisins, potato, onion, beans, egg-yolk, ground almonds, to find what food materials are present (Figure 5). The solid samples are crushed in a mortar and shaken with warm water to extract the soluble products. Separate samples of the watery mixture of crushed food are tested for starch, glucose or protein as described above. To test for fats, the food must first be crushed with ethanol, not water, and then filtered. The clear filtrate is poured into water to see if it goes cloudy, indicating the presence of fats.

Test for vitamin C. 2 cm^3 fresh lemon juice is drawn up into a plastic syringe. This juice is added drop by drop to 2 cm^3 of a 0.1 per cent solution of PIDCP (a blue dye) in a test-tube. The PIDCP will become colourless quite suddenly as the juice is added. The amount of juice added from the syringe should be noted down. The experiment is now repeated but with orange juice in the syringe. If it takes more orange juice than lemon juice to decolourize the PIDCP, the orange juice must contain less vitamin C.

BASIC FACTS

- **Carbohydrates, such as starch and sugar, provide us with energy.**
- **Carbohydrates can be turned into fat and stored.**
- **Fats, such as animal fat or margarine, provide us with energy. They can be stored in the fat depots.**
- **Proteins, such as meat and fish, are used to build new cells and tissues. They cannot be stored.**
- **A balanced diet must contain (a) enough food to provide our energy needs, (b) carbohydrates, fats and proteins, (c) mineral salts, (d) vitamins, (e) water and (f) roughage.**

12 Teeth

FUNCTIONS OF TEETH

Before food can be swallowed, pieces have to be bitten off which are small enough to pass down the gullet into the stomach. Digestion is made easier if these pieces of food are crushed into even smaller particles by the action of chewing. Biting and chewing are actions carried out by our teeth, jaws and muscles.

Carnivore's teeth

Figure 3 on page 149 shows the human skull with the muscles which close the jaws, but the function of the different kinds of teeth is best understood by first looking at a carnivore (meat-eating animal) such as a dog. In the wild, the dog would use its teeth and jaws for catching, holding and killing its prey. The teeth also have to cut up the flesh and break the bones into pieces small enough to be swallowed. In Figure 1, which shows a dog's skull and teeth, it can be seen that the teeth in different parts of the jaw vary greatly in shape and size.

The small front teeth, or **incisors**, meet at their tips to get the first grip on the prey and, later, to pull off small pieces of flesh close to the bones (Figure 1). The long pointed **canine** teeth (Figure 2) pass each other and may kill the prey or hold it securely while the dog shakes it to death. At the back of the jaws are the **molar** teeth. These have broad surfaces with blunt projections on them. When the jaw closes, the molars meet and crush flesh and bone between them.

Between the canines and molars are the **premolars**, which are narrower than the molars and have sharp cutting edges. The last

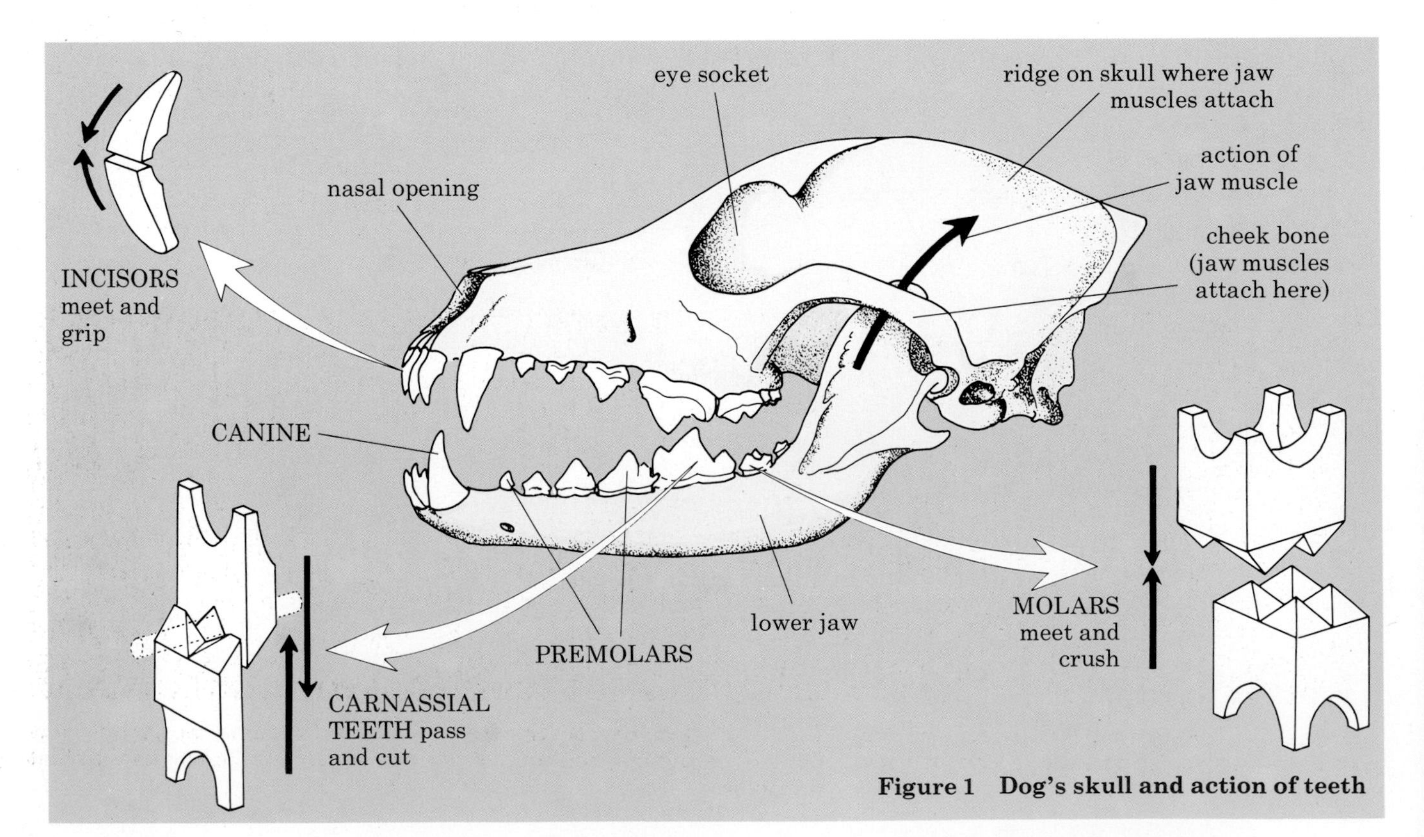

Figure 1 Dog's skull and action of teeth

Figure 2 Lioness, clearly showing the six incisors and the two canines in each jaw.

premolar of the top jaw and the first molar of the bottom jaw are called **carnassial** teeth. They are very large and their cutting edges pass each other like the blades of shears, cutting flesh off a carcass or cracking bones (Figure 1). A dog eating a bone will often move the bone to the side of its jaws so that the carnassial teeth can cut off the meat or crack off bits of bone.

Human teeth

Our teeth are not used for catching, holding, killing and tearing up prey, and we cannot cope with bones. Thus, although we have incisors, canines, premolars and molars, they do not show such big variations in size and shape as the dog's. Figures 3 and 5 show the position of teeth in the upper jaw and Figure 4 shows how they appear in both jaws when seen from the side.

Our top incisors pass in front of our bottom incisors (Figure 6) and cut pieces off the food as when biting into an apple or taking a bite out of a piece of toast. The canines are more pointed than the incisors but are not much larger and they function like extra incisors. The premolars and molars are similar in shape and function. Their knobbly surfaces meet when the jaws are closed, and crush the food into small pieces. Small particles of food are easier to digest than large chunks.

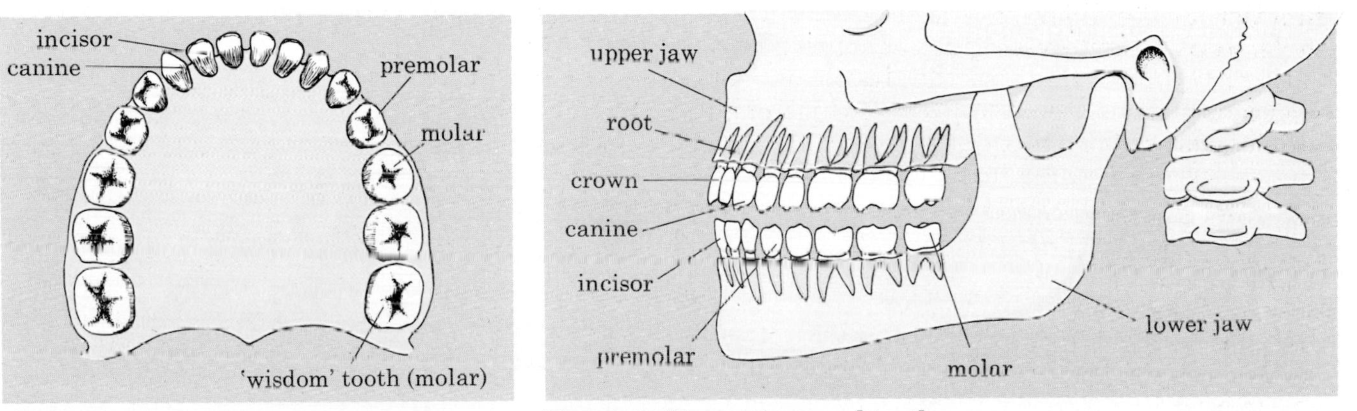

Figure 3 Teeth in man's upper jaw

Figure 4 Human jaws and teeth

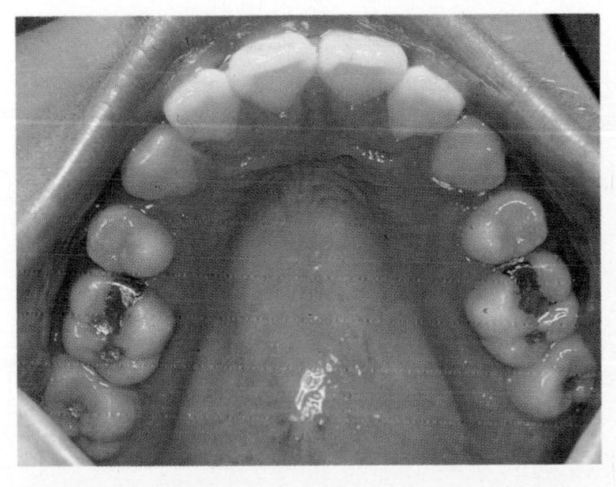

Figure 5 Human teeth, top set. One premolar and one molar on each side are missing, and all the molars have fillings.

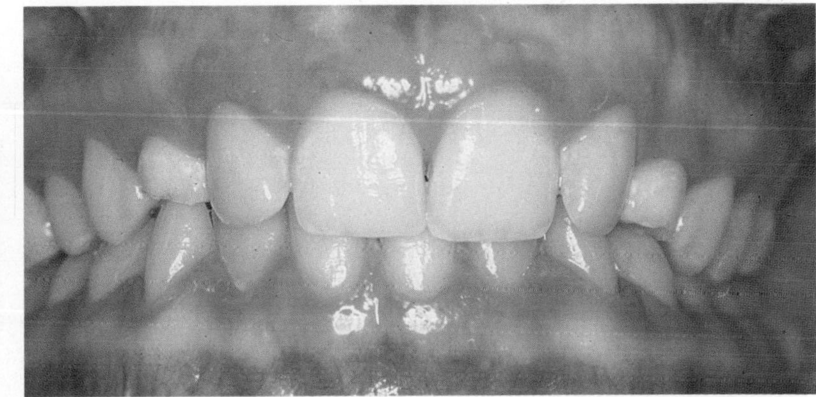

Figure 6 Healthy teeth, front view. Notice how the top incisors overlap the bottom incisors. You can also see the top canines and the first premolar.

QUESTIONS

1 How many incisors, canines, premolars and molars are there in a human upper jaw (see Figure 3)? How many teeth are there in a full set?

2 The words 'biting, crushing and chewing' have been used several times in this section. Say exactly what you think each word means.

3 Look at the position of the jaw muscles in Figure 1 and in Figure 3 on page 149. Which teeth do you think exert the greatest force on the food?

STRUCTURE OF TEETH

Figure 7*a* shows the structure of an incisor or canine tooth as it would appear in vertical section (i.e. cut from top to bottom), and Figure 7*b* shows a section through a molar tooth.

Enamel covers the exposed part, or **crown**, of the tooth and makes a hard, biting surface. Enamel is a non-living substance containing calcium salts.

Dentine is rather like bone and softer than enamel. It is a living tissue with threads of cytoplasm running through it.

Pulp. In the centre of the tooth is soft connective tissue. It contains cells which make the dentine and keep the tooth alive. In the pulp are blood vessels which bring food and oxygen, so that the tooth can grow at first and then remain alive when growth has stopped. There are also sensory nerve endings in the pulp, which are sensitive to heat and cold but give only the sensation of pain. If you plunge your teeth into an ice-cream, they do not feel cold but they do hurt.

Cement is a bone-like substance which covers the root of the tooth. In the cement are embedded tough fibres which pass into the bone of the jaw and hold the tooth in place.

Milk teeth and permanent teeth

Mammals have two sets of teeth in their lifetimes (Figure 8). In humans, the first set, or **milk teeth**, grow through the gum during the first year of life and consist of four incisors, two canines and four molars in each jaw. Between the ages of six and twelve years, these milk teeth gradually fall out (Figure 9) and are replaced by the **permanent teeth**, including six molars in each jaw. The last of these molars, the 'wisdom teeth', may not grow until the age of 17 or later. If these permanent teeth are lost for any reason, they do not grow again.

Dental health

The most likely reasons for the loss of teeth are dental decay or gum disease. Dental decay is

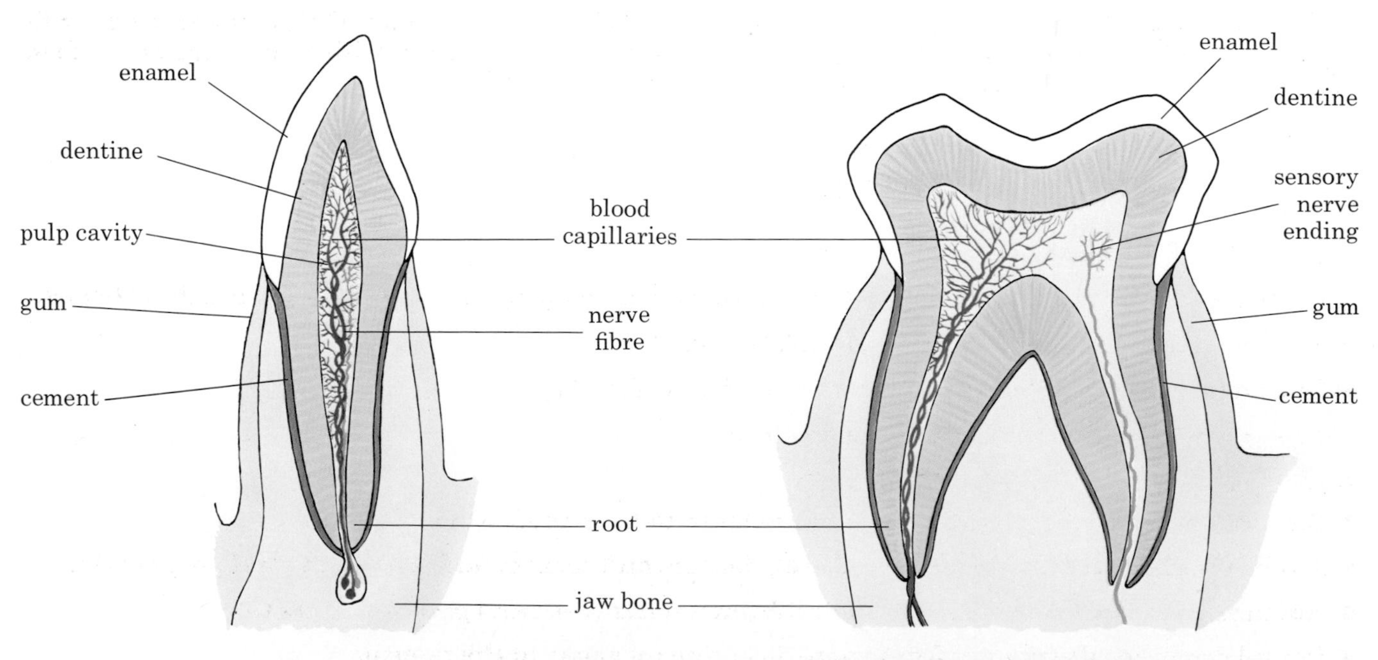

(*a*) INCISOR TOOTH

(*b*) MOLAR TOOTH (nerve shown on one side, blood vessels on the other)

Figure 7 Sections through incisor and molar teeth

caused by certain bacteria in the mouth. The chemical activities of these bacteria dissolve the enamel and dentine of the teeth and so produce cavities (holes) (Figure 5). The bacteria grow and reproduce actively in the film of moisture over the teeth, particularly if they can obtain continuous supplies of sucrose (sugar) from sweets, sweetened drinks and ice-creams. The film of moisture with its population of bacteria

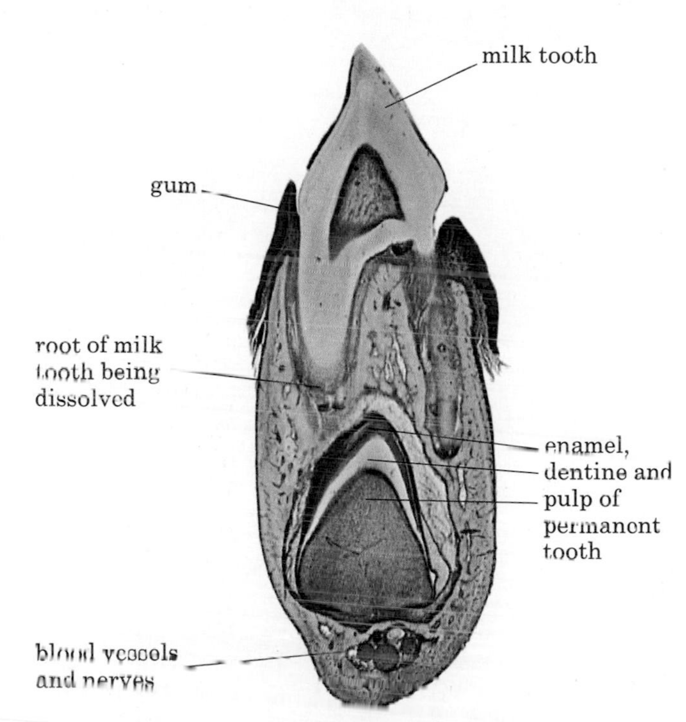

Figure 8 Section through a kitten's lower jaw (×15). A permanent tooth is growing beneath the milk tooth, whose root is being dissolved away.

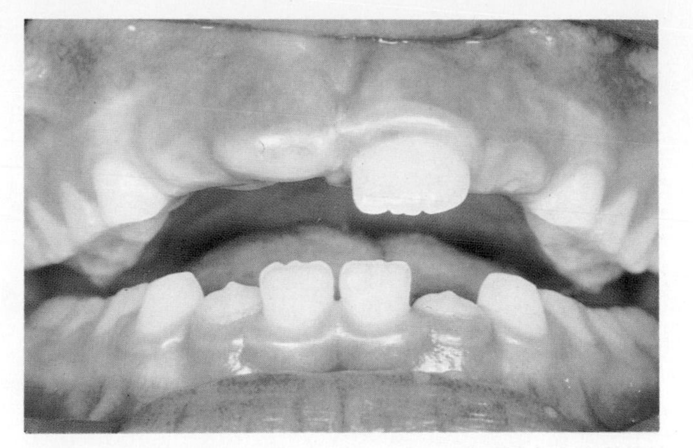

Figure 9 Permanent teeth coming through. In the bottom jaw, the two outer incisors are emerging. Only one of the top incisors has come through.

and the substances they produce, is called **plaque.** If plaque is not removed by efficient daily teeth-brushing, the bacteria may invade the gum where the tooth emerges and so cause gum disease. Reducing the intake of sugar helps to prevent decay, and regular, thorough brushing reduces the chances of gum disease.

QUESTIONS

4 It is thought that one reason our teeth stop growing is that the hole in the root gets smaller and pinches the blood vessels. Why should this stop the growth of the tooth?

5 Why do you think that decayed teeth with cavities are more sensitive to temperature than healthy teeth?

BASIC FACTS

- **Teeth do different jobs in different animals.**
- **A dog has teeth with very different shapes, for holding its prey, cutting flesh and crushing bones.**
- **Our incisors and canines bite off mouthfuls of food.**
- **Our molars and premolars crush food ready for swallowing.**
- **The crown of a tooth projects into the mouth.**
- **The root of a tooth is held in the jawbone.**
- **Blood vessels bring food and oxygen to the teeth to keep them alive.**
- **A tooth is made up from layers of enamel, dentine and cement and it has a central pulp cavity.**
- **Humans have two sets of teeth in their lifetime—milk teeth and permanent teeth.**
- **Dental decay results mainly from bacteria acting on sugar in the mouth.**

13 Digestion, absorption and use of food

Feeding, in animals, involves taking food into the mouth, chewing it and swallowing it down into the stomach. This satisfies the animal's hunger, but for food to be of any use to the whole body it has first to be **digested** and **absorbed**. This means the food is dissolved, passed into the blood stream and carried by the blood all round the body. In this way, the blood delivers dissolved food to the living cells in all parts of the body such as the muscles, brain, heart and kidneys. This chapter describes how the food is digested and absorbed. Chapter 14 describes how the blood carries it round the body.

THE ALIMENTARY CANAL

Alimenta is Latin for food, and the **alimentary canal** is a tube running through the body, in which food is digested and absorbed. Figure 1 shows a simple diagram of an alimentary canal.

The inside of the alimentary canal is lined with layers of cells forming what is called an **epithelium**. New cells in the epithelium are being produced all the time to replace the cells worn away by the movement of the food. There are also cells in the lining which produce **mucus**. Mucus is a slimy liquid that lubricates the lining of the canal and protects it from wear and tear. Mucus also protects the lining from attack by **digestive enzymes**, powerful chemicals that are released into it.

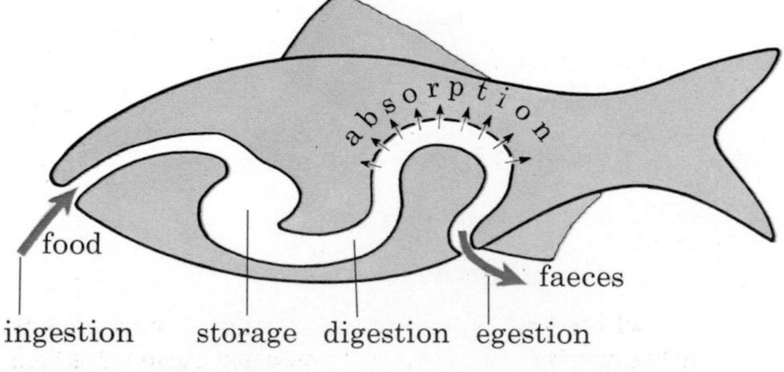

Figure 1 The alimentary canal (generalized)

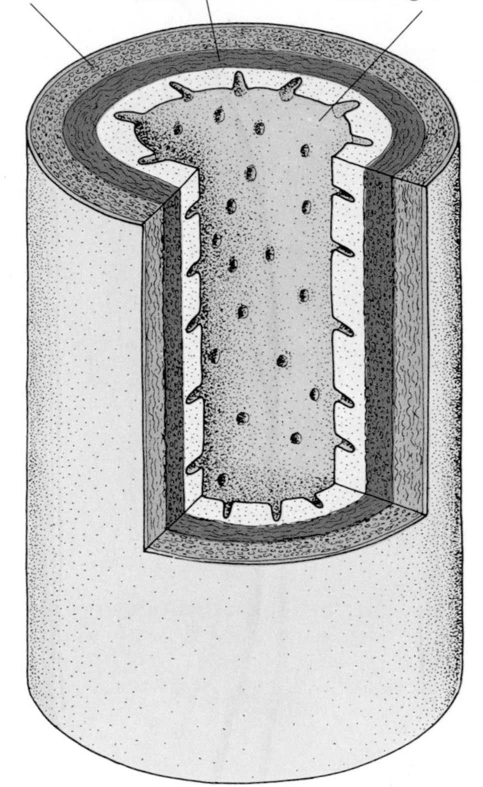

Figure 2 The general structure of the alimentary canal

Some of the digestive enzymes are produced by cells in the lining of the alimentary canal, as in the stomach lining. Others are produced by **glands** which are outside the alimentary canal but pour their enzymes through tubes (called **ducts**) into the alimentary canal. The **salivary glands** (Figure 6) and the **pancreas** (Figure 8) are examples of such digestive glands.

The alimentary canal has a great many blood vessels in its walls, close to the lining. These bring oxygen needed by the cells and take away the carbon dioxide they produce. They also absorb the digested food from the alimentary canal.

Peristalsis

The alimentary canal has layers of muscle in its walls. The inner layer of muscle has fibres which run round the canal (**circular muscle**). The fibres of the outer layer run along the length of the canal (**longitudinal muscle**) (Figure 2). When the circular muscle in one region contracts, it makes the alimentary canal narrow in that region. Food is forced along the alimentary canal by contractions of the circular muscles. The circular muscles just behind the food contract and squeeze the food forward. The muscles just in front of the food relax and allow it to pass along. The region of contraction moves along the alimentary canal, pushing the food in front of it. The function of the longitudinal muscle is not clear.

The wave of contraction, called **peristalsis**, is illustrated in Figure 3. If you feed a swan with bread, you can see the effect of peristalsis in its long neck as the bread passes down the gullet.

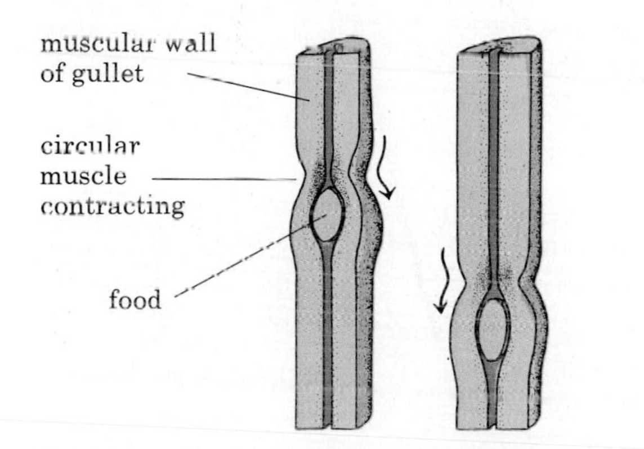

Figure 3 Diagram to illustrate peristalsis

QUESTIONS

1 What three functions of the alimentary canal are shown in Figure 1?

2 Into what parts of the alimentary canal do (a) the pancreas, (b) the salivary glands, pour their digestive juices?

3 Starting from the inside, name the layers of tissue that make up the alimentary canal.

DIGESTION

Digestion is a chemical process and consists of breaking down large molecules to small molecules. The large molecules are usually not soluble in water, while the smaller ones are. The small molecules can pass through the epithelium of the alimentary canal, through the walls of the blood vessels and into the blood.

Some food can be absorbed without digestion. The glucose in a soft drink, for example, could pass through the walls of the alimentary canal and enter the blood vessels. Most food, however, is solid and cannot get into blood vessels. Digestion is the process by which solid food is dissolved to make a solution. The chemicals which dissolve the food are **enzymes**, described on page 12.

Digestive enzymes speed up the rate at which food can be dissolved. A protein might take 50 years to dissolve if just placed in water but is completely digested by enzymes in a few hours. All the solid starch in foods such as bread and potatoes is digested to **glucose** which is soluble in water. The solid proteins in meat, egg-white and beans are digested to soluble substances called **amino acids**. Fats are digested to two soluble products called **glycerol** and **fatty acids**.

The chemical breakdown usually takes place in stages. For example, the starch molecule is made up of hundreds of carbon, hydrogen and oxygen atoms. The first stage of digestion breaks it down to a 12-carbon sugar molecule called **maltose**. The last stage of digestion breaks the maltose molecule into two 6-carbon sugar molecules called glucose (Figure 4). Protein

Figure 4 Enzymes acting on starch

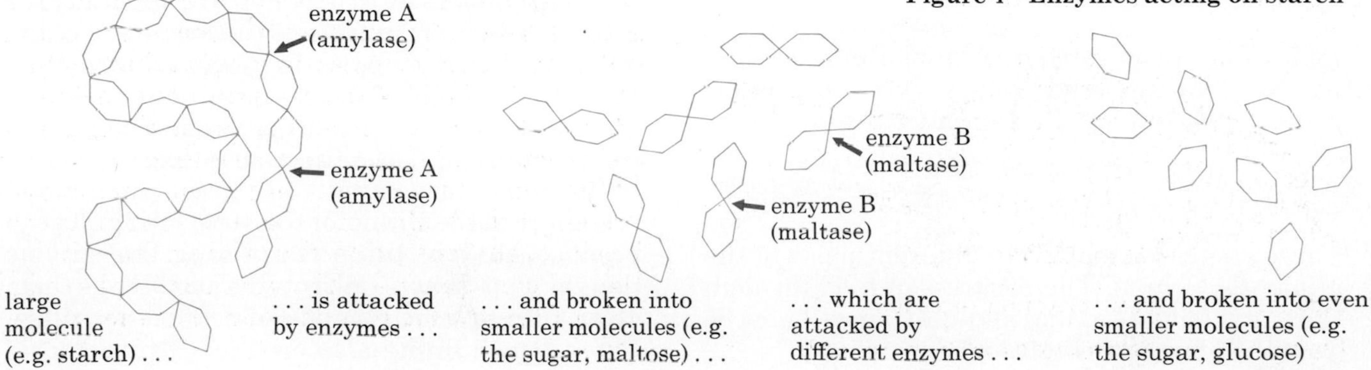

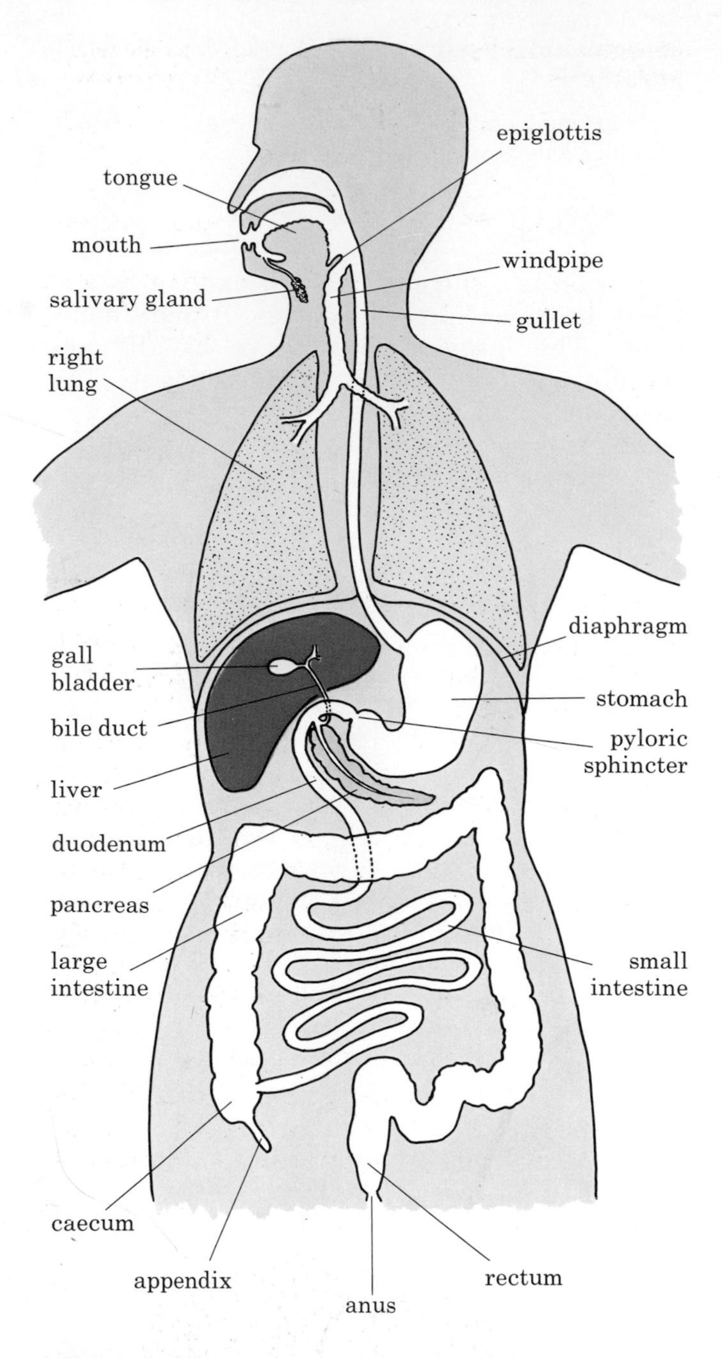

Figure 5 The human alimentary canal

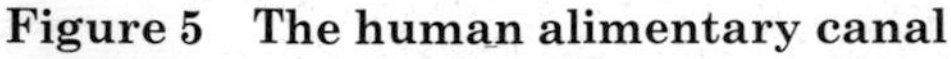

molecules are digested first to smaller molecules called **peptides** and finally into completely soluble molecules called amino acids.

starch→ maltose→glucose

protein→peptide→ amino acid

These stages take place in different parts of the alimentary canal. The progress of food through the canal and the stages of digestion will now be described for man (Figures 5 and 6).

Digestion in the mouth

Ingestion is the act of taking food into the mouth. In the mouth, the food is chewed and mixed with **saliva**. The chewing breaks the food into pieces which can be swallowed and it also increases the surface area for the enzymes to work on later. Saliva is a digestive juice produced by three pairs of glands whose ducts lead into the mouth (Figure 6). It helps to lubricate the food and make the small pieces stick together. Saliva contains one enzyme, **salivary amylase**, which acts on cooked starch and begins to break it down into maltose.

Swallowing

By studying Figure 6*a*, it can be seen that for food to enter the gullet, it has to pass over the windpipe. All the complicated actions which occur during swallowing ensure that food does not enter the windpipe and cause choking.

(a) The tongue presses upwards and back against the roof of the mouth, forcing the food to the back of the mouth.

(b) The soft palate closes the nasal cavity at the back.

(c) The 'Adam's apple' cartilage round the top of the windpipe is pulled upwards by muscles so that the opening of the windpipe lies under the back of the tongue.

(d) This opening is also partly closed by the contraction of a ring of muscle.

(e) The **epiglottis**, a flap of cartilage (gristle) prevents the food from going down the windpipe instead of the gullet.

The food is then forced down the gullet by peristalsis.

Digestion in the stomach

The stomach has elastic walls which extend as the food collects in it. The **pyloric sphincter** is a circular band of muscle at the lower end of the stomach which stops solid pieces of food from passing through. The main function of the stomach is to store the food from a meal and release it in small quantities at a time to the rest of the alimentary canal.

Glands in the lining of the stomach (Figure 7) produce **gastric juice** containing the enzyme **pepsin**. Pepsin acts on proteins and breaks them down into soluble compounds called **peptides**. The stomach lining also produces hydrochloric

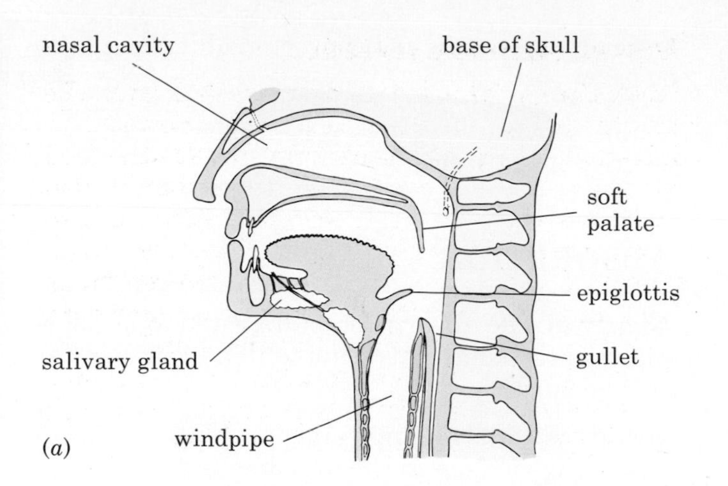

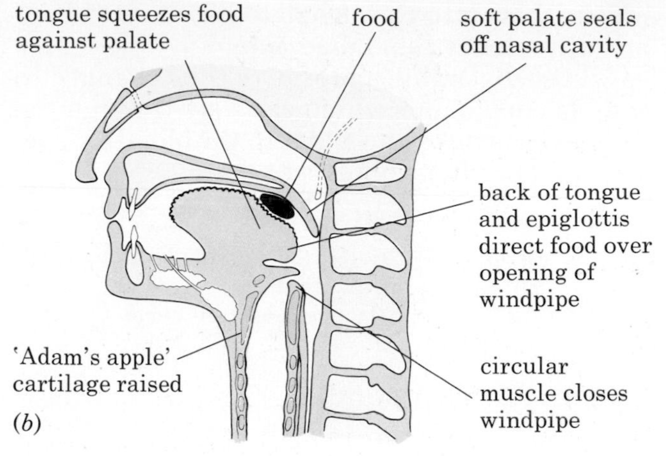

Figure 6 Section through head to show swallowing action

acid which makes a weak solution in the gastric juice. The acid provides the best degree of acidity for pepsin to work in (see page 13) and probably kills many of the bacteria taken in with the food.

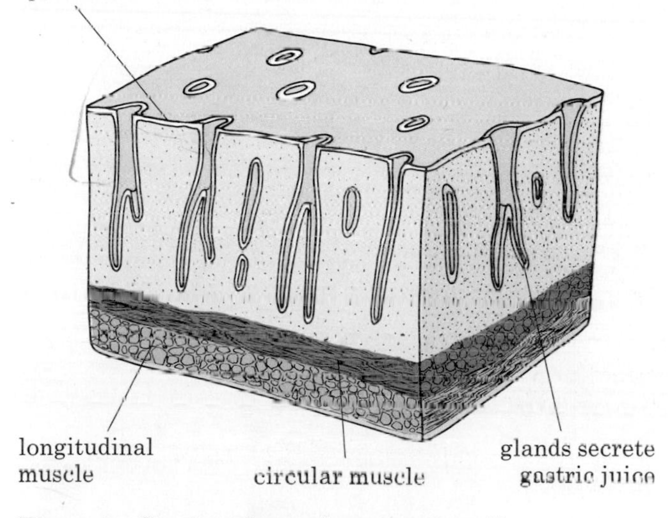

Figure 7 Section through stomach wall

The regular, peristaltic movements (Figure 3) of the stomach, about once every 20 seconds, mixes up the food and gastric juice into a creamy liquid. How long food remains in the stomach depends on its nature. Water may pass through in a few minutes; a meal of carbohydrate such as porridge may be held in the stomach for less than an hour, but a mixed meal containing protein and fat may be in the stomach for one or two hours.

The pyloric sphincter lets the liquid products of digestion pass, a little at a time, into the first part of the small intestine called the **duodenum**.

Digestion in the small intestine

A digestive juice from the pancreas (**pancreatic juice**) and bile from the liver are poured into the duodenum to act on food there. The pancreas is a digestive gland lying below the stomach (Figure 8). It makes several enzymes, which act on all classes of food. One breaks down proteins and peptides to **amino acids**. Another attacks starch and converts it to maltose while a third, called **lipase**, digests fats to fatty acids and glycerol. Pancreatic juice is alkaline and partly neutralizes the acid liquid from the stomach, because the enzymes of the pancreas do not work well in acid conditions.

Bile is a green, watery fluid made in the liver, stored in the gall bladder and delivered to the duodenum by the bile duct (Figure 8). It is not an enzyme but acts like a detergent on fats, and

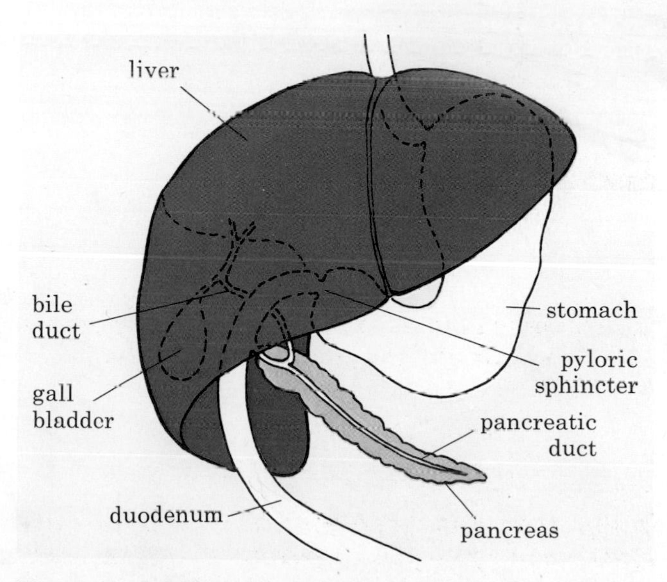

Figure 8 Diagram to show the relationship between the stomach, liver and pancreas

breaks them into tiny droplets which spread out and allow more rapid digestion.

All the digestible material is thus changed to soluble compounds which can pass through the lining of the intestine and into the blood stream. The final products of digestion are:

food	*final products*
starch	→glucose (see 'Basic facts')
proteins	→amino acids
fats	→fatty acids and glycerol

Caecum and appendix

In humans, these are small structures which may have no digestive function. In grass-eating animals (herbivores) like the rabbit and the horse, however, they are much larger and it is here that most digestion of the cellulose in plant cell walls takes place, largely as a result of bacterial activity.

QUESTIONS

4 Why can you not breathe while you are swallowing?

5 Why is it necessary for an animal's food to be digested? Why do plants not need a digestive system? (See page 24.)

6 Write down the menu for your breakfast and lunch (or supper). State the main food substances present in each item of the meal. State the final digestion product of each.

7 In which parts of the alimentary canal is (a) starch, (b) protein digested?

8 Study the classes and characteristics of enzymes on pages 12–13.

(a) Suggest a more logical name for pepsin.

(b) In what ways does pepsin show the characteristics of an enzyme?

ABSORPTION

Nearly all the absorption of digested food takes place in the small intestine which is efficient at this because

(a) it is fairly long and presents a large absorbing surface to the digested food,

(b) its internal surface is greatly increased by thousands of tiny projections called **villi** (singular=villus) (Figures 9 and 10). These are about one millimetre long and may be finger-like or flattened in shape.

(c) the lining epithelium is very thin and the fluids can pass rapidly through it,

(d) there is a dense network of blood capillaries (tiny blood vessels, see page 116) in each villus (Figure 10*b*).

The small molecules of the digested food, mainly glucose and amino acids, pass through the epithelium and the capillary walls and enter the blood stream. They are then carried away in the capillaries which join up to form veins. These veins then unite to form one large vein, the **hepatic portal vein** (see Figure 8 on page 114). This vein carries all the blood from

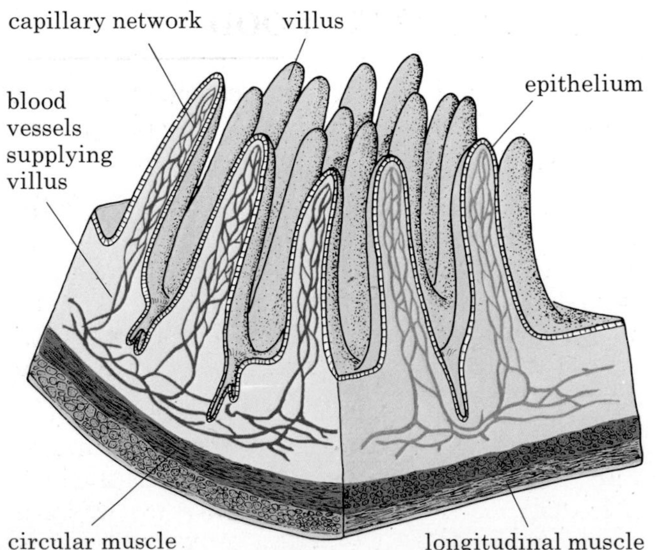

(*a*) Diagram to show structure

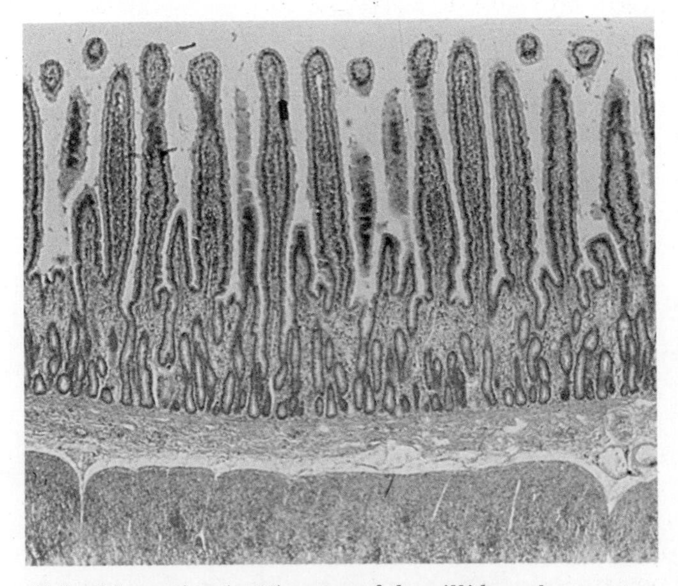

(*b*) In this section (×20), some of the villi have been cut so that their tops appear as isolated circles. A lacteal can be seen in the central villus.

Figure 9 Structure of the small intestine

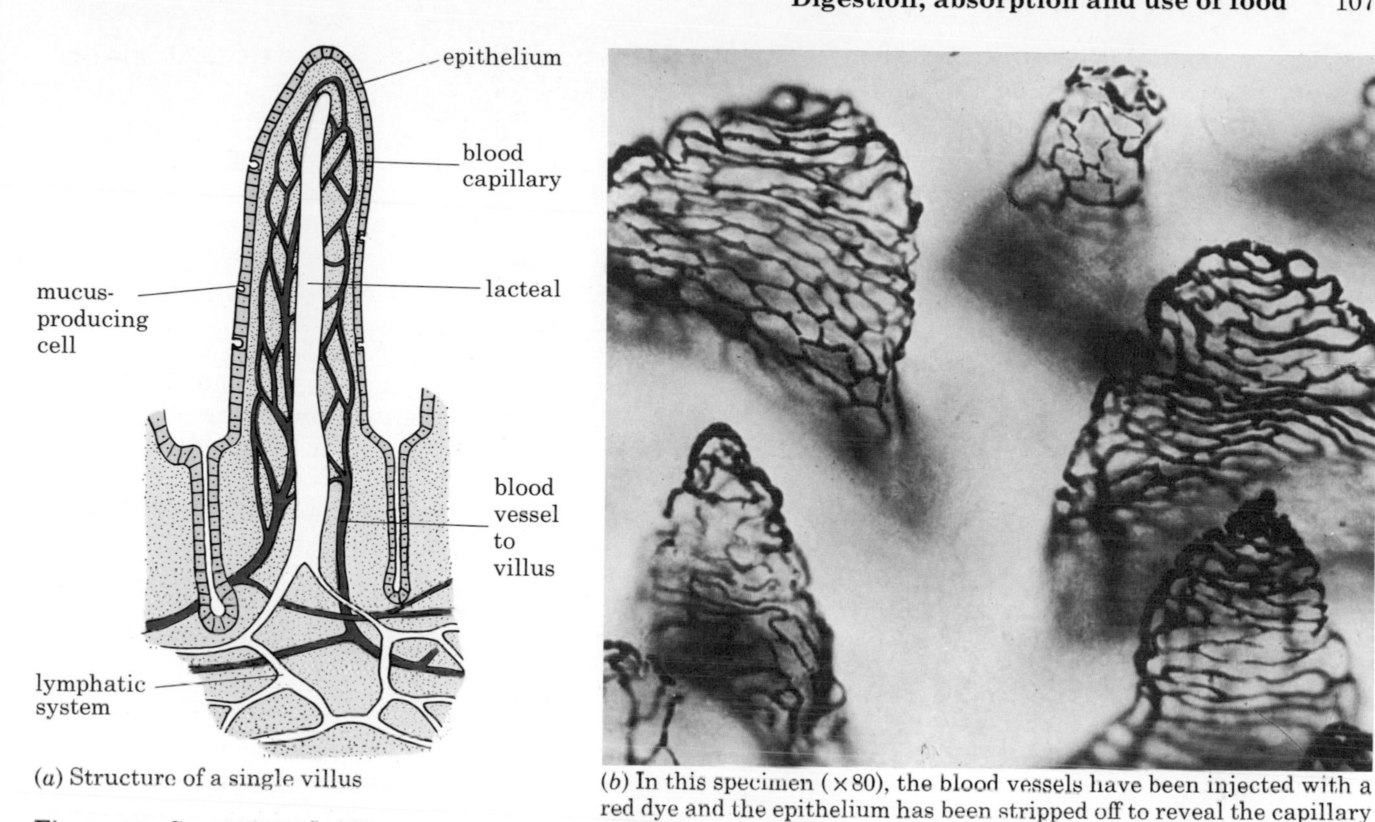

(*a*) Structure of a single villus

(*b*) In this specimen (×80), the blood vessels have been injected with a red dye and the epithelium has been stripped off to reveal the capillary network in each villus

Figure 10 Structure of villi

the intestine to the liver, which may store or alter any of the digestion products. When the digested food is released from the liver, it enters the general blood circulation.

Some of the fatty acids and glycerol from the digestion of fats enter the blood capillaries of the villi. However, a large proportion of the fatty acids and glycerol may be combined to form fats again in the lining of the intestine. These fats then pass into the **lacteals** (Figure 10). The fluid in the lacteals flows into the **lymphatic system** which forms a network all over the body and eventually empties its contents into the blood stream (see page 118).

The large intestine

The material passing into the large intestine consists of water and those parts of the food which cannot be digested, largely cellulose and vegetable fibres (called roughage). The large intestine produces no digestive enzymes and can absorb very little digested food. It does, however, absorb much of the water from the undigested remains called the **faeces**. This semi-solid waste is passed into the rectum by peristalsis and is expelled at intervals through the **anus**. This act is called **egestion** or **defecation**.

QUESTIONS

9 Which regions of the alimentary canal absorb (a) water, (b) amino acids, (c) fats?

10 What characteristics of the small intestine enable it to absorb digested food efficiently?

USE OF DIGESTED FOOD

The products of digestion are carried round the body in the blood. From the blood, cells absorb and use glucose, fats and amino acids.

(a) Glucose. During respiration in the cells, glucose is oxidized to carbon dioxide and water (see page 14). This reaction provides energy to drive the many chemical processes in the cells which result in, for example, the building up of proteins, contraction of muscles or electrical changes in nerves.

(b) Fats. These are built into cell membranes and other cell structures. Any fats not used for growth or maintenance in this way are oxidized to carbon dioxide and water, providing energy for the vital processes of the cell.

(c) Amino acids are absorbed by the cells and built up into proteins. These proteins may form structures such as the cell membrane or they may become enzymes which control the chemical activity within the cell.

QUESTIONS

11 State briefly what happens to a protein from the time it is swallowed, to the time its products are built up into the cytoplasm of a muscle cell.

12 List the chemical changes which a starch molecule undergoes from the time it reaches the duodenum to the time its carbon atoms become part of carbon dioxide molecules. Say where in the body these changes occur.

STORAGE OF DIGESTED FOOD

If more food is taken in than the body needs for energy or for building tissues, such as bone or muscle, it is stored in one of the following ways:

(a) Glucose. The sugar not required immediately for energy in the cells is changed in the liver to **glycogen**. The glycogen molecule is built up by combining many glucose molecules into a long chain molecule similar to that of starch. Some of this insoluble glycogen is stored in the liver and the rest in the muscles. When the blood sugar falls below a certain level, the liver changes its glycogen back to glucose and releases it into the circulation. The muscle glycogen is not returned to the circulation but is used by muscle cells as a source of energy during muscular activity.

The glycogen in the liver is a 'short-term' store, sufficient for only about six hours. Excess glucose not stored as glycogen is converted to fat and stored in the fat depots.

(b) Fat. Unlike glycogen, there is no limit to the amount of fat stored and because of its high energy value (page 93) it is an effective 'long-term' store. The fat is stored in tissues in the abdomen, round the kidneys and under the skin. These are the **fat depots**. Figures 1 and 3 on pages 133 and 134 show the fatty tissue of the skin.

(c) Amino acids. Amino acids are not stored in the body. Those not used in protein formation are **de-aminated** (see below). The protein of the liver and other tissues can act as a kind of protein store to maintain the protein level in the blood, but absence of protein in the diet soon leads to serious disorders.

Body weight

The rate at which glucose is oxidized or changed into glycogen and fat is controlled by hormones (page 170). When intake of carbohydrate and fat is more than enough to meet the energy requirements of the body, the surplus will be stored mainly as fat. Some people never seem to get fat no matter how much they eat, while others start to lay down fat when their intake only just exceeds their needs. Putting on weight is certainly the result of eating more food than the body needs, but it is not clear why people should differ so much in this respect. The explanation probably lies in the balance of hormones which, to some extent, is determined by heredity. A slimming diet designed to reduce food intake must always include the essential proteins, mineral salts and vitamins.

QUESTIONS

13 List the ways in which the body can store an excess of carbohydrates taken in with the diet.

14 If you were deprived of food for several days, how would your body meet the demands for energy by your heart and other organs?

THE LIVER

The liver has been mentioned several times in connection with the digestion, use and storage of food. This is only one aspect of its many important functions, some of which are listed below. It is a large, reddish-brown organ which lies just beneath the diaphragm and partly overlaps the stomach. All the blood from the blood vessels of the alimentary canal passes through the liver, which adjusts the composition of the blood before releasing it into the general circulation.

(a) Regulation of blood sugar. After a meal, the liver removes excess glucose from the blood and stores it as glycogen. In the periods between meals, when the glucose concentration in the blood starts to fall, the liver converts some of its stored glycogen into glucose and releases it into the blood stream. In this way, the concentration of sugar in the blood is kept at a fairly steady level.

(b) Production of bile. Cells in the liver make bile continuously and this is stored in the gall bladder until needed to assist digestion of fats in the small intestine. The bile breaks up the liquid fats into small droplets (an *emulsion*) which makes them easier to digest and absorb.

(c) De-amination. The amino acids not needed for making proteins are converted to glycogen in the liver. During this process, the nitrogen-containing, amino part (NH_2) of the amino acid is removed and changed to **urea**, which is later excreted by the kidneys (see page 128).

(d) Storage of iron. Millions of red blood cells (page 111) break down every day. The iron from their haemoglobin (page 111) is stored in the liver.

(e) Detoxication. Most poisonous compounds which are absorbed in the intestine are made harmless when the blood passes through the liver on its way to the general circulation.

Note Contrary to statements made in previous impressions (prior to 1994), the liver is not a net exporter of heat. Although some of its chemical reactions release heat energy, many require an input of energy.

QUESTION

15 Explain how the liver exercises control over the substances coming from the intestine and entering the general blood circulation.

EXPERIMENTS WITH DIGESTIVE ENZYMES

1 The action of saliva on starch

Rinse the mouth with water to remove traces of food. Collect saliva in two test-tubes, labelled A and B, to a depth of about 15 mm (see Figure 11). Heat the saliva in tube B over a small Bunsen flame until it boils for about 30 seconds and then cool the tube under the tap. Add about 2 cm^3 of a 2 per cent starch solution to each tube; shake each tube and leave them for five minutes.

Share the contents of tube A between two clean test-tubes. To one of these add some iodine solution. To the other add some Benedict's solution and boil with care as described on page 96. Test the contents of tube B in exactly the same way.

Results. The contents of tube A fail to give a blue colour with iodine, showing that the starch has gone. The other half of the contents, however, gives a red precipitate with Benedict's solution showing that sugar is present.

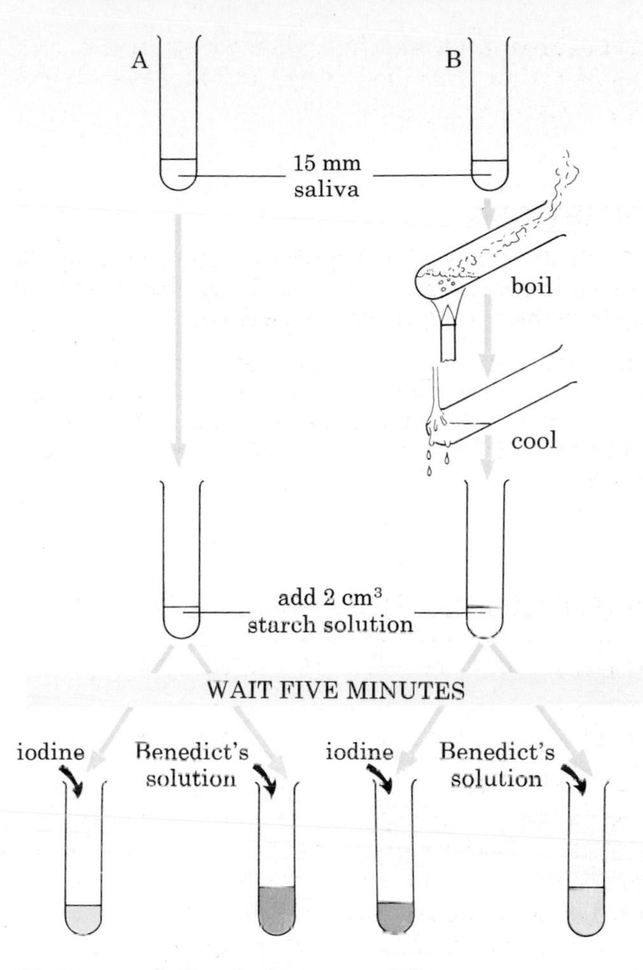

Figure 11 Saliva acting on starch

The contents of tube B still give a blue colour with iodine but do not form a red precipitate on boiling with Benedict's solution.

Interpretation. The results with tube A suggest that something in saliva has converted starch into sugar. The fact that the boiled saliva in tube B fails to do this, suggests that it was an enzyme in saliva which brought about the change (see page 12), because enzymes are proteins and are destroyed by boiling. If the boiled saliva had changed starch to sugar, it would have ruled out the possibility of an enzyme being responsible.

2 The action of pepsin on egg-white protein

A cloudy suspension of egg-white is prepared by stirring the white of one egg into 500 cm^3 tap water, heating it to boiling point and filtering it through glass wool to remove the larger particles.

Label four test-tubes A, B, C and D and place 2 cm^3 egg-white suspension in each of them.

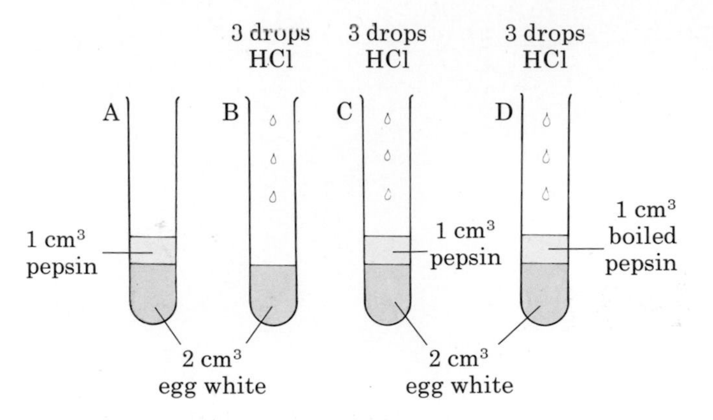

Figure 12 Pepsin acting on egg white

Then add pepsin solution and/or dilute hydrochloric acid to the tubes as follows (Figure 12).

A Egg-white suspension + 1 cm^3 pepsin solution (1 %)

B Egg-white suspension + 3 drops dilute hydrochloric acid (HCl)

C Egg-white suspension + 1 cm^3 pepsin + 3 drops HCl

D Egg-white suspension + 1 cm^3 boiled pepsin + 3 drops HCl

Place all four tubes in a beaker of warm water at 35 °C for 10–15 minutes.

Result. The contents of tube C go clear. The rest remain cloudy.

Interpretation. The change from a cloudy suspension to a clear solution shows that the solid particles of egg protein have been digested to soluble products. The failure of the other three tubes to give clear solutions shows that:

A Pepsin will only work in acid solutions.

B It is the pepsin and not the hydrochloric acid which does the digestion.

D Pepsin is an enzyme, because its activity is destroyed by boiling.

BASIC FACTS

- **Digestion is the process which changes insoluble food into soluble substances.**
- **Digestion takes place in the alimentary canal.**
- **The changes are brought about by chemicals called digestive enzymes.**

Region of alimentary canal	Digestive gland	Digestive juice produced	Enzymes in the juice	Class of food acted upon	Substances produced
mouth	salivary glands	saliva	salivary amylase	starch	maltose
stomach	glands in stomach lining	gastric juice	pepsin	proteins	peptides
duodenum	pancreas	pancreatic juice	trypsin	proteins and peptides	amino acids
			amylase	starch	maltose
			lipase	fats	fatty acids and glycerol

- **Maltose and sucrose are changed to glucose by enzymes in the epithelium of the villi.**
- **The small intestine absorbs amino acids, glucose and fats.**
- **These are carried in the blood stream first to the liver and then to all parts of the body.**
- **The digested food is used or stored in the following ways:**
  Glucose is (a) oxidized for energy or (b) changed to glycogen or fat and stored.
  Amino acids are (a) built up into proteins or (b) de-aminated to urea and glycogen and used for energy.
  Fats are (a) oxidized for energy or (b) stored.
- **The liver stores glycogen and changes it to glucose and releases it into the blood stream to keep a steady level of blood sugar.**

14 The blood circulatory system

The previous chapter explained how food is digested to amino acids, glucose, etc. which are absorbed in the small intestine. These substances are needed in all living cells in the body such as the brain cells, leg muscle cells and kidney cells. The substances are carried from the intestine to other parts of the body by the blood system. In a similar way, the oxygen taken in by the lungs is needed by all the body cells and is carried round the body in the blood.

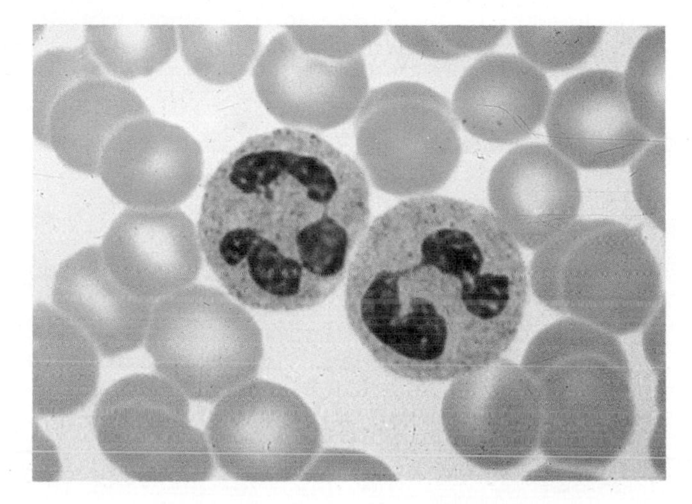

Figure 2 Red and white cells (×2000). The two white cells are stained mauve and their lobed nuclei are stained purple. The red cells are stained pink.

COMPOSITION OF BLOOD

Blood is a mixture of red and white cells floating in a liquid called **plasma**. There are between five and six litres of blood in the body of an adult.

Red cells

These are tiny, disc-like cells (Figures 1*a* and 2) which do not have nuclei. They are made of spongy cytoplasm enclosed in an elastic cell membrane. In their cytoplasm is the red pigment, **haemoglobin**, a protein combined with iron. Haemoglobin combines with oxygen in places where there is a high concentration of oxygen to form **oxy-haemoglobin**. Oxy-haemoglobin is an unstable compound. It breaks down and releases its oxygen in places where the oxygen concentration is low (Figure 3). This makes haemoglobin very useful in carrying oxygen from the lungs to the tissues.

Each red cell lives for about four months, after which it dies and is broken down in the liver which stores the iron from the haemoglobin. About 200 000 million red cells die and are replaced each day. They are made by the red bone-marrow of certain bones in the skeleton; in the ribs and breast-bone for example.

Figure 1 Blood cells

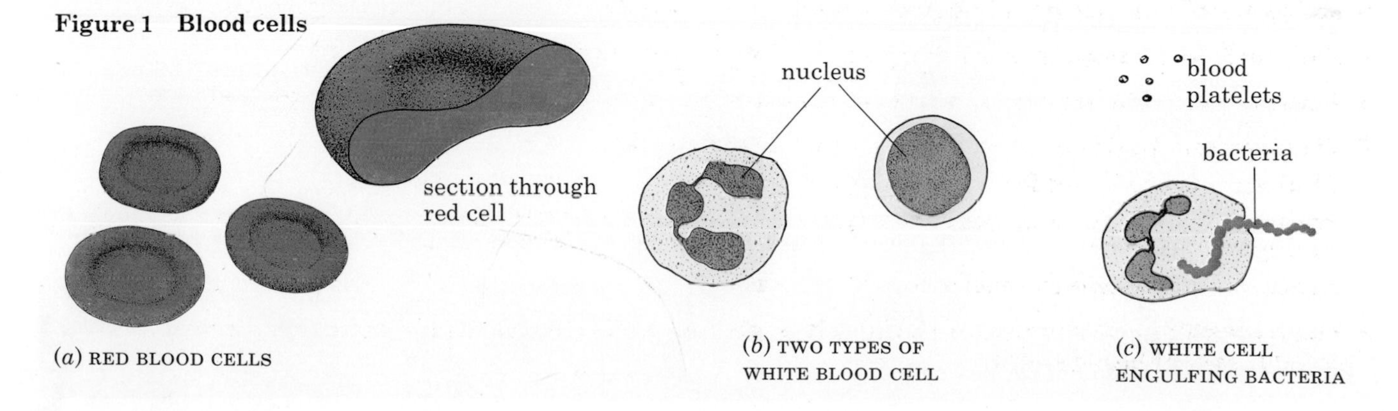

White cells

There are several different kinds of white cell (Figures 1*b* and 2). Most are larger than the red cells, and they all have a nucleus. Some of them can move about by a flowing action of their cytoplasm and can escape from the blood capillaries into the tissues by squeezing between the cells of the capillary walls. They collect at the site of an infection, engulfing (**ingesting**) and digesting harmful bacteria (Figure 1*c*). In this way they prevent the spread of infection through the body. There is one white cell to every 600 red cells.

Platelets

These are small pieces of cytoplasm budded off from larger cells. They help to clot the blood at wounds and so stop the bleeding. White cells and platelets are also made in the red marrow of bones.

Plasma

The liquid part of the blood is called plasma. It is water with a large number of substances dissolved in it—salts, for example, and the products of digestion such as amino acids and glucose. There are also nitrogen-containing waste products such as urea (see page 128), hormones like adrenalin (page 170), and the plasma proteins. The plasma proteins have important functions in the blood (clotting and antibodies for example, see page 120) but many of the other substances are simply being carried from one place to another, e.g. food from the intestine to the muscles.

QUESTIONS

1 In what ways are white cells different from red cells in (a) their structure, (b) their function?

2 (a) Where, in the body, would you expect haemoglobin to be combining with oxygen to form oxy-haemoglobin?

(b) In what parts of the body would you expect oxy-haemoglobin to be breaking down to oxygen and haemoglobin?

3 Why is it important for oxy-haemoglobin to be an unstable compound, i.e. easily changed to oxygen and haemoglobin?

4 What might be the effect on a person whose diet contained too little iron?

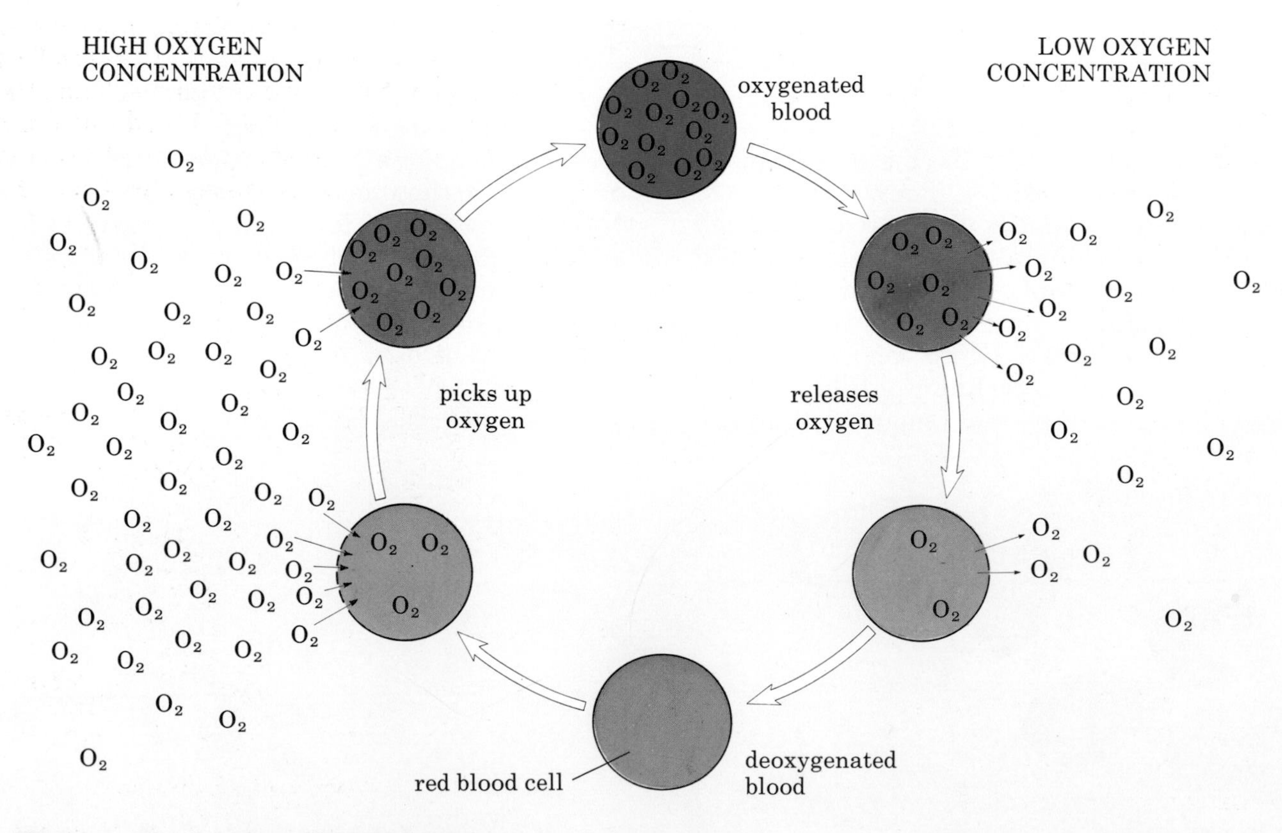

Figure 3 The function of the red cells

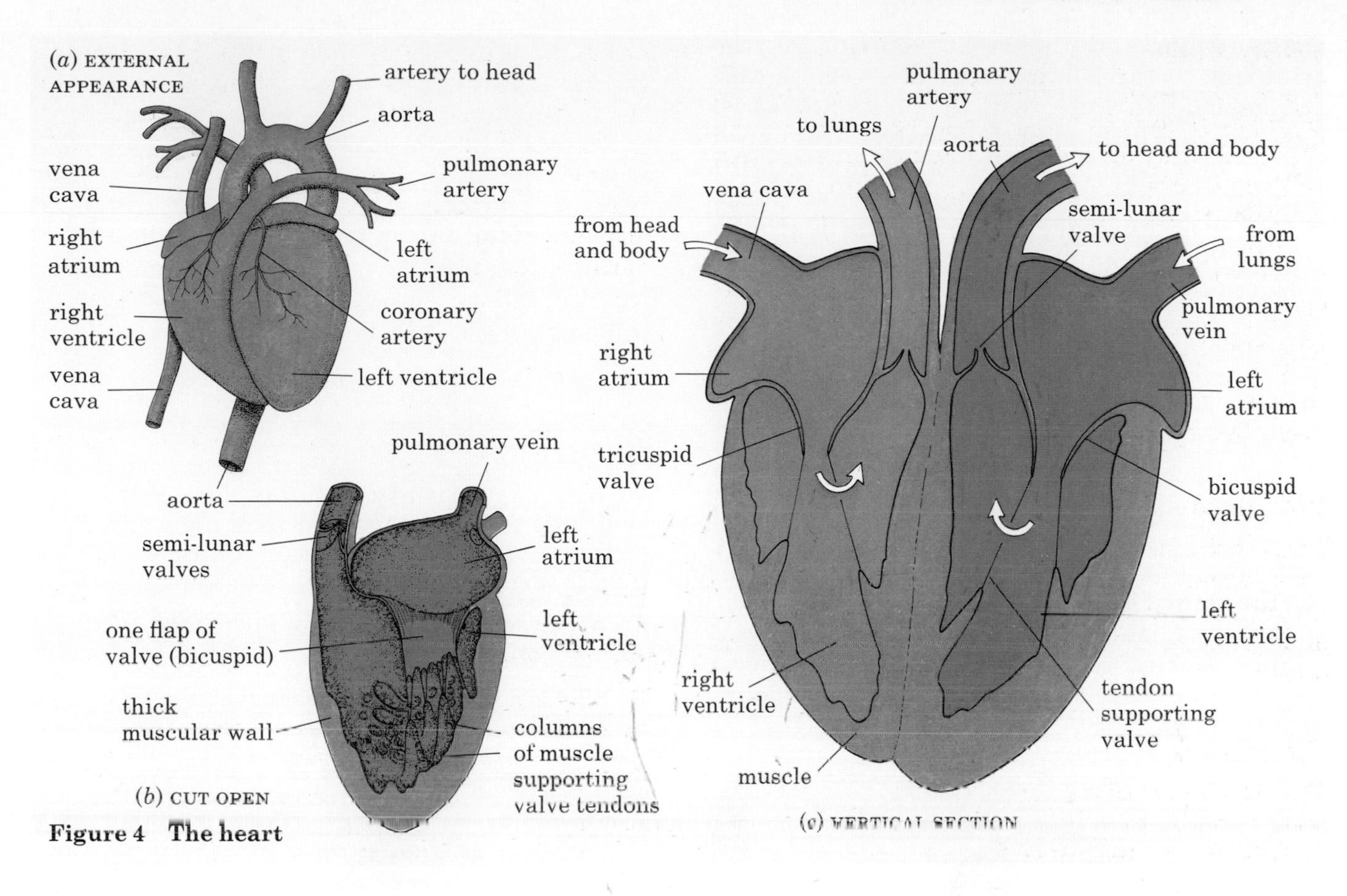

Figure 4 The heart

THE HEART

The heart pumps blood through the circulatory system all round the body. Figure *4a* shows its appearance from the outside, Figure *4b* shows the left side cut open, while Figure *4c* is a diagram of a vertical section to show its internal structure. Since the heart is seen as if in a dissection of a person facing you, the left side is drawn on the right.

If you study Figure *4c* you will see that there are four chambers. The upper, thin-walled chambers are the **atria** (singular=atrium) and each of these opens into a thick-walled chamber, the **ventricle**, below.

Blood enters the atria from large veins. The **pulmonary vein** brings oxygenated blood from the lungs into the left atrium. The **vena cava** brings deoxygenated blood from the body tissues into the right atrium. The blood passes from each atrium to its corresponding ventricle, and the ventricle pumps it out into the arteries. The artery carrying oxygenated blood to the body from the left ventricle is the **aorta**. The **pulmonary artery** carries deoxygenated blood from the right ventricle to the lungs.

In pumping the blood, the muscle in the walls of the atria and ventricles contracts and relaxes (Figure 5). The walls of the atria contract first and force blood into the two ventricles. Then the ventricles contract and send blood into the arteries. The blood is stopped from flowing backwards by four sets of valves. Between the

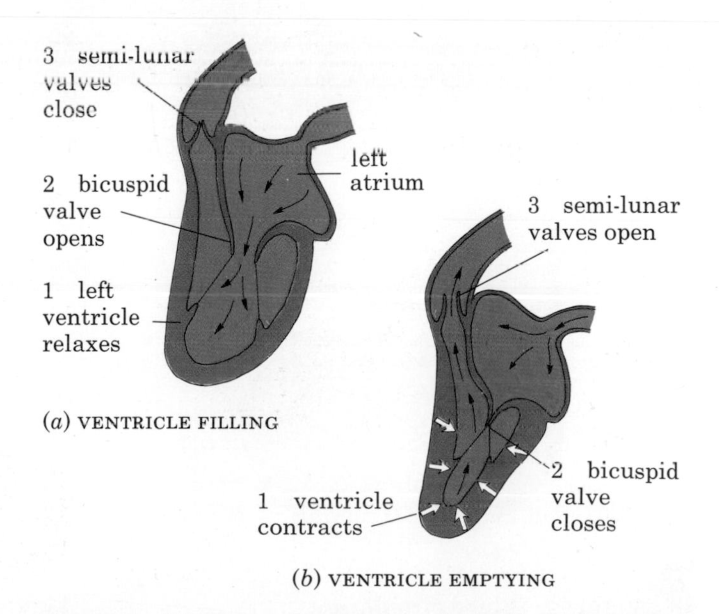

Figure 5 Diagram of heart beat (only the left side is shown)

right atrium and the right ventricle is the **tricuspid** (=three flaps) valve. Between the left atrium and left ventricle is the **bicuspid** (=two flaps) valve. The flaps of these valves are shaped rather like parachutes, with 'strings' called **tendons** to prevent their being turned inside out. In the pulmonary artery and aorta are the **semi-lunar** (=half-moon) valves. These each consist of three pockets which are pushed flat against the artery walls when blood flows one way. If blood tries to flow the other way, the 'pockets' fill up and meet in the middle to stop the flow of blood (Figure 6).

When the ventricles contract, blood pressure closes the bicuspid and tricuspid valves and these prevent blood returning to the atria. When the ventricles relax, the blood pressure in the arteries closes the semi-lunar valves so preventing the return of blood to the ventricles.

The heart contracts and relaxes 60–80 times a minute. During exercise, this rate goes up to over 100 and increases the supply of oxygen and food to the tissues.

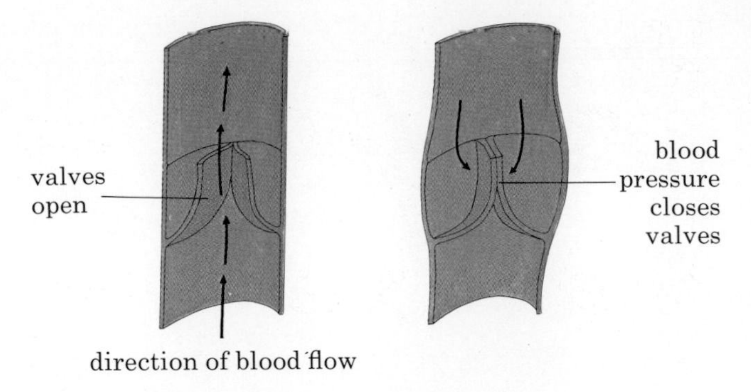

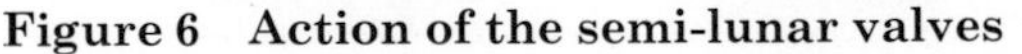

Figure 6 Action of the semi-lunar valves

QUESTIONS

5 Why do you think that (a) the walls of the ventricles are more muscular than the walls of the atria and (b) the muscle of the left ventricle is thicker than that of the right ventricle? (Consult Figure 8.)

6 Which important veins are not shown in Figure 4*a*?

7 Why is a person whose heart valves are damaged by disease unable to take part in active sport?

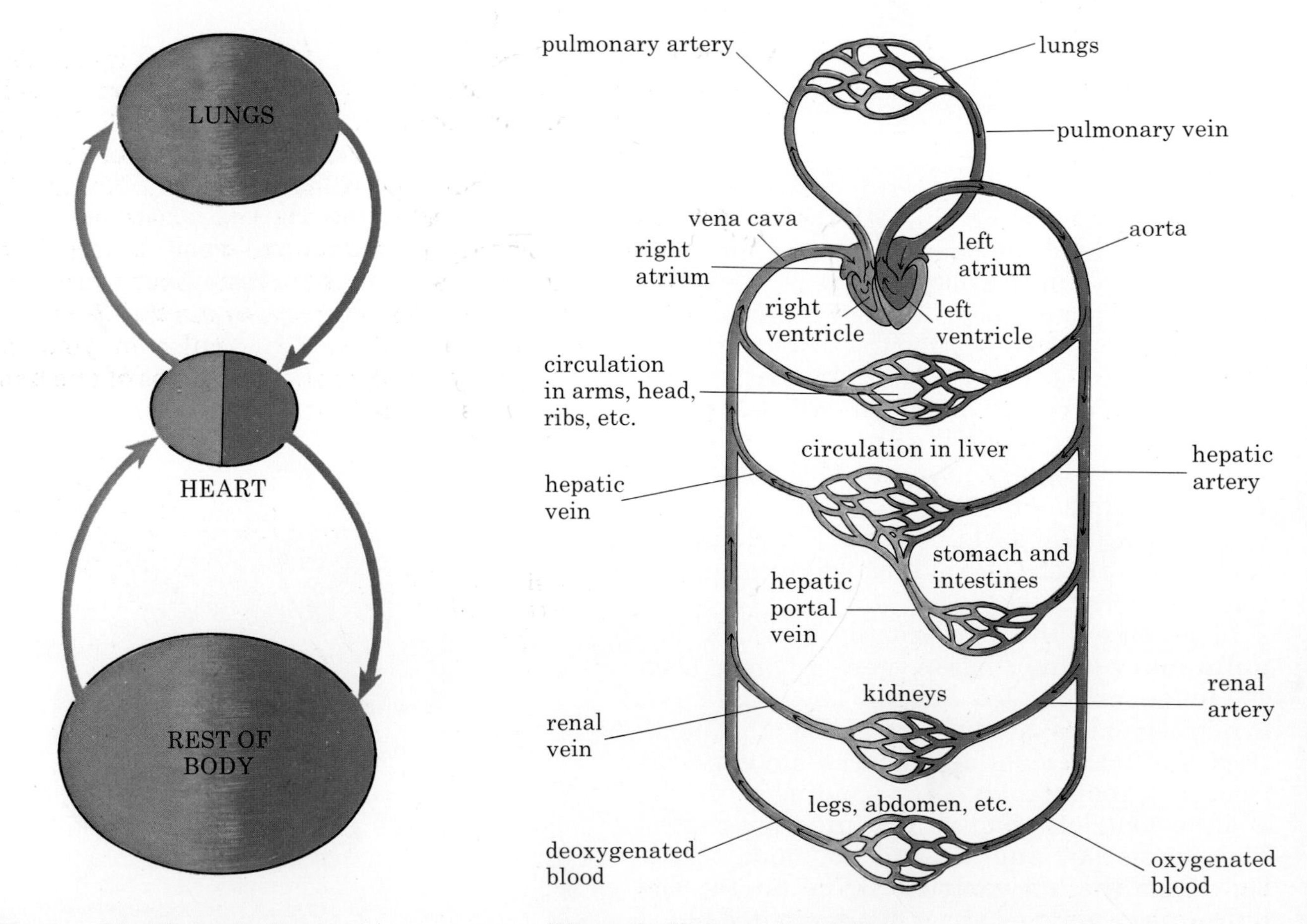

Figure 7 Blood circulation

Figure 8 Human circulation

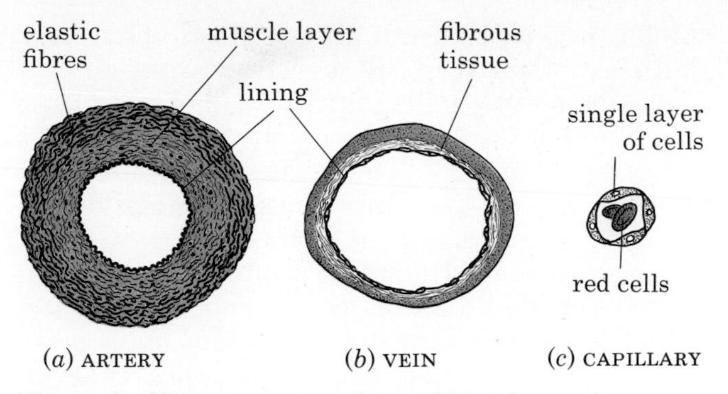

Figure 9 Transverse sections of blood vessels

THE CIRCULATION

Pumped by the heart, the blood travels all round the body in blood vessels. It leaves the heart in arteries and returns in veins. Figure 7 shows the route of the circulation as a diagram. The blood passes twice through the heart during one complete circuit; once on its way to the body and again on its way to the lungs. On average, a red cell would go round the circulation in 45 seconds. Figure 8 is a more detailed diagram of the circulation.

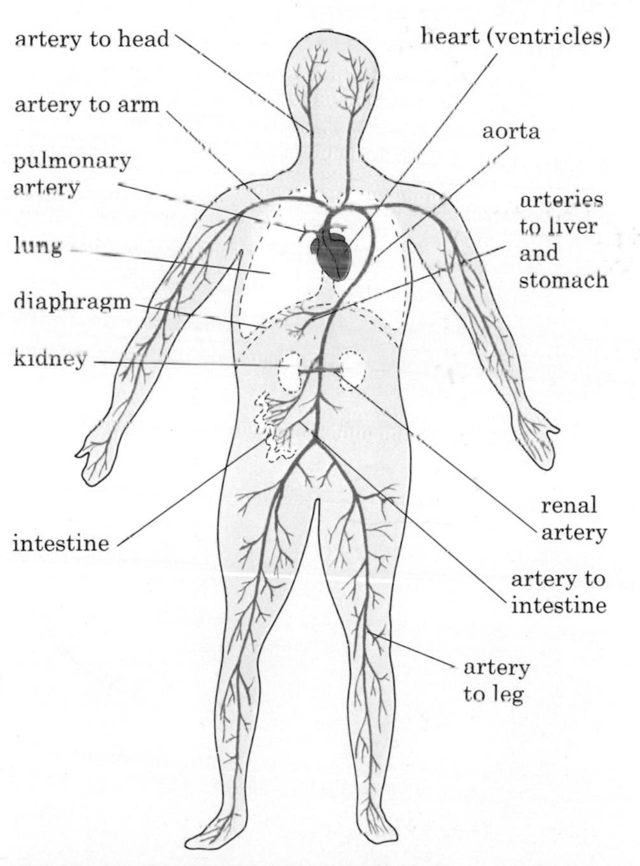

Figure 10 Human arteries. (The veins, which are not shown, would mostly run alongside the arteries, carrying the blood back to the heart.)

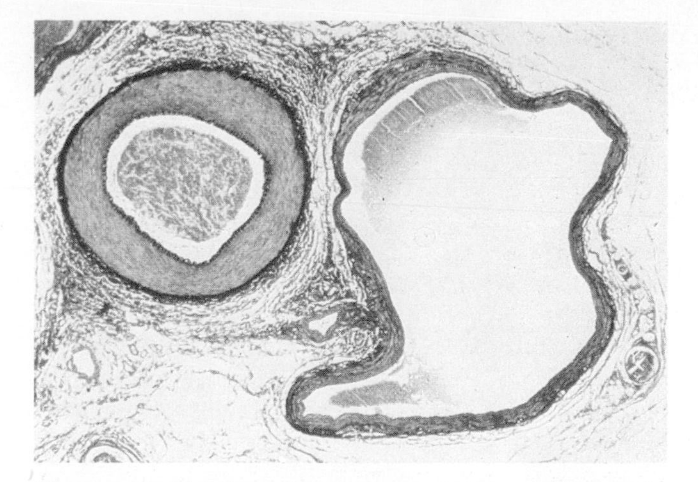

Figure 11 Section through an artery and a vein (×25). The artery, on the left, has a thick wall and contains some clotted blood. The vein, on the right, is larger than the artery and has a much thinner wall. Its irregular shape results from distortion when the section was cut.

Arteries

These are fairly wide vessels (Figures 9*a* and 11) which carry blood from the heart to the limbs and organs of the body (Figure 10). They have elastic tissue and muscle fibres in their thick walls which must stand up to the surges of high pressure caused by the heart beat.

The ripple of pressure which passes down an artery as a result of the heart beat can be felt as a 'pulse' when the artery is near the surface of the body. You can feel the pulse in your radial artery by pressing the finger-tips of one hand on the wrist of the other (Figure 12).

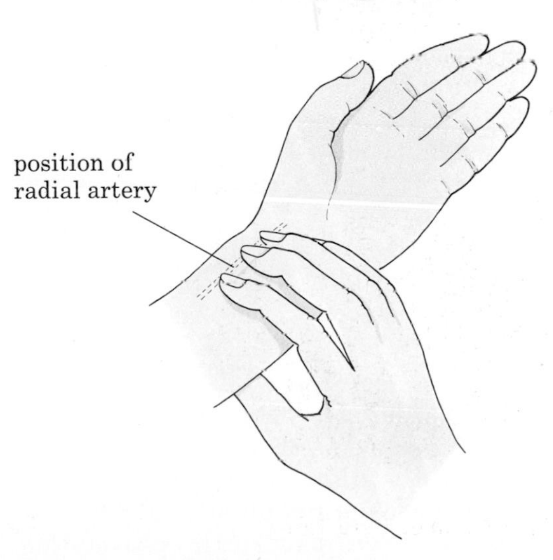

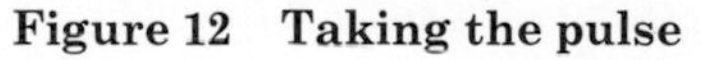

Figure 12 Taking the pulse

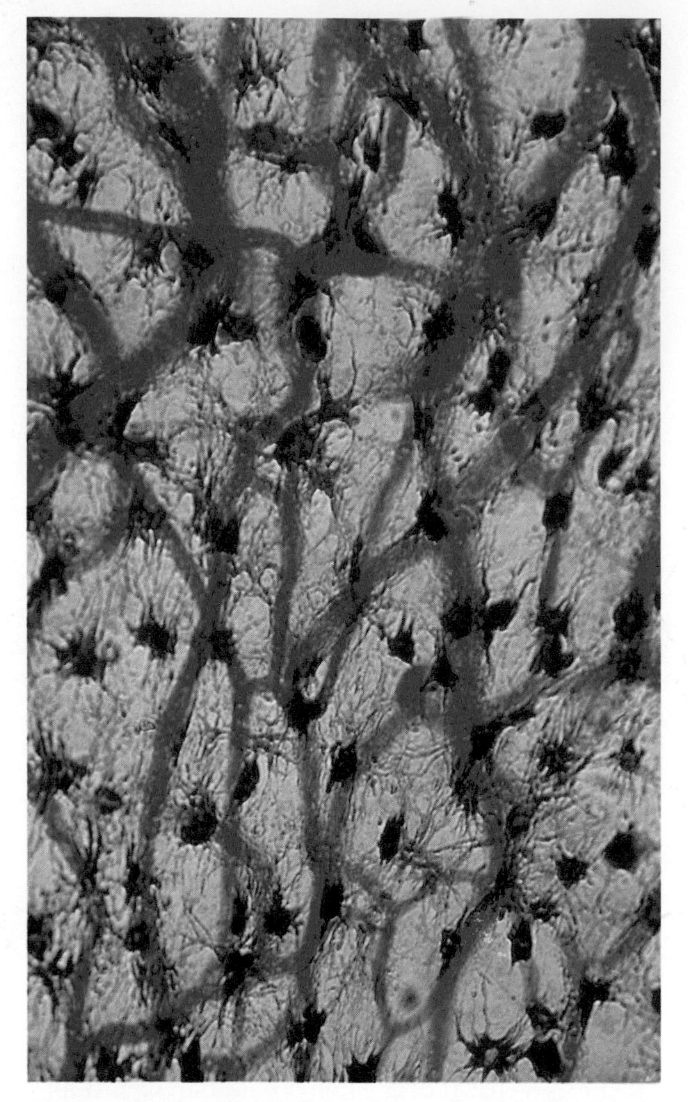

Figure 13 Capillary network in a frog's skin (×60). Oxygen from the air or water is absorbed by the blood in these capillaries.

The arteries divide into smaller vessels called **arterioles**, which themselves divide repeatedly (Figure 13) to form a branching network of microscopic vessels passing between the cells of every living tissue. These final branches are called **capillaries**.

Capillaries

These are tiny vessels with walls only one cell thick (Figures 10*c* and 14). Although the blood as a whole cannot escape from the capillary, the thin capillary walls allow some liquid to pass through, i.e. they are permeable. Blood pressure in the capillaries forces part of the plasma out through the walls. The fluid which escapes is not blood, nor plasma but **tissue fluid**. This fluid bathes all the living cells of the body and since it contains amino acids, glucose and oxygen from the blood, it supplies the cells with their needs (Figures 15 and 16). The tissue fluid eventually seeps back into the capillaries, having given up its oxygen and dissolved food to the cells, but it has now received the waste products of the cells, namely carbon dioxide and nitrogenous waste, which are carried away by the blood stream.

The capillary network is so dense that no living cell is far from a supply of oxygen and food. The capillaries join up into larger vessels, which then combine to form **veins**, and these return blood to the heart.

Veins

Veins return blood from the tissues to the heart. The blood pressure in them is steady and is less than that in the arteries. They are wider and their walls are thinner, less elastic and less muscular than those of the arteries (Figures 10*b* and 11). They also have valves in them similar to the semi-lunar valves (Figure 6).

The blood in most veins contains less oxygen and food, but more carbon dioxide and nitrogenous waste, than the blood in most arteries.

QUESTIONS

8 Why is it not correct to say that all arteries carry oxygenated blood and all veins carry deoxygenated blood?

9 How do veins differ from arteries in (a) their function, (b) their structure?

10 How do capillaries differ from other blood vessels in (a) their structure, (b) their function?

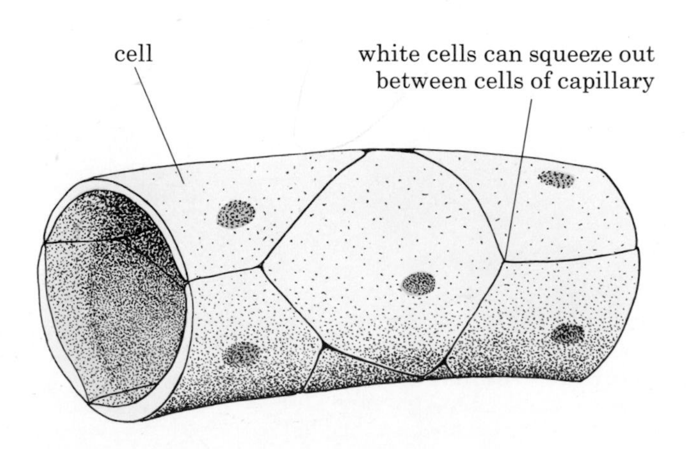

Figure 14 Structure of a blood capillary

11 Describe the path taken by a molecule of glucose, from the time it is absorbed in the small intestine, and the path taken by a molecule of oxygen absorbed in the lungs, to the time when they both meet in a muscle cell of the leg (use Figure 8).

THE LYMPHATIC SYSTEM

Not all the tissue fluid returns to the capillaries. Some of it enters blindly-ending, thin-walled vessels called **lymphatics** (Figure 15). The lymphatics from all parts of the body join up to make two large vessels which empty their contents into the blood system as shown in Figure 17. The lacteals from the villi in the small intestine (page 107) join up with the lymphatic system, so most of the fats absorbed in the intestine reach the circulation by this route. The fluid in the lymphatic vessels is called **lymph** and is similar in composition to tissue fluid.

Lymphatic vessels contract rhythmically when filled with lymph. Also, they are squashed when the body muscles around them contract in

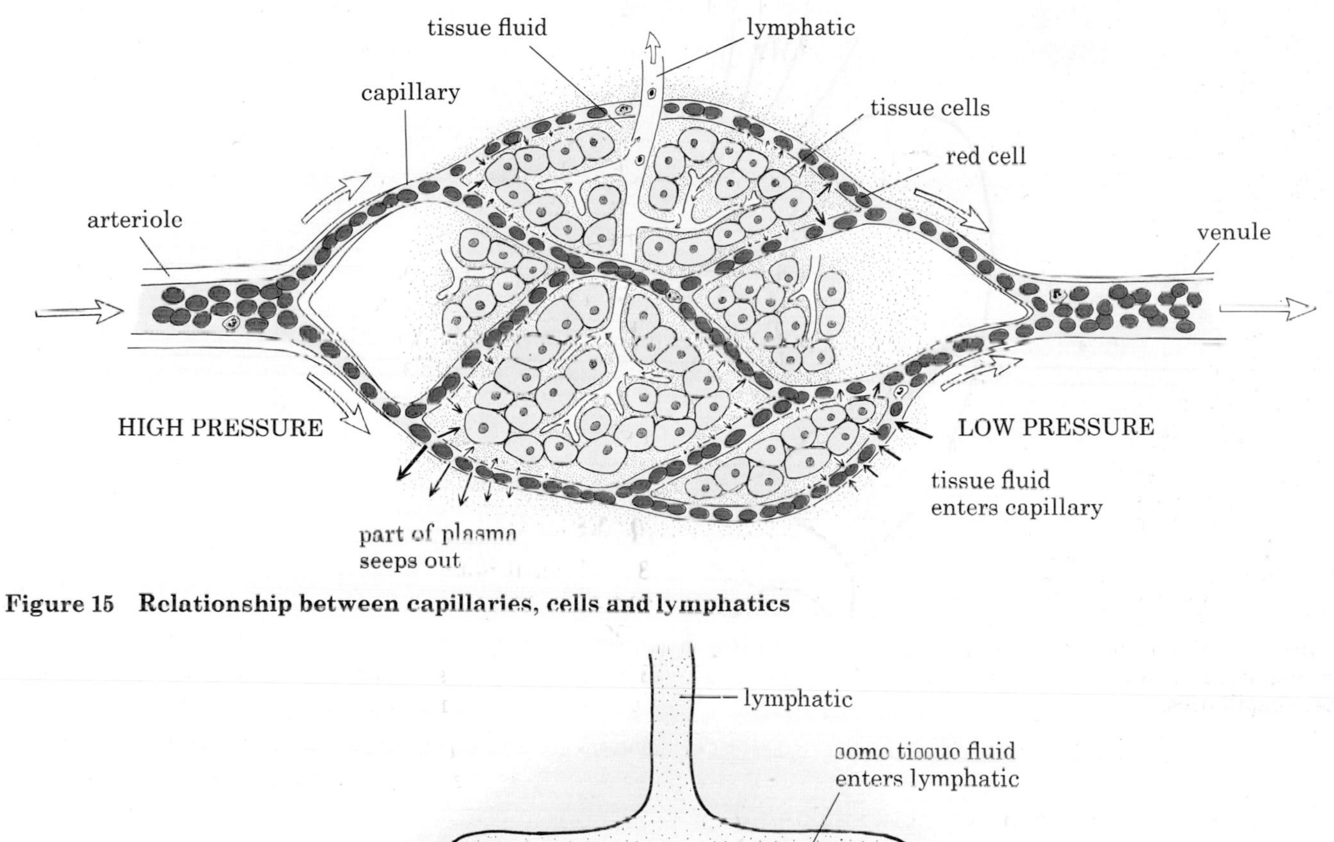

Figure 15 Relationship between capillaries, cells and lymphatics

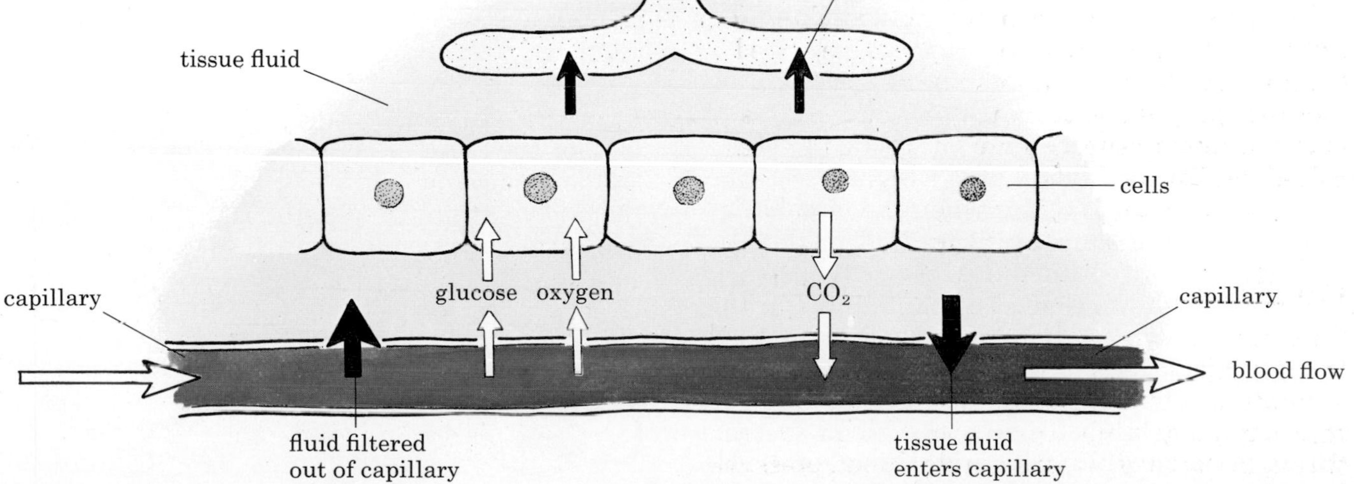

Figure 16 Tissue fluid and lymph

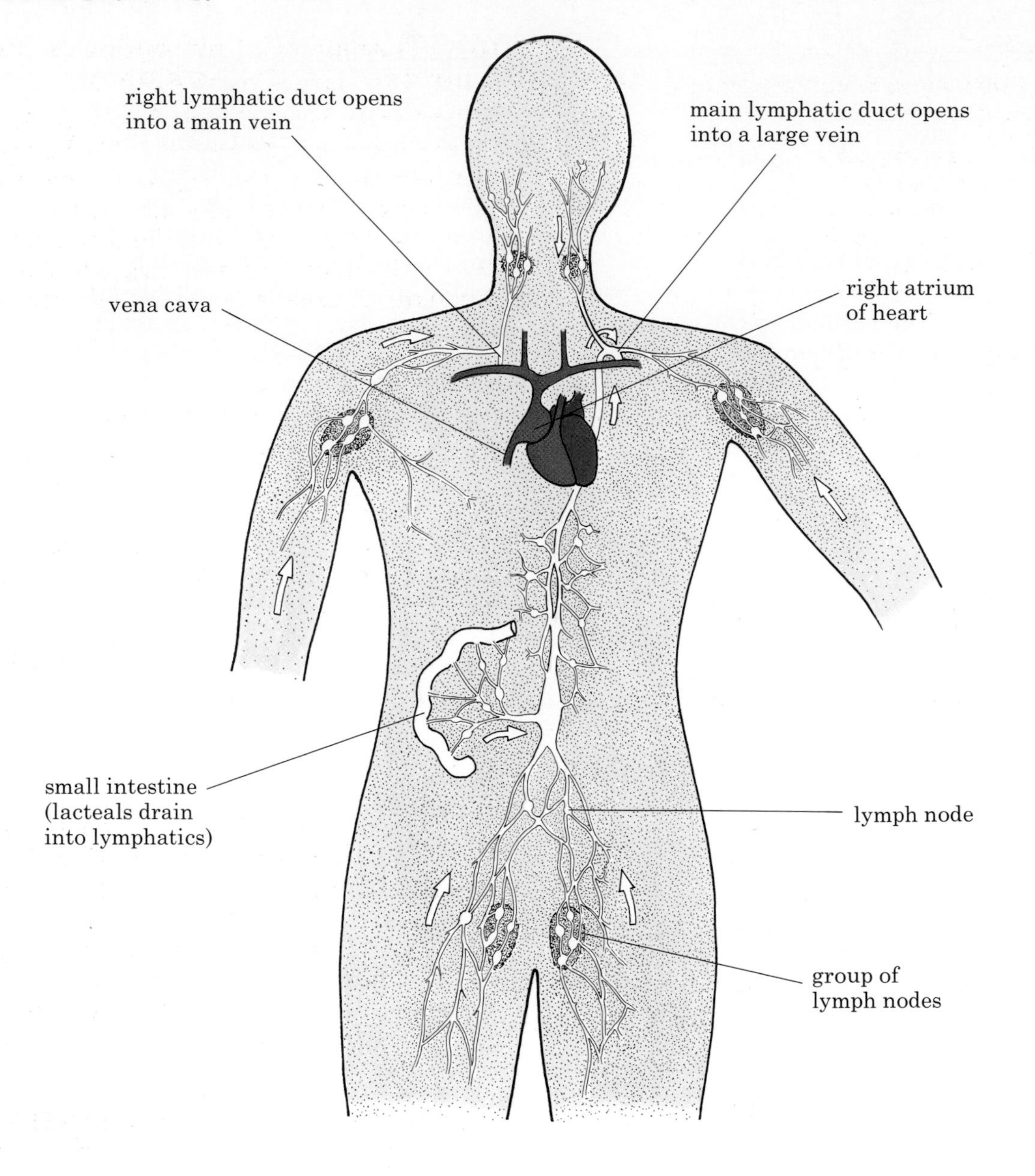

Figure 17 Main drainage routes of lymphatic system

movements such as walking or breathing. There are valves in the lymphatics (Figure 18), like those in the veins, so that when the lymphatics contract or are squashed, the fluid in them is forced in one direction only, towards the heart.

Figure 17 shows that at certain points in the lymphatic vessels, there are swellings called **lymph nodes**. Certain kinds of white cell are made in the lymph nodes and released into the lymph to reach, eventually, the blood system. There are also white cells which remain in the lymph nodes. If bacteria enter a wound and are not ingested by the white cells of the blood or lymph, they will be carried in the lymph to a lymph node and white cells there will ingest them. The lymph nodes thus form part of the body's defence system against infection.

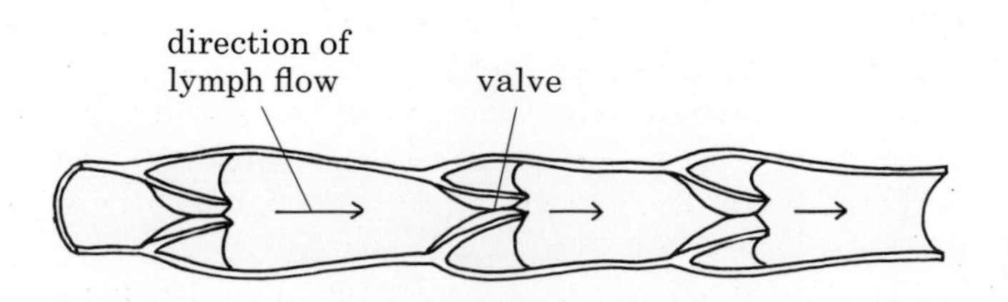

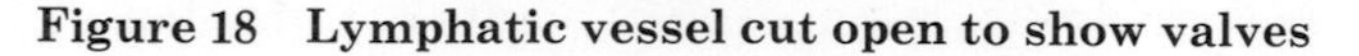

Figure 18 Lymphatic vessel cut open to show valves

QUESTIONS

12 List the things you would expect to find if you analysed a sample of lymph.

13 Describe the course taken by a molecule of fat from the time it is absorbed in the small intestine to the time it reaches the liver to be oxidized for energy. (Use Figure 8 on page 114 and Figure 10 on page 107.)

FUNCTIONS OF THE BLOOD

All the living tissues of the body are bathed in the fluid which escapes from the blood capillaries. This tissue fluid supplies the cells with food and oxygen and removes their waste products. The blood circulating round the body replaces this tissue fluid and so acts as a transport system.

(a) Transport of oxygen from the lungs to the tissues

In the lungs, the concentration of oxygen is high and so the oxygen combines with the haemoglobin in the red cells, forming oxy-haemoglobin. The blood is now said to be **oxygenated**. When this oxygenated blood reaches tissues where oxygen is being used up, the oxy-haemoglobin breaks down and releases its oxygen to the tissues. Oxygenated blood is a bright red colour; **deoxygenated** blood is dark red.

(b) Transport of carbon dioxide from the tissues to the lungs

The blood picks up carbon dioxide from actively respiring cells and carries it to the lungs. In the lungs, the carbon dioxide escapes from the blood and is breathed out (see page 125).

(c) Transport of digested food from the intestine to the tissues

The soluble products of digestion pass into the capillaries of the villi lining the small intestine (page 106). They are carried in solution by the plasma and, after passing through the liver, enter the main blood system. Glucose and amino acids diffuse out of the capillaries and into the cells of the body. Glucose may be oxidized in a muscle, for example, and provide the energy for contraction; amino acids will be built up into new proteins and make new cells and tissues.

(d) Transport of nitrogenous waste from the liver to the kidneys

When the liver changes amino acids into glycogen (see pages 109 and 128), the amino part of the molecules ($-NH_2$) is changed into the nitrogenous waste product, urea. This substance is carried away in the blood circulation. When the blood passes through the kidneys, much of the urea is removed and excreted (page 130).

(e) Transport of hormones

Hormones are chemicals made by certain glands in the body (see page 170). The blood carries these chemicals from the glands which make them, to the organs where they affect the rate of activity. For example, a hormone called **insulin**, made in the pancreas, is carried by the blood to the liver and stimulates it to convert glucose to glycogen. This helps to keep a steady concentration of blood sugar (page 108).

(f) Transport of heat

The limbs and head lose heat to the surrounding air. Chemical activity in the body such as in the contracting muscles produces heat. The blood carries the heat from the warm places to the cold places and so helps to keep an even temperature in all regions. Also by opening or closing blood vessels in the skin, the blood system helps to control the body temperature (see page 136).

TRANSPORT BY THE BLOOD SYSTEM

	From	To
oxygen	lungs	whole body
carbon dioxide	whole body	lungs
urea	liver	kidneys
digested food	intestine	whole body
heat	liver and muscles	whole body

Note that the blood is not directed to a particular organ. A molecule of urea may go round the circulation many times before it enters the renal artery, by chance, and is removed by the kidneys.

QUESTION

14 What substances would the blood (a) gain, (b) lose, on passing through (i) the kidneys, (ii) the lungs, (iii) an active muscle? Remember that respiration (page 14) is taking place in all these organs.

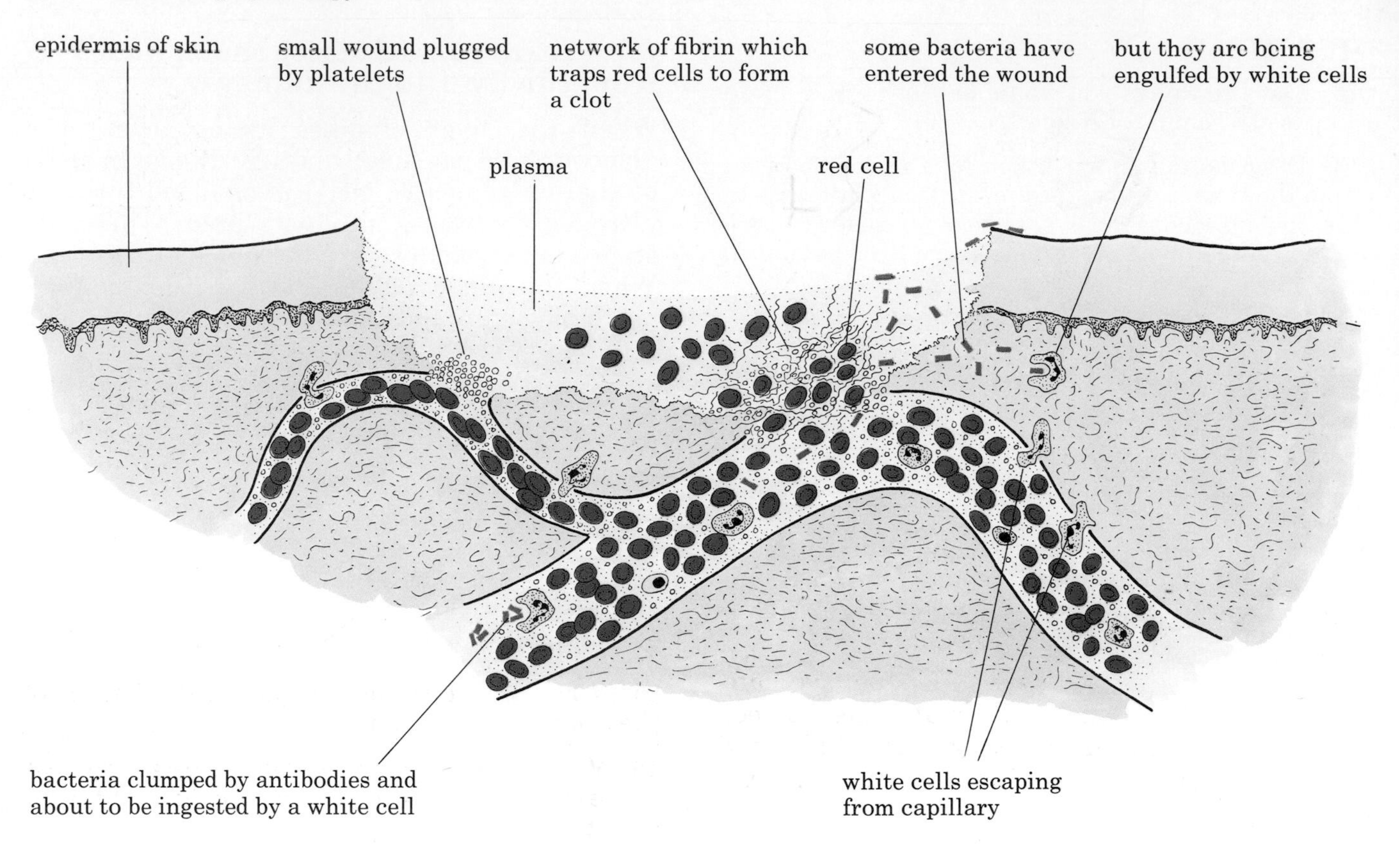

Figure 19 The defence against infection. An area of skin has been damaged and two capillaries broken open.

DEFENCE AGAINST INFECTION

(a) Clotting

When tissues are damaged and blood vessels cut, platelets clump together and block the smaller capillaries. The platelets and damaged cells at the wound also produce a substance which acts on the plasma protein called **fibrinogen**. As a result of this action, the fibrinogen is changed into **fibrin**, which forms a network of fibres across the wound. Red cells become trapped in this network and so form a blood clot. The clot not only stops further loss of blood, but also prevents the entry of harmful bacteria into the wound (Figure 19).

(b) White cells

White cells at the site of the wound, in the blood capillaries or in lymph nodes (page 118) may ingest harmful bacteria and so stop them entering the general circulation. White cells can squeeze through the walls of capillary vessels and so attack bacteria which get into the tissues, even though the capillaries themselves are not damaged.

(c) Antibodies

Certain types of white cell produce chemicals called **antibodies**, which attack bacteria and other foreign particles which get into the body. Antibodies are proteins, released into the plasma by the white cells. They may clump the bacteria together, make them easier for the white cells to ingest or simply neutralize the poisons produced by the bacteria.

Each antibody is very **specific**. This means that an antibody which attacks a typhoid bacterium will not affect a pneumonia bacterium. Figure 20 illustrates this in the form of a diagram.

Once an antibody has been made by the blood, it may remain in the circulation for some time. This means that the body has become **immune** to the disease, because the antibody will attack the germs immediately they get into the body. Even if the antibodies do not remain for long in the circulation, the white cells can usually make them again very quickly, so giving the person some degree of immunity. This explains why, once you have had measles or chicken pox, for example, you are very unlikely to catch the same disease again.

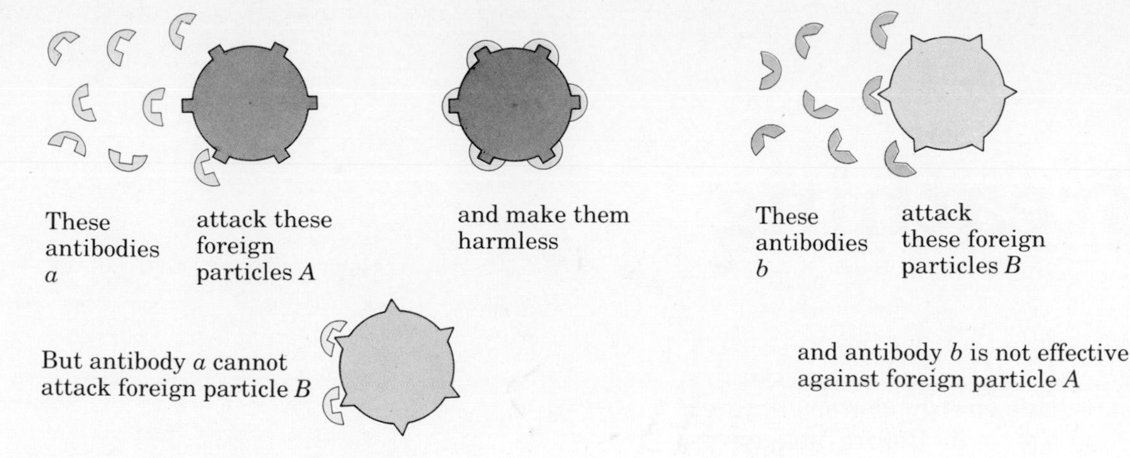
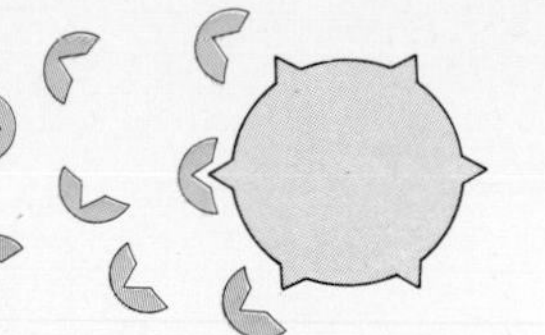
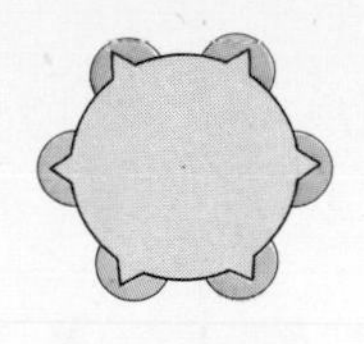

Figure 20 Antibodies are specific

When you are **inoculated** against a disease, a harmless form of the germ is introduced into your body. The white cells make the correct antibodies, so that if the real germs get into the blood, the antibody is already present or very quickly made by the blood.

QUESTIONS

15 What part do white cells play in the defence of the body against infection?

16 Why is it necessary to inoculate a person against a disease before he catches it rather than wait until he catches it?

BASIC FACTS

- **Blood consists of red cells, white cells and platelets suspended in plasma.**
- **The red cells carry oxygen. The white cells attack bacteria.**
- **The heart is a muscular pump with valves, which sends blood round the circulatory system.**
- **Arteries carry blood from the heart to the tissues.**
- **Veins return blood to the heart from the tissues.**
- **Capillaries form a network of tiny vessels in all tissues. Their thin walls allow dissolved food and oxygen to pass from the blood into the tissues, and carbon dioxide and other waste substances to pass back into the blood.**
- **Lymph vessels return tissue fluid to the lymphatic system and finally into the blood system.**
- **The function of the blood is to carry substances round the body, e.g. oxygen from lungs to body, food from intestine to body and urea from the body to the kidneys.**
- **Antibodies are chemicals made by white cells in the blood. They attack any bacteria which get into the circulation.**

15 Breathing

All the processes carried out by the body, such as movement, growth and reproduction, require energy. In animals, this energy can be obtained only from the food they eat. Before the energy can be used by the cells of the body, it must be set free from the chemicals of the food by a process called respiration (see page 14). Respiration needs a supply of oxygen and produces carbon dioxide as a waste product. All cells, therefore, must be supplied with oxygen and must be able to get rid of carbon dioxide.

In man and other mammals, the oxygen is obtained from the air by means of the lungs. In the lungs, the oxygen dissolves in the blood and is carried to the tissues by the circulatory system (page 111).

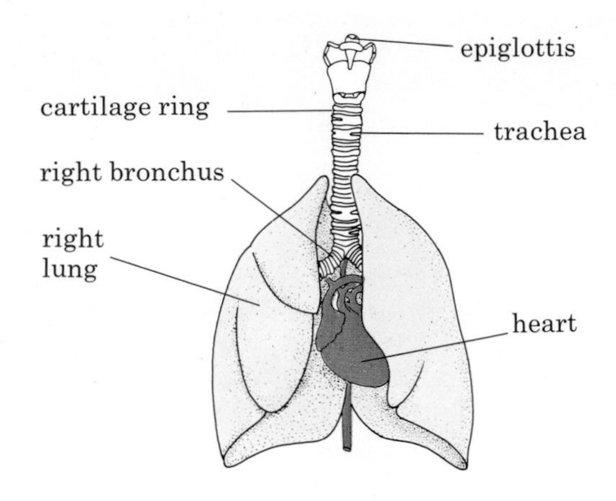

Figure 1 Diagram of lungs to show position of heart

LUNG STRUCTURE

The lungs are enclosed in the thorax (chest region) (see Figure 5 on page 104). They have a spongy texture and can be expanded and compressed by movements of the thorax in such a way that air is sucked in and blown out. The lungs are joined to the back of the mouth by the windpipe or **trachea** (Figure 1). The trachea divides into two smaller tubes called

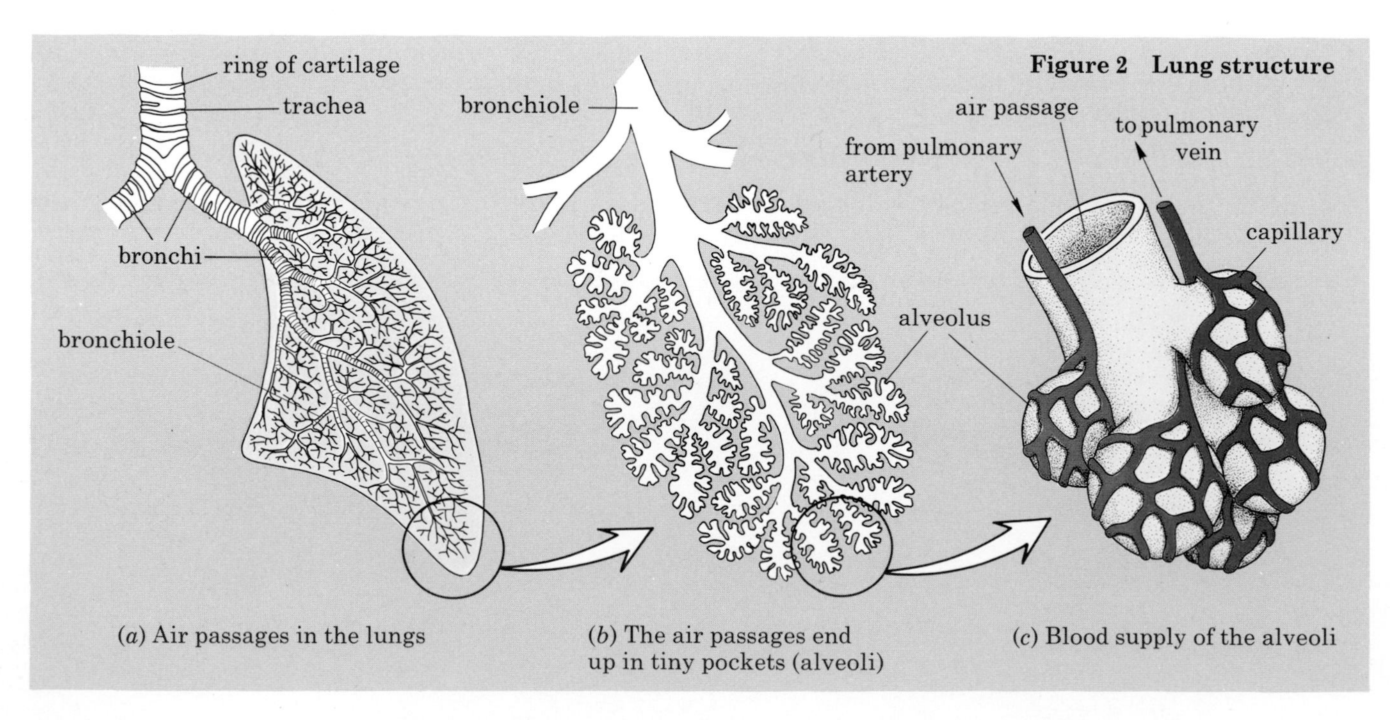

Figure 2 Lung structure

(*a*) Air passages in the lungs

(*b*) The air passages end up in tiny pockets (alveoli)

(*c*) Blood supply of the alveoli

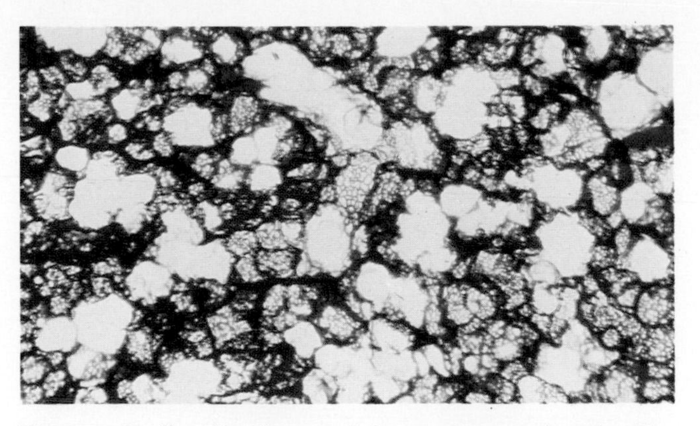

Figure 3 Small piece of lung tissue (×60). The capillaries have been injected with red and blue dye. The networks surrounding the alveoli can be seen.

bronchi (singular = bronchus), which enter the lungs and divide into smaller branches, the smallest of which are called bronchioles (Figure 2*a*). The **bronchioles** end up in a mass of tiny, thin-walled, pouch-like air sacs called alveoli (Figures 2*b* and *c*).

Rings of gristle (cartilage) stop the trachea and bronchi collapsing when we breathe in. The **epiglottis** and other structures at the top of the trachea stop food and drink from entering the air passages when we swallow (see page 105).

The alveoli have thin elastic walls of a single cell layer or **epithelium**. Beneath the epithelium is a dense network of capillaries (Figures 2*c* and 3) supplied with deoxygenated blood (page 119). This blood from which the body has taken oxygen is pumped from the right ventricle, through the pulmonary artery (see Figure 8 on page 114). In man, there are about 350 million alveoli, with a total absorbing surface of about 90 square metres.

QUESTIONS

1 Place the following structures in the order in which air will reach them when breathing in: bronchus, trachea, nasal cavity, alveolus.

2 One function of the small intestine is to absorb food (page 106). One function of the lungs is to absorb oxygen. Point out the basic similarities in these two structures which help to speed up the process of absorption.

VENTILATION OF THE LUNGS

The movement of air into and out of the lungs, called **ventilation**, renews the oxygen supply in the lungs and removes the surplus carbon dioxide from them. The lungs contain no muscle fibres and are made to expand and contract by movements of the ribs and diaphragm.

The diaphragm is a sheet of muscular tissue which separates the thorax from the abdomen (see Figure 5 on page 104). When relaxed, it is domed slightly upwards. The ribs are moved by the **intercostal muscles** which run from one rib to the next (Figure 4). Figure 5 shows how the contraction of the intercostal muscles makes the ribs move upwards.

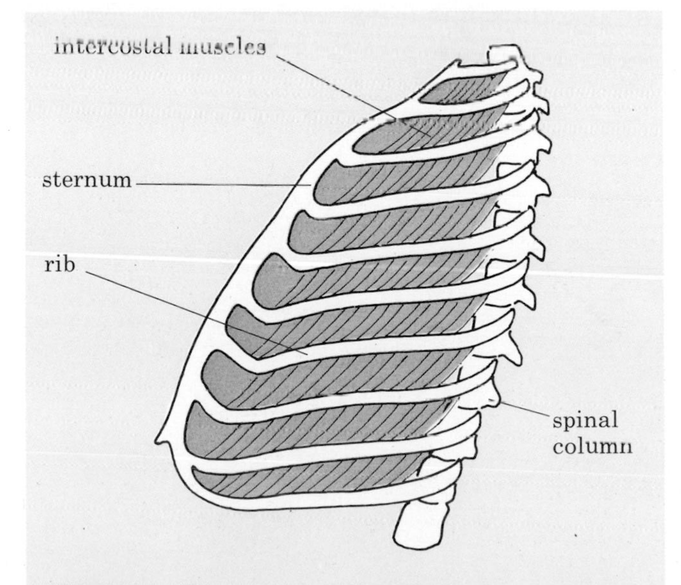

Figure 4 Rib cage seen from left side showing intercostal muscles

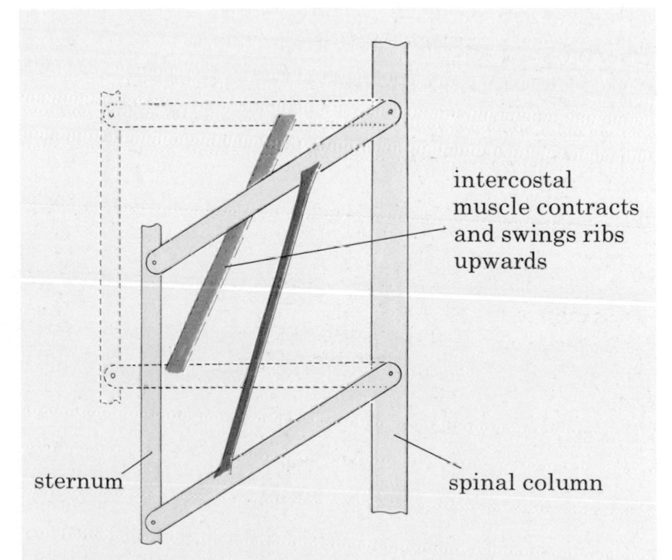

Figure 5 Model to show action of intercostal muscles

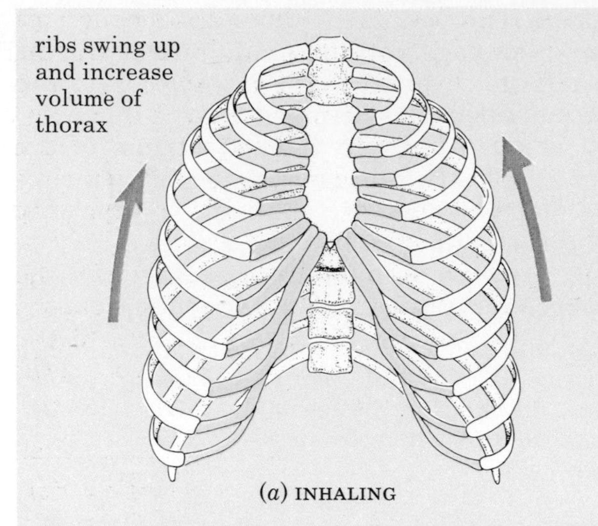

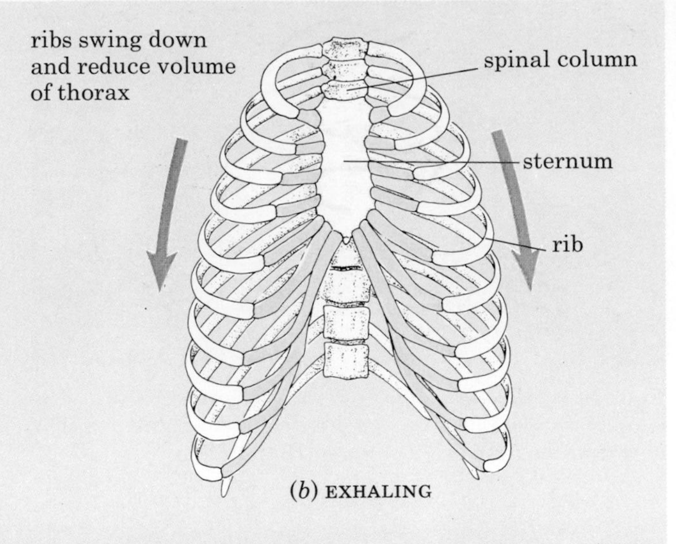

Figure 6 Movement of rib cage during breathing

Inhaling

(a) The diaphragm muscles contract and pull it down (Figure 7*a*).

(b) The intercostal muscles contract and pull the rib cage upwards (Figure 6).

These two movements make the space in the thorax bigger, so forcing the lungs to expand and draw air in through the nose and trachea.

Exhaling

(a) The diaphragm muscles relax, allowing the diaphragm to return to its domed shape (Figure 7*b*).

(b) The intercostal muscles relax, allowing the ribs to move downwards under their own weight.

The lungs are elastic and shrink back to their relaxed size, forcing air out again.

The outside of the lungs and the inside of the thorax are lined with a smooth membrane called the **pleural membrane**. This produces a thin layer of liquid called **pleural fluid** which stops the lungs rubbing against the inside of the thorax.

Breathing rate

When you are at rest and breathing quietly, you normally inhale and exhale about 16 times a minute. Only about 500 cm^3 air is exchanged,

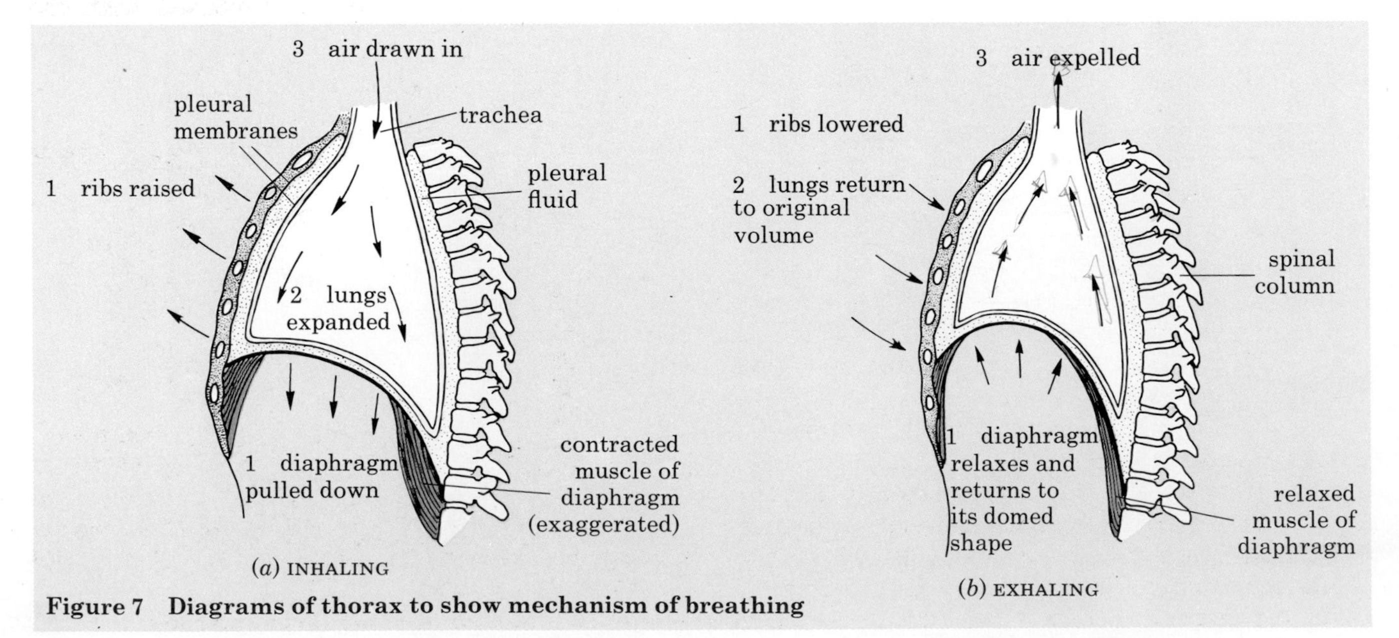

Figure 7 Diagrams of thorax to show mechanism of breathing

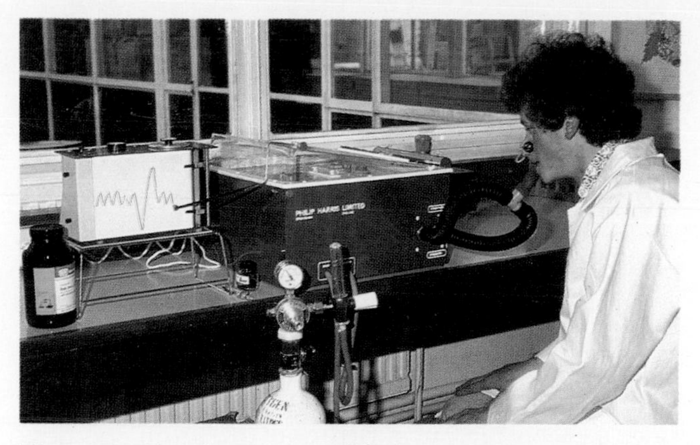

Figure 8 A spirometer. This instrument measures the volume of air breathed in and out of the lungs. The first part of the recording chart on the left shows quiet breathing. In the middle, the chart shows a deep breath in and a deep breath out.

even though your lungs have a total capacity of about five litres (5000 cm^3) (Figure 8). When you take exercise, the rate and depth of breathing increases. As a result of this, more oxygen will dissolve in the blood and supply the active muscles. The extra carbon dioxide which the muscles put into the blood will also be removed by the faster, deeper breathing.

QUESTIONS

3 Place the following in the correct order: lungs expand, ribs rise, air enters lungs, intercostal muscles contract, thorax expands.

4 During and after exercise, you breathe faster. This is because the extra carbon dioxide in your blood affects a sensitive region of the brain. The brain sends out nerve impulses which make you breathe faster. To which structures will the nerve impulses go and what effect will they have on these structures?

GASEOUS EXCHANGE

Ventilation refers to the movement of air into and out of the lungs. Gaseous exchange refers to the exchange of oxygen and carbon dioxide which takes place between the air and the blood vessels in the lungs.

Even when relaxed, your lungs contain 1.5 litres of air which cannot be expelled, no matter how hard you try to breathe out. As a result, the air in the alveoli is not exchanged during ventilation and the oxygen has to reach the blood capillaries by the slower process of diffusion. Figure 9 shows how oxygen reaches the red blood cells and how carbon dioxide escapes from the blood. The capillaries carrying oxygenated blood from the alveoli join up to form the pulmonary vein (see Figure 8 on page 114), which returns to the left atrium of the heart. From here, blood enters the left ventricle and is pumped all round the body, so supplying the tissues with oxygen.

The process of gaseous exchange in the alveoli does not remove all the oxygen from the air. The air breathed in contains about 21 per cent of oxygen; the air breathed out still contains 16 per cent of oxygen.

	Changes in the composition of air	
	Inhaled %	Exhaled %
oxygen	21	16
carbon dioxide	0.04	4
water vapour	variable	saturated

The lining of the alveoli is coated with a film of moisture in which the oxygen dissolves. Some of this moisture evaporates into the alveoli and saturates the air with water vapour. The air you breathe out, therefore, always contains a great deal more water vapour than the air you breathe in. The exhaled air is warmer as well, so in cold and temperate climates, you lose heat to the atmosphere by breathing.

Sometimes the word **respiration** or **respiratory** is used in connection with breathing. The lungs, trachea and bronchi are called the **respiratory system**; a person's rate of breathing may be called his **respiration rate**. This use of the word should not be confused with the

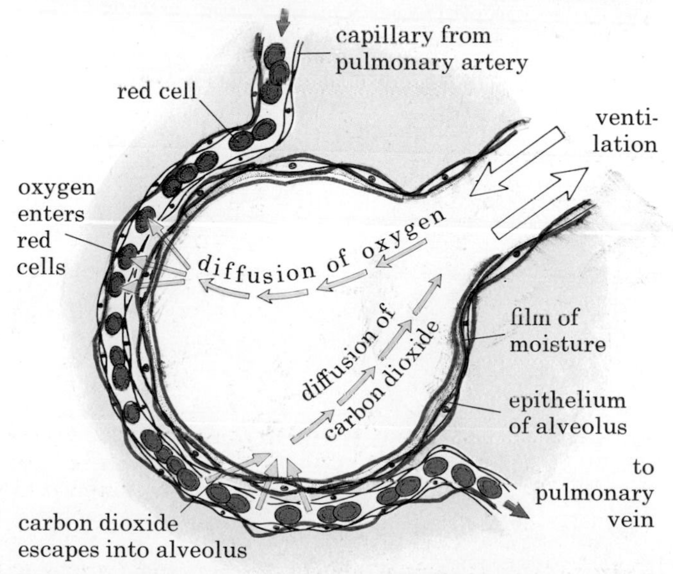

Figure 9 Gaseous exchange in the alveolus

biological meaning of respiration, namely, the release of energy in cells (page 14). This chemical process is sometimes called **tissue respiration** or **internal respiration** to distinguish it from breathing.

QUESTIONS

5 Try to make a clear distinction between **respiration** (page 14), **gaseous exchange** and **ventilation**. Say how one depends on the other.

6 Describe the path taken by a molecule of oxygen from the time it is breathed in through the nose, to the time it enters the heart in some oxygenated blood.

7 Figure 9 shows oxygen and carbon dioxide diffusing across an alveolus. What causes them to diffuse in opposite directions? (See page 18.)

8 In 'mouth to mouth' resuscitation, air is breathed from the rescuer's lungs into the lungs of the person who has stopped breathing. How can this 'used' air help to revive the person?

PRACTICAL WORK

1 Oxygen in exhaled air

Place a large coffee jar on its side in a bowl of water (Figure 10*a*). Put a rubber tube in the mouth of the jar and then turn the jar upside-down, still full of water and with the rubber tube still in it. Start breathing out and when you feel your lungs must be about half empty, breathe the last part of the air down the rubber tubing so that the air collects in the upturned jar and fills it (Figure 10*b*). Put the screw top back on the jar under water, remove the jar from the bowl and place it upright on the bench.

Light the candle on the special wire holder (Figure 10*c*), remove the lid of the jar, lower the burning candle into the jar and count the number of seconds the candle stays alight. Now take a fresh jar, with ordinary air, and see how long the candle stays alight in this.

Results. The candle will burn for about 15–20 seconds in a large coffee jar of ordinary air. In exhaled air it will go out in about five seconds.

Interpretation. Burning needs oxygen. When the oxygen is used up, the flame goes out. It looks as if exhaled air contains much less oxygen than atmospheric air.

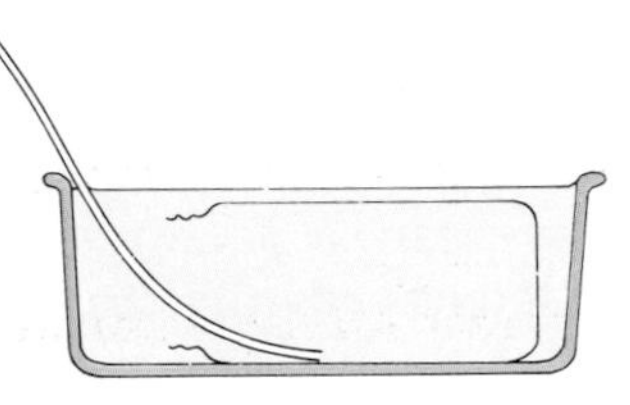

(*a*) Lay the jar on its side under the water

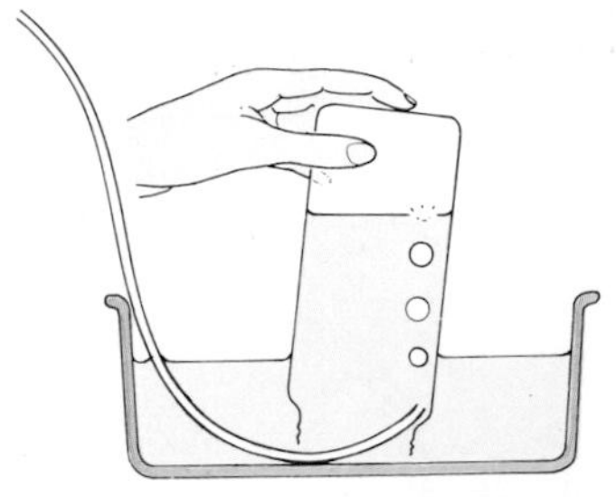

(*b*) Breathe out through the rubber tube and trap the air in the jar

(*c*) Lower the burning candle into the jar until the lid is pressed against the rim

Figure 10 Testing exhaled gas for oxygen

2 Carbon dioxide in exhaled air

Prepare two large test-tubes as shown in Figure 11, each with a little clear lime water in. Put the ends of both rubber tubes at the same time in your mouth and breathe in and out gently through the tubes for about 15 seconds. Notice which tube is bubbling when you breathe out and which one bubbles when you breathe in.

If after 15 seconds there is no difference in the appearance of the lime water in the two tubes, continue breathing through them for another 15 seconds.

Results. The lime water in tube B goes milky. The lime water in tube A stays clear.

Interpretation. Carbon dioxide turns lime water milky. Exhaled air passes through tube B. Inhaled air passes through tube A. Exhaled air must, therefore, contain more carbon dioxide than inhaled air.

3 Volume of air in the lungs

Use a large plastic bottle with lines marked on the side to show litres and half litres. Fill the bottle with water and put on the stopper. Put about 50 mm depth of water in a large plastic bowl. Hold the bottle upside-down with its neck under water and remove the screw top. Some of the water will run out but this does not matter. Push a rubber tube into the mouth of the bottle

(Figure 12) to position A shown on the diagram. Take a deep breath and then exhale as much air as possible down the tubing into the bottle. The final water level inside the bottle will tell you how much air you can exchange in one deep breath.

Now push the rubber tubing further into the bottle, to position B (Figure 12), and blow out any water left in the tube. Support the bottle with your hand and breathe quietly in and out through the tube, keeping the water level inside and outside the bottle the same. This will give you an idea of how much air you exchange when breathing normally.

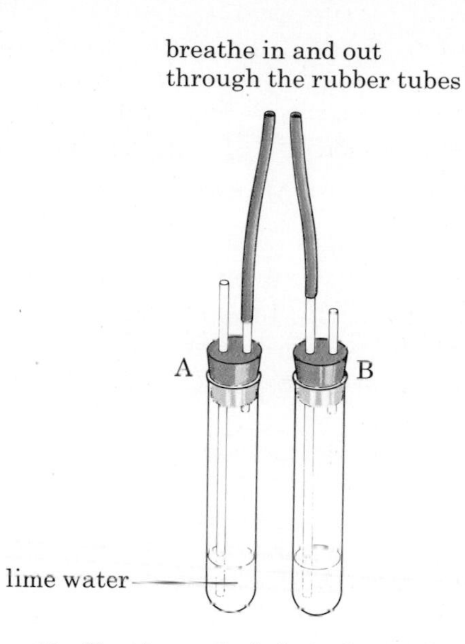

Figure 11 Testing exhaled air for carbon dioxide

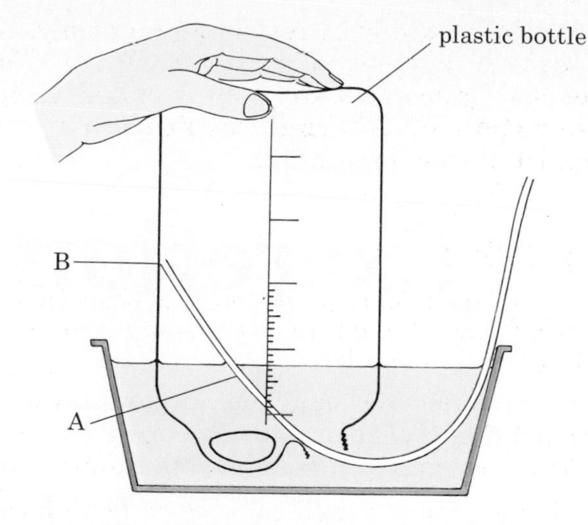

Figure 12 Measuring the volume of air exhaled from the lungs

BASIC FACTS

- **Ventilation is inhaling and exhaling air.**
- **The ribs, rib muscles and diaphragm make the lungs expand and contract. This causes inhaling and exhaling.**
- **The vast number of air pockets (alveoli) give the lungs an enormous internal surface area. This surface is moist and lined with capillaries.**
- **The blood in the capillaries picks up oxygen from the air in the alveoli and gives out carbon dioxide. This is called gaseous exchange.**
- **The oxygen is carried round the body by the blood and used by the cells for their respiration.**

16 Excretion

A great number of chemical reactions take place inside the cells of an organism in order to keep it alive. The products of some of these reactions are poisonous and must be removed from the body. For example, the breakdown of glucose during respiration (page 14) produces carbon dioxide. This is carried away by the blood and removed in the lungs. Excess amino acids are deaminated in the liver to form glycogen and **urea**, as explained on page 109. The urea is removed from the tissues by the blood, to be expelled by the kidneys.

Urea and similar waste products, like **uric acid**, from the breakdown of proteins, contain the element nitrogen. For this reason they are often called **nitrogenous waste products**.

During feeding, more water and salts are taken in with the food than are needed by the body. So these excess substances need to be removed as fast as they build up.

Excretion is the name given to the removal from the body of (a) the waste products of its chemical reactions (carbon dioxide and nitrogenous waste) and (b) the excess water and salts taken in with the diet. Excretion also includes the removal of drugs or other foreign substances taken into the alimentary canal and absorbed by the blood. The term 'excretion' should not usually be applied to the passing out of faeces (page 107), because most of the contents of the faeces have not taken part in reactions in the body.

QUESTION

1 Write a list of the substances that are likely to be excreted from the body during the day.

EXCRETORY ORGANS

The lungs supply the body with oxygen, but they are also excretory organs because they get rid of carbon dioxide. They also lose a great deal of water vapour, but this loss is unavoidable and is not a method of controlling the water content of the body.

The kidneys remove urea and other nitrogenous waste from the blood. They also expel excess water, salts, hormones (page 170) and drugs.

QUESTION

2 When your body becomes overheated, you sweat through the pores in your skin. Sweat contains water, salts and urea. Do you think the skin should be classed as an excretory organ for this reason? Justify your answer.

THE KIDNEYS

Structure

The two kidneys are fairly solid, oval structures. They are red-brown, enclosed in a transparent membrane and attached to the back of the abdominal cavity (Figure 1). The **renal artery** branches off from the aorta and brings oxy-

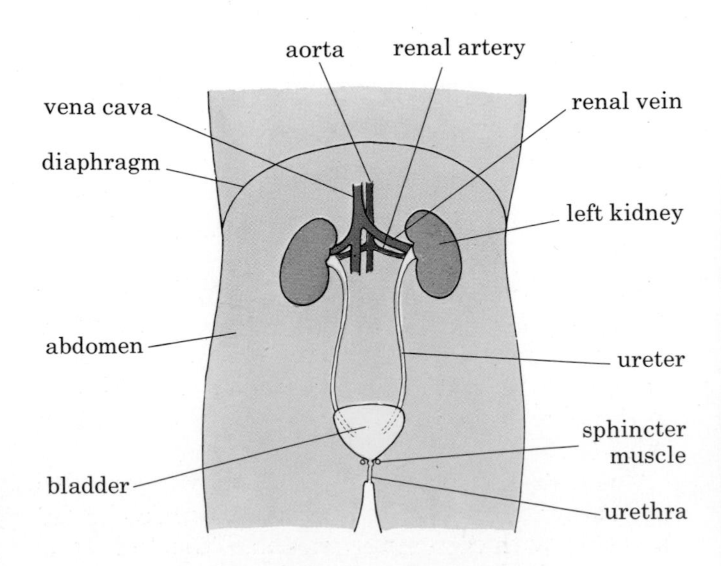

Figure 1 Position of kidneys in the body

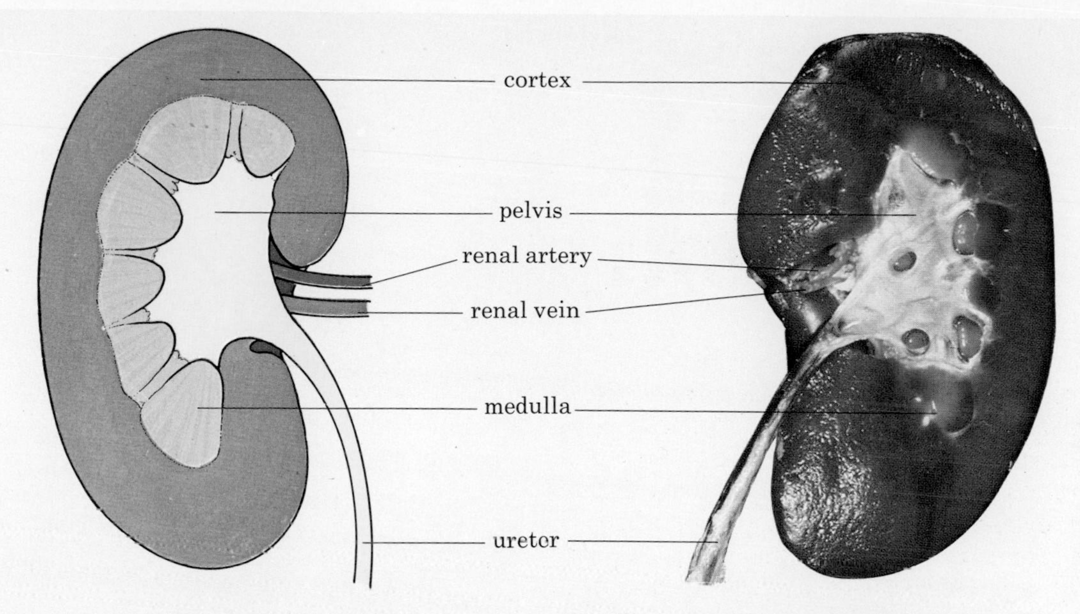

(*a*) Section through kidney to show regions

(*b*) Kidney cut open to show internal structure

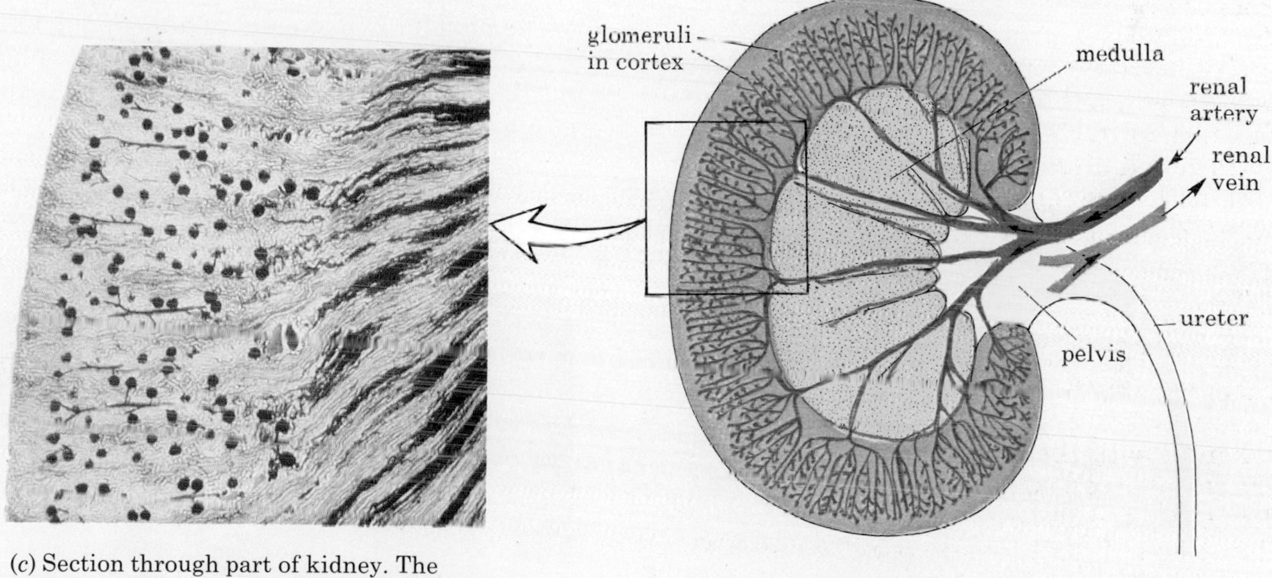

(*c*) Section through part of kidney. The blood vessels have been injected so that the glomeruli show up clearly (×8).

(*d*) Diagram of section through kidney to show distribution of glomeruli

Figure 2 Structure of the kidney

genated blood to them. The **renal vein** takes deoxygenated blood away from the kidneys to the vena cava (see Figure 8 on page 114). A tube, called the **ureter**, runs from each kidney to the bladder in the lower part of the abdomen (Figure 2).

The kidney tissue consists of many capillaries and tiny tubes, called **renal tubules**, held together with connective tissue. If the kidney is cut down its length (sectioned), it is seen to have a darker, outer region called the **cortex** and a lighter, inner zone, the **medulla**. Where the ureter joins the kidney there is a space called the pelvis (Figures 2*a* and *b*).

The renal artery divides up into a great many arterioles and capillaries, mostly in the cortex. Each arteriole leads to a **glomerulus** (Figure 2*d*). This is a capillary repeatedly divided and coiled, making a knot of vessels (Figure 2*c*). Each glomerulus is almost entirely surrounded

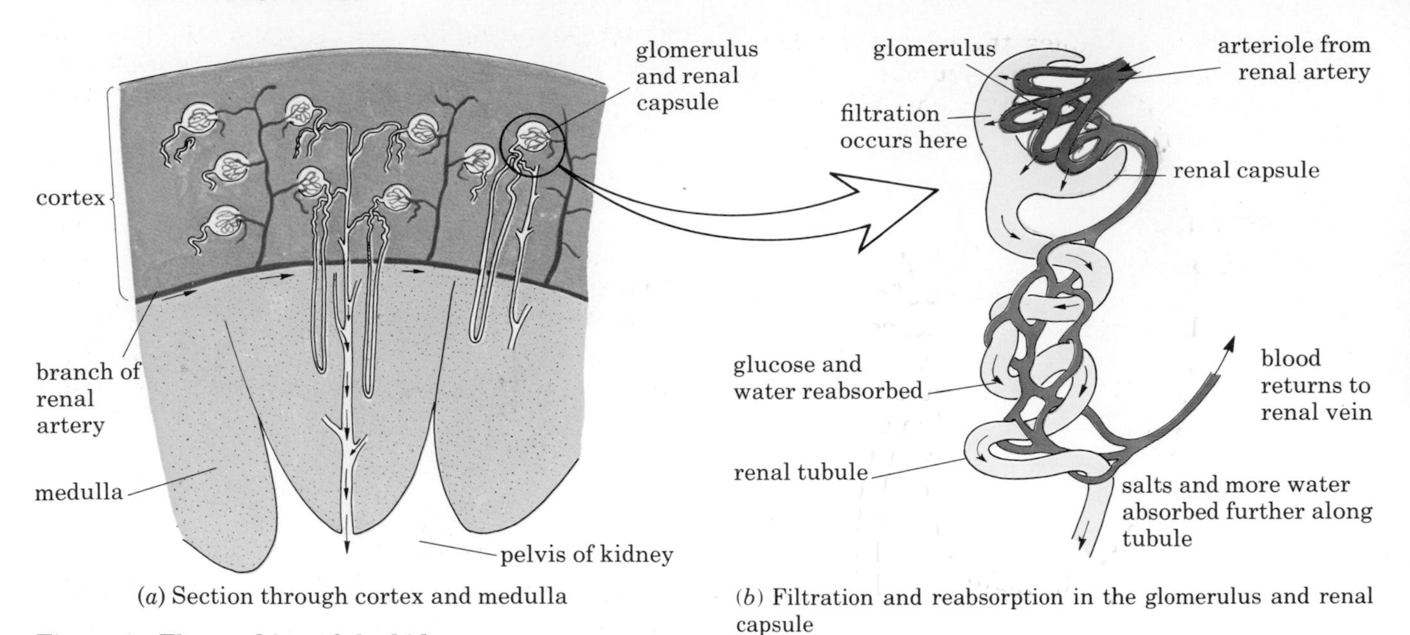

(*a*) Section through cortex and medulla

(*b*) Filtration and reabsorption in the glomerulus and renal capsule

Figure 3 The working of the kidney

by a cup-shaped organ called a **renal capsule**, which leads to a coiled renal tubule. This tubule, after a series of coils and loops, joins other tubules and passes through the medulla to open into the pelvis (Figure 3*a*). There are thousands of glomeruli in the kidney cortex and the total surface area of their capillaries is very great.

Function of the kidneys

The blood pressure in a glomerulus causes blood fluid to leak through the capillary walls. The red blood cells and the plasma proteins are too big to pass out of the capillary, so the fluid that does filter through is plasma without the proteins (see page 112). The fluid thus consists mainly of water with dissolved salts, glucose, urea and uric acid. This filtrate from the glomerulus collects in the renal capsule and trickles down the renal tubule (Figure 3*b*). As it does so, the capillaries which surround the tubule absorb back into the blood those substances which the body needs. First, all the glucose and amino acids are reabsorbed with some of the water. Then some of the salts are taken back to keep the correct concentration in the blood. Salts in excess of these needs, plus urea and uric acid are left to pass on down the kidney tubule. In this way, nitrogenous waste products, excess salts and water continue down the renal tubules and into the pelvis of the kidney. From here the fluid, now called **urine**, passes down the ureter to the bladder.

The bladder can expand to hold about 400 cm^3 urine. The urine cannot escape from the bladder because a band of circular muscle, called a sphincter, is contracted, so shutting off the exit. When this sphincter muscle relaxes, the muscular walls of the bladder expel the urine through the **urethra**. Adults can control this sphincter muscle and relax it only when they

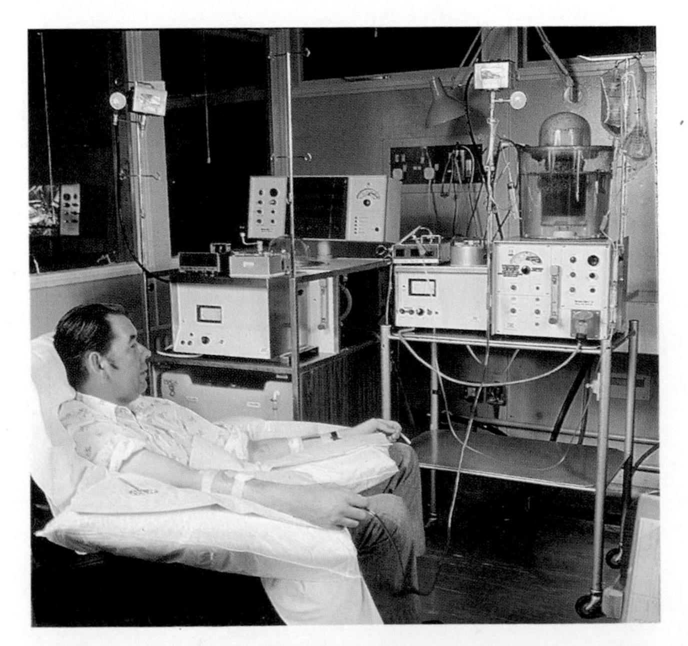

Figure 4 Kidney machine. In cases of kidney failure, the patient's blood is passed through a coil of cellophane tubing, immersed in a container of warm salt solution (top right). The urea and excess salts diffuse out of the blood through the tubing and into the water. The blood is then returned to the patient. This is done about twice a week.

want to urinate. In babies, the sphincter relaxes by a reflex action (page 166), set off by pressure in the bladder. At 3 years old, most children can control the sphincter voluntarily.

Osmo-regulation

As well as excreting unwanted substances, the kidneys also regulate the amount of water in the blood. If the blood is too dilute (i.e. contains too much water), less water is reabsorbed from the renal tubules, leaving more to enter the bladder. If the blood is too concentrated, more water is absorbed back into the blood from the kidney tubules, so that less passes to the bladder. This regulatory process helps to keep the blood at a steady concentration and is called **osmo-regulation** because it regulates the osmotic strength (see page 19) of the blood. Osmo-regulation is one example of the process of **homeostasis** which is described below.

QUESTIONS

3 How does the composition of urine differ from the composition of blood plasma? (See page 112 for the composition of blood plasma.)

4 Where, in the urinary system, do the following take place (answer as precisely as possible): filtration, reabsorption, storage of urine, transport of urine, osmo-regulation?

5 In hot weather when you sweat a great deal, you urinate less often and the urine is a dark colour. In cold weather when you sweat little, urination occurs more often and the urine is pale in colour. Use your knowledge of kidney function to explain these observations.

6 Trace the path taken by a molecule of urea from the time it is produced in the liver, to the time it leaves the body in the urine (see also page 114).

HOMEOSTASIS

Homeostasis means 'staying the same'. It refers to the fact that the composition of the tissue fluid (page 116) in the body is kept very steady. Its concentration, acidity and temperature are being adjusted all the time to prevent any big changes.

On page 12 it was explained that in living cells, all the chemical reactions are controlled by enzymes. The enzymes are very sensitive to the conditions in which they work. A slight fall

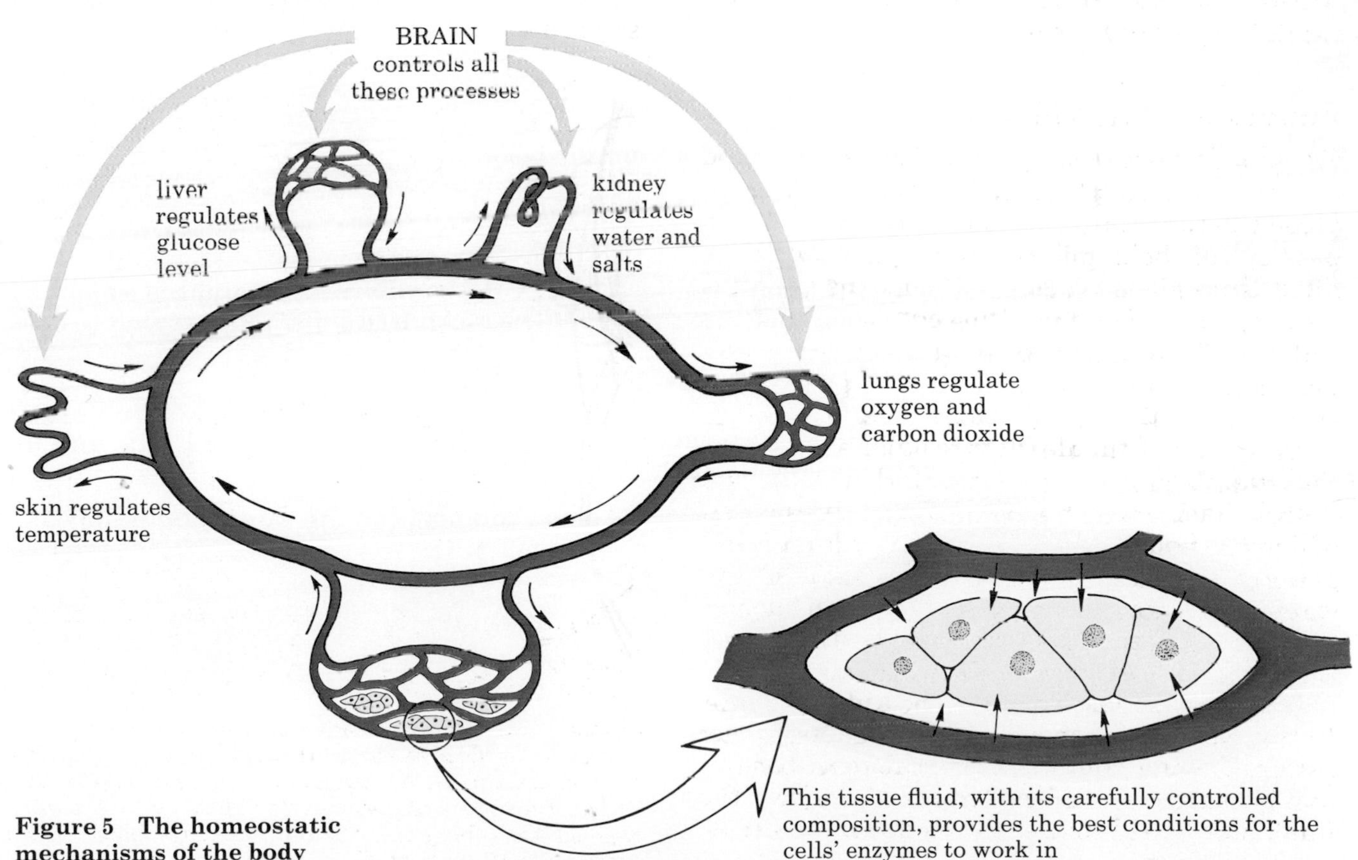

Figure 5 The homeostatic mechanisms of the body

in temperature or a rise in acidity may slow down or stop an enzyme from working and thus prevent an important reaction from taking place in the cell.

The cell membrane controls the substances which enter and leave the cell, but it is the tissue fluid which supplies or removes these substances, and it is, therefore, important to keep the composition of the tissue fluid as steady as possible. If the tissue fluid were too concentrated, it would withdraw water from the cells by osmosis (page 19). If the tissue fluid were too weak, the cells would take up too much water from it by osmosis and the tissues would become waterlogged and swollen.

Many systems in the body contribute to homeostasis (Figure 5). The obvious example is the kidney, which removes substances that might poison the enzymes and controls the level of salts, water and acids in the blood. The composition of the blood affects the tissue fluid (page 116) which, in turn, affects the cells.

Another example of a homeostatic organ is the liver, which regulates the level of glucose in the blood (page 108). The liver stores any excess glucose as glycogen, or turns glycogen back into glucose if the concentration in the blood gets too low. The brain cells are very sensitive to the glucose concentration in the blood and if the level drops too far, they stop working properly, and the person becomes unconscious and will die unless glucose is injected into his blood system. This shows how important homeostasis is to the body.

The lungs (page 122) play a part in homeostasis by keeping the concentrations of oxygen and carbon dioxide in the blood at the best level for the cell's chemical reactions, especially respiration.

The next chapter describes the way in which the skin regulates the temperature of the blood. If the cells were to get too cold, the chemical reactions would become too slow to maintain life. If they became too hot, the enzymes would be destroyed (page 12).

The brain has overall control of the homeostatic processes in the body. It checks the composition of the blood flowing through it and if it is too warm, too cold, too concentrated or has too little glucose, nerve impulses or hormones are sent to the organs concerned, causing them to make the necessary adjustments.

QUESTION

7 Where will the brain send nerve impulses or hormones if the blood flowing through it (a) has too much water, (b) contains too little glucose, (c) is too warm, (d) has too much carbon dioxide?

BASIC FACTS

- **Excretion is getting rid of unwanted substances from the body.**
- **The lungs excrete carbon dioxide.**
- **The kidneys excrete urea, unwanted salts and excess water.**
- **Part of the blood plasma entering the kidney is filtered out by the capillaries. Substances which the body needs, like glucose, are absorbed back into the blood. The unwanted substances are left to pass down the ureters into the bladder.**
- **The bladder stores urine, which is discharged at intervals.**
- **The kidneys help to keep the blood at a steady concentration by excreting excess salts and by adjusting the amount of water (osmo-regulation).**
- **The kidneys, lungs, liver and skin all help to keep the blood composition the same (homeostasis).**

17 The skin and temperature control

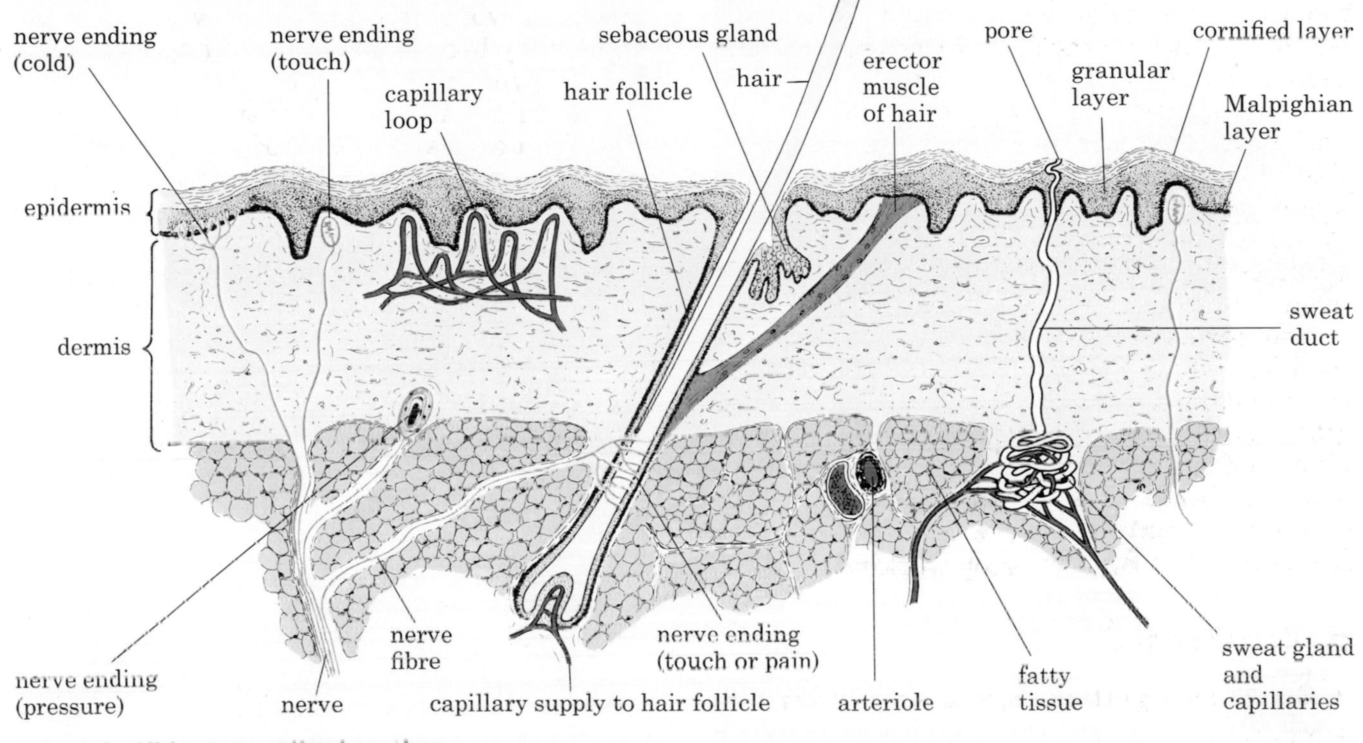

Figure 1 Skin: generalized section

The skin forms a continuous layer over the entire body. In the words of a once popular song it 'keeps the outside out and the inside in', which is true enough, but skin does more than that.

FUNCTIONS OF THE SKIN

Protection

(a) The brown or black pigment in the skin absorbs the harmful ultra-violet rays from sunlight. If the skin contains only a little pigment, it may be damaged by these rays. This happens when a fair-skinned person becomes sunburned.

(b) The layer of dead cells at the surface of the skin, stops harmful bacteria getting into the living tissues beneath.

(c) The skin greatly reduces the evaporation of water from the body and so helps to maintain the composition of the body fluids.

Sensitivity

Scattered through the skin are a large number of tiny sense organs which give rise to sensations of touch, pressure, heat, cold and pain. These make us aware of changes in our surroundings and enable us to take action to avoid damage, to recognize objects by touch and to manipulate objects with our hands.

Temperature regulation

The way in which the skin helps to keep the body temperature constant is described on pages 136–7.

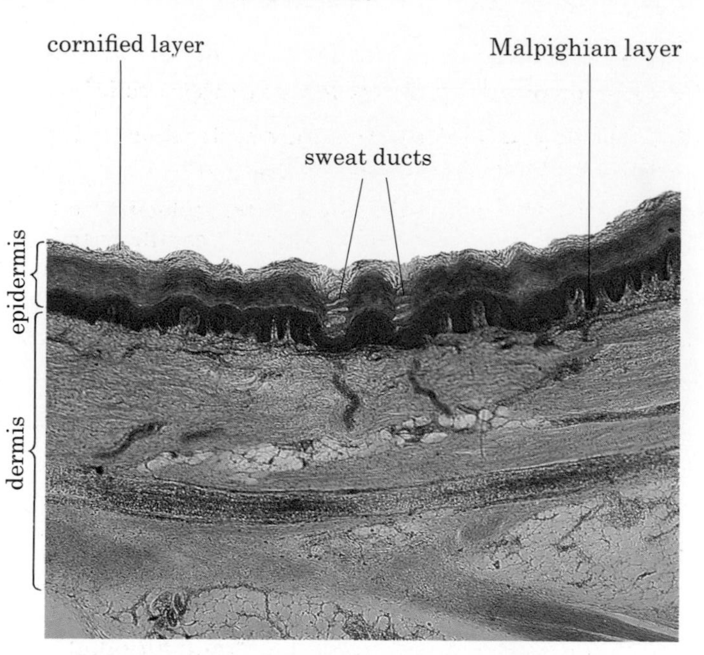

Figure 2 Section through non-hairy skin (×25). The epidermis is much thicker than that in Figure 3.

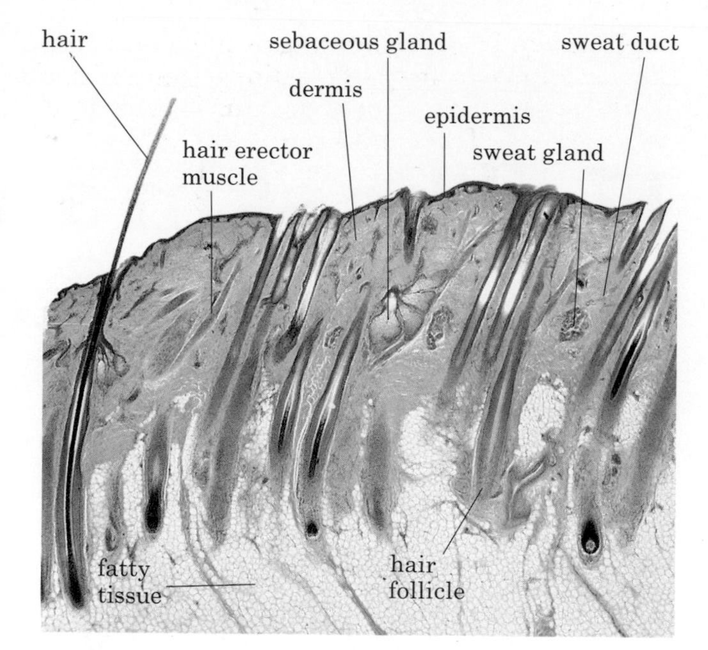

Figure 3 Section through hairy skin (×25). About 15 follicles are visible in this section.

QUESTION

1 To what dangers is the body exposed if (a) a small area of skin is damaged, (b) a large area of skin is damaged?

STRUCTURE OF THE SKIN

There are two main layers in the skin, an outer **epidermis** and an inner **dermis**. The thickness of these two layers depends on which part of the body they are covering. The skin on the palms of the hands and soles of the feet has a very thick epidermis and no hairs (Figure 2). Over the rest of the body, the epidermis is thinner and has hairs (Figure 3). The diagram of a section through the skin (Figure 1) is a generalized one. It shows all the structures that are in skin as a whole, even though some may be absent in a particular area.

Epidermis

(a) Malpighian layer. This is the innermost layer of cells in the epidermis. These cells contain the pigment that gives the skin its colour and helps to absorb ultra-violet light from the sun. The cells of the Malpighian layer keep dividing and producing new cells which are pushed towards the outside of the skin, forming the granular layer.

(b) Granular layer. The cells produced by the Malpighian layer move through the granular layer and then die, so forming the cornified layer.

(c) The cornified layer consists of dead cells. This is the outermost layer of the epidermis which helps to cut down evaporation and keep bacteria out. The cells are worn away all the time and replaced from below by the Malpighian and granular layers (Figure 4).

Dermis

The dermis is a layer of connective tissue containing capillaries, sensory nerve endings, lymphatics, sweat glands and hair follicles.

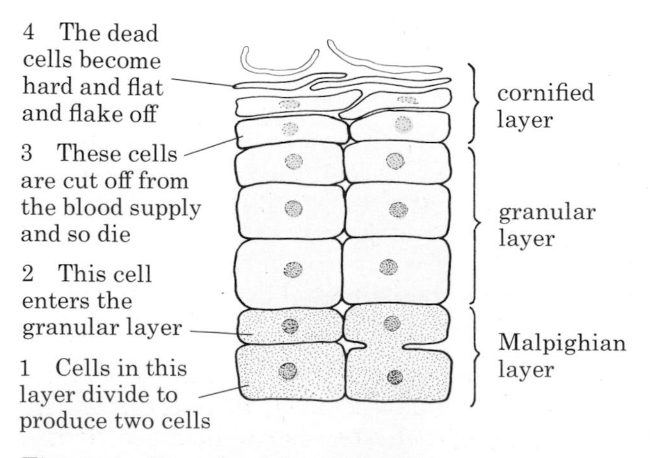

Figure 4 Growth of the epidermis

Capillaries. These bring oxygen and food to the skin and remove its carbon dioxide and nitrogenous waste. They supply the hair follicles and sweat glands. The capillary loops close to the surface, i.e. just beneath the epidermis, play an important part in regulating heat loss from the body (see page 136).

Sweat glands. A sweat gland is a coiled tube deep in the dermis. When the body temperature is too high, the sweat gland takes up water from the capillaries around it. The water collects in the gland, travels up the **sweat duct**, comes out of a **pore** in the epidermis and on to the skin surface. When the sweat evaporates, it takes heat from the body and so cools it down. Unless the sweat evaporates, it has no cooling effect.

Sweat is mainly water but there are some dissolved salts and urea in it. Some people regard this as a form of excretion (page 128), but sweating occurs in response to a rise in temperature and not because there is too much water or salt in the body. For this reason, the skin is not included in the list of excretory organs on page 128.

Hair follicles. A hair follicle is a deep pit lined with granular and Malpighian cells. The Malpighian cells keep dividing and adding cells to the base of the hair, making it grow. A hair is a lot of cornified cells formed into a tube. The hair follicle has nerve endings which respond when the hair is touched, or give a sensation of pain if the hair is pulled.

In furry mammals, the hairs trap a layer of air close to the body. Air is a bad conductor of heat and so this layer of air insulates the body against heat loss. When the **hair erector muscle** contracts, it pulls the hair more upright. In mammals, this makes the fur stand up more and so provides a thicker layer of insulating air in cold weather. Most of our body is covered with short hairs, but they trap very little air and when the hair erector muscles contract, they produce only 'goose pimples'.

Sebaceous glands. The sebaceous glands open into the top of the hair follicles and produce an oily substance that keeps the epidermis waterproof and stops it drying out.

Sensory nerve endings. These are described more fully on page 154.

Fat layer. The fat stored beneath the skin not only provides a store of food, it also forms an insulating layer and reduces the heat lost from the body.

QUESTIONS

2 Is a hair made by the epidermis or the dermis?

3 Why do you think dead cells are better than living cells in reducing water loss from the skin?

4 Describe the path taken by a water molecule that enters the skin in the blood plasma of a capillary and ends up in sweat on the surface of the skin.

'WARM-BLOODED' ANIMALS

Birds and mammals, whose body temperature is kept at a level often above that of their surroundings, are called 'warm-blooded'. A fish in a tropical sea or a lizard basking in the sun may also have warm blood, but if they move to a cooler place their body temperature will fall. Such animals are often called 'cold-blooded'. Figure 5 shows that the cat's body temperature stays the same even though the outside temperature rises from 10 °C to 40 °C. The 'cold-blooded' lizard's temperature changes as the external temperature rises or falls. The spiny ant-eater controls its body temperature better than the lizard but not so well as the cat.

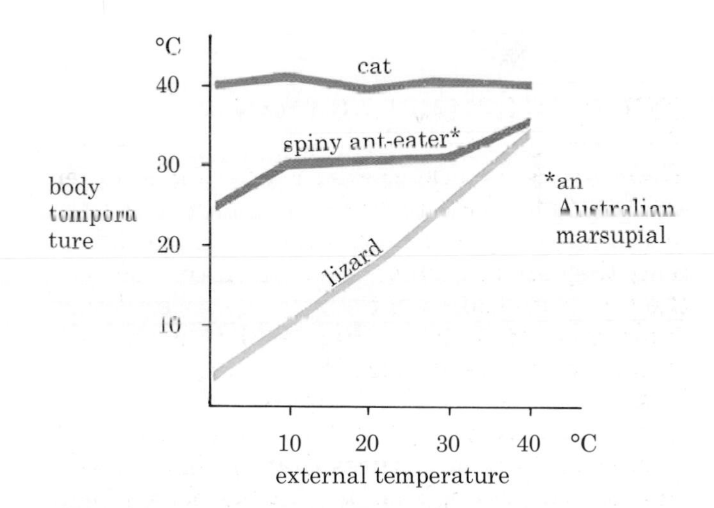

Figure 5 Changes in body temperature when the external temperature changes (after C. J. Martin, 1930).

Chemical reactions in cells will take place faster if the temperature is high. So a warm animal can be more active than a cold animal. A tropical lizard in the daytime may have a body temperature of about 30 °C and be able to dart about catching insects at great speed. As soon as the sun goes down and the air temperature drops, the lizard's movements will slow down.

Birds and mammals keep their temperature at a constant level so that even if the temperature of their surroundings goes down, they can carry out their activities at a fast rate. The essential difference between 'warm-blooded' and 'cold-blooded' animals is that the 'warm-blooded' ones can control their body temperature and keep it constant, while the 'cold-blooded' ones cannot. It is better to use the terms **constant temperature** and **variable temperature** animals, rather than 'warm-blooded' and 'cold-blooded'. The tropical lizard whose temperature is 30 °C can hardly be called 'cold-blooded'. Keeping the body temperature within narrow limits is an example of homeostasis (page 131).

QUESTION

5 Why do you think that 'warm-blooded' animals have fur or feathers while 'cold-blooded' ones do not?

HEAT BALANCE

The body gains or loses heat in the following ways:

Heat gain

(a) Internal. A great many of the chemical reactions in the cells release heat. The chief heat-producers are the abdominal organs, the brain and active muscle. Any increase in muscular activity of the body will result in more heat being produced.

(b) External. Direct heat from the sun will be absorbed by the body. Hot food and drink also add heat to the body.

Heat loss

Heat is lost to the air from the exposed surfaces of the body by conduction, convection and radiation. Evaporation from the skin and lungs takes place all the time and is a cause of heat loss. The cold air breathed into the lungs and cold food or drink taken into the stomach all absorb heat from the body.

To a large extent, the heat lost from the body is balanced by the heat absorbed or produced. However, changes in the temperature of the surroundings or in the rate of activity by the animal may upset this balance. In man, any change in the temperature balance is regulated mainly by changes in the skin.

QUESTIONS

6 What conscious action does man take to reduce the heat lost from his body?

7 Discuss whether drinking (a) a large whisky and (b) a cup of hot tea is likely to 'warm you up' if you are feeling cold.

TEMPERATURE REGULATION

Skin structure

The structures in the skin which play a part in temperature control are the small blood vessels, the sweat glands and, in furry mammals, the hairs and hair muscles.

The arterioles and capillaries near the surface of the skin can increase or decrease in width and so increase or reduce the amount of blood flowing through them. An increase in width of the vessel is called **vaso-dilation**; a reduction in width is **vaso-constriction**.

The sweat glands are not active when the loss and gain of heat are balanced, but as soon as the body temperature starts to rise, the sweat glands produce sweat as described above.

Over-heating

If the body gains or produces heat faster than it is losing it, the following processes occur:

(a) Vaso-dilation. The widening of the blood vessels in the dermis allows more warm blood to flow near the surface of the skin and so lose more heat (Figure 6*a*).

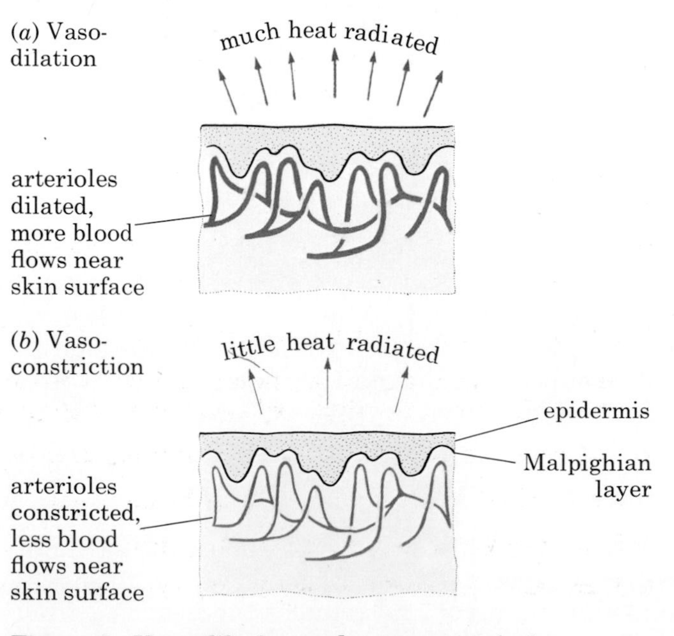

Figure 6 Vaso-dilation and vaso-constriction

(b) Sweating. The sweat glands pour sweat on to the skin surface. When this layer of liquid evaporates it takes heat from the body and so cools it down (Figure 7).

Figure 7 The athlete Brendan Foster in a sweat. During the race, the sweat evaporates from the skin and helps to cool the runner. After the race the continued evaporation of sweat may over-cool the body unless it is towelled off.

Over-cooling

If the body begins to lose heat faster than it can produce it, the following changes occur:

(a) Sweat production stops, so reducing the heat lost by evaporation.

(b) Vaso-constriction of the blood vessels in the skin reduces the amount of warm blood flowing near the surface (Figure 6*b*).

(c) Shivering. Uncontrollable bursts of rapid muscular contraction in the limbs release heat as a result of the chemical changes in the muscles.

As a result of these processes, the body temperature of an adult does not usually vary by more than 1 °C either side of 37 °C even when the external temperature changes. 36.7 °C is the average temperature recorded by a thermometer under the tongue and is called the 'core' temperature of the body. The temperature of the skin can change through very wide limits without affecting the core temperature. It is the sensory nerve endings in your dermis which make you *feel* hot or cold. You cannot consciously detect the small changes in your core temperature.

QUESTIONS

8 Draw up a balance sheet to show all the possible ways the human body can gain or lose heat. Make two columns, with 'Gains' on the left and 'Losses' on the right.

9 (a) Which structures in the skin of a furry mammal help to reduce heat loss?
(b) What changes take place in the skin of man to reduce heat loss?

10 (a) What do you think happens in your skin when you blush?
(b) Why may your face also feel hot when you blush?

11 Why do you think people look pale when they are feeling cold?

12 If your body temperature hardly changes at all, why do you sometimes *feel* hot or *feel* cold?

13 Sweating only cools you down if the sweat can evaporate.
(a) In what conditions might the sweat be unable to evaporate from your skin?
(b) What conditions might speed up the evaporation of sweat and so make you feel very cold?

BASIC FACTS

- **Skin consists of an outer layer of epidermis and an inner dermis.**
- **The epidermis is growing all the time and has an outer layer of dead cells.**
- **The dermis contains the sweat glands, hair follicles, sense organs and capillaries.**
- **Skin (a) protects the body from bacteria and drying out, (b) contains sense organs which give us the sense of touch, warmth, cold and pain, and (c) controls the body temperature.**
- **Chemical activity in the body and muscular contractions produce heat.**
- **Heat is lost to the surroundings by conduction, convection, radiation and evaporation.**
- **If the body temperature rises too much, the skin cools it down by sweating and vaso-dilation.**
- **If the body loses too much heat, vaso-constriction and shivering help to keep it warm.**

18 Sexual reproduction

Sexual reproduction in both animals and plants involves the production of male and female reproductive cells called **gametes**. A process such as mating or pollination brings the male and female gametes together. The two gametes **fuse**, that is, they join together to make a single cell called a **zygote**.

Fertilization is the name given to the fusion of two gametes to form a zygote (Figure 1 or Figure 15 on page 77).

In man and other animals, the male gamete is the **sperm** and the female gamete is the egg or **ovum** (plural= ova) (Figure 5). The gametes are produced in the male and female reproductive organs.

REPRODUCTIVE ORGANS

Female

The female reproductive organs (Figures 2*a* and *b*) are the **ovaries**, two whitish oval bodies, 3–4 cm long. They lie in the lower half of the abdomen, one on each side of the **uterus**. Close to each ovary is the expanded, funnel-shaped opening of the **oviduct**, the tube down which the ova pass when released from the ovary.

The oviducts are narrow tubes that open into a wider tube, the uterus or womb, lower down in the abdomen. When there is no embryo developing in it, the uterus is only about 80 mm long. It leads to the outside through a muscular tube, the **vagina**. The **cervix** is a ring of muscle closing the lower end of the uterus where it joins the vagina. The urethra, from the bladder, opens into the **vulva** just in front of the vagina.

Male

The male reproductive organs (Figures 3*a* and *b*) are the **testes** (singular=testis) which lie outside the abdominal cavity, in a special sac called the **scrotum**. In this position they are kept at a temperature slightly below the rest of the body. This is the best temperature for sperm production. The testes consist of a mass of sperm-producing tubes (Figures 4*a* and *b*). These tubes join to form ducts leading to the **epididymis**, a coiled tube about six metres long on the outside of each testis. The epididymis, in turn, leads into a muscular **sperm duct**. The two sperm ducts, one from each testis, open into the top of the urethra just after it leaves the bladder. A short, coiled tube called the **seminal vesicle** branches from each sperm duct just before it enters the **prostate gland**, which surrounds the urethra at this point. Either urine

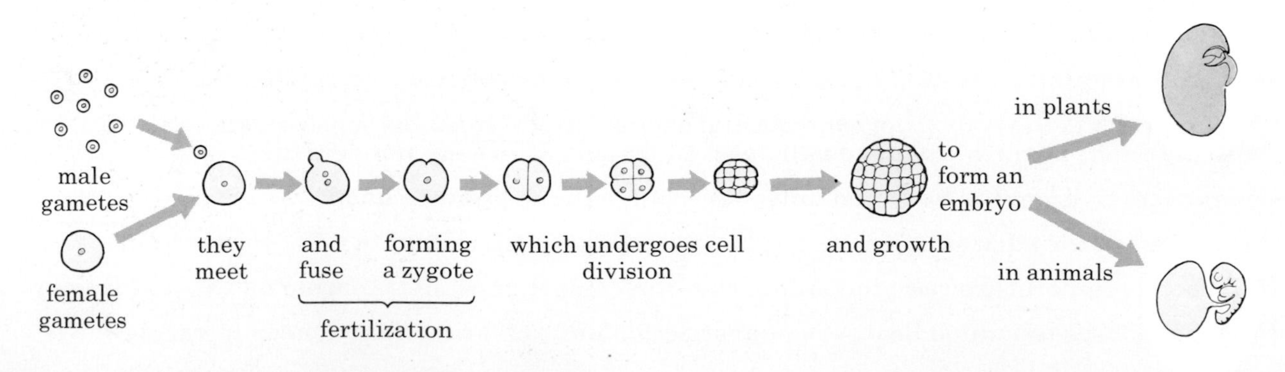

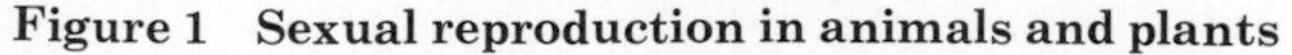

Figure 1 Sexual reproduction in animals and plants

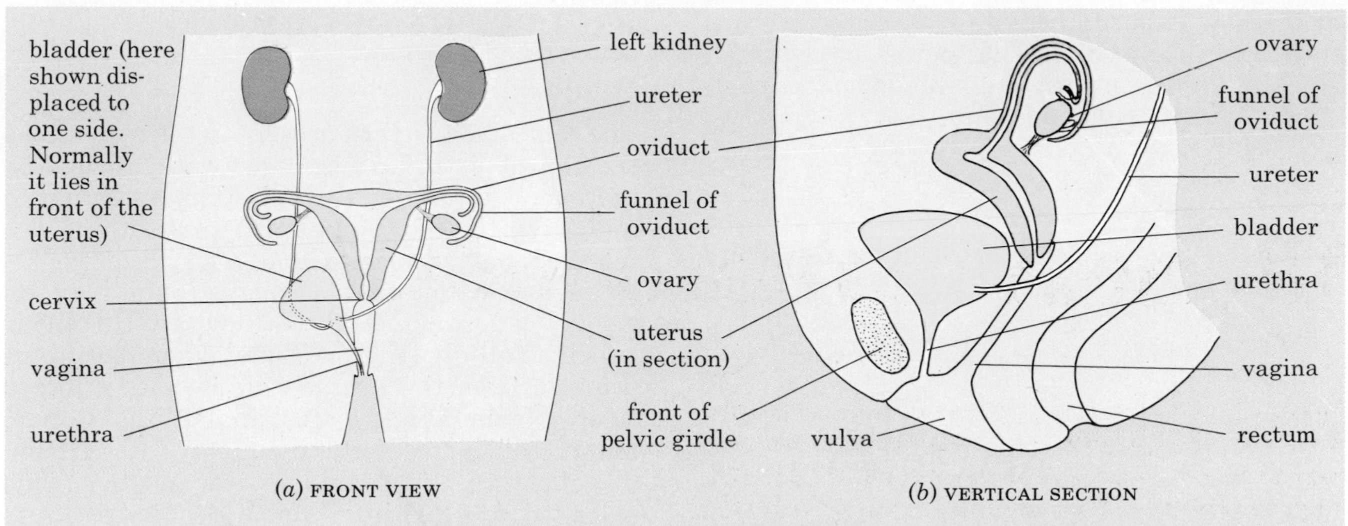

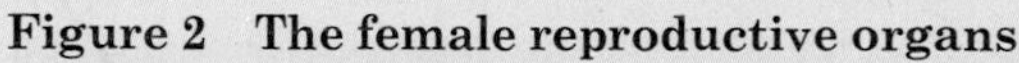
Figure 2 The female reproductive organs

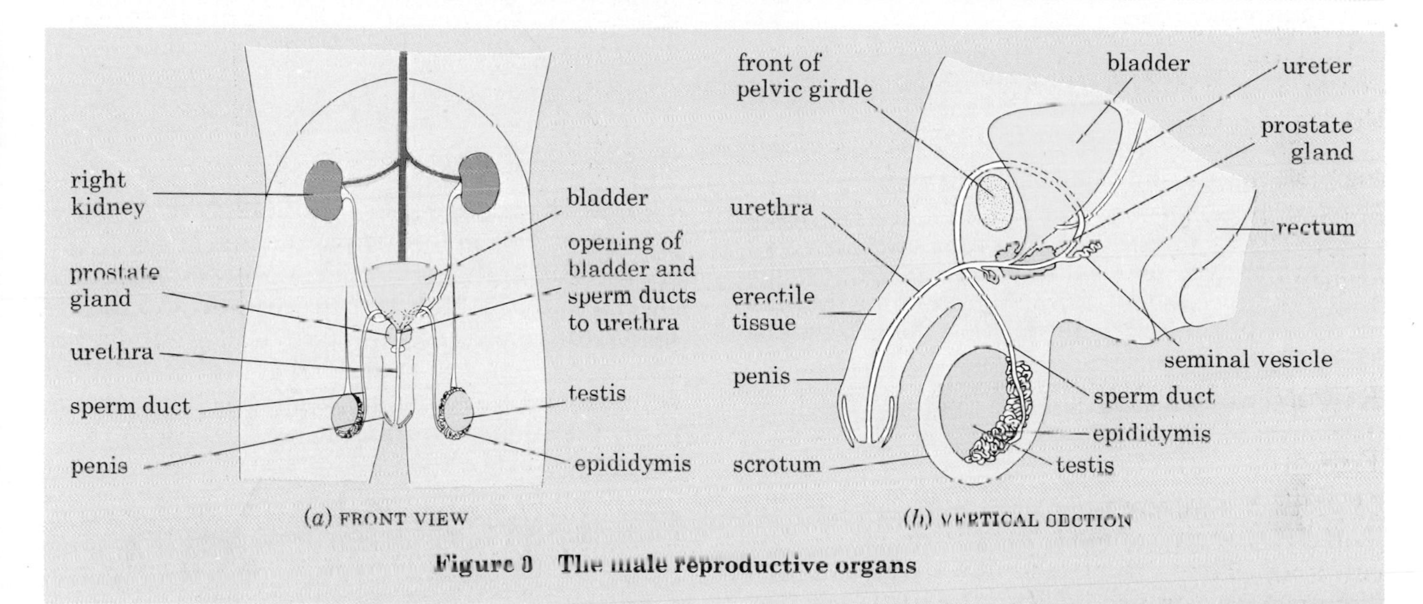

Figure 3 The male reproductive organs

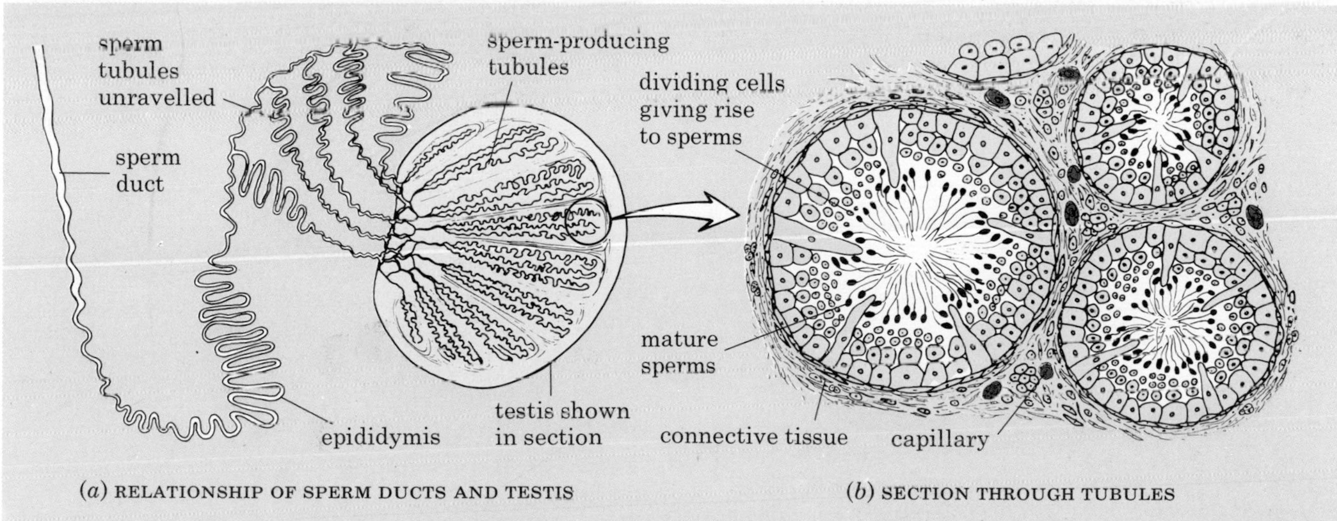

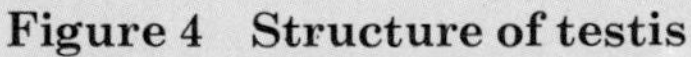
Figure 4 Structure of testis

or sperms can pass down the urethra, which runs through the **penis**. The penis consists of connective tissue with many blood spaces in it. This is called **erectile tissue**.

QUESTIONS

1 How do sperms differ from ova in their structure (see Figure 5)?

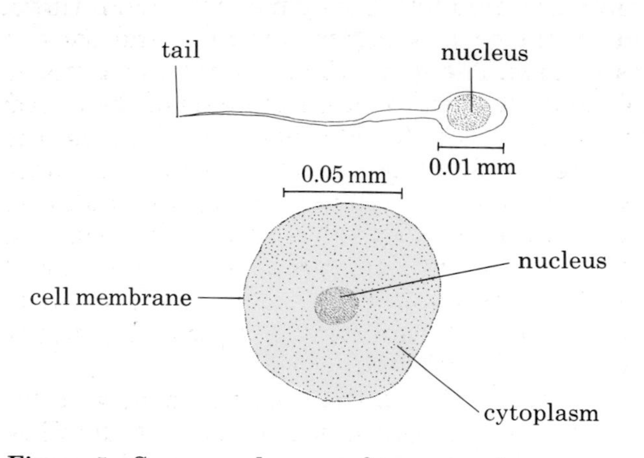

Figure 5 Sperm and ovum of a mammal

2 List the structures, in the correct order, through which the sperms must pass from the time they are produced in the testis, to the time they leave the urethra.

3 What structures are shown in Figure 3*b* which are not shown in Figure 3*a*?

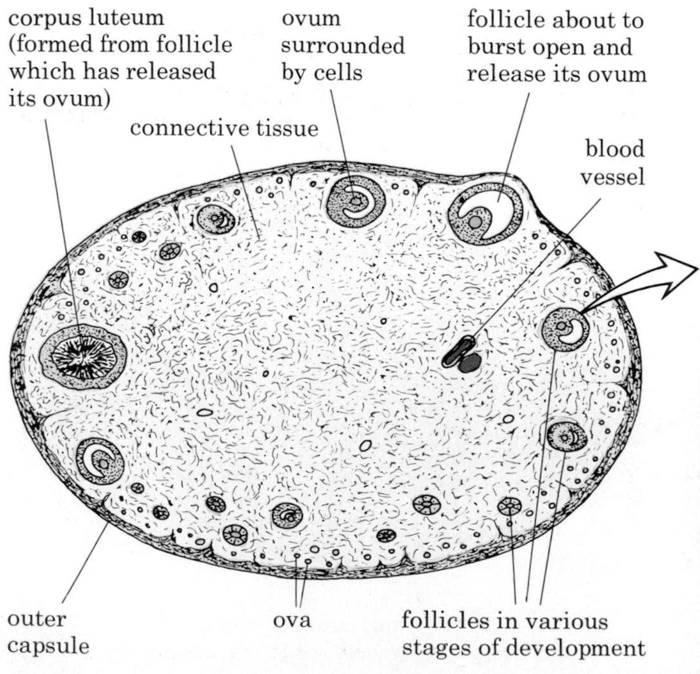

Figure 6 Section through an ovary

PRODUCTION OF GAMETES

Ovulation

The egg cells (ova) are present in the ovary from the time of birth. No more are formed during the lifetime, but between the ages of about 10 and 14 (puberty), the egg cells start to ripen and are released, one at a time about every four weeks. As each ovum matures, the cells round it divide rapidly and produce a fluid-filled sac. This sac is called a **follicle** (Figure 7) and when mature, it projects from the surface of the ovary like a small blister (Figure 6). Finally, the follicle bursts and releases the ovum with its coating of cells into the funnel of the oviduct. This is called **ovulation**. From here, the ovum is wafted down the oviduct by the action of cilia (page 6) in the lining of the tube. If it meets sperm cells in the oviduct, it may be fertilized by one of them.

Sperm production

The testes produce sperm cells (Figure 8) which collect in the epididymis. During mating, the epididymis and sperm ducts contract and force sperms out through the urethra. The prostate gland and seminal vesicle add fluid to the sperm. This fluid plus the sperms it contains is called **semen**, and the ejection of sperms through the penis is called **ejaculation**.

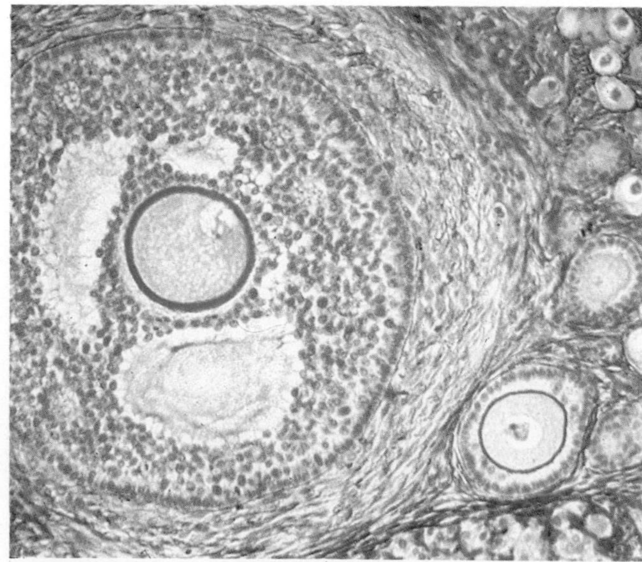

Figure 7 Mature follicle (×150). The ovum is stained pink with its nucleus visible at the top right. It is enclosed in an outer membrane stained deep blue, and surrounded by follicle cells (purple) and fluid (pale blue). On the right are several immature follicles.

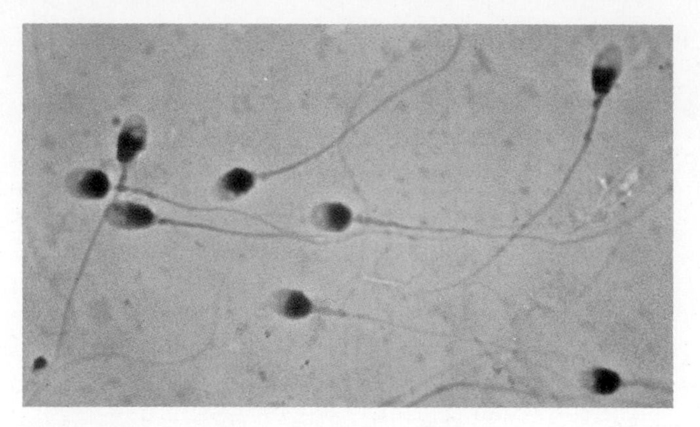

Figure 8 Human sperm cells (×900)

QUESTION

4 If a woman starts ovulating at 13 and stops at 50, (a) how many ova will she produce, (b) about how many of these are likely to be fertilized?

MATING AND FERTILIZATION

Mating

As a result of sexual stimulation, the male's penis becomes erect. This is due to blood flowing into the erectile tissue round the urethra. In the female, the lining of the vagina produces mucus which makes it possible for the penis to enter. The sensory stimulus (sensation) produced by mating causes a reflex action (page 166) in the male which results in the ejaculation of semen into the top of the vagina.

Fertilization

The sperms swim through the cervix by wriggling movements of their tails. They continue up the uterus (how they do so is not certain) and enter the oviduct. If they meet an ovum there, one of the sperms may bump into it and stick to its surface. The sperm then enters the cytoplasm of the ovum and the male nucleus of the sperm fuses with the female nucleus. This is the moment of fertilization and is shown in more detail in Figure 9. Although a single ejaculation may contain about five hundred million sperms, only one will fertilize the ovum. The function of the others is not understood but it is likely that a great many do not manage to travel from the vagina to the oviduct.

The released ovum is thought to survive for 8–24 hours; the sperms might be able to fertilize an ovum for up to 3 days. So there is only a short period of 3–4 days each month when fertilization might occur.

QUESTIONS

5 List, in the correct order, the parts of the female reproductive system through which sperms must pass before reaching and fertilizing an ovum.

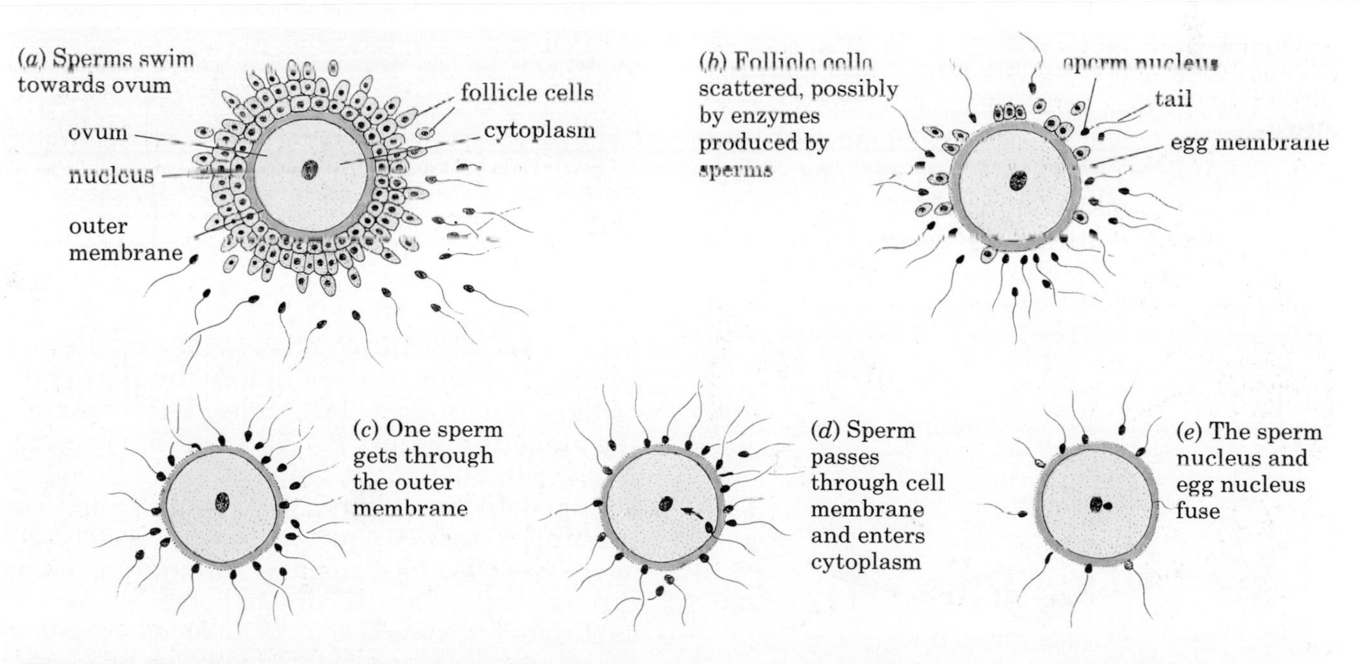

Figure 9 Fertilization of human ovum. (The diagrams show what is thought to happen, but the events are not known for certain.)

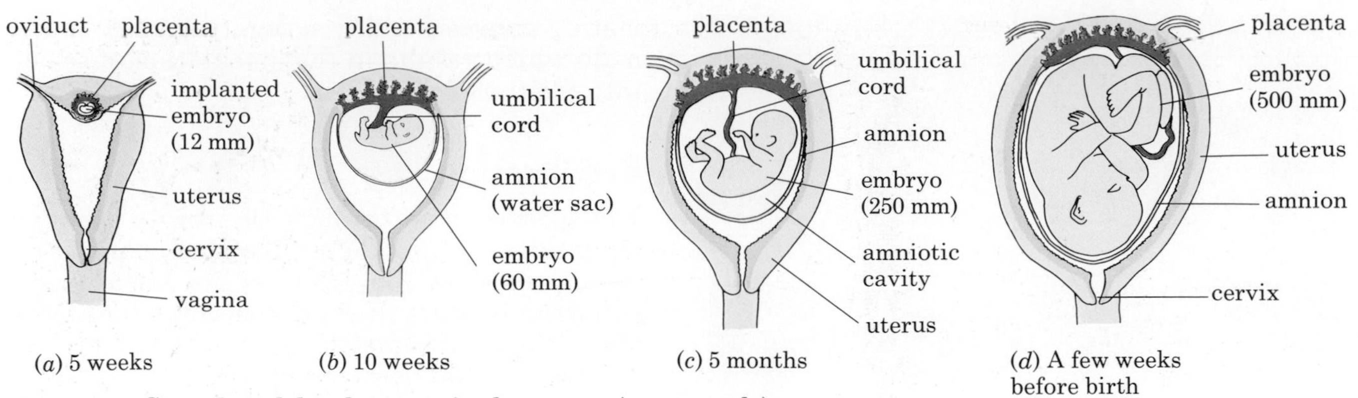

Figure 10 Growth and development in the uterus (not to scale)

6 State exactly what happens at the moment of fertilization.

7 If mating takes place (a) two days before ovulation, (b) two days after ovulation, is fertilization likely to occur? Explain your answers.

PREGNANCY AND DEVELOPMENT

The fertilized ovum first divides into two cells. Each of these divides again so producing four cells. The cells continue to divide in this way to produce a solid ball of cells (Figure 1), an early stage in the development of the **embryo**. This early embryo travels down the oviduct to the uterus. Here it sinks into the lining of the uterus, a process called **implantation** (Figure 10*a*). The embryo continues to grow and produces new cells which form tissues and organs (Figure 11). After 8 weeks, when all the organs are formed, the embryo is called a **fetus**. The blood vessels and heart, which pumps blood round the body of the embryo, are one of the first organ systems to form.

As the embryo grows, the uterus enlarges to contain it. Inside the uterus the embryo becomes enclosed in a fluid-filled sac called the **amnion** or water sac, which protects it from damage and prevents unequal pressures from acting on it (Figures 10*b* and *c*). The oxygen and food needed to keep the embryo alive and growing are obtained from the mother's blood by means of a structure called the placenta.

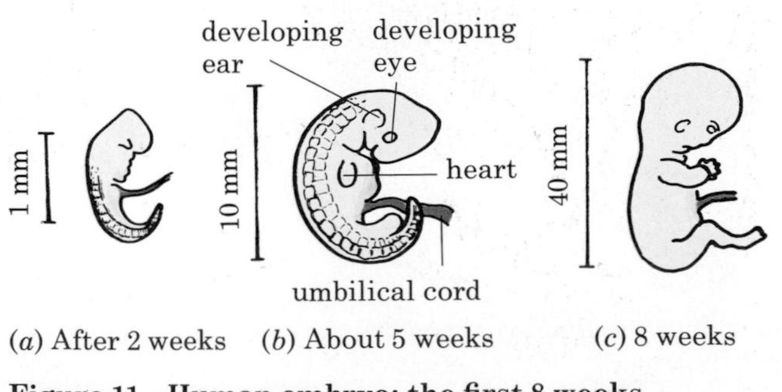

Figure 11 Human embryo: the first 8 weeks

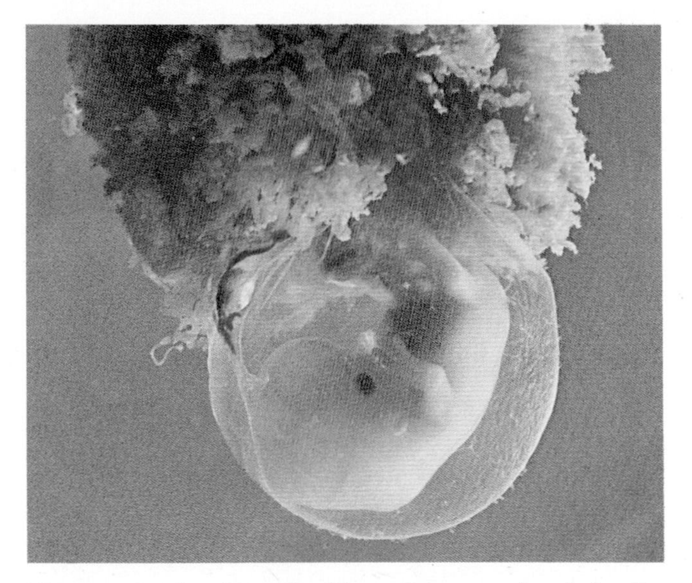

Figure 12 An 8-week-old embryo, enclosed in the amnion and attached to the placenta (×1.8)

Placenta

Soon after the ball of cells reaches the uterus, some of the cells, instead of forming the organs of the embryo, grow into a disc-like structure, the **placenta** (Figure 12). The placenta becomes closely attached to the lining of the uterus and is attached to the embryo by a tube called the **umbilical cord**. After a few weeks, the embryo's heart has developed and is circulating blood through the umbilical cord and placenta as well as through its own tissues. The blood vessels in the placenta are very close to the blood vessels in the uterus so that oxygen, glucose, amino acids and salts can pass from the mother's blood

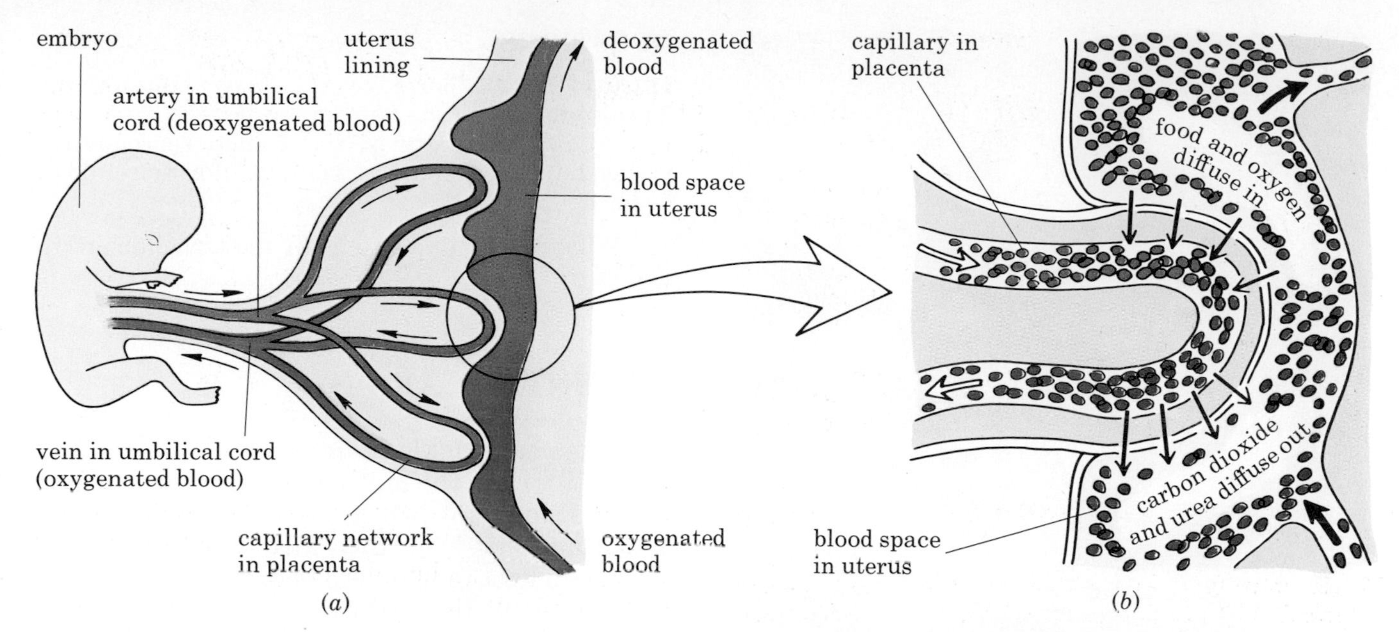

Figure 13 The exchange of substances between the blood of the embryo and the mother

to the embryo's blood (Figure 13*a*). So the blood flowing in the umbilical vein from the placenta carries food and oxygen to be used by the living, growing tissues of the embryo. In a similar way, the carbon dioxide and urea in the embryo's blood escape from the vessels in the placenta and are carried away by the mother's blood in the uterus (Figure 13*b*). In this way the embryo gets rid of its excretory products.

There is no direct communication between the mother's blood system and the embryo's. The exchange of substances takes place across the thin walls of the blood vessels. In this way, the mother's blood pressure cannot damage the delicate vessels of the embryo and it is possible for the placenta to select the substances allowed to pass into the embryo's blood. The placenta can prevent some harmful substances in the mother's blood from reaching the embryo. It cannot prevent all of them, however, as is shown by the reduced birth weight of babies if the mother smokes cigarettes during her pregnancy.

QUESTIONS

8 In a pregnant woman what changes will take place in the composition of the blood as it passes through the uterus?

9 What is the main function of the umbilical cord?

10 An embryo is surrounded with fluid, its lungs are filled with fluid and it cannot breathe. Why doesn't it suffocate?

BIRTH

From fertilization to birth takes about nine months in humans. This is called the **gestation** period. A few weeks before birth, the embryo has come to lie head downwards in the uterus, with its head just above the cervix (Figures 10*d* and 14). When the birth starts, the uterus begins to

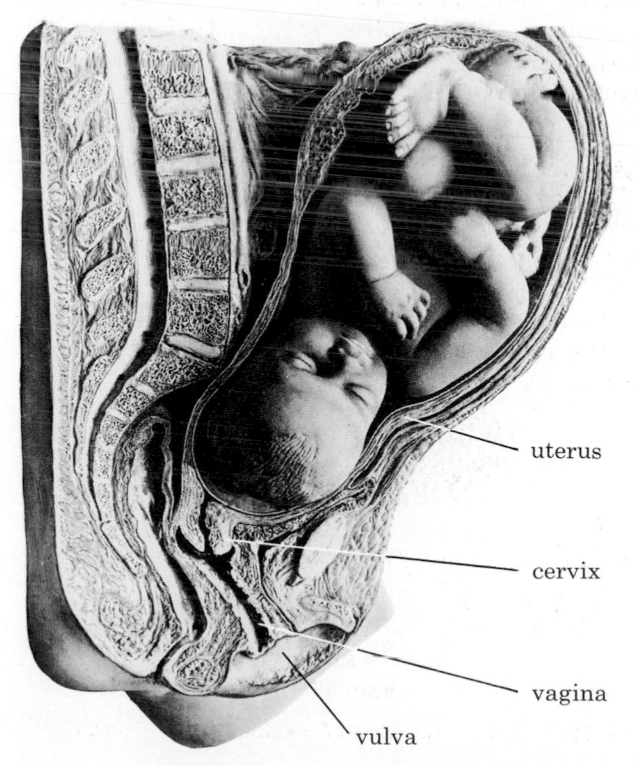

Figure 14 Model of a human fetus just before birth

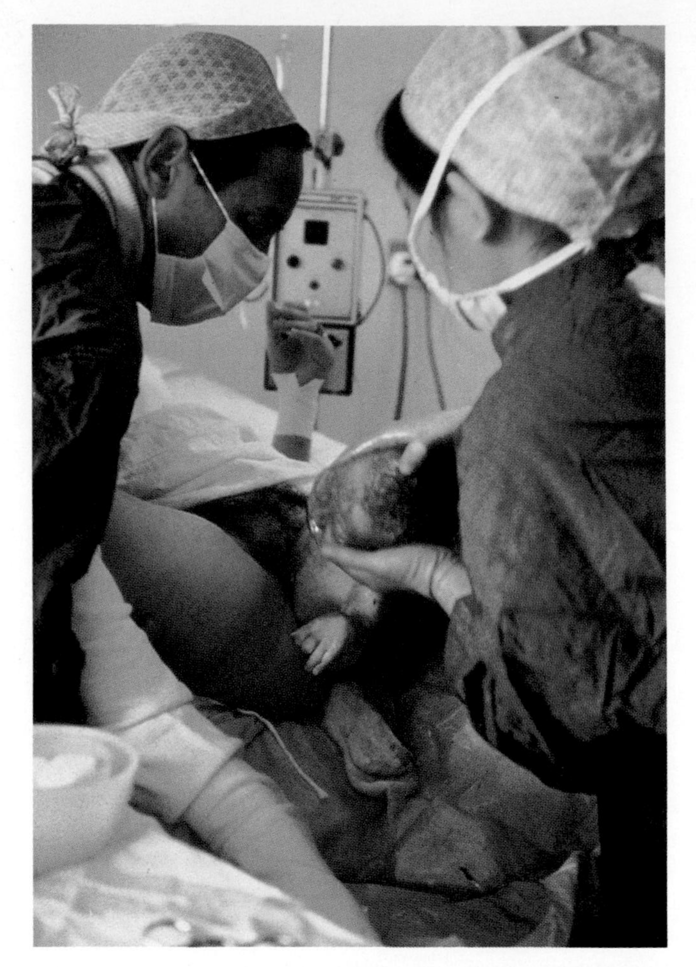

Figure 15 Delivery of a baby. The nurse supports the baby's head while the mother's contractions push out the rest of the body.

contract rhythmically. This is the beginning of what is called 'labour'. These regular rhythmic contractions become stronger and more frequent. The opening of the cervix gradually widens enough to let the baby's head pass through and the contractions of the uterus are assisted by muscular contractions of the abdomen. The water sac breaks at some stage in labour and the fluid escapes through the vagina. Finally, the muscular contractions of the uterus and abdomen push the baby head-first through the widened cervix and vagina (Figure 15). The umbilical cord which still connects the child to the placenta is tied and cut. Later, the placenta breaks away from the uterus and is pushed out separately as the 'after-birth'.

The sudden fall in temperature felt by the newly born baby stimulates it to take its first breath and it usually cries. In a few days, the remains of the umbilical cord attached to the baby's abdomen shrivel and fall away, leaving a scar in the abdominal wall, called the navel.

QUESTIONS

11 (a) Which parts of the body contain the muscles that push the baby out at birth?

(b) Which parts of the female reproductive system have to increase in width to allow the child to be born?

12 What do you understand by the term 'labour' in childbirth?

PARENTAL CARE

All mammals suckle their young (i.e. the young suck milk from the mother) and protect them in various ways until they are old enough to move about and find their own food. Many mammals prepare a nest in which to bear their young, and in this way the babies are protected from animals of prey and from temperature changes.

Figure 16 Two-day-old rats. Some mammals, such as rats and mice, are born naked, blind and almost helpless. Others, such as guinea pigs or foals, are born covered with fur and with their eyes open, and they are active within a few hours of birth.

A nest reduces the chances of the young wandering away and getting lost, and prevents them from injuring themselves. Some young mammals are born without fur (Figure 16), but the mother's body prevents them losing heat.

The food at first is just milk, sucked from the mother's mammary glands. The milk contains nearly all the food, vitamins and salts that the young need for tissue-building and energy but there is no iron present for making haemoglobin (page 111). All the iron needed for the first weeks or months is stored in the body of the embryo during the period of gestation in the uterus. The parents' milk supply increases as the young animals grow larger, and solid food is brought to the nest. In carnivorous (meat-eating) mammals, the prey is torn into pieces small enough for the young to swallow. Often the parents are more aggressive when they have young and react violently to intruders. When young animals, like mice, are old enough to find their own food and escape from predators (hunting animals), they leave the nest and scatter. Other young mammals, such as deer, may remain in a family group or herd for a longer period.

Human parental care is similar to that in other mammals. The mother may choose to feed the baby from a bottle but it is thought that breast-feeding is better (Figure 17). This is because (a) the composition of human milk is best suited to human babies, (b) there is less chance of harmful bacteria getting into the milk, (c) the milk contains antibodies (page 120) which protect the baby against diseases, and (d) there may be an emotional advantage for both mother and child in the close relationship involved in suckling.

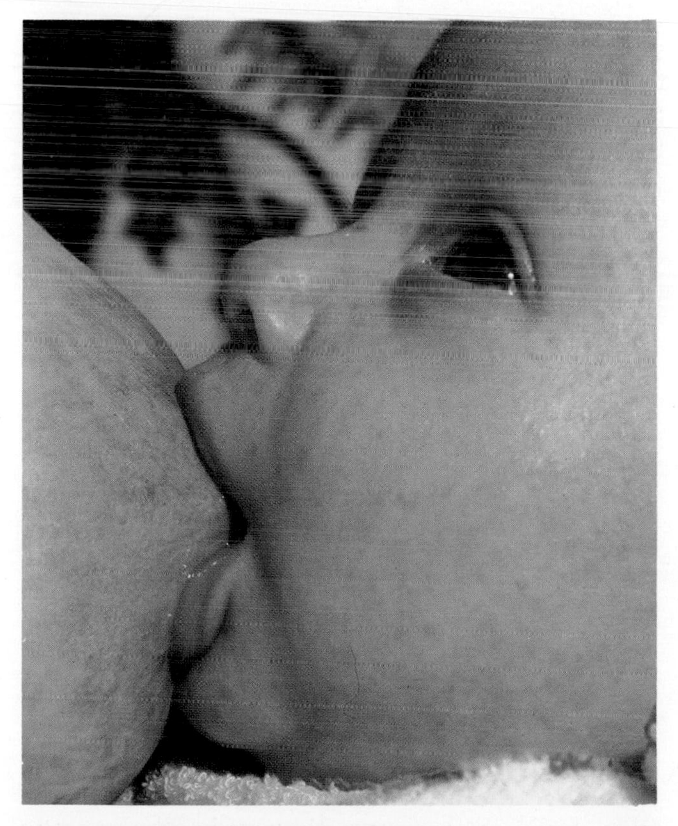

Figure 17 Breast-feeding. Human milk has the right composition for a human baby and there is an emotional benefit to both mother and baby.

Most young mammals are independent of their parents after a few weeks or months even though they may stay together as a family group. In man, however, the young are dependent on their parents for food, clothing and shelter for many years. During this long period of dependence, the young learn to talk, read and write and learn a great variety of other skills that help them to survive and be self-sufficient (Figure 18).

Figure 18 Parental care. Human parental care includes a long period of education.

QUESTION

13 Apart from learning to talk, what other skills might a young human develop that would help him when he is no longer dependent on his parents in (a) an agricultural society, (b) an industrial society?

PUBERTY AND THE MENSTRUAL CYCLE

Puberty

Although the ovaries of a young girl contain all the ova she will ever produce, they do not start to be released until she reaches an age of about 10–14 years. This stage in her life is known as puberty. At about the same time as the first ovulation, the ovary also releases female sex hormones into the blood stream. These hormones are called **oestrogens** and when they circulate round the body, they bring about the development of **secondary sexual characters**. In the girl these are the increased growth of the breasts, a widening of the hips and the growth of hair in the pubic region and in the armpits.

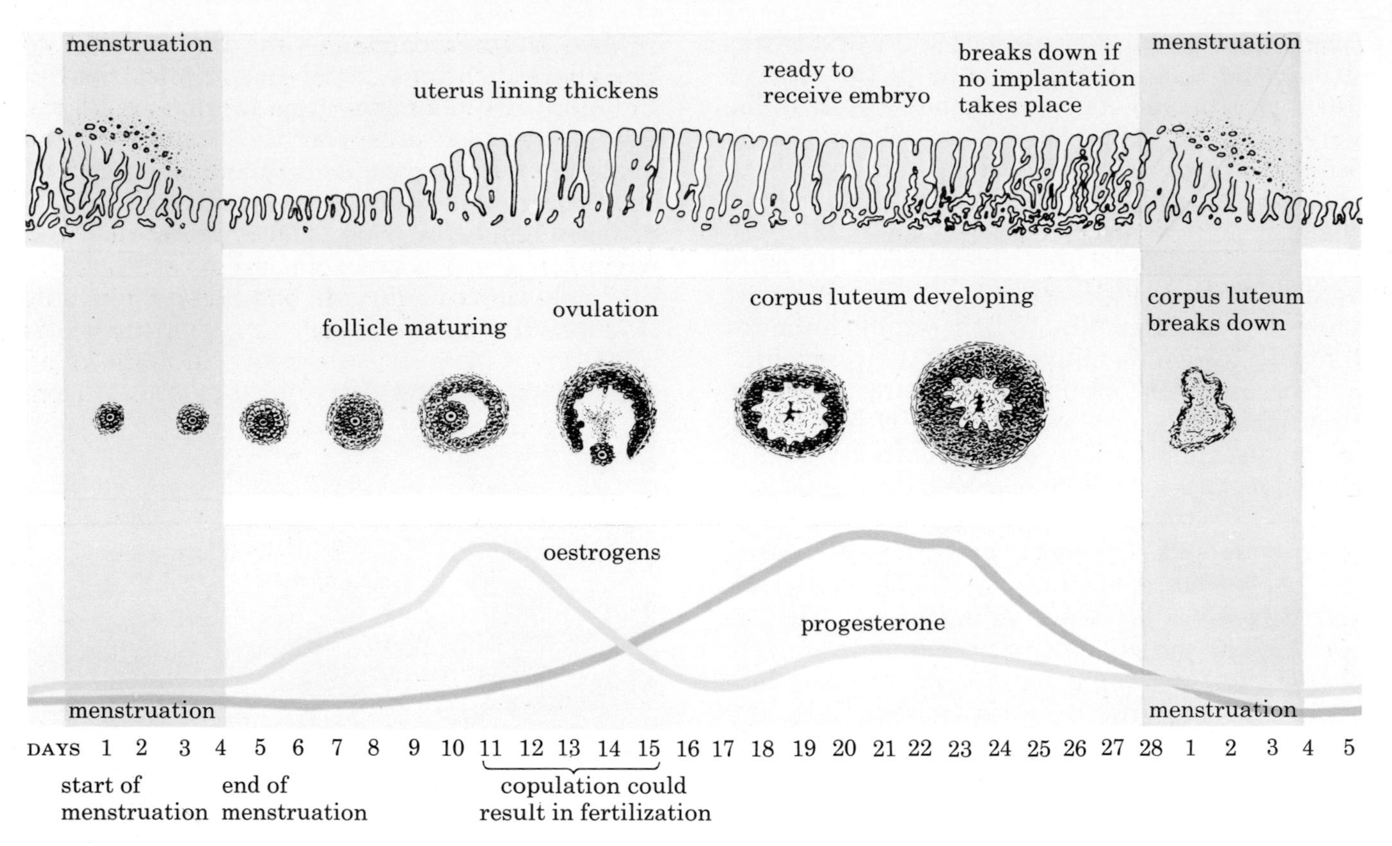

Figure 19 The menstrual cycle (after G. W. Corner, *The hormones in human reproduction*, Princeton).

There is also an increase in the size of the uterus and vagina. Once all these changes are complete, the girl is capable of having a baby.

Puberty in boys occurs at about the same age as in girls. The testes start to produce sperms for the first time and also release a hormone, called **testosterone**, into the blood stream. The male secondary sexual characters which begin to appear at puberty are enlargement of the testes and penis, deepening of the voice, growth of hair in the pubic region, armpits, chest and, later on, the face. In both sexes there is a rapid increase in the rate of growth during puberty.

The menstrual cycle

The ovaries release an ovum about every four weeks. As each follicle develops, the amount of oestrogens produced by the ovary increases. The oestrogens act on the uterus and cause its lining to become thicker and develop more blood vessels. These are changes which help an early embryo to implant as described on page 142.

Once the ovum has been released, the follicle which produced it develops into a solid body called the **corpus luteum**. This produces a hormone called **progesterone**, which affects the uterus lining in the same way as the oestrogens, making it grow thicker and produce more blood vessels.

If the ovum is fertilized, the corpus luteum continues to release progesterone and so keep the uterus in a state suitable for implantation. If the ovum is not fertilized, the corpus luteum breaks down and stops producing progesterone. As a result, the thickened lining of the uterus breaks down and loses blood which escapes through the cervix and vagina. This is known as a **menstrual period**. The appearance of the first menstrual period is one of the signs of puberty in girls. The events in the menstrual cycle are shown in Figure 19.

QUESTION

14 (a) List the changes that occur in a girl at puberty.

(b) How are these changes related to child-bearing?

BIRTH CONTROL

When couples want to limit the size of their families, they use some form of birth control. One method is to have sexual intercourse only when an ovum is unlikely to be in the oviduct, i.e. during the seven days after one menstrual period or the twelve days before the start of the next period (see Figure 19). On its own this is not very reliable because ovulation is not always regular. More effective methods are:

(a) The sheath or the diaphragm. A thin rubber sheath is worn on the penis during sexual intercourse and traps the sperms so that they cannot reach the cervix. The diaphragm is a small rubber disc which is placed in the vagina before sexual intercourse. It stops sperms from entering the uterus through the cervix.

(b) The loop. A small plastic strip bent into a loop or coil is placed in the uterus and left there. It is not known whether it stops fertilization happening or stops the embryo implanting, but it is very efficient.

(c) The pill. The contraceptive pill contains chemicals similar to the female sex hormones, oestrogen and progesterone. If a pill is taken each day, these chemicals stop the release of ova from the ovaries. This method is almost 100 per cent effective.

World population

Because of the great improvement in drugs and medical knowledge, fewer and fewer people die from infectious diseases such as typhoid and cholera. But the birth rate in many countries has not gone down, so their population is doubling every 50 years or less. In these countries, the rapid growth of population leads to shortages of food and living space. It may also result in environmental damage, e.g. the soil erosion which occurs when forests are cleared to plant crops.

There must be a limit to the number of people who can live on the Earth and, therefore, it is necessary to try and make people understand the need to prevent the population increase from getting out of hand (Figure 20).

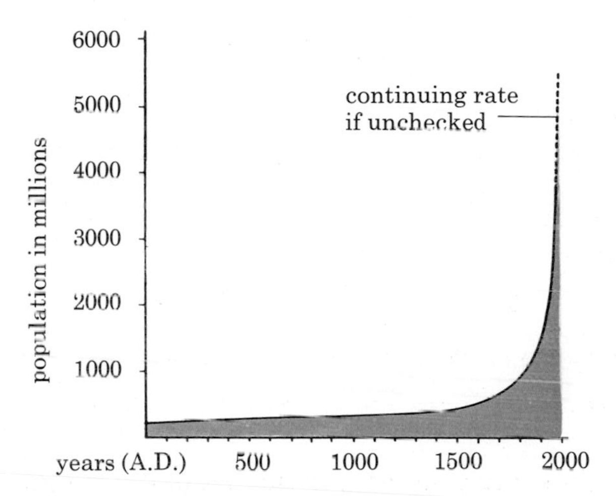

Figure 20 World population growth in the last 2000 years

BASIC FACTS

- **The male reproductive cells (gametes) are sperms. They are produced in the testes and expelled through the urethra and penis during mating.**
- **The female reproductive cells (gametes) are ova (eggs). They are produced in the ovaries. One is released each month. If sperms are present, the ovum may be fertilized as it passes down the oviduct to the uterus.**
- **Fertilization happens when a sperm enters an ovum and the sperm and egg nuclei join up (fuse).**
- **The fertilized ovum (zygote) divides into many cells and becomes embedded in the lining of the uterus. Here it grows into an embryo.**
- **The embryo gets its food and oxygen from its mother.**
- **The embryo's blood is pumped through blood vessels in the umbilical cord to the placenta, which is attached to the uterus lining. The embryo's blood comes very close to the mother's blood so that food and oxygen can be picked up and carbon dioxide and nitrogenous waste can be got rid of.**
- **When the embryo is fully grown, it is pushed out of the uterus through the vagina by contractions of the uterus and abdomen (birth).**
- **Each month, the uterus lining thickens up in readiness to receive the fertilized ovum. If an ovum is not fertilized, the lining and some blood is lost through the vagina (menstrual period).**
- **The production of ova and the development of an embryo are under the control of hormones like oestrogen and progesterone.**

19 The skeleton, muscles and movement

STRUCTURE OF THE SKELETON

The vertebral column

The **vertebral column** forms the central supporting structure of the skeleton. The legs are joined to the vertebral column by the **pelvic girdle (pelvis)** and the arms are connected, by means of muscles, through the shoulder blade (**scapula**) and collar bone (**clavicle**). All these are shown in Figures 1*a* and 4.

The vertebral column (Figure 1*b*) is made up of 33 separate bones called **vertebrae** (singular=vertebra) and one of these is shown in Figure 1*c*. Each vertebra is made up of a solid **centrum** with a **neural arch**. The spinal cord runs under the neural arch (Figure 2). The bony projections sticking out from the vertebrae help

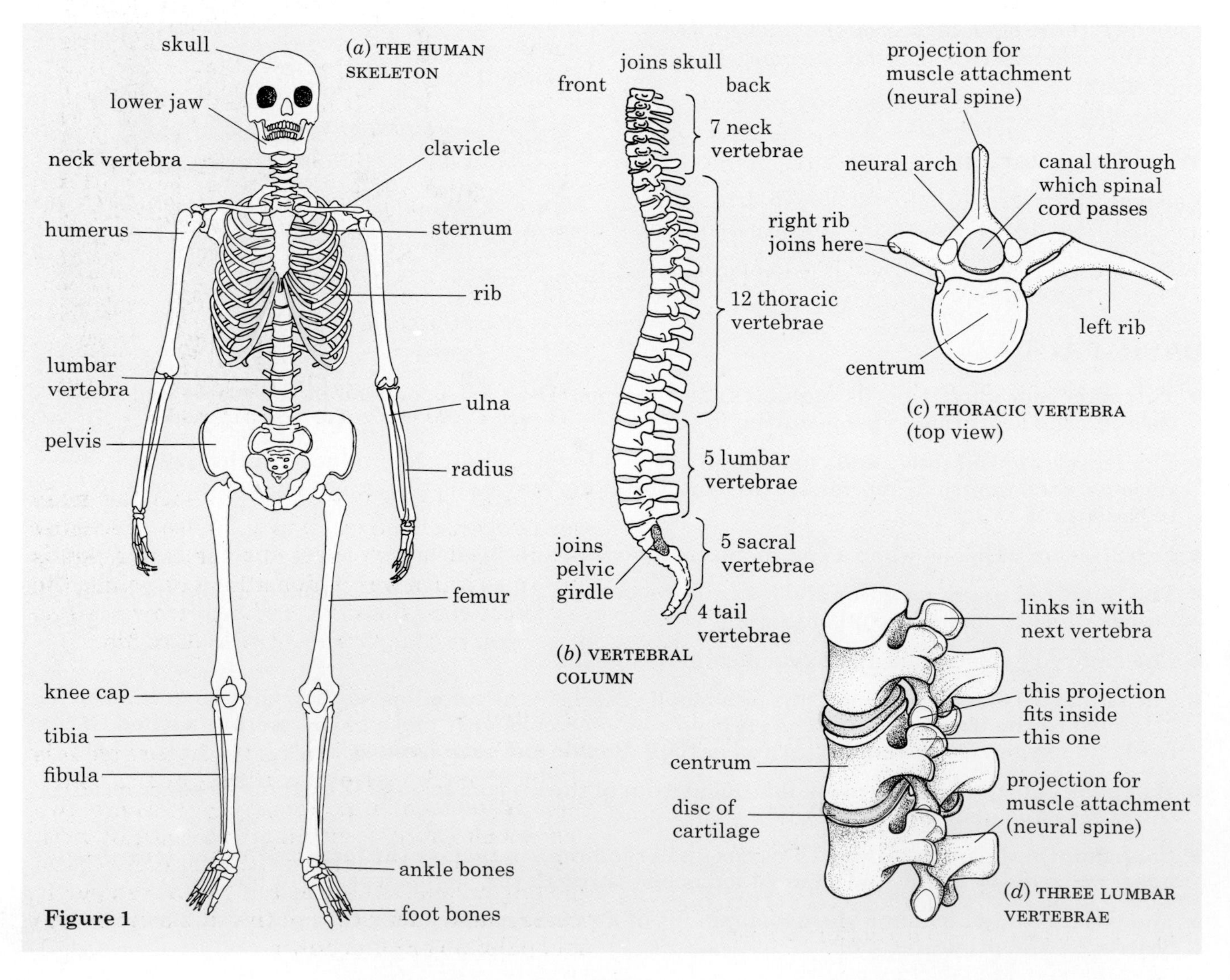

Figure 1

to lock the vertebrae together and restrict their movement (see Figure 1*d*). These projections also provide points for the attachment of muscles which bend or straighten the whole vertebral column.

In each region of the vertebral column, the vertebrae have slightly different functions and, therefore, they vary in their structure. The **thoracic vertebrae** have projections where the ribs are attached (Figure 1*c*) and the five **sacral vertebrae** are joined together to form a rigid structure to which the pelvic girdle is attached (Figure 1*b*).

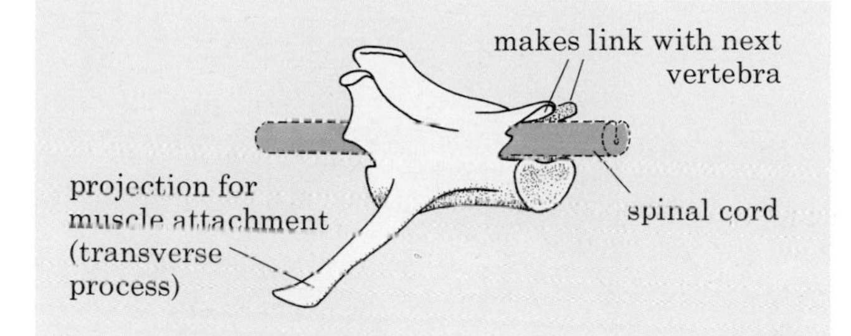

Figure 2 Lumbar vertebra of rabbit (side view)

Between the vertebrae are discs of **cartilage** (gristle and connective tissue) which allow the vertebrae to move slightly and so enable the vertebral column to bend forwards and backwards or from side to side. These discs, called **inter-vertebral** discs (Figure 1*d*), also act as shock-absorbers when you walk, run or jump.

The skull

This is made up of many bony plates joined together (Figure 3). It encloses and protects the brain and also carries and protects the main sense organs, the eyes, ears and nose. The upper jaw is fixed to the skull but the lower jaw is hinged to it in a way which allows chewing. The joints between the skull and top two vertebrae allow you to nod your head or look round.

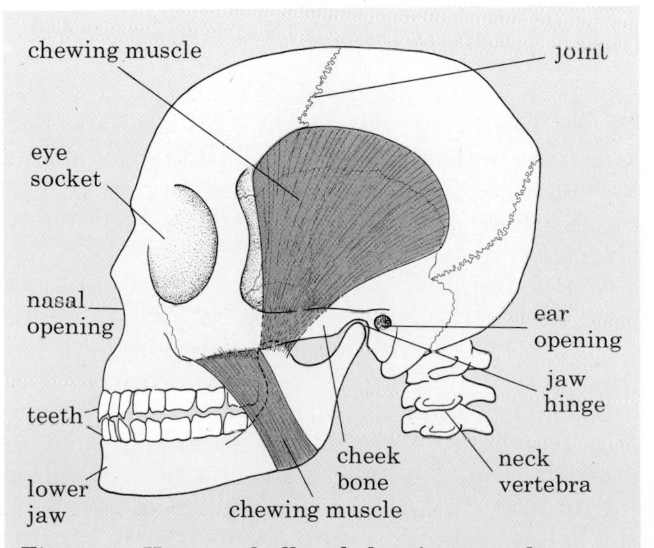

Figure 3 Human skull and chewing muscles

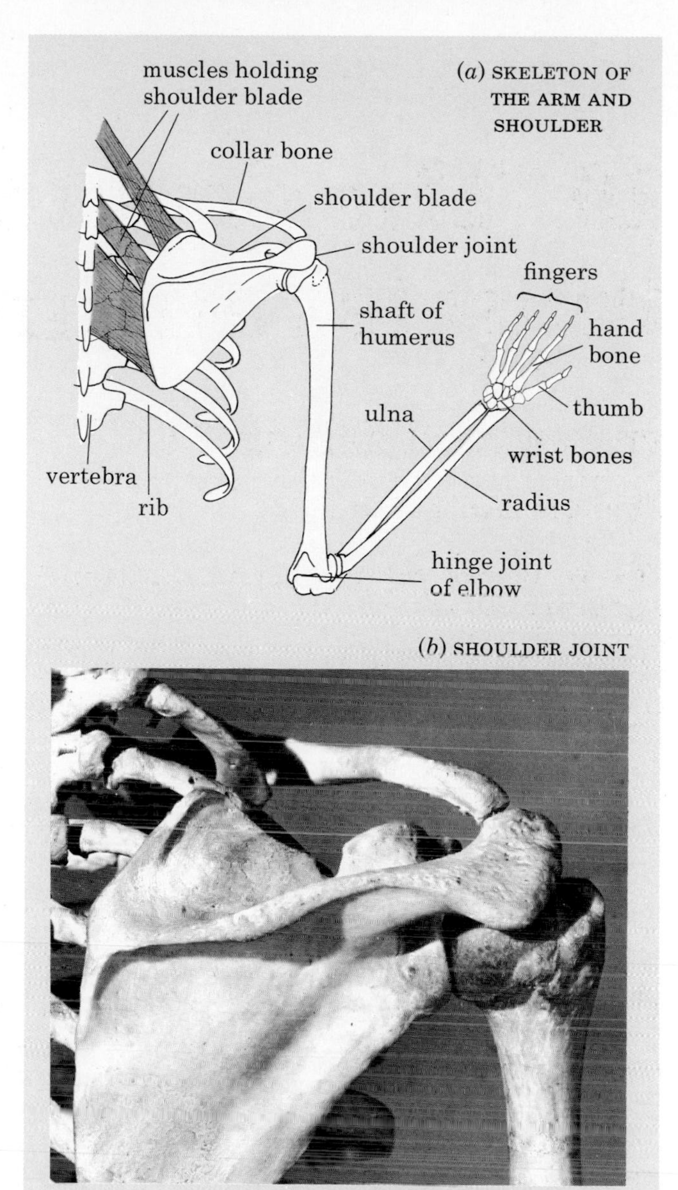

Figure 4 Skeleton of arm and shoulder

The limbs

Arm. The upper arm bone is the **humerus**. It is attached by a hinge joint to the lower arm bones, the **radius** and **ulna** (Figure 4*a*). These two bones make a joint with a group of small wrist bones which in turn join to a series of five hand and finger bones. The ulna and radius can partly rotate round each other so that the hand can be held palm up or palm down.

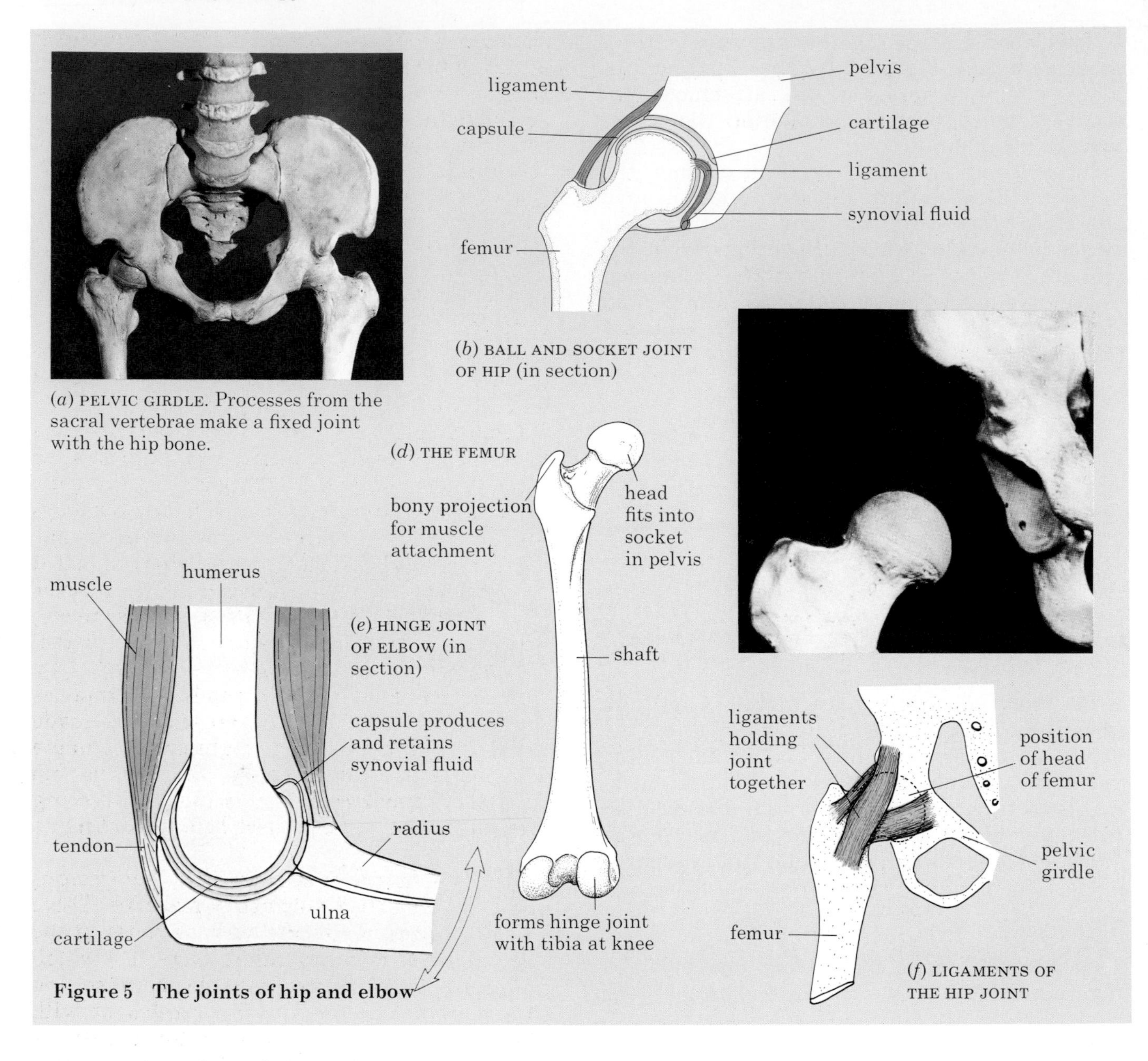

(*a*) PELVIC GIRDLE. Processes from the sacral vertebrae make a fixed joint with the hip bone.

(*b*) BALL AND SOCKET JOINT OF HIP (in section)

(*d*) THE FEMUR

(*e*) HINGE JOINT OF ELBOW (in section)

(*f*) LIGAMENTS OF THE HIP JOINT

Figure 5 The joints of hip and elbow

Leg. The thigh bone or **femur** (Figure 5*d*) is attached at the hip to the pelvic girdle by a ball joint and at the knee it makes a hinge joint with the **tibia**. The **fibula** runs parallel to the tibia but does not form part of the knee joint. The ankle, foot and toe bones are similar to those of the wrist, hand and fingers.

Joints

Where two bones meet they form a joint. It may be a **fixed joint** as in the junction of the pelvic girdle and the sacral vertebrae (Figure 5*a*) or a **movable joint** as in the knee. Two important types of movable joint already mentioned are the **ball and socket joints** of the hip (Figures 5*b* and *c*) and the shoulder (Figure 4), and the **hinge joints** of the elbow (Figure 5*e*), and knee. The ball and socket joint allows movement forwards, backwards and sideways, while the hinge joint allows movement in only one direction.

Where the surfaces of the bones in a joint rub over each other, they are covered with smooth cartilage which reduces the friction between them. Friction is also reduced by a thin layer of lubricating liquid called **synovial fluid** (Figure 5*b*). The bones forming the joint are held in place by tough bands of fibrous tissues called **ligaments** (Figures 5*b* and *f*). Ligaments keep the bones together but do not stop their various movements.

QUESTIONS

1 Study Figure 1*a* and then write the biological names of the following bones: upper arm bone, upper leg bone, hip bone, breastbone, 'backbone', lower arm bones.

2 How many vertebrae are there in your vertebral column (not counting the 'tail')?

3 Apart from the elbow and knee, what other joints in the body function like hinge joints?

4 What types of joints are visible in Figure 5*a*? Name the bones which form each type of joint that you mention.

MOVEMENT AND LOCOMOTION

Muscles and movement

Muscles are bundles or layers of long cells which are able to contract and thus shorten the muscle as a whole. The ends of the limb muscles are drawn out into **tendons** which attach each end of the muscle to the skeleton (Figure 5*e*).

Figure 6*a* shows how a muscle is attached to a limb to make it bend at the joint. The tendon at one end is attached to a non-moving part of the skeleton while the tendon at the other end is attached to the movable bone close to the joint. When the muscle contracts it pulls on the bones and makes one of them move. The position of the attachment means that a small contraction of the muscle will produce a large movement at the end of the limb. Figure 6*b* is a model which shows how the shortening of muscle can move a limb. Figure 6*a* shows how a contraction of the **biceps muscle** bends (or **flexes**) the arm at the elbow, while the **triceps** straightens (or **extends**) the arm.

The non-moving end of the biceps is attached to the scapula, while the moving end is attached to the ulna, near the elbow joint.

Limb muscles are usually arranged in pairs having opposite effects. This is because muscles can only shorten or relax, they cannot elongate, so the triceps is needed to pull the relaxed biceps back to its elongated shape after it has contracted. Pairs of muscles like this are called **antagonistic** muscles. Antagonistic muscles are also important in holding the limbs steady, both muscles keeping the same state of tension.

The contraction of muscles is controlled by nerve impulses. The brain sends out impulses in the nerves so that the muscles are made to contract or relax in the right order to make a movement. For example, in picking up an object, the biceps and triceps will bring the hand close to the object before the fingers grasp it.

Muscles which can be controlled by conscious decisions are called **voluntary muscles**. There are also layers of muscle such as those in the wall of the alimentary canal (page 102) which are called **involuntary muscle**, because you cannot make them contract or relax at will. Their contractions take place automatically as they do in peristalsis. Arterioles (page 116) also contain layers of involuntary muscle which make them get narrower when the muscle contracts or wider when the muscle relaxes, as described on page 136.

There are many muscular activities which bring about movements but do not result in locomotion. Chewing, breathing, throwing, swallowing and blinking are examples of such movements.

All muscles need energy to make them contract. This energy comes from respiration (page 14). In the muscle cells, glucose and oxygen, both brought by the blood stream, are made to react together and provide the energy which causes the long muscle cells to get shorter.

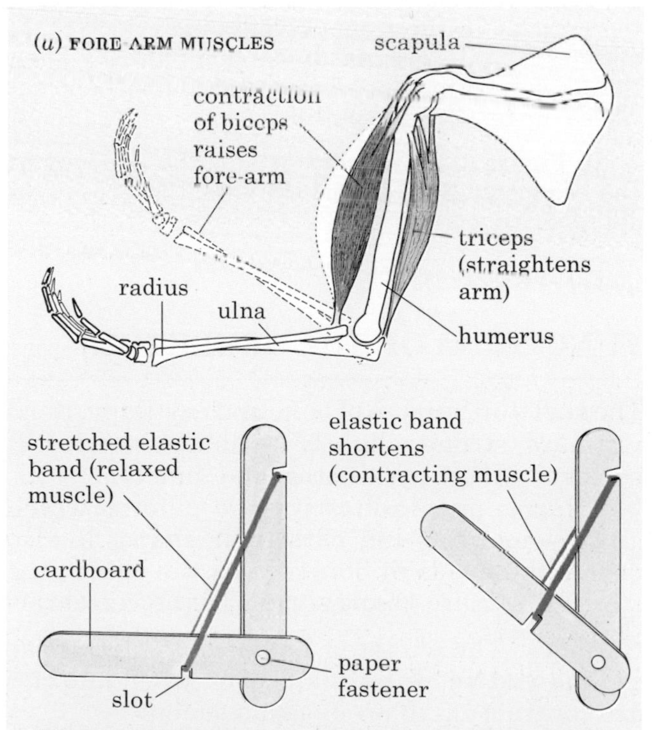

Figure 6 Antagonistic muscles of the fore-arm

Locomotion

Locomotion is brought about by the limb muscles contracting and relaxing in an orderly (i.e. co-ordinated) manner. Figure 8 shows a sprinter at the start of a relay race, and Figure 9 shows how some of his leg muscles are acting on the bones to thrust him forward. When muscle A contracts, it pulls the femur backwards. Contraction of muscle B straightens the leg at the knee. Muscle C contracts and pulls the foot down at the ankle. When these three muscles contract at the same time, the leg is pulled back and straightened and the foot is extended, pushing the foot downwards and backwards against the ground. If the ground is firm, the straightening of the leg pushes upwards against the pelvic girdle which in turn pushes the vertebral column and so lifts the whole body upwards and forwards.

Figure 8 A sprint start. The muscles of the right leg are contracting to straighten the leg and push the sprinter forward. Muscles B and C in Figure 9 can be made out in the picture but there are many others at work.

Limbs as levers

Figure 7 shows how the lower arm works as a lever. The muscle supplies the **effort**; the **load** is an object held in the hand, and the elbow joint is the **fulcrum** or pivot. Most limb bones act as levers which greatly magnify the movement of the muscles which act on them.

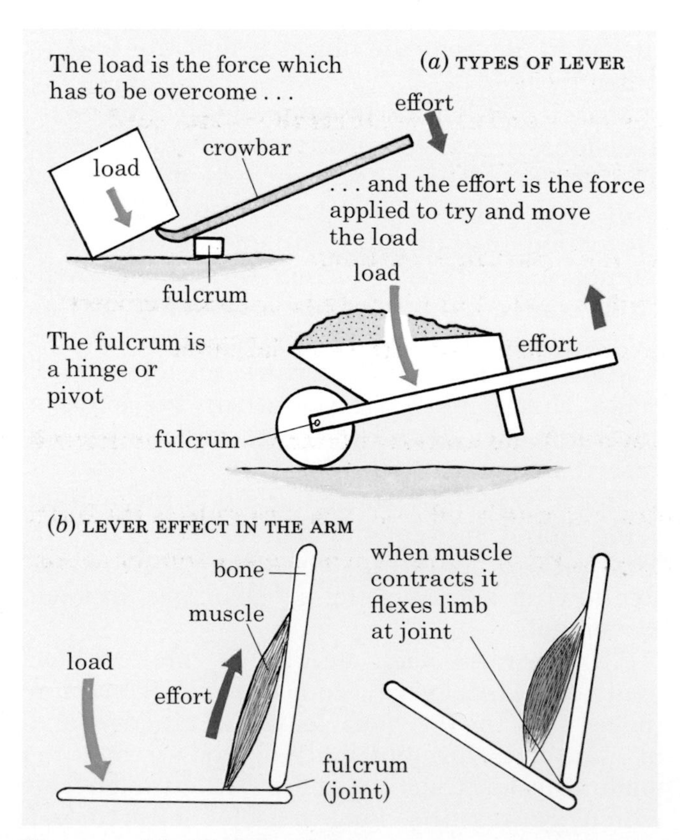

Figure 7 The limb as a lever

QUESTIONS

5 What is the difference between the functions of a ligament and a tendon?

6 What is the main action of (a) your calf muscle, (b) the muscles in the front of your thigh and (c) the muscles in your forearm? If you don't already know the answer, try making the muscles contract and feel where the tendons are pulling.

7 In Figure 9, (a) to what bone is the non-moving end of muscle B attached? (b) Which muscle is the antagonistic partner to C?

FUNCTIONS OF THE SKELETON

The skeleton is made of bone and cartilage. Bone is a hard, strong, slightly flexible substance. It consists of protein fibres and calcium salts. Cartilage is much softer; it is the material which gives your nose and ears their shape. It also covers the heads of bones, joins the ribs to the sternum (Figure 1) and separates the vertebrae.

Support. The skeleton holds the body off the ground and keeps its shape even when muscles are contracting to produce movement.

Protection. The brain is protected from injury by being enclosed in the skull. The heart, lungs and liver are protected by the rib cage, and the

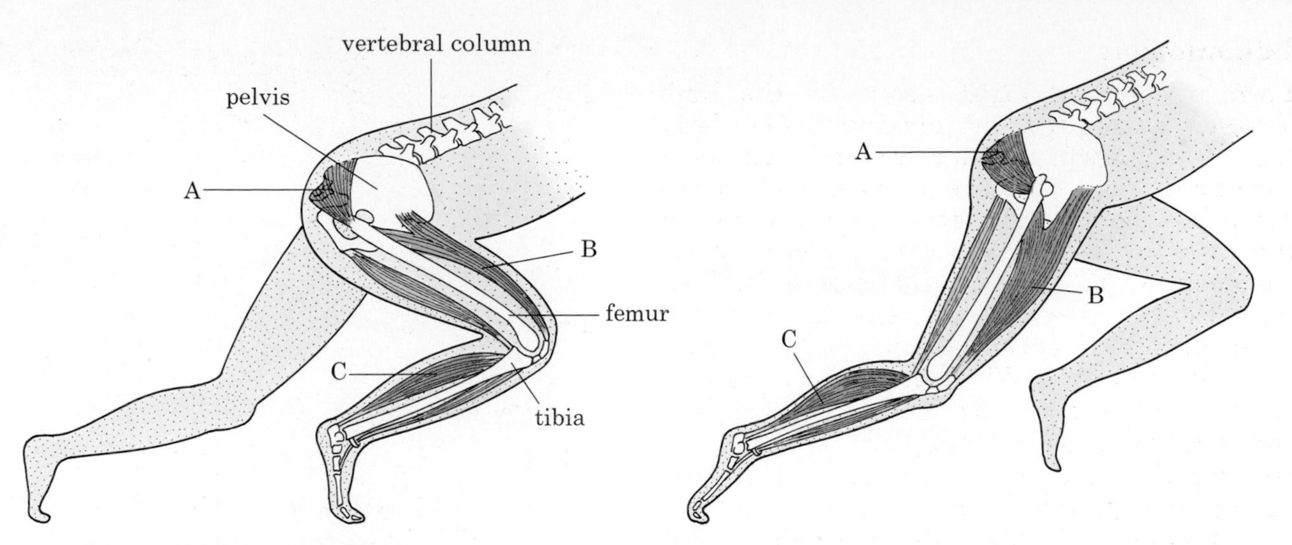

Figure 9 Action of the leg muscles. Muscles A, B and C are all contracting to straighten the leg and the foot. There are many other muscles which are not shown here.

spinal cord is enclosed inside the neural arches of the vertebrae.

Movement. Many bones of the skeleton act as levers. When muscles pull on these bones, they produce movements such as the raising of the ribs during breathing (see page 123), or the chewing action of the jaws. For a muscle to produce movement, both its ends need to have a firm attachment. The skeleton provides suitable points of attachment for the ends of the muscles.

QUESTION

8 Which parts of the skeleton are concerned with both protection and movement?

BASIC FACTS

- **The vertebral column ('backbone') is made up of 33 vertebrae.**
- **The vertebral column forms the main support for the body and also protects the spinal cord.**
- **The legs are attached to the vertebral column by the pelvic girdle.**
- **The skull protects the brain, eyes and ears.**
- **The ribs protect the lungs, heart and liver, and also play a part in breathing.**
- **The limb joints are either ball and socket (e.g. hip and shoulder) or hinge (e.g. knee and elbow).**
- **The surfaces of the joints are covered with cartilage and lubricated with synovial fluid.**
- **Limb muscles are attached to the bones by tendons.**
- **When the muscles contract, they pull on the bones and so bend and straighten the limb or move it forwards and backwards.**
- **Most limb muscles are arranged in antagonistic pairs, e.g. one bends and one straightens the limb.**
- **By straightening the leg and thrusting it against the ground, an animal can propel itself forwards.**

20 The senses

Our senses make us aware of changes in our surroundings and in our own bodies. We have sense cells which respond to stimuli (singular =stimulus). A **stimulus** is a change in light, temperature, pressure, etc. which produces a reaction in a plant or animal. Structures which respond to stimuli are called **receptors**. Some of these receptors are scattered through the skin while others are concentrated into special organs such as the eye and the ear.

SKIN SENSES

There are a great many sensory nerve endings in the skin which respond to the stimuli of touch (Figure 1), pressure, heat and cold, and some which cause a feeling of pain. These sensory nerve endings are very small; they can be seen only in sections of the skin when studied under the microscope (Figure 2), and some have not yet been identified.

When the nerve ending receives a stimulus, it sends a nerve impulse to the brain which makes

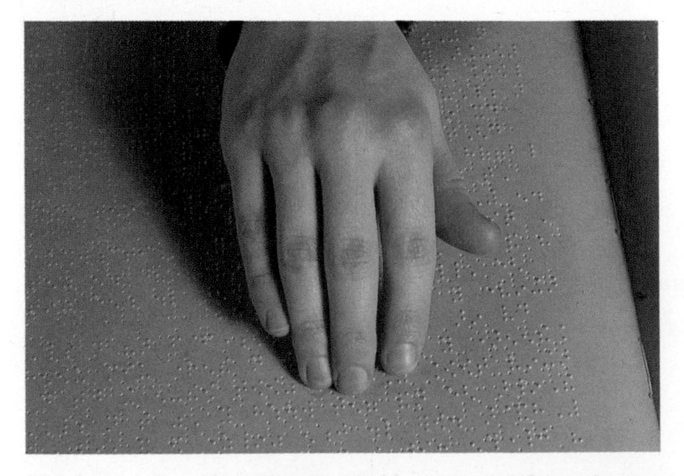

Figure 1 Reading Braille. The sensory endings in our finger-tips provide information about texture and temperature. In this picture, a blind person is using his sense of touch to read Braille by feeling the raised dots with his fingers. The patterns of dots represent letters of the alphabet.

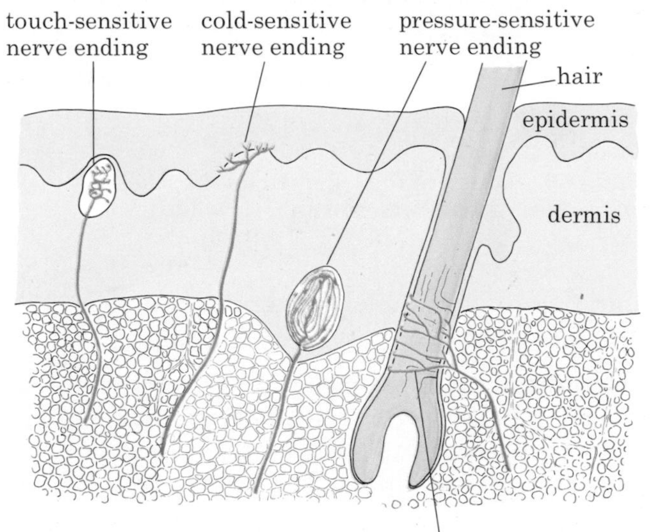

Figure 2 The sense organs of the skin (generalized diagram).

us aware of the sensation. Generally, each type of nerve ending responds to only one kind of stimulus. For example, a heat receptor would send off a nerve impulse if its temperature were raised but not if it were touched.

QUESTIONS

1 What sensation would you expect to feel if a warm pin-head was pressed on to a touch receptor in your skin? Explain.

2 If a piece of ice is pressed on to the skin, which receptors are likely to send impulses to the brain?

TASTE AND SMELL

In the lining of the nasal cavity and on the tongue are groups of sensory cells which respond to chemicals. On the tongue, these groups are called **taste buds** and they lie mostly in the grooves round the bases of the little

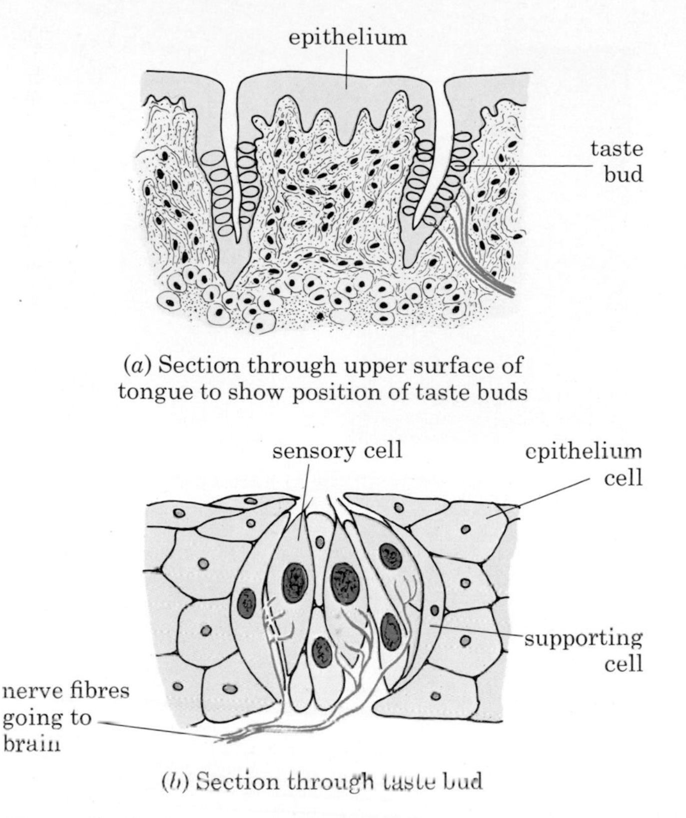

(*a*) Section through upper surface of tongue to show position of taste buds

(*b*) Section through taste bud

Figure 3 Sensory system of tongue

projections on the upper surface of the tongue (Figure 3). The receptor cells in the taste buds can recognize only four classes of chemicals. These are the chemicals which give the taste sensations of sweet, sour, salt or bitter. Nearly all acids, for example, give the taste sensation we call 'sour'. Generally, the taste cells are sensitive to only one or two of these classes of chemical. For a substance to produce a sensation of taste, it must be able to dissolve in the film of water covering the tongue. The smell receptors in the nasal cavity, however, can recognize a much wider variety of chemicals in the air than the tongue can recognize in our food.

The sensation we call flavour (as distinct from 'taste') is the result of the vapours from the food reaching the sensory cells in the nasal cavity from the back of the mouth (see Figure 6 on page 105).

QUESTIONS

3 Apart from the cells which detect chemicals, what other types of receptor must be present in the tongue?

4 What is the difference between taste, smell and flavour?

SIGHT

If you are not already familiar with the way in which lenses work, you are advised to study Figure 4 before reading the next section.

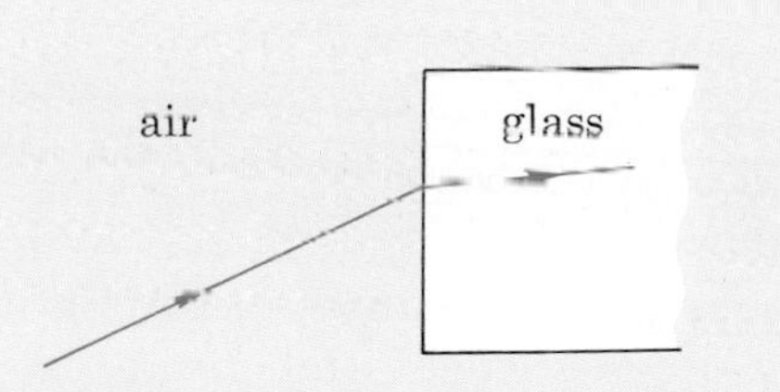

(*a*) When a ray of light passes, at an angle, from air to glass, or from air to water, the ray is bent slightly, as shown. The bending is called 'refraction'

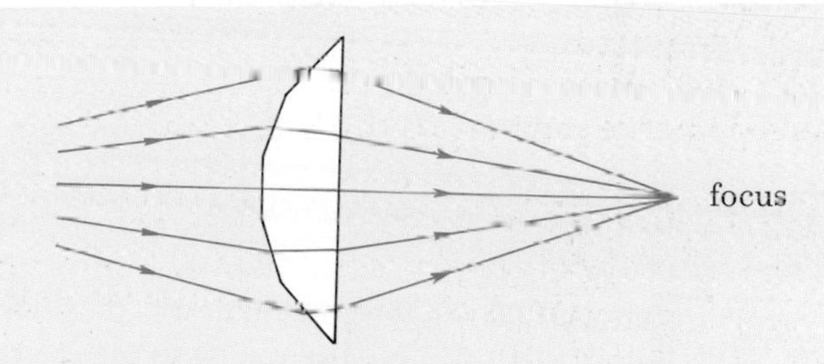

(*b*) If the rays pass through the curved glass surface shown here (a lens) they are bent towards each other and come to a focus

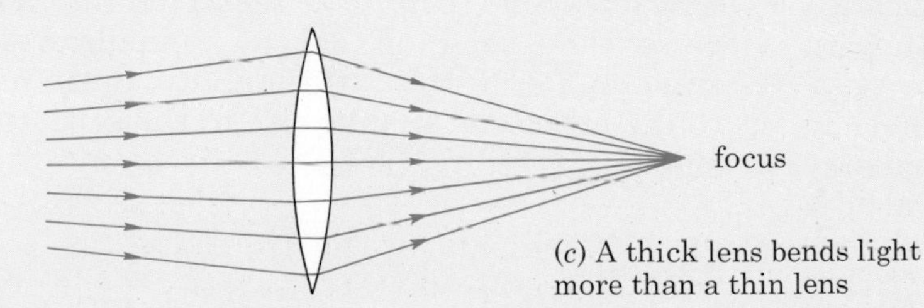

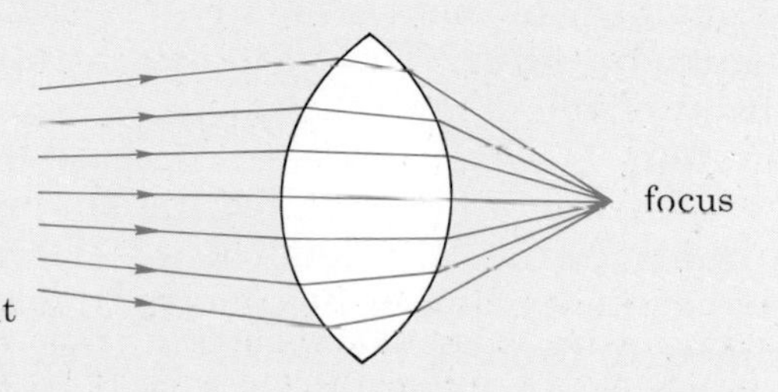

(*c*) A thick lens bends light more than a thin lens

Figure 4 How a convex lens works

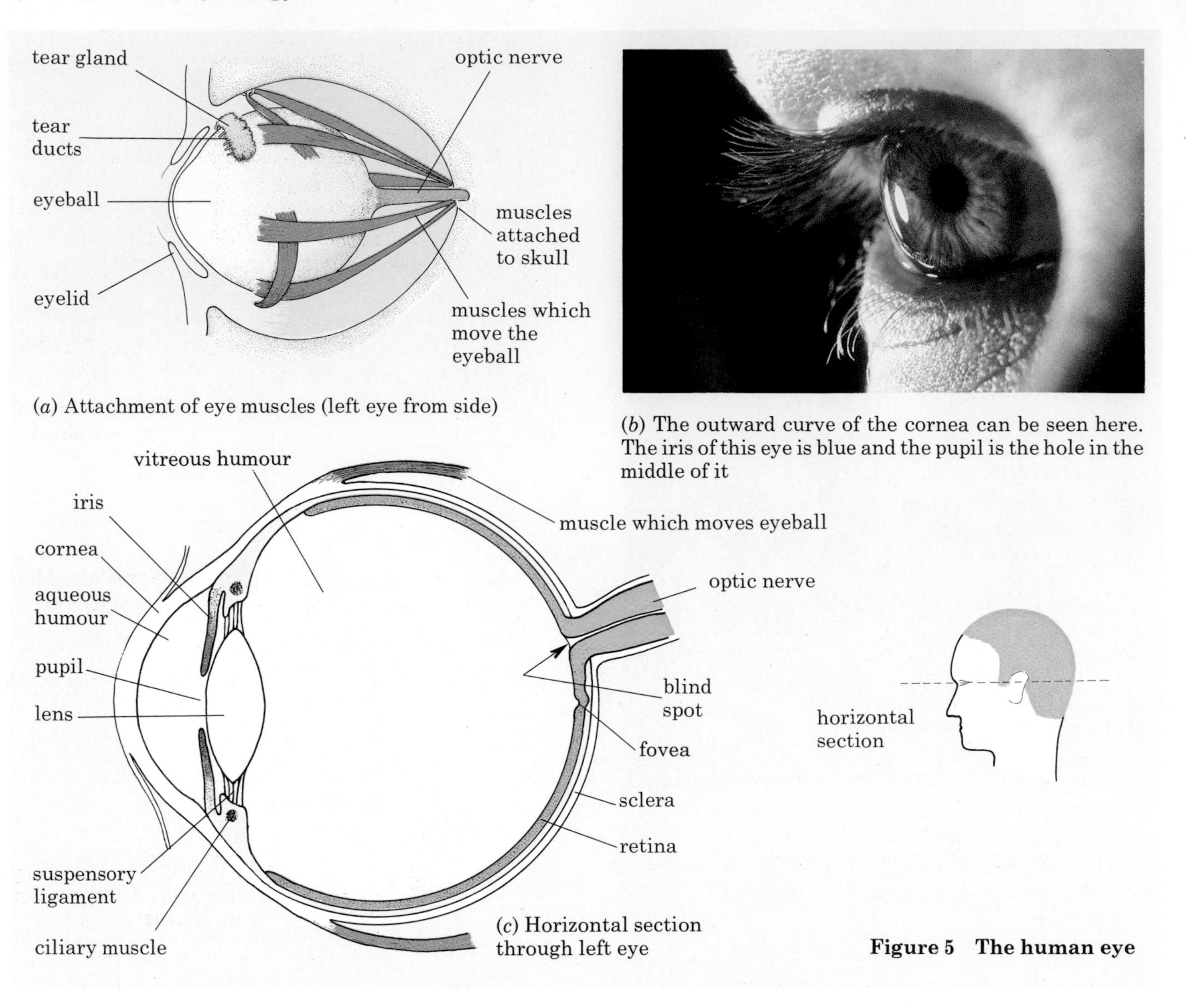

(*a*) Attachment of eye muscles (left eye from side)

(*b*) The outward curve of the cornea can be seen here. The iris of this eye is blue and the pupil is the hole in the middle of it

(*c*) Horizontal section through left eye

Figure 5 The human eye

The eye

The structure of the eye is shown in Figures 5*c* and 9. The **sclera** is the tough, white outer coating. The front part of the sclera is clear and allows light to enter the eye. This part is called the **cornea**. The eye contains a clear liquid whose outward pressure on the sclera keeps the spherical shape of the eyeball. The liquid in the back chamber of the eye is jelly-like and called **vitreous humour**. The **aqueous humour** in the front chamber is watery.

The **lens** is a transparent structure, held in place by a ring of fibres called the **suspensory ligament**. Unlike the lens of a camera or a telescope, the eye lens is flexible and can change its shape. In front of the lens is a disc of tissue called the **iris**. It is the iris we refer to when we describe the colour of the eye as blue or brown. There is a hole in the centre of the iris called the **pupil**. This lets in light to the rest of the eye. The pupil looks black because all the light entering the eye is absorbed and none is reflected. Muscles in the iris can make the pupil wider if the light gets dimmer and make it narrower if the light gets brighter.

The internal lining at the back of the eye is the **retina** and it consists of many thousands of cells which respond to light. When light falls on these cells, they send off nervous impulses which travel in nerve fibres, through the **optic nerve**, to the brain and so give rise to the sensation of sight.

Vision

Figures 6 and 7 explain how light from an object produces a focused **image** on the retina (like a 'picture' on a cinema screen). The curved

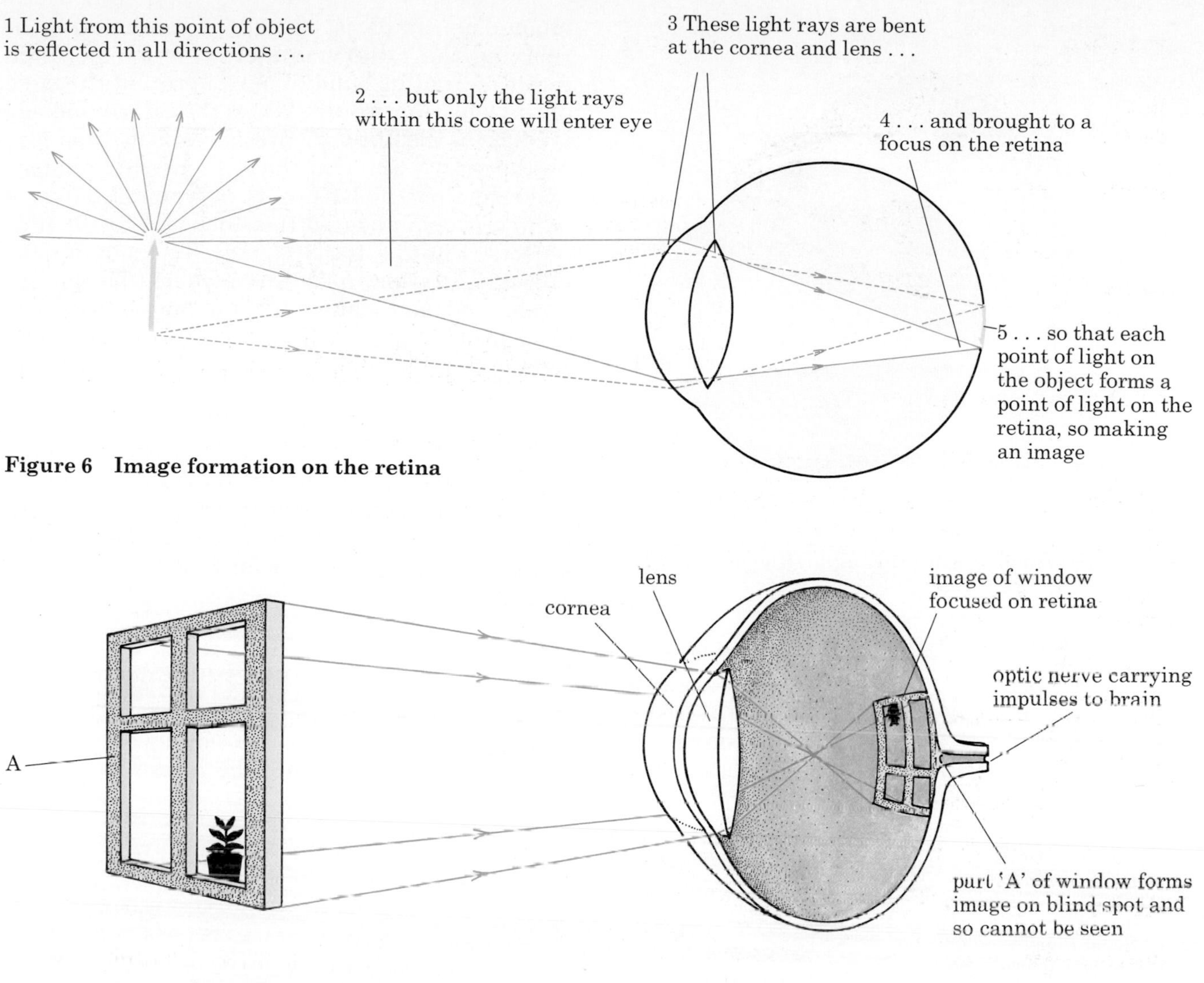

Figure 6 Image formation on the retina

Figure 7 Image formation in the eye

surfaces of the cornea and lens both 'bend' the light rays which enter the eye, in such a way that each 'point of light' from the object forms a 'point of light' on the retina. These points of light will form an image, upside-down and smaller than the object.

The pattern of sensory cells affected by the image will produce a pattern of nerve impulses sent to the brain. The brain interprets this pattern, using its past experience and learning, and so forms an impression of the real size, distance and upright nature of the object.

At the point where the optic nerve leaves the retina, there are no sensory cells and so no information reaches the brain about that part of the image which falls on this **blind spot** (see Figure 8).

The millions of light-sensitive cells in the retina are of two kinds, the **rods** and the **cones** (according to shape). The cones enable us to distinguish colours. There are thought to be three types of cone cell. One type responds to red light, one to green and one to blue. If all three types are equally stimulated we get the sensation of white. The cone cells are concentrated in a central part of the retina, called the **fovea** (Figure 5), and when you study an object closely, you are making its image fall on the fovea.

Figure 8 The blind spot. Hold the book about 60 cm away. Close the left eye and concentrate on the cross with the right eye. Slowly bring the book closer to the face. When the image of the dot falls on the blind spot it will seem to disappear.

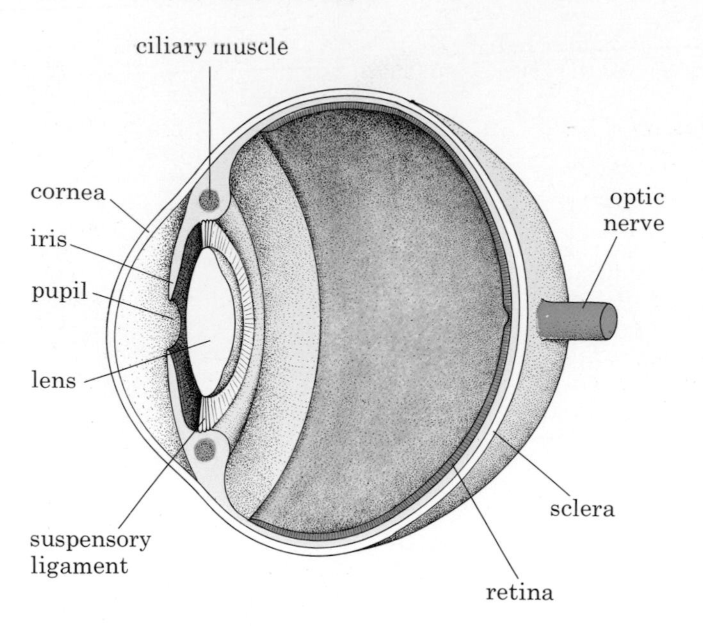

Figure 9 Vertical section through left eye

Accommodation (focusing)

The eye can produce a focused image of either a near object or a distant object. To do this the lens changes its shape, becoming thinner for distant objects and fatter for near objects. This change in shape is caused by contracting or relaxing the **ciliary muscle** (Figure 9) which forms a circular band of muscle round the edge of the lens (Figure 10). When the ciliary muscle is relaxed, the outward pressure of the humours on the sclera pulls on the suspensory ligament and stretches the lens to its thin shape. The eye is now accommodated (i.e. focused) for distant objects (Figures 10*a* and 11*a*). To focus a near object, the ciliary muscle contracts to a smaller circle and this takes the tension out of the suspensory ligament (Figures 10*b* and 11*b*). The lens is flexible and so is able to change to its fatter shape. This shape is better at bending the light rays from a close object.

As we become older, the lens becomes less flexible and will not change shape so much when the ciliary muscle contracts. This makes it more difficult to accommodate for close objects and we have to use spectacles for reading and other close work.

3-D vision and distance judgement

When we look at an object, each eye forms its own image of the object. So, two sets of impulses are sent to the brain. The brain somehow

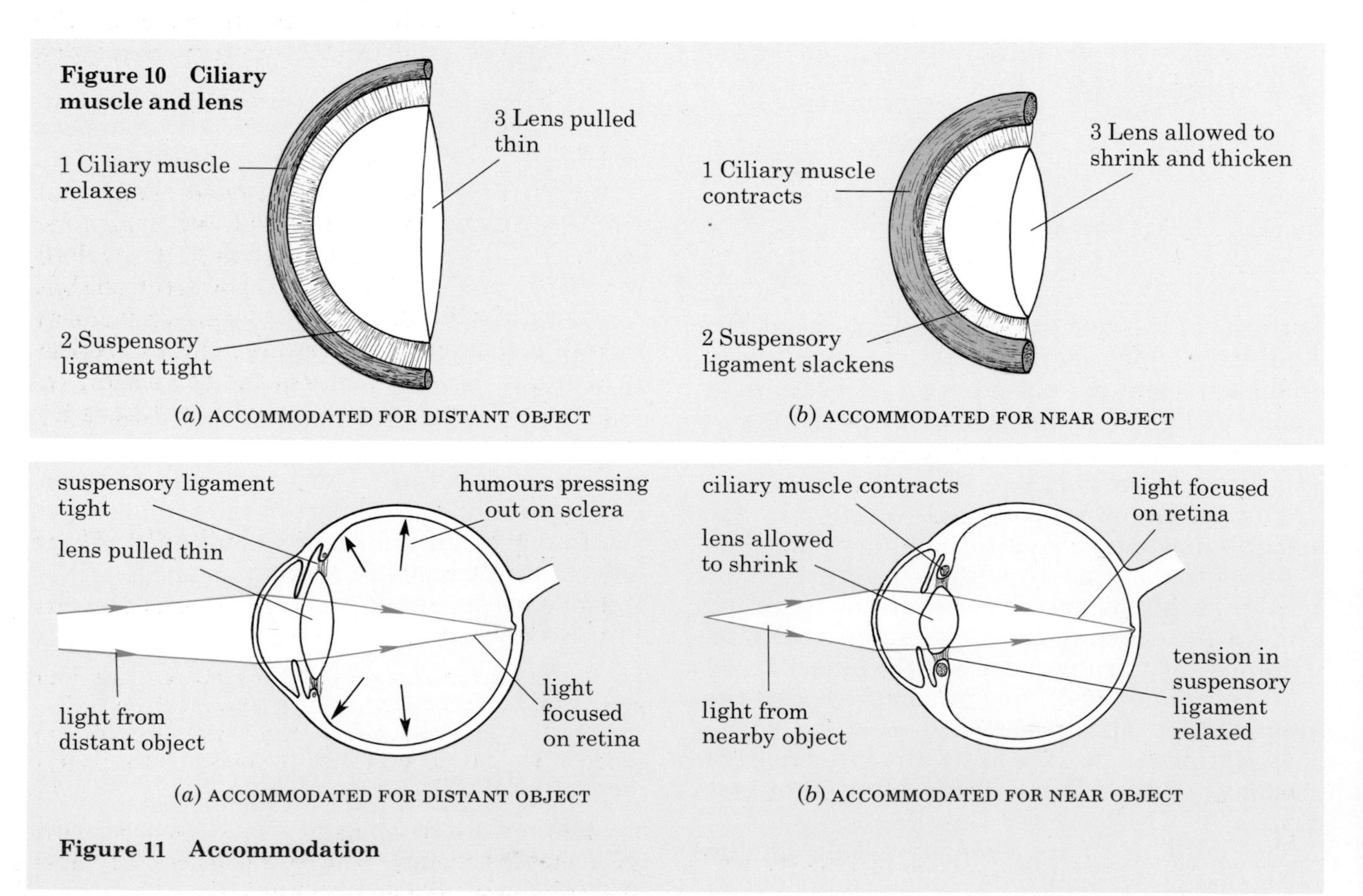

Figure 10 Ciliary muscle and lens

Figure 11 Accommodation

Figure 12 Long and short sight

(*a*) LONG SIGHT

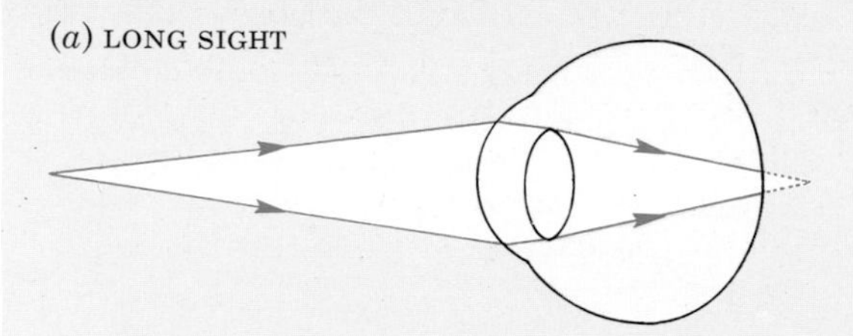

Long sight is caused by small or 'short' eyeballs. Light from a close object would be brought to a focus behind the retina, so the image on the retina is blurred

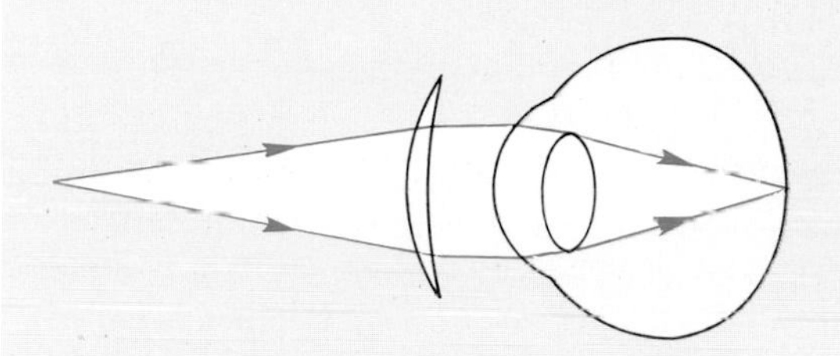

Long sight can be corrected by wearing converging (convex) lenses

(*b*) SHORT SIGHT

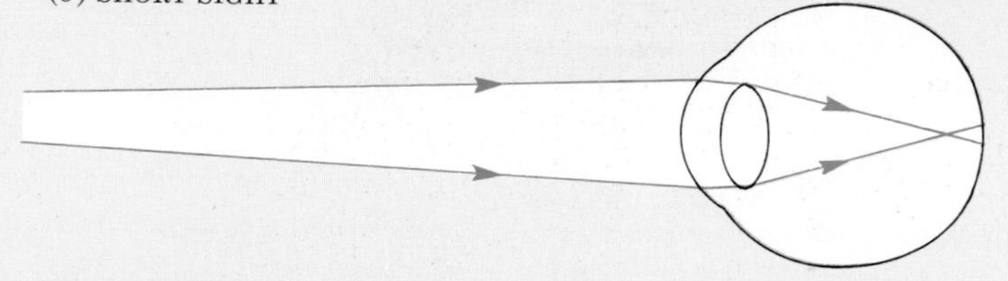

Short sight is usually caused by large or elongated eyeballs. Light from a distant object is focused in front of the retina, so the image on the retina is blurred

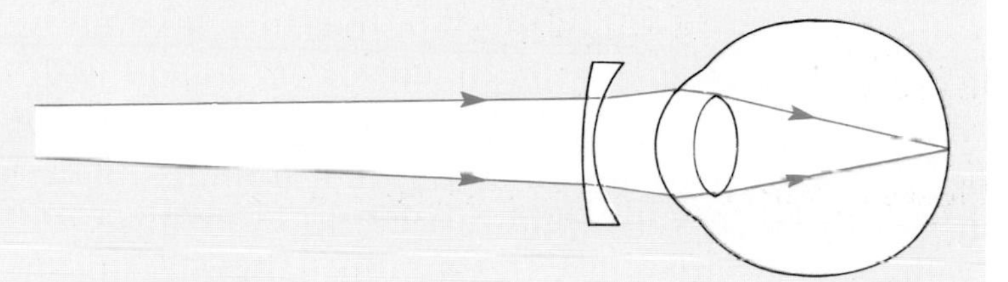

Short sight can be corrected by wearing diverging (concave) lenses

combines these impulses so that we see only one object and not two. However, because the eyes are spaced apart, they do not have the same view of the object. In Figure 13, the left eye sees more of the left side of the box and the right eye sees more of the right side. When the brain combines the information from both eyes, it gives the impression that the object is three-dimensional (3-D) rather than flat.

In order to look at a nearby object, the eyes have to turn inwards slightly. (Trying to focus on a very close object will make you look 'cross-eyed'.) Stretch receptors in the eye muscles will send impulses to the brain and make it aware of how much the eyes are turning inwards. This is one way by which the brain may be able to judge how close an object is, but it is not likely to be much use for objects beyond 15 metres. We probably judge greater distances by comparing

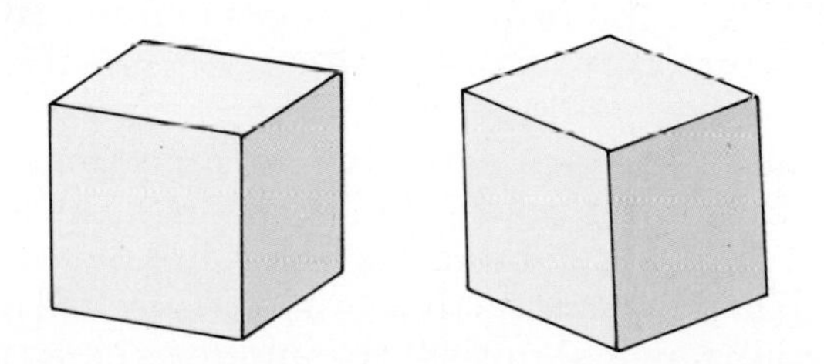

Figure 13 Different views of a cube seen by the left and right eyes

the sizes of familiar objects. The further away an object is, the smaller will be its image on the retina. We use many other clues to help us judge distances.

Long and short sight

In some people, the size of the eyeball does not exactly match the strength of the lens. For example, in short-sighted people, the eyeball may be too long or the lens too powerful, so that the image of a distant object comes to a focus in front of the retina and the image on the retina itself is blurred. The causes of and correction for long and short sight are explained in Figure 12.

QUESTIONS

5 In Figure 7, what structures of the eye are *not* shown in the diagram?

6 In Figure 6, explain what the broken lines are meant to represent.

7 (a) If your ciliary muscles are relaxed, are your eyes focused on a near or a distant object? Explain.

(b) If you dissected a cow's eye, would you expect the lens to be at its thinnest or its fattest shape? Explain.

8 Many people over the age of 50 have to wear spectacles for reading. What sort of lenses will these spectacles need? Explain your answer.

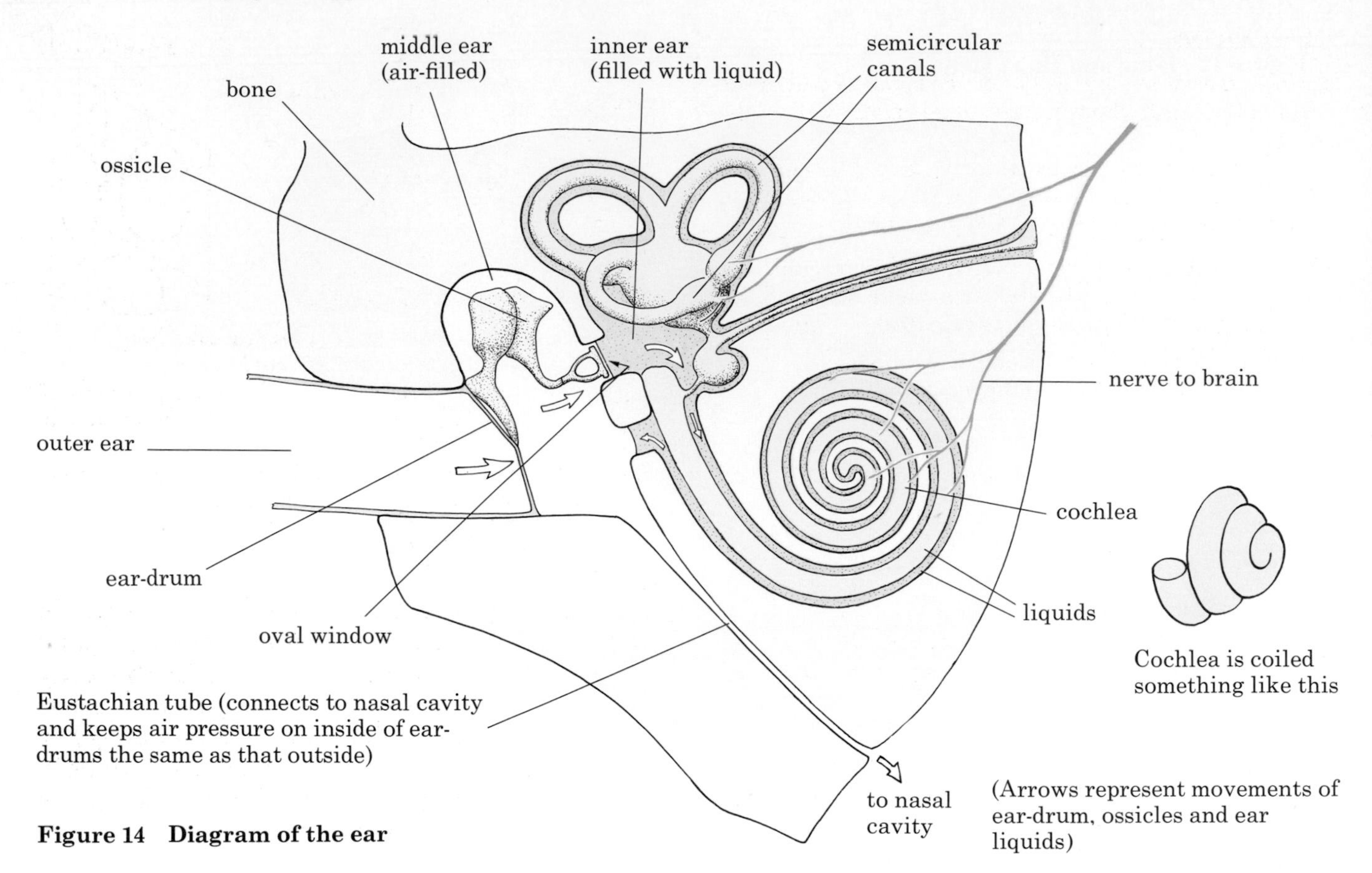

Figure 14 Diagram of the ear

HEARING

The hearing apparatus is protected in two bony boxes attached to the skull, one on each side. They each open to the outside world by a wide tube. Figure 14 shows the structure of the ear.

Outer ear

Sound is the name we give to the sensation we get as a result of vibrations in the air. These vibrations are pulses of compressed air. They enter the tube of the outer ear and hit the **ear-drum**, a thin membrane like a drum-skin across the inner end of the tube. The air vibrations cause the ear-drum to vibrate backwards and forwards. If there are 200 pulses of compressed air every second, the ear-drum will move backwards and forwards at the same rate.

Middle ear

This is a cavity with air in it. It contains a chain of tiny bones or **ossicles**. The first of these ossicles is attached to the ear-drum and the inner ossicle fits into a small hole in the skull called the **oval window**. When the ear-drum vibrates back and forth, it forces the ossicles to vibrate in the same way, so that the innermost ossicle moves rapidly backwards and forwards like a tiny piston in the oval window. The way the ossicles are attached to each other increases the force of the vibrations.

Inner ear

This is where the vibrations are changed into nerve impulses. The inner ear contains liquid, and the vibrations of the ossicles are passed to this liquid. The sensitive part of the inner ear is the **cochlea**, a coiled tube with sensory nerve endings in it. When the liquid in the cochlea is made to vibrate, the nerve endings send off impulses to the brain. The nerve endings at the inner (top) end of the cochlea are sensitive to low-frequency vibration (low notes) and those at the outer end are sensitive to high-frequency vibration (high notes). So, if the brain receives nerve impulses coming from the first part of the cochlea, it interprets this as a high-pitched noise or a high musical note. If the impulses come from the top end of the cochlea, the brain recognizes them as being caused by a low note.

QUESTIONS

9 (a) How does the function of a sensory cell in the retina differ from the function of a sensory cell in the cochlea?

(b) If nerve impulses from the cochlea were fed into the optic nerve, what sensations would you expect if somebody clapped their hands?

10 Sometimes, as a result of catching a cold, the middle ear becomes filled with a clear sticky fluid. Why do you think this causes deafness? (The fluid usually drains away through the Eustachian tube when the cold goes.)

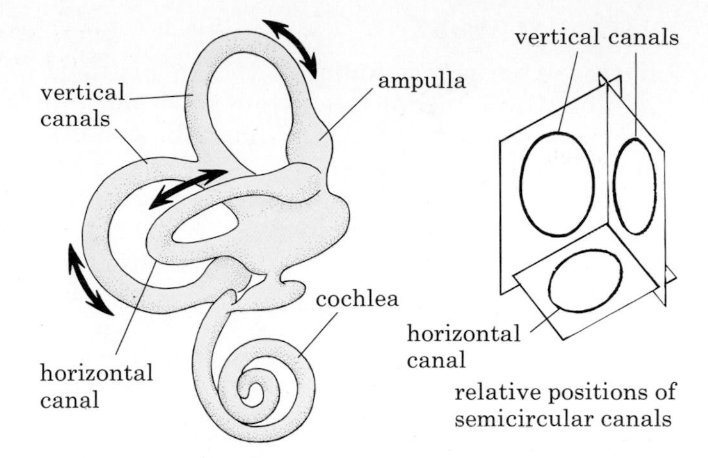

Figure 15 Semicircular canals. (Arrows show direction of rotation which stimulates each canal.)

BALANCE

Semicircular canals

These are in the inner ear (see Figure 14), but do not play a part in hearing. There are two vertical (upright) canals at right angles to each other and one horizontal canal. Each canal has a swelling at one end called an **ampulla** (Figure 15). In each ampulla there is a structure called a **cupula**. This is rather like a swing door which can be pushed either way by the liquid in the semicircular canal (Figure 16). When the body spins round, the liquid in the canals lags slightly behind and pushes the cupula to one side. The cupula pulls on sensory hairs in the ampulla and these send nerve impulses to the brain.

The impulses to the brain inform it of the direction and speed of rotation. The horizontal canal responds best to rotations in the horizontal plane, e.g. twisting the body round while in an upright position. The vertical canals respond particularly to tilting movements of the body forwards, backwards or sideways. The information from the semicircular canals and other sense organs in the inner ear enables us to hold our position when standing still and to keep our balance while moving about.

QUESTION

11 (a) Which semicircular canals are likely to be stimulated the most while you are turning a somersault?

(b) What other sensory information will be reaching the brain while you are doing this?

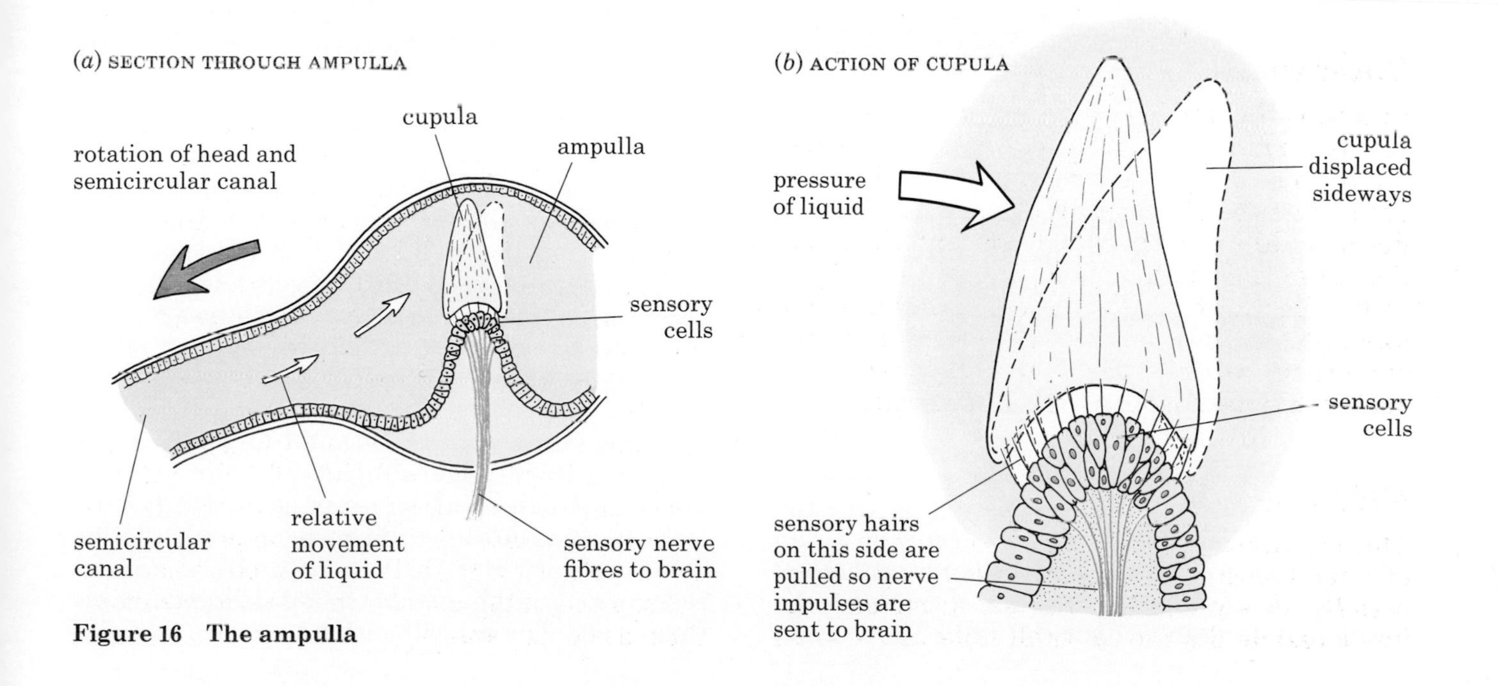

Figure 16 The ampulla

PRACTICAL WORK

All the experiments suggested here are best done by three people working together: an **experimenter** who applies or offers the stimuli, the **subject** who is given the stimuli and a **recorder** who watches and makes a note of the responses made by the subject.

1 Sensitivity to touch

The experimenter marks a regular pattern of dots on the back of the subject's hand. To do this he uses an ink pad and a rubber stamp like the one in Figure 17. He also stamps the same pattern on to a piece of paper for the recorder to use. The experimenter now tests the sensitivity of the subject's skin by pressing a fine bristle on to each dot in turn. The bristle, e.g. a horsehair, is glued to a wooden handle or held in forceps and pressed on to each mark until the bristle just starts to bend. The subject must not look, and simply says 'yes' if he feels the stimulus. The recorder marks on his pattern how many spots in each row were sensitive to touch.

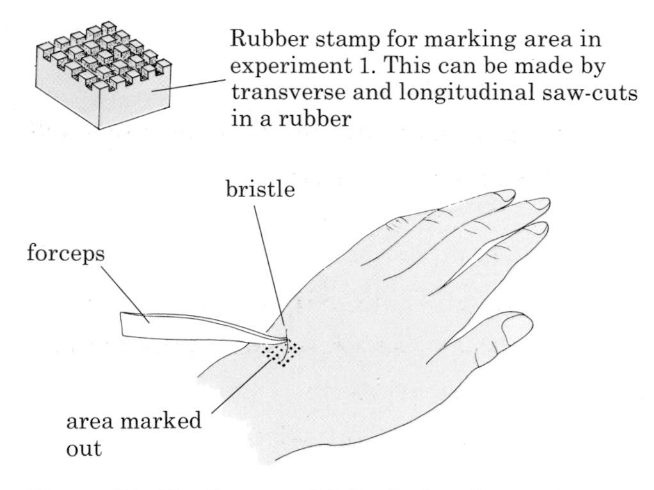

Figure 17 Testing sensitivity to touch

The sensitivity of the back of the hand can now be compared with the sensitivity of the finger-tips or the back of the neck by repeating the experiment in these regions.

2 Ability to distinguish between two touch stimuli

A piece of wire is bent to a hairpin shape like the one in Figure 18*a*. The experimenter adjusts the hairpin so that the points are exactly 5 mm apart. He then presses *one* of the points or *both at once* (Figure 18*b*) on to the skin of the subject's hand, just enough to dent the skin. The subject must not watch the experiment and simply says 'one' or 'two' if he thinks he is being touched by one or two points. The recorder notes how many times the subject is correct. The experimenter must use one point or two points in a random way so that the subject does not recognize a pattern. The recorder should make up a programme of ten stimuli—for example, 1.1.2.1.1.2.2.2.1.2, with five single and five double stimuli. The experimenter carries out this programme and the recorder ticks the correctly recognized stimuli on his plan.

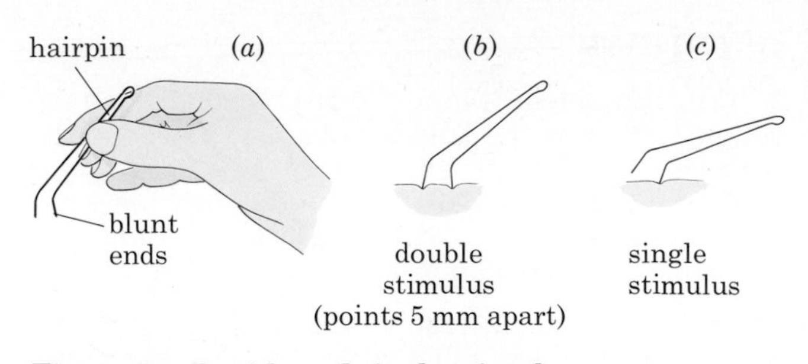

Figure 18 Double and single stimulus

If the subject gets them all right, the experimenter should move the points closer together and try again. If the subject makes a lot of mistakes, the points should be opened out to 10 mm and the experiment repeated. The idea is to find the least distance the points have to be apart for the subject to recognize every time whether one or two points are touching him. Once this distance has been discovered for the back of the hand, the experiment is repeated for the finger-tips and the back of the neck.

3 Sensitivity of the tongue to different tastes

(CAUTION. It is normally forbidden to taste chemical substances in a laboratory. The substances in the next two experiments are harmless but should be made up exactly as described on page 260 and the test-tubes and other apparatus used must be very clean.)

Sweet, sour, salt and bitter solutions are made up as described on page 260. The subject sticks his tongue out and the experimenter picks up a drop of one of the solutions in the end of a drinking straw (Figure 19) and touches it on the subject's tongue. The subject must not know in advance which solution is being used, but must leave his tongue out while he decides whether he can recognize the taste or not. If he cannot recognize it, he shakes his head, still leaving his tongue out. The experimenter now tries a drop of

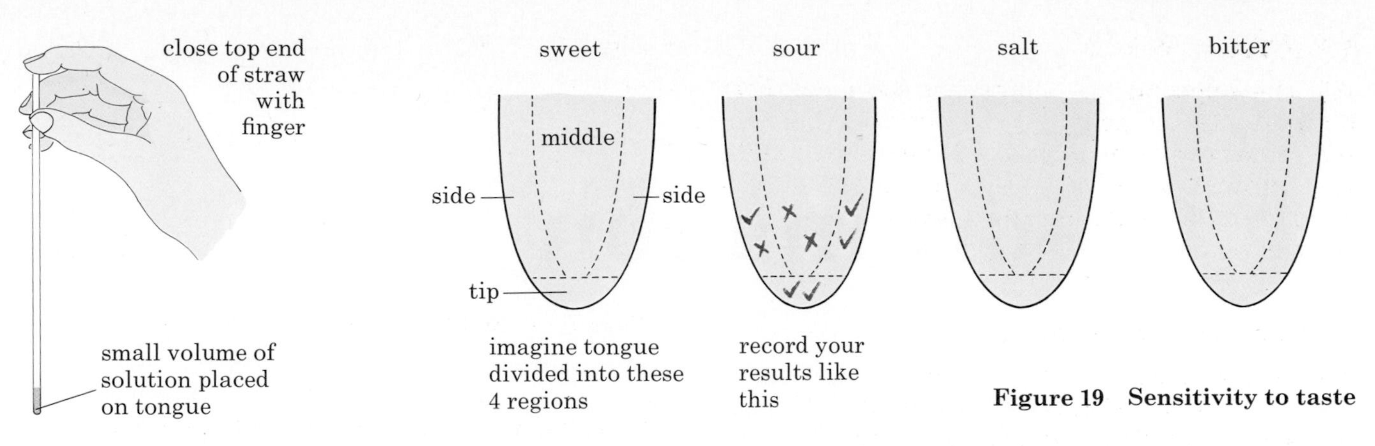

Figure 19 Sensitivity to taste

the same solution on a different part of the tongue and keeps doing this until the subject recognizes the taste. At this point, the subject has to pull his tongue in to say what the taste is.

The recorder makes four charts of the tongue as shown in Figure 19 and puts a tick where each taste was recognized and a cross where it was not. The experimenter now changes to a different solution (and a different drinking straw) and repeats the experiment. He should try to test all four regions of the tongue equally with all four solutions and may have to go back several times to any one of the solutions during the course of the experiment if the subject recognizes the taste after only one or two trials.

It should be possible to build up a picture of which regions of the tongue are particularly sensitive or insensitive to the four tastes.

BASIC FACTS

- **Our senses detect changes in ourselves and in our surroundings.**
- **The skin is a sense organ which detects heat, cold, touch and pressure.**
- **Sensory endings in the nose respond to chemical substances in the air and give us the sense of smell.**
- **The tongue responds to chemicals in food and drink and gives the sense of taste.**

THE EYE

- **The lens focuses light from the outside world to form a tiny image on the retina.**
- **The sensory cells of the retina are stimulated by the light and send nerve impulses to the brain.**
- **The brain interprets these nerve impulses and so gives us the sense of vision.**
- **The eye can focus on near or distant objects by changing the thickness of the lens.**

THE EAR

- **The ear-drum is made to vibrate to and fro by sound waves in the air.**
- **The vibrations are passed on to the inner ear by the three tiny ear bones.**
- **Sensory nerve endings in the inner ear respond to the vibrations and send impulses to the brain.**
- **The brain interprets these impulses as sound.**

SEMICIRCULAR CANALS

- **These are tubes filled with fluid, which detect turning movements of the head.**
- **When the head rotates, the fluid in the tubes lags behind slightly and pushes a cupula to one side.**
- **The movement of the cupula sends nerve impulses to the brain.**
- **As a result of these impulses, the brain makes your body move in a way that helps to keep your balance.**

21 Co-ordination

Co-ordination is the way all the organs and systems of the body are made to work efficiently together (Figure 1). If the leg muscles are being used for running, they will need extra supplies of glucose and oxygen. To meet this demand, the liver turns some of its stored glycogen into glucose (page 108), the lungs breathe faster and deeper to obtain the extra oxygen and the heart pumps more rapidly to get the oxygen and glucose to the muscles more quickly.

There has to be some way of 'telling' the liver, diaphragm and heart that the leg muscles need more oxygen and glucose, and this is the function of the nervous system and endocrine system. The **nervous system** co-ordinates the activities of the leg muscles, lungs and heart. The **endocrine system** co-ordinates the activities of the liver with these other systems.

The nervous system works by sending electrical impulses along nerves. The endocrine system depends on the release of chemicals, called **hormones**, from **endocrine glands**. Hormones are carried by the blood stream. For example, insulin (page 171), is carried from the pancreas to the liver by the circulatory system.

THE NERVOUS SYSTEM

Figure 2 is a diagram of the human nervous system. The brain and spinal cord together form the **central nervous system**. Nerves carry electrical impulses from the central nervous system to all parts of the body, making muscles contract for movement or glands produce enzymes or hormones. The nerves also carry impulses back to the central nervous system from the sense organs of the body. These impulses from the eyes, ears, skin, etc. make us aware of changes in our surroundings or in ourselves. Nerve impulses from the sense organs *to* the central nervous system are called **sensory**

Figure 1 Co-ordination. The brain receives a steady stream of sensory information from the eyes, semicircular canals and stretch receptors. From this information it computes a pattern of muscular movements to make the arm hit the ball accurately with the racket.

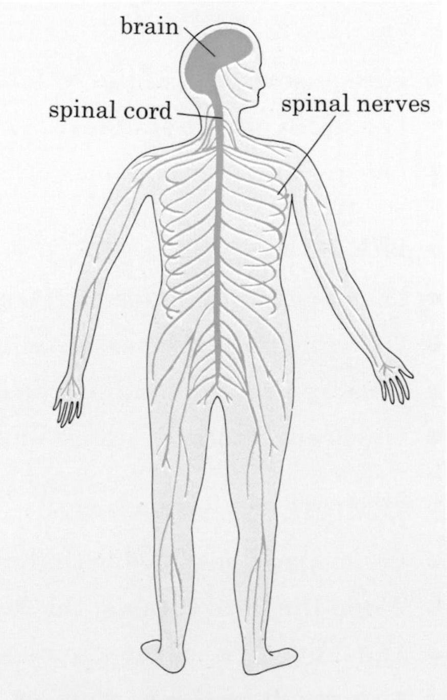

Figure 2 Nervous system of man

impulses; those *from* the central nervous system to the muscles, resulting in action of some kind, are called **motor impulses**.

Nerve cells (neurones)

The central nervous system and the nerves are made up of nerve cells, called **neurones**. Figure 3 shows three types of neurone. The **motor neurones** carry impulses from the central nervous system to muscles and glands. The **sensory neurones** carry impulses from the sense organs to the central nervous system. The **multi-polar neurones** are neither sensory nor motor but make connections to other neurones inside the central nervous system.

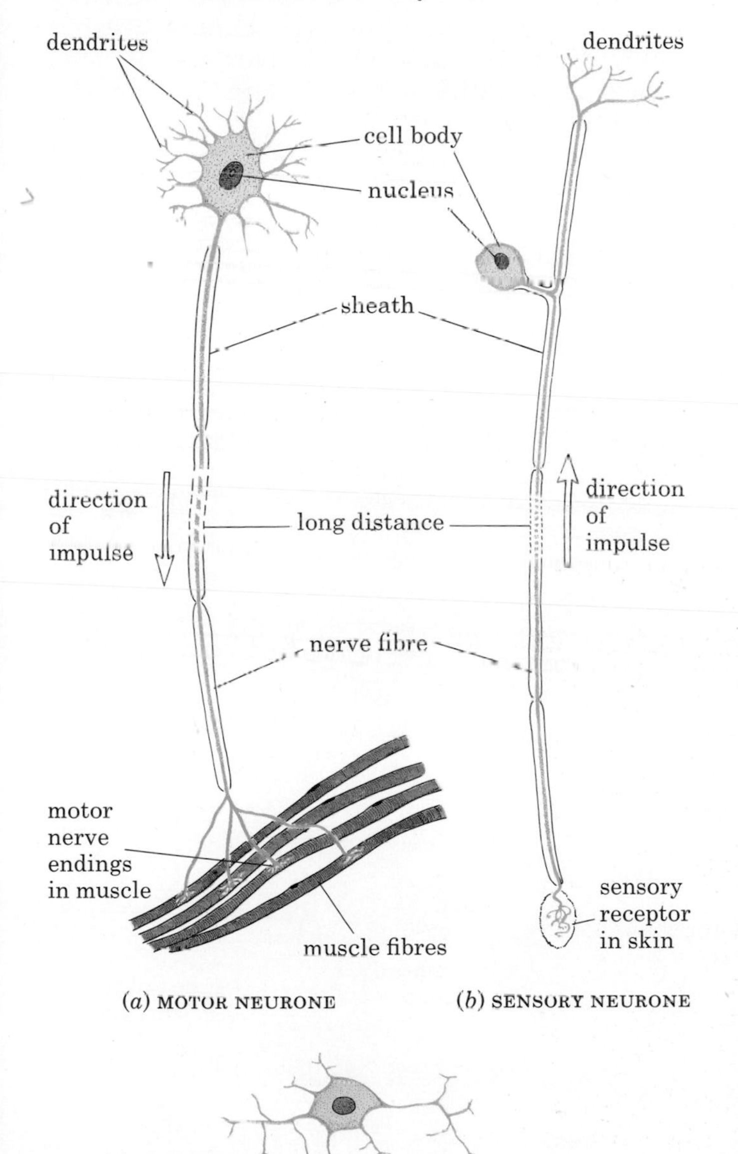

Figure 3 Nerve cells (neurones)

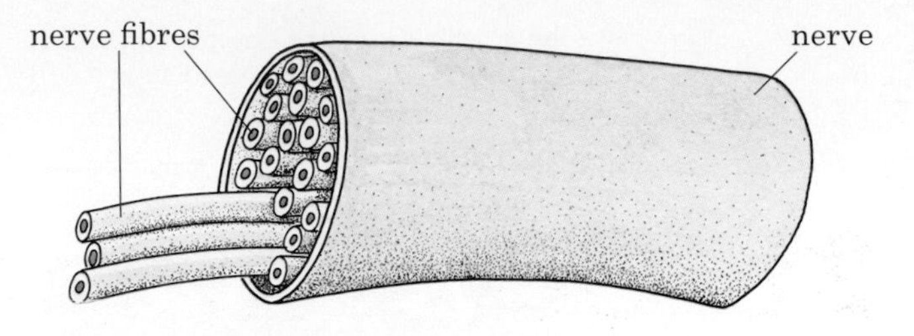

Figure 4 Nerve fibres grouped into a nerve

Each neurone has a **cell body** consisting of a nucleus surrounded by a little cytoplasm. Branching fibres, called **dendrites**, from the cell body make contact with other neurones. A long filament of cytoplasm, surrounded by an insulating sheath, runs from the cell body of the neurone. This filament is called a **nerve fibre** (Figures 3*a* and *b*). The cell bodies of the neurones are mostly located in the brain or spinal cord and it is the nerve fibres which run in the nerves. A **nerve** is easily visible, white, tough, and stringy and consists of hundreds of microscopic nerve fibres bundled together (Figure 4). Most nerves contain a mixture of sensory and motor fibres. So a nerve will carry sensory and motor impulses in opposite directions in the separate fibres it contains.

Some of the nerve fibres are very long. The nerve fibres to the foot have their cell bodies in the spinal cord and the fibres run inside the nerves, without a break, down to the skin of the toes or the muscles of the foot. Thus a single nerve cell may have a fibre about one metre long.

QUESTIONS

1 What is the difference between a nerve and a nerve fibre?

2 In what ways are sensory neurones and motor neurones similar (a) in structure, (b) in function? How do they differ?

3 Can (a) a nerve fibre, (b) a nerve, carry both sensory and motor impulses? Explain your answers.

Synapse

Although nerve fibres are insulated, it is necessary for impulses to pass from one neurone to another. An impulse from the finger-tips has to pass through at least three neurones before reaching the brain and so produce a conscious sensation. The regions where impulses are able to cross from one neurone to the next are called **synapses**.

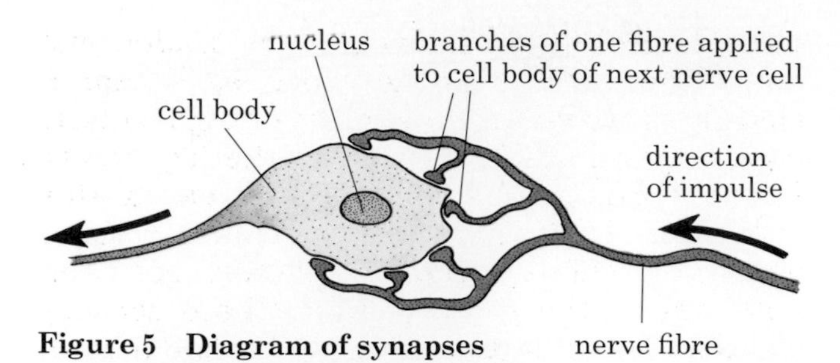

Figure 5 Diagram of synapses

At a synapse, a branch at the end of one fibre is in close contact with the cell body or dendrite of another neurone (Figure 5). When an impulse arrives at the synapse, it releases a chemical which sets off an impulse in the next neurone. Sometimes several impulses have to arrive at the synapse before an impulse is fired off in the next neurone.

The nerve impulse

The nerve fibres do not carry sensations like pain or cold. These sensations occur only when a nerve impulse reaches the brain. The impulse itself is a series of electrical pulses which travel down the fibre rather like morse code. All nerve impulses are similar; there is no difference between nerve impulses from the eyes, ears or hands. If you could 'tune in' to nerve impulses as they passed down a nerve fibre, you would hear something like the 'number engaged' tone on a telephone, but with much faster 'pips'.

We are able to tell where the sensory impulses have come from and what caused them only because the impulses are sent to different parts of the brain.

QUESTIONS

4 Look at Figure 6*b*. (a) How many cell bodies are drawn? (b) How many synapses are shown?
Look at Figure 8 and answer the same questions.

5 If you could intercept and 'listen to' the nerve impulses travelling in the spinal cord, could you tell which ones came from pain receptors and which from cold receptors? Explain your answer.

The reflex arc

One of the simplest situations where impulses cross synapses to produce action is in the reflex arc. A **reflex action** is an automatic response which you cannot consciously control. When a particle of dust touches the cornea of your eye, you will blink; you cannot prevent yourself from blinking. A particle of food touching the lining of the windpipe will set off a coughing reflex which cannot be suppressed. When a bright light shines in the eye, the pupil contracts (see page 156). You cannot stop this reflex and you are not even aware that it is happening.

The nervous pathway for such reflexes is called a **reflex arc**. Figure 7 shows such a nervous pathway from a receptor in the hand to a muscle in the arm. The organ which responds to the motor impulse is sometimes called the **effector**. In this case, the biceps muscle is the effector. Figure 6 shows a reflex arc in more detail. Here the spinal cord is drawn in transverse section. The spinal nerve divides into two 'roots' at the point where it joins the spinal cord. All the sensory fibres enter through the **dorsal root** and the motor fibres all leave

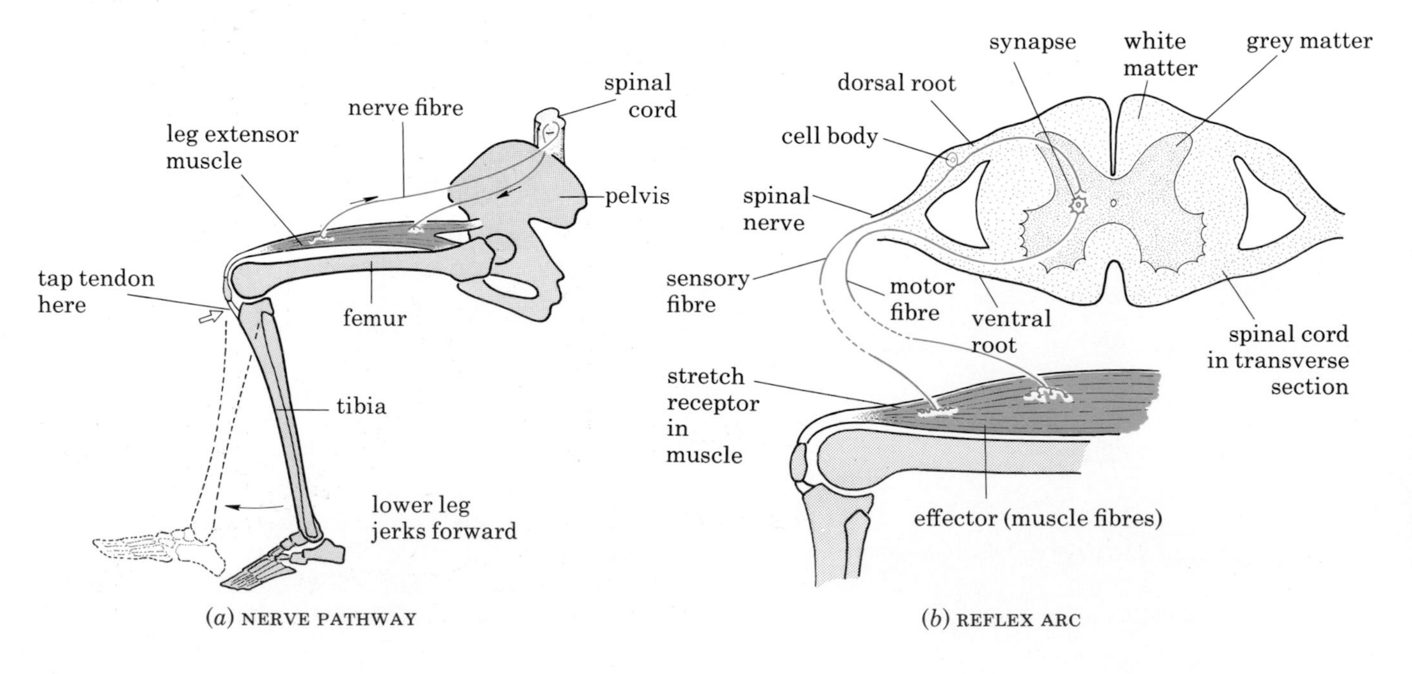

(*a*) NERVE PATHWAY

(*b*) REFLEX ARC

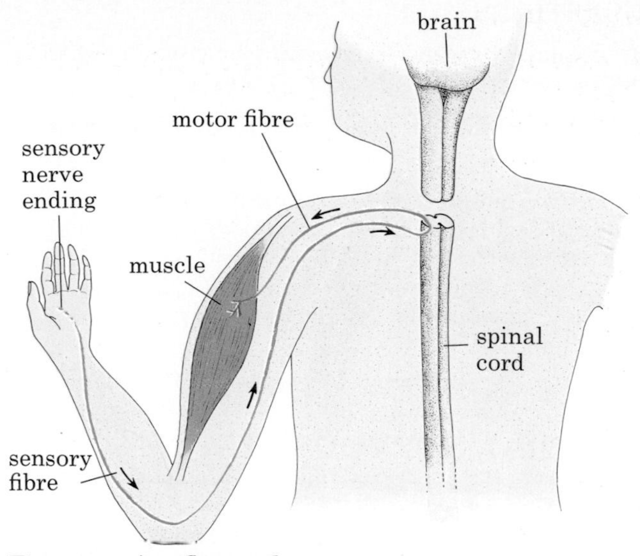

Figure 7 A reflex pathway

through the **ventral root**, but both kinds of fibre are contained in the same spinal nerve. This is like a group of insulated wires in the same electric cable. The cell bodies of all the sensory fibres are situated in the dorsal root and they make a bulge called a **ganglion** (Figure 8).

A well-known reflex is the 'knee jerk' (Figure 6*a*). One leg is crossed over the other and the muscles are totally relaxed. If the tendon just below the kneecap of the upper leg is tapped sharply, a reflex arc makes the thigh muscle contract and the lower part of the leg swings forward. Figure 6*b* traces the pathway of this reflex arc. Hitting the tendon makes a sense organ, called a stretch receptor, send off impulses in the sensory fibre. These sensory impulses travel in the nerve to the spinal cord. In the central region of the spinal cord, the sensory fibre passes the impulse across to a motor neurone which conducts the impulse down the fibre, back to the thigh muscle. The arrival of the impulses at the muscle makes it contract, and jerk the lower part of the limb forward. You are aware that this is happening (which means that sensory impulses must be reaching the brain), but there is nothing you can do to stop it.

In even the simplest reflex action, many more nerve fibres, synapses and muscles are involved than are described here. Figure 8 shows the reflex arc which would result in the hand being removed from a painful stimulus. On the left side of the spinal cord, an incoming sensory fibre makes its first synapse with a **relay neurone** (sometimes called an **association neurone**).

Figure 8 Reflex arc (withdrawal reflex)

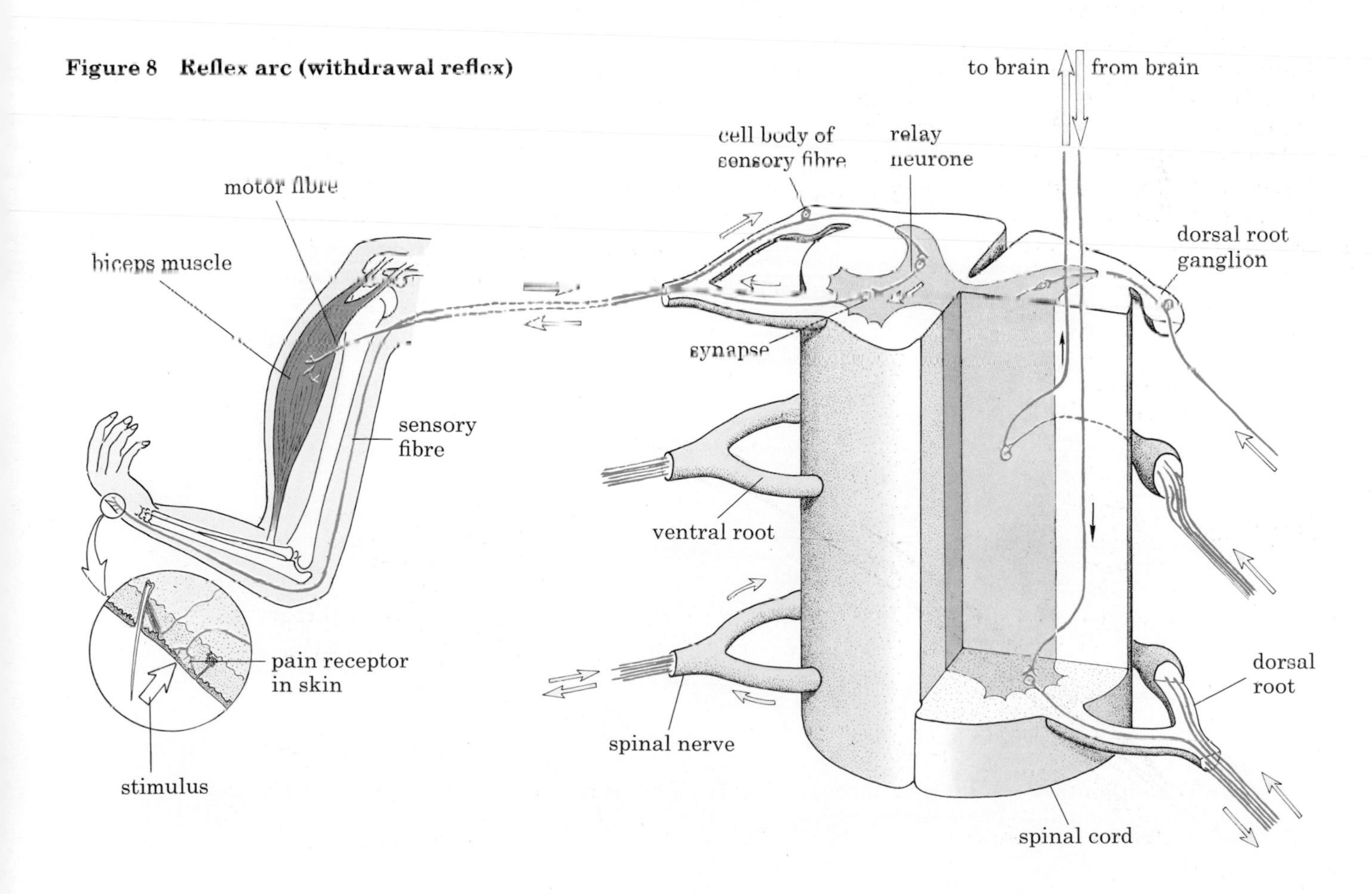

This can pass the impulse on to many other motor neurones, although only one is shown in the diagram. On the right side of the spinal cord, some of the incoming sensory fibres are shown making synapses with neurones which send nerve fibres to the brain, thus keeping the brain informed about events in the body. Also, nerve fibres from the brain make synapses with motor neurones in the spinal cord so that 'commands' from the brain can be sent to muscles of the body.

The reflex just described is a **spinal reflex**. The brain, theoretically, is not needed for it to happen. Responses which take place in the head, such as reflex blinking, reflex coughing and iris contraction, have their reflex arcs in the brain, but still cannot be consciously controlled.

Voluntary actions

A voluntary action starts in the brain. It may be the result of external events, such as seeing a book on the floor, but any resulting action, such as picking up the book, is entirely voluntary; it does not have to happen. The brain sends motor impulses down the spinal cord in the nerve fibres. These make synapses with motor fibres which enter spinal nerves and make connections to the sets of muscles needed to produce effective action. Many sets of muscles in the arms, legs and trunk would be brought into play in order to stoop and pick up the book, and impulses passing between the eye, brain and arm would direct the hand to the right place and 'tell' the fingers when to close on the book.

QUESTIONS

6 Explain why the tongue may be considered to be both a receptor and an effector organ.

7 Discuss whether coughing is a voluntary or reflex action.

8 Draw a simple diagram, similar to Figure 6*b*, for a reflex arc which would make you stand up quickly if you sat on a thistle. Refer to Figure 2 on page 154 for the sensory organs involved and to Figure 9 on page 153 for the muscles (effector organs) involved in standing up.

CENTRAL NERVOUS SYSTEM

Spinal cord

Like all other parts of the nervous system, the spinal cord consists of thousands of nerve cells. Figure 8 shows its structure as a diagram. Figure 6*b* shows it in transverse section and Figure 9 is a photograph of such a section. The diagrams on pages 148 and 149 show how it is protected by being enclosed in the spinal column.

All the cell bodies, apart from those in the dorsal root ganglion, are concentrated in the central region called the **grey matter**. The **white matter** consists of nerve fibres. Some of these will be passing from the grey matter to the spinal nerves and others will be running along the spinal cord connecting the spinal nerve fibres to the brain. The spinal cord is thus concerned with (a) reflex actions involving body structures below the neck, (b) conducting

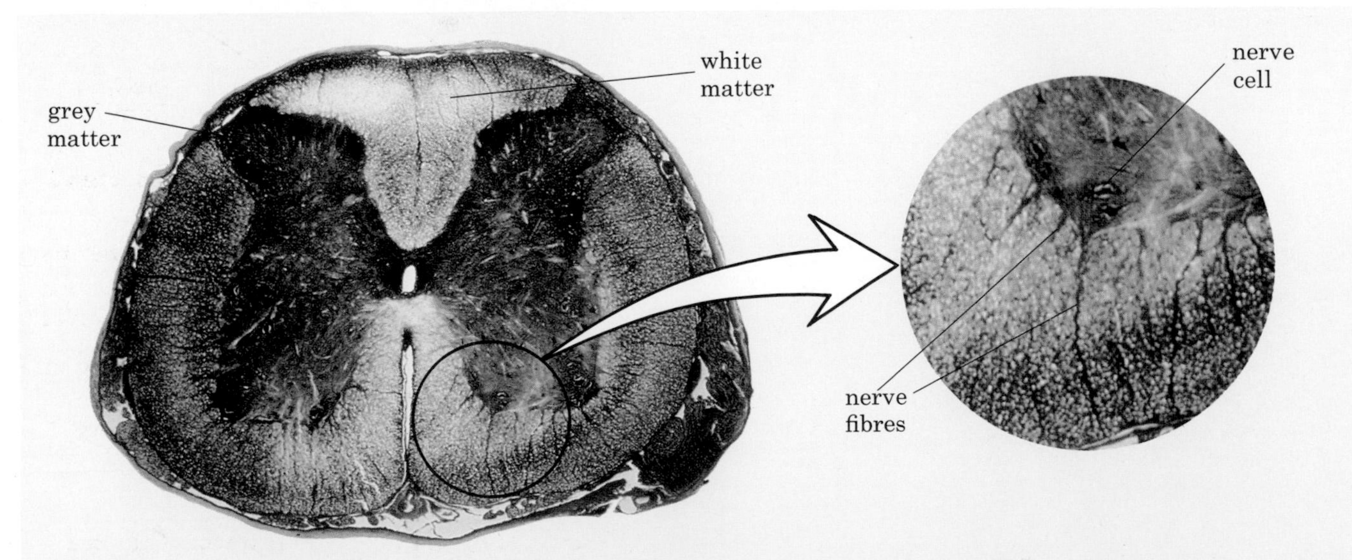

Figure 9 Section through the spinal cord (×10). The grey matter is here stained brown. The black blobs in the grey matter are the cell bodies of neurones.

sensory impulses from the skin and muscles to the brain and (c) carrying motor impulses from the brain to the muscles of the limbs and trunk.

The brain

The brain may be thought of as the expanded front end of the spinal cord (Figure 10). Certain areas are greatly enlarged to deal with all the information arriving from the ears, eyes, tongue, nose and semicircular canals. Figure 11 gives a simplified diagram of the main regions of the brain as seen in vertical section. The **medulla** is concerned with regulation of the heart beat, body temperature and breathing rate. The **cerebellum** controls balance and movement. The mid-brain deals with reflexes involving the eye. The largest part of the brain, however, consists of the **cerebral hemispheres**. These are very large and highly developed in mammals, especially man, and are thought to be the regions concerned with intelligence, memory, reasoning ability and acquired skills.

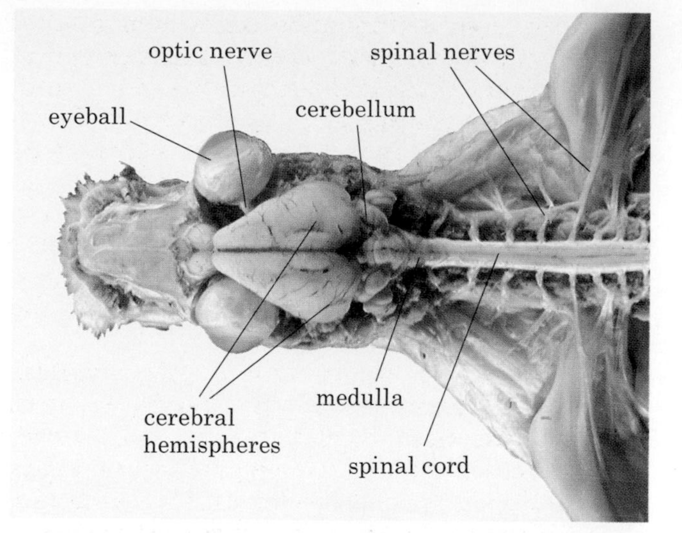

Figure 10 Dissection of the brain and spinal cord of a rabbit, seen from above. (Dissection by Griffin and George Ltd., Gerrard Biological Centre.)

In the cerebral hemispheres and the cerebellum, there is an outer layer of grey matter, the **cortex**, with hundreds of thousands of multi-

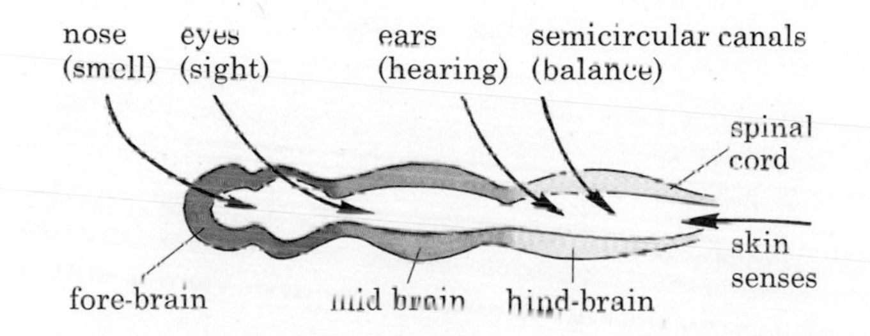

(*a*) The front end of the spinal cord develops three bulges: the fore-, mid- and hind-brain. Each region receives sensory impulses mainly from sense organs in the head.

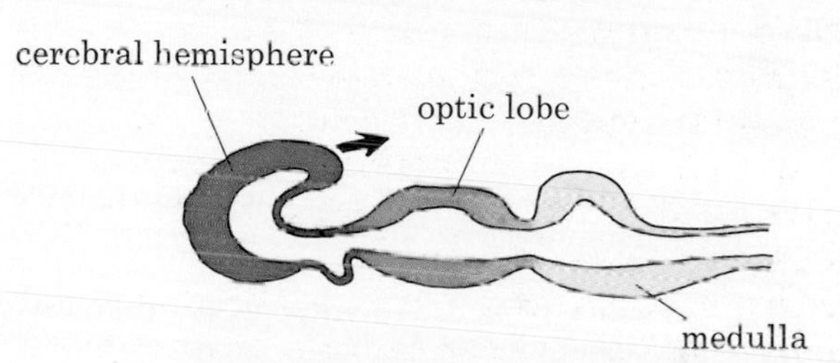

(*b*) The roofs of the fore-, mid- and hind-brain become thicker and form the cerebral hemispheres, optic lobes and cerebellum. The floor of the hind-brain thickens to form the medulla.

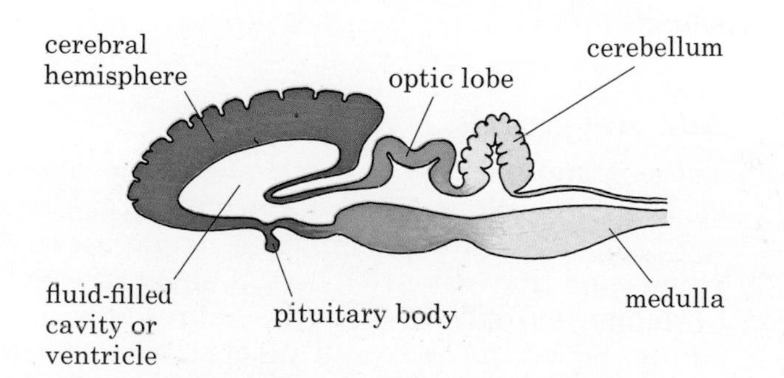

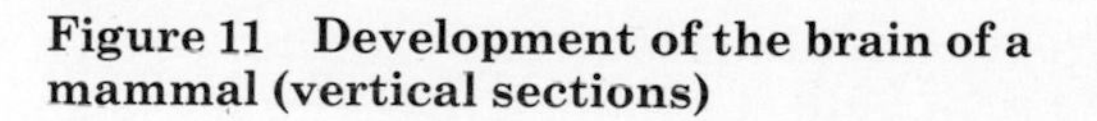

(*c*) A rabbit's brain would look something like this in vertical section.

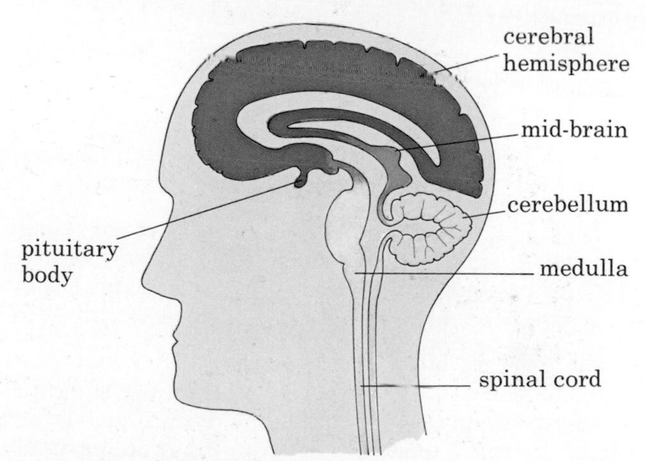

(*d*) The same regions are present in a man's brain but because of his upright position, the brain is bent through 90°.

Figure 11 Development of the brain of a mammal (vertical sections)

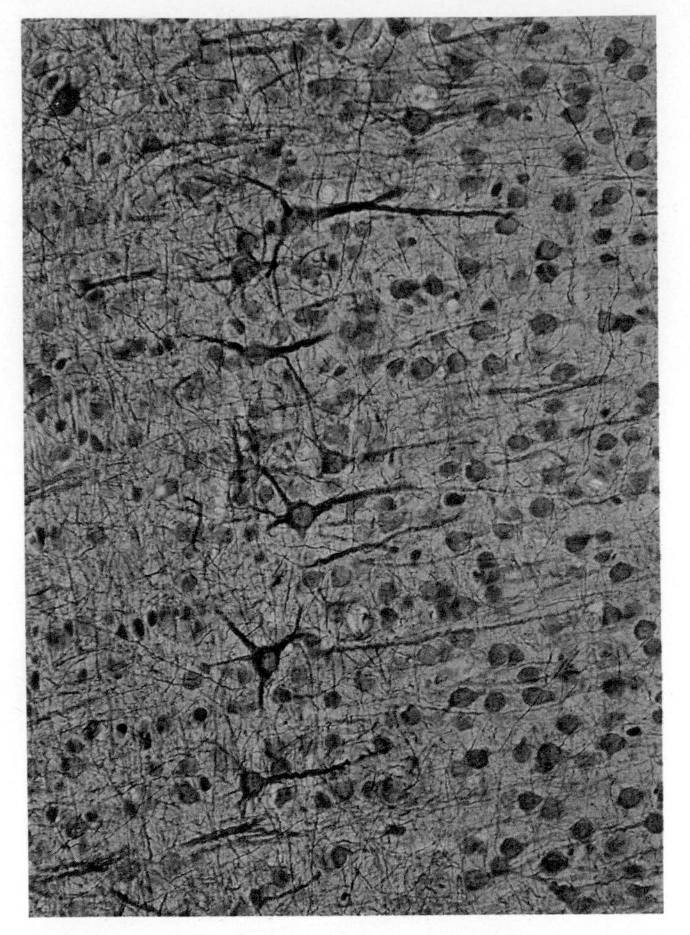

Figure 12 Multi-polar neurones in the brain cortex (×300). There are hundreds of cell bodies and nerve fibres in this small piece of brain tissue. Five of them have taken up the stain particularly well and you can see the nerve fibres branching from their cell bodies.

polar neurones (Figure 3*c*) forming the outer layers and making possible an enormous number of synapse connections between the dendrites (Figure 12).

QUESTIONS

9 Would you expect synapses to occur in grey matter or in white matter? Explain your answer.

10 Look at Figure 2. If the spinal cord were damaged at a point about one-third of the way up the vertebral column, what effect would you expect this to have on the bodily functions?

11 Suggest, in simple terms, a way in which the brain and nervous system might co-ordinate the rate and depth of breathing with the body's need for oxygen during exercise. For your answer, assume that there is a region of the medulla particularly sensitive to the oxygen concentration in the blood.

12 Describe the biological events involved when you hear a sound and turn your head towards it.

THE ENDOCRINE SYSTEM

The positions of the glands in the endocrine system are shown in Figure 13. These **endocrine glands** do not produce their effects by sending out nerve impulses but by releasing chemicals, called **hormones**, into the blood stream. The hormones circulate round the body in the blood and when they reach certain organs (**target organs**) they speed up or slow down or alter the activity of that organ in some way.

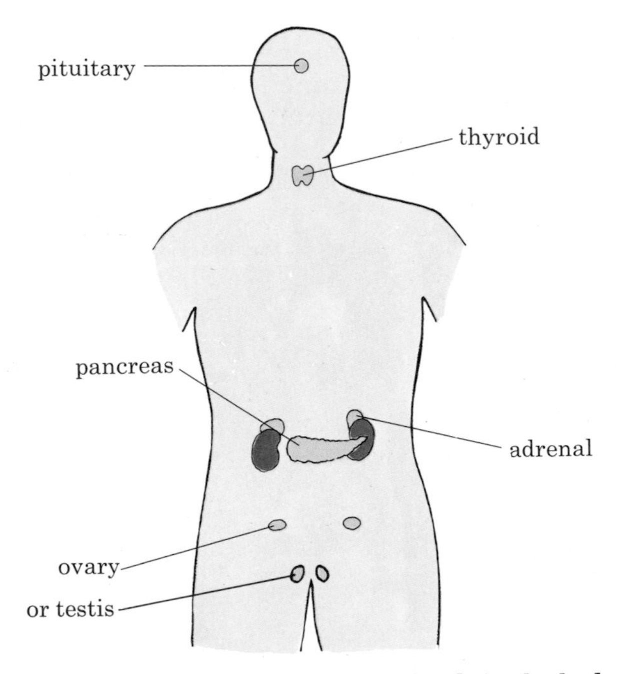

Figure 13 Position of endocrine glands in the body

When the hormones have done their job, they are changed by the liver into inactive compounds and excreted by the kidneys.

Adrenal glands

These glands are in the abdomen, one above each kidney. When we are alarmed or excited, the brain sends nerve impulses to the adrenal glands and they release into the blood stream a hormone called **adrenalin**. Adrenalin circulates round the body and affects the heart, the blood vessels and the muscles as shown in the table on page 171.

The changes listed in column 3 of the table show that the effect of adrenalin is to co-ordinate the heart, blood system and muscles to support effective action such as running away or putting up a fight. Column 4 lists the sensations that these changes produce, which we describe

1 Organ	2 Effects of adrenalin	3 Biological advantage	4 Effect or sensation
heart	beats faster	sends more glucose and oxygen to the muscles	thumping heart
blood vessels of the skin	constricts them (see page 137)	less blood to the skin means more blood is available for the muscles	person goes pale
blood vessels of the alimentary canal	constricts them	less blood going to the stomach and intestines means more is available for the muscles	dry mouth; slack, 'hollow' feeling in stomach
muscles	makes them tense	ready for immediate action	tense feeling or even shivering

The action of adrenalin on some systems of the body

as 'fear' or 'excitement'. If you feel like this before the start of a race, you are likely to perform to the best of your ability. The effects of adrenalin may also help you to give a sparkling, vigorous performance on stage, in spite of the first feelings of 'stage fright'. However, the effects can be a disadvantage for some people. Playing solo flute with a dry mouth, shaking hands and trembling knees is not likely to result in a good performance.

Adrenalin is not the only hormone produced by the adrenal gland and the functions mentioned above are only a selection from a much larger list (see question 14).

Thyroid gland

This endocrine gland is in the neck, in front of the windpipe. It releases a hormone called **thyroxine** into the blood stream. Thyroxine affects nearly all the cells in the body because it speeds up the rate of respiration (page 14). Too little thyroxine results in a low rate of respiration and sometimes causes sluggishness and fatness. Too much thyroxine in the body may make a person overactive.

Pancreas

Most of the pancreas cells produce digestive enzymes (page 105) but some of them, in little groups called **islets**, produce a hormone called **insulin**. When insulin circulates in the blood stream it controls the level of glucose in the blood and the amount of glucose which is stored in the liver as glycogen (see page 108). A person whose pancreas produces too little insulin is **diabetic**. After a meal, his liver fails to remove the extra sugar from the blood and the blood sugar level rises so high that glucose is excreted by the kidneys into the urine. There are many other effects which result from a shortage of insulin but they can be kept under control by regular injections of the hormone, and by reducing the quantity of carbohydrates in the diet.

Reproductive organs

These produce hormones as well as gametes (sperms and ova) and their effects have been described on page 145.

The hormones from the ovary, **oestrogen** and **progesterone**, both prepare the uterus, for the implantation of the embryo, by making its lining thicker and increasing its blood supply.

The hormones **testosterone** (from the testes) and oestrogen (from the ovaries) play a part in the development of the secondary sexual characters as described on page 145.

Pituitary gland

This gland is attached to the base of the brain (Figure 11*d*). It produces many hormones. One of these acts on the kidneys and regulates the amount of water reabsorbed in the kidney tubules (page 131). Another pituitary hormone affects the growth rate of the body as a whole and the skeleton in particular. Several of the pituitary hormones act on the other endocrine glands and stimulate them to produce their own hormones. For example, the pituitary releases into the blood a **follicle-stimulating hormone** (FSH) which, when it reaches the ovaries, makes one of the follicles start to mature and to produce oestrogen.

Feedback

When the level of oestrogen in the blood rises, it affects the pituitary gland, making it produce less follicle-stimulating hormone. A low level of FSH in the blood reaching the ovary will cause the ovary to slow down its production of oestrogen. With less oestrogen in the blood, the pituitary is able to resume its production of FSH which, in turn, makes the ovary start to produce oestrogen again (Figure 14). This cycle of events takes about one month and is the basis of the monthly menstrual cycle.

The oestrogen and progesterone in the female contraceptive pill act on the pituitary and suppress the production of FSH. If there is not enough FSH, none of the follicles in the ovary will grow to maturity and so no ovum will be released.

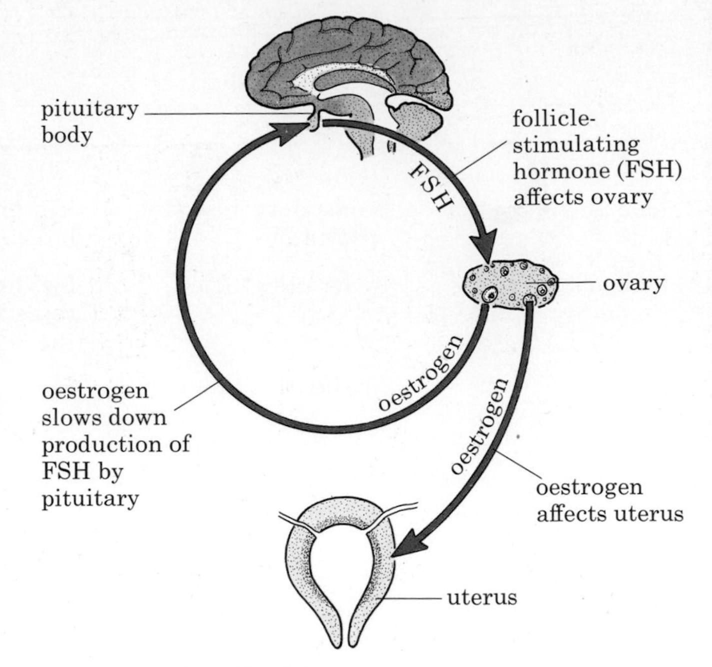

Figure 14 'Feedback'

QUESTIONS

13 One of the results of too little insulin is that the cells of the body cannot take up glucose efficiently from the blood or use it in normal respiration. What effect do you think this has on a diabetic patient?

14 Some effects of adrenalin not listed in the table are, to make the pupils of the eyes wider, to make the

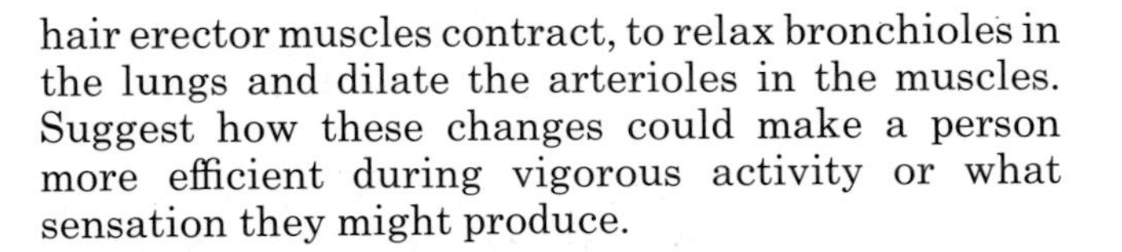
hair erector muscles contract, to relax bronchioles in the lungs and dilate the arterioles in the muscles. Suggest how these changes could make a person more efficient during vigorous activity or what sensation they might produce.

BASIC FACTS

- **The body systems are made to work efficiently together by the nervous system and the endocrine system.**
- **The nervous system consists of the brain, the spinal cord and the nerves.**
- **The nerves consist of bundles of nerve fibres.**
- **Each nerve fibre is a thin filament which grows out of a nerve cell body.**
- **The nerve cell bodies are mostly in the brain and spinal cord.**
- **Nerve fibres carry electrical impulses from sense organs to the brain or from the brain to muscles and glands.**
- **A reflex is an automatic nervous reaction that cannot be consciously controlled.**
- **A reflex arc is the nervous pathway which carries the impulses causing a reflex action.**
- **The simplest reflex involves a sensory nerve cell and a motor nerve cell, connected by synapses in the spinal cord.**
- **The brain and spinal cord contain millions of nerve cells.**
- **The millions of possible connections between the nerve cells in the brain allow complicated actions, learning, memory and intelligence.**
- **The thyroid, adrenal and pituitary are all endocrine glands.**
- **The testes, ovaries and pancreas are also endocrine glands in addition to their other functions.**
- **The endocrine glands release hormones into the blood system.**
- **When the hormones reach certain organs they change the rate or kind of activity of the organ.**

SECTION FOUR

Genetics and Evolution

22 Heredity and genetics

GENETICS

We often talk about people inheriting certain characteristics: 'John has inherited his father's curly hair', or 'Mary has inherited her mother's blue eyes.' We expect tall parents to have tall children. The inheritance of such characteristics is called **heredity** and the branch of biology which studies how heredity works is called **genetics**.

Genetics also tries to forecast what sort of offspring are likely to be produced when plants or animals reproduce sexually. What will be the eye colour of children whose mother has blue eyes and whose father has brown eyes? Will a mating between a black mouse and a white mouse produce grey mice, black and white mice or some black and some white mice?

To understand the method of inheritance, we need to look once again at the process of sexual reproduction and fertilization. In sexual reproduction, a new organism starts life as a single cell called a **zygote** (page 138). This means that *you* started from a single cell. Although you were supplied with oxygen and food in the uterus, all your tissues and organs were produced by cell division from this one cell. So, the 'instructions' that dictated which cells were to become liver, or muscle, or bone must all have been present in this first cell. The 'instructions' which decided that you should be tall or short, dark or fair, male or female must also have been present in the zygote.

To understand how these 'instructions' are passed from cell to cell, we need to look in more detail at what happens when the zygote divides and produces an organism consisting of thousands of cells. This type of cell division is called **mitosis**. It does not take place only in a zygote but occurs in all growing tissues.

QUESTION

1 (a) What are gametes? What are the male and female gametes of (i) plants and (ii) animals called, and where are they produced?

(b) What happens at fertilization?

(c) What is a zygote and what does it develop into?

(The information needed to answer these questions is given on pages 77 and 138.)

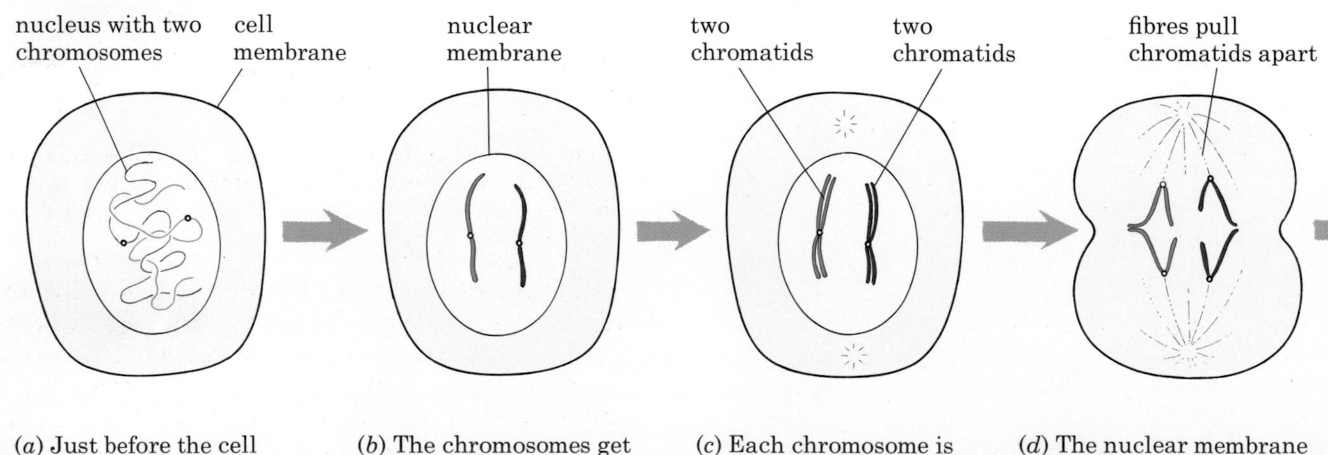

Figure 1 Mitosis. Three of the stages described here are shown in Figure 2.

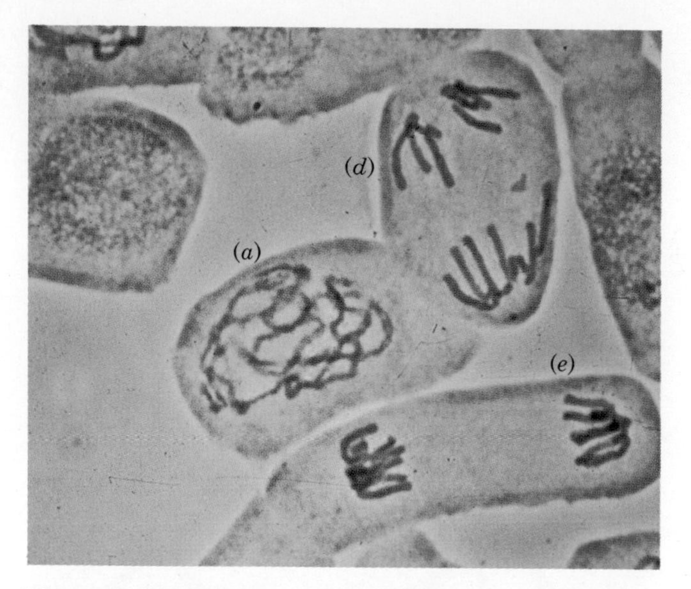

Figure 2 Mitosis in crocus cells. The letters refer to the stages described in Figure 1. (The tissue has been squashed to separate the cells.)

CHROMOSOMES AND MITOSIS

When a cell is not dividing, there is not much detailed structure to be seen in the nucleus even if it is treated with special dyes called stains. Just before cell division, a number of long, thread-like structures appear in the nucleus and show up very clearly when the nucleus is stained (Figures 1*a* and 2). These thread-like structures are called **chromosomes**. Although they are present in the nucleus all the time, they only show up clearly at cell division because at this time they get shorter and thicker.

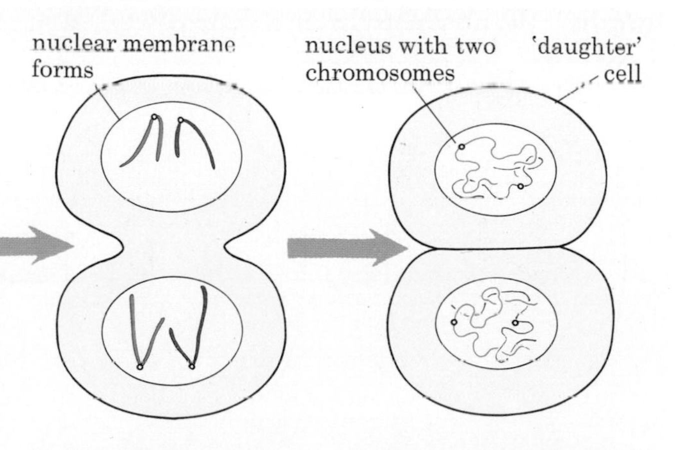

(*e*) A nuclear membrane forms round each set of chromatids, and the cell starts to divide

(*f*) Cell division complete giving two 'daughter' cells, each containing the same number of chromosomes as the parent cell

Each chromosome is
parallel strands, called
nucleus divides into t
each chromosome g
nucleus. The chroma
become chromosomes
copies of themselves
division. The proce
replication because
replica (copy) of itself.
mitosis, showing only two chromosomes,
there are always more than this. Human cells contain 46 chromosomes.

QUESTIONS

2 In the nucleus of a human cell just before cell division, how many chromatids will there be?

3 Why can chromosomes not be seen when a cell is not dividing?

4 Look at Figure 12 on page 64. Where would you expect mitosis to be occurring most often?

5 In which human tissues would you expect mitosis to be going on in (a) a five-year-old, (b) an adult?

The function of chromosomes

When a cell is not dividing, its chromosomes become very long and thin. Along the length of the chromosome is a series of chemical structures called **genes** (Figure 3). The chemical

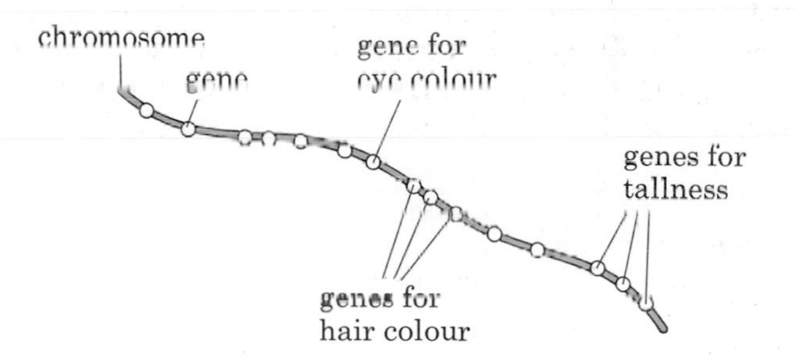

Figure 3 Relationship between chromosomes and genes. The drawing does not represent real genes or a real chromosome. There are probably thousands of genes on a chromosome.

which forms the genes is called DNA (which is short for deoxyribonucleic acid). Each gene controls some part of the chemistry of the cell. It is these genes which provide the 'instructions' mentioned at the beginning of the chapter. For example, one gene may 'instruct' the cell to make the pigment which is formed in the iris of brown eyes. On one chromosome there will be a gene which causes the cells of the stomach to make the enzyme pepsin. When the chromosome

…uilds up an exact replica of itself, …e (Figure 4). When the chromatids …t mitosis, each cell will receive a full … genes. In this way, the chemical in…tions in the zygote are passed on to all cells …he body. All the chromosomes, all the genes …nd, therefore, all the 'instructions' are faithfully reproduced by mitosis and passed on complete to all the cells.

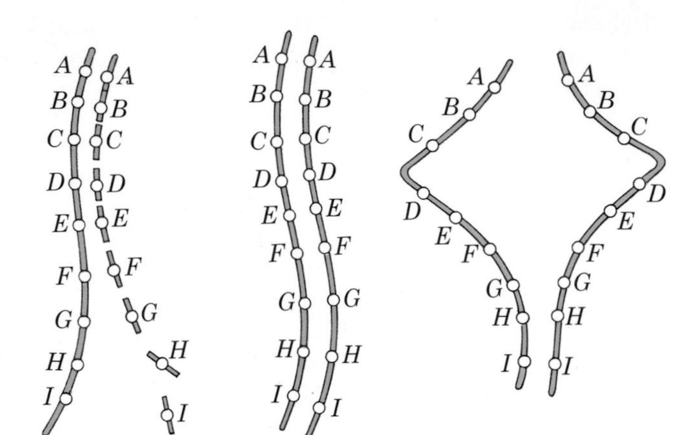

(*a*) A chromosome builds up a replica of itself

(*b*) The original and the replica are called chromatids

(*c*) Mitosis separates the chromatids. Each new cell gets a full set of genes

Figure 4 Replication. *A*, *B*, *C*, etc. are genes.

Which of the 'instructions' are used depends on where a cell finally ends up. The gene which causes brown eyes will have no effect in a stomach cell and the gene for making pepsin will not function in the cells of the eye. So the gene's chemical instructions are carried out only in the correct situation.

QUESTION

6 Genes are too small to be seen even with a powerful microscope. If it *were* possible to see them, where would you look for them?

Number of chromosomes

(a) There is a fixed number of chromosomes in each species. Man's body cells each contain 46 chromosomes, mouse cells contain 40, and garden pea cells 14 (see Figure 5).

(b) The number of chromosomes in a species is the same in all of its body cells. There are 46 chromosomes in each of your liver cells, in every nerve cell, skin cell and so on.

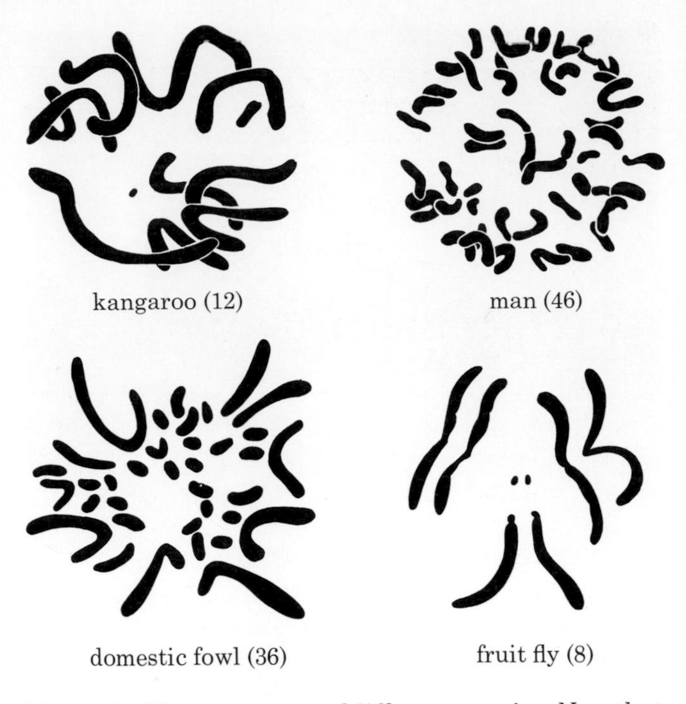

Figure 5 Chromosomes of different species. Note that chromosomes are always in pairs.

(c) The chromosomes have different shapes and sizes and can be recognized by a trained observer.

(d) The chromosomes are always in pairs (Figure 5), e.g. two long ones, two short ones, two medium ones. This is because when the zygote is formed, one of each pair comes from the male gamete and one from the female gamete. Your 46 chromosomes consist of 23 from your mother and 23 from your father.

(e) The number of chromosomes in each body cell of a plant or animal is called the **diploid number**. Because the chromosomes are in pairs, it is always an even number.

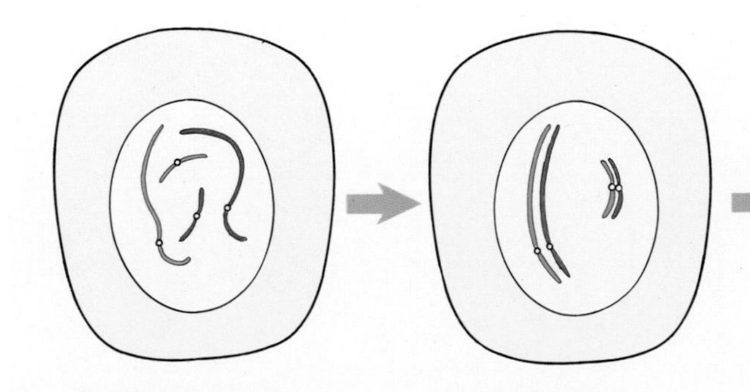

(*a*) The chromosomes appear. The red ones are from the organism's mother, and the blue ones are from its father

(*b*) Corresponding chromosomes lie alongside each other

QUESTIONS

7 How many chromosomes would there be in the nucleus of (a) a human muscle cell, (b) a mouse kidney cell, (c) a human skin cell that has just been produced by mitosis?

8 What is the diploid number in humans?

GAMETE PRODUCTION AND CHROMOSOMES

The genes on the chromosomes carry the 'instructions' which turn a single-cell zygote into a bird, or a rabbit or an oak tree. The zygote is formed when a male gamete fuses with a female gamete. Each gamete brings a set of chromosomes to the zygote. The gametes, therefore, must each contain only half the diploid number of chromosomes, otherwise the chromosome number would double each time an organism reproduced sexually. Each human sperm cell contains 23 chromosomes and each human ovum has 23 chromosomes. When the sperm and ovum fuse at fertilization (page 141), the diploid number of 46 (23 + 23) chromosomes is produced (Figure 6).

The process of cell division which gives rise to gametes is different from mitosis because it results in the cells containing only half the diploid number of chromosomes. This number is called the **haploid number** and the process of cell division which gives rise to gametes is called **meiosis**.

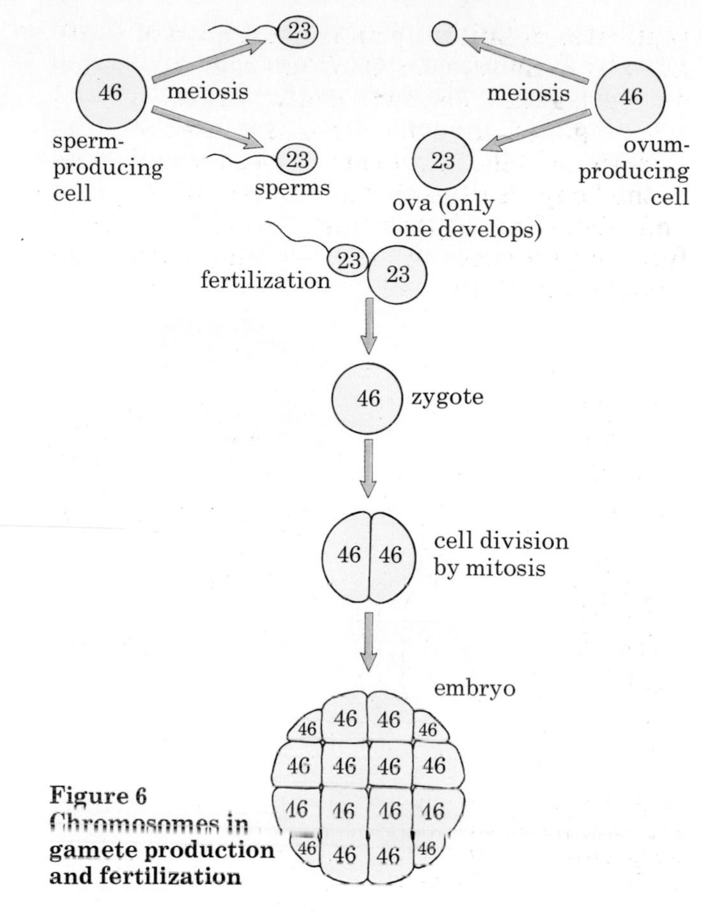

Figure 6
Chromosomes in gamete production and fertilization

Meiosis

In a cell which is going to divide and produce gametes the diploid number of chromosomes appears. The pairs of corresponding chromosomes, e.g. the two long ones and the two short ones in Figure 7*b* lie alongside each other and, when the nucleus divides for the first time, it is the chromosomes and not the chromatids which are separated. This results in only half the total number of chromosomes going to each daughter cell. In Figure 7*d* the diploid number of four chromosomes is reduced to two chromosomes after the first cell division.

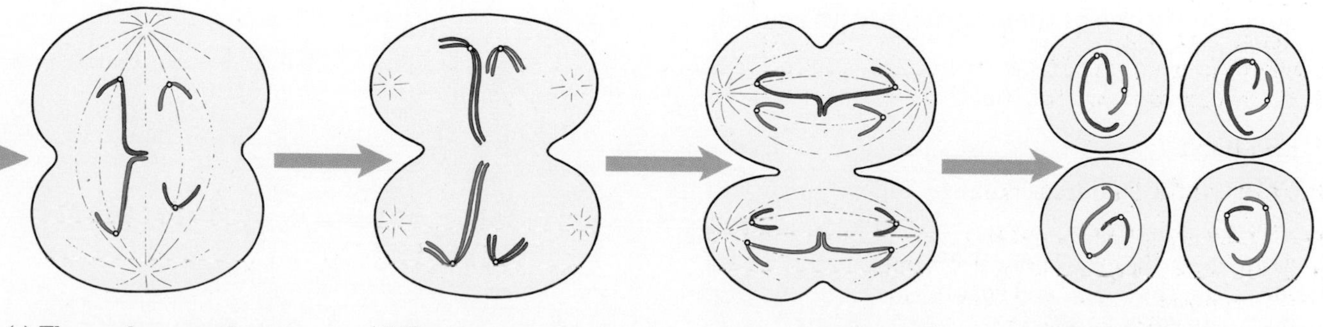

(*c*) The nuclear membrane disappears, and corresponding chromosomes move apart to opposite ends of the cell

(*d*) By now each chromosome has become two chromatids

(*e*) A second division takes place to separate the chromatids

(*f*) Four gametes are formed. Each contains only two chromosomes

Figure 7 Meiosis

By now, each chromosome is seen to consist of two chromatids and there is a second division of the nucleus which separates the chromatids into four distinct nuclei. This gives rise to four gametes, each with the haploid number of chromosomes. In the anther of a plant, four haploid pollen grains are produced when a cell divides by meiosis (Figure 8). In the testis of an animal, meiosis of each sperm-producing cell forms four sperms. In the cells of the ovule of a flowering plant or in the ovary of a mammal, meiosis gives rise to only one female gamete instead of four. Although four gametes may be produced, only one of them turns into an egg cell which can be fertilized.

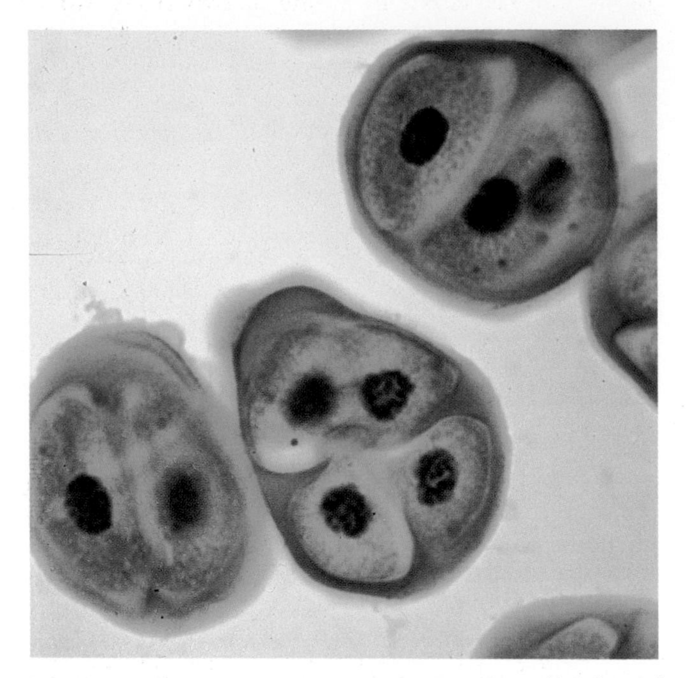

Figure 8 Meiosis in an anther. The last division of meiosis in the anther of a flower produces four pollen grains. The cell in the middle is viewed 'end on' so that the four grains are clearly seen. The other two cells are seen 'sideways on' so that only two or three of the four nuclei are visible.

QUESTIONS

9 What is the haploid number for (a) man, (b) mouse?

10 How does meiosis differ from mitosis (a) in its first division, (b) in the end result?

11 Where in the body of (a) a human male, (b) a human female and (c) a plant, would you expect meiosis to be taking place?

12 How many chromosomes would be present in (a) a mouse sperm cell, (b) a mouse ovum?

HEREDITY

A knowledge of mitosis and meiosis allows us to explain, at least to some extent, how heredity works. The gene in a mother's body cells which causes her to have brown eyes may be present on one of the chromosomes in each ovum she produces. If the father's sperm cell contains a gene for brown eyes on the corresponding chromosome, the zygote will receive a gene for brown eyes from each parent. These genes will be reproduced by mitosis in all the embryo's body cells and when the embryo's eyes develop, the genes will make the cells of the iris produce brown pigment and the child will have brown eyes.

In a similar way, the child may receive genes for curly hair. Figure 9 shows this happening, but it does not, of course, show all the other chromosomes with thousands of genes for producing the enzymes, making different types of cell and all the other processes which control the development of the organism.

Single factor inheritance

Because it is impossible to follow the inheritance of the thousands of characteristics controlled by genes, it is usual to start with the study of a single gene which controls one characteristic. We have used eye colour as an example so far. Probably more than one gene pair is involved, but the simplified example will serve our purpose. It was explained above how a gene for brown eyes from each parent would result in the child having brown eyes. Suppose, however, that the mother has blue eyes and the father brown eyes. The child might receive a

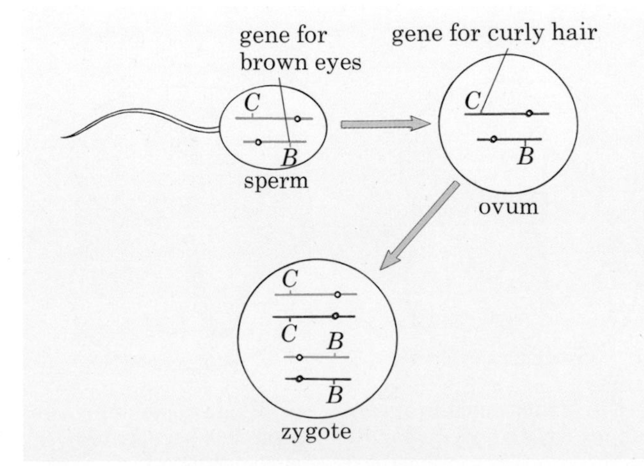

Figure 9 Fertilization. Fertilization restores the diploid number of chromosomes and combines the genes from the mother and father.

gene for blue eyes from its mother and a gene for brown eyes from its father (Figure 10). If this happens, the child will, in fact, have brown eyes. The gene for brown eyes is said to be **dominant** to the gene for blue eyes. Although the gene for blue eyes is present in all the child's cells, it does not contribute to the eye colour. It is said to be **recessive** to brown.

This example illustrates the following important points:

(a) There is a pair of genes for each characteristic, one gene from each parent. These genes are called **alleles**.

(b) The alleles control the same characteristic, e.g. eye colour, but they may have different effects. For instance, one tries to produce blue eyes, the other tries to produce brown eyes.

(c) Often one allele is dominant over the other.

(d) The alleles of each pair are on corresponding chromosomes and occupy corresponding positions, e.g. in Figure 9 the alleles for eye colour are shown in the corresponding position on the two short chromosomes and the alleles for hair curliness are in corresponding positions on the two long chromosomes.

In diagrams and explanations of heredity

(i) genes are represented by letters,

(ii) genes controlling the same characteristic are called alleles and are given the same letter;

(iii) the dominant allele is given the capital letter.

For example, in rabbits, the dominant allele for black fur is labelled *B*. The recessive allele for white fur is labelled *b* to show that it corresponds to *B* for black fur. If it were labelled *w*, we would not see any connection between *B* and *w*. *B* and *b* are obvious 'partners' (alleles). In the same way *L* could represent the allele for long fur and *l* the allele for short fur.

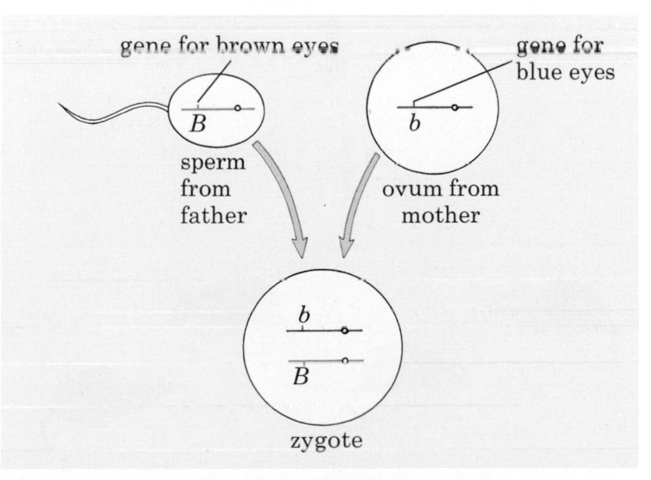

Figure 10 Combination of genes in the zygote (only one chromosome is shown). The zygote has both genes for eye colour; the child will have brown eyes.

QUESTIONS

13 Some plants occur in one of two sizes, tall or dwarf. This characteristic is controlled by one pair of alleles. Tallness is dominant to shortness.
Choose suitable letters for the alleles.

14 Why are there two genes controlling one characteristic? Do the two genes affect the characteristic in the same way as each other?

15 The allele for red hair is recessive to the allele for black hair. What colour hair will a person have if he inherits the allele for red hair from his mother and the allele for black hair from his father?

Breeding true

A white rabbit must have both the recessive alleles *b* and *b*. If it had *B* and *b*, the dominant allele for black (*B*) would override the allele for white (*b*) and produce a black rabbit. A black rabbit, on the other hand, could be either *BB* or *Bb* and, by just looking at the rabbit, you could not tell the difference. When the male black rabbit *BB* produces sperms by meiosis, each one of the pair of chromosomes carrying the *B* alleles will end up in different sperm cells. Since the alleles are the same, all the sperms will have the *B* allele for black fur (Figure 11*a*).

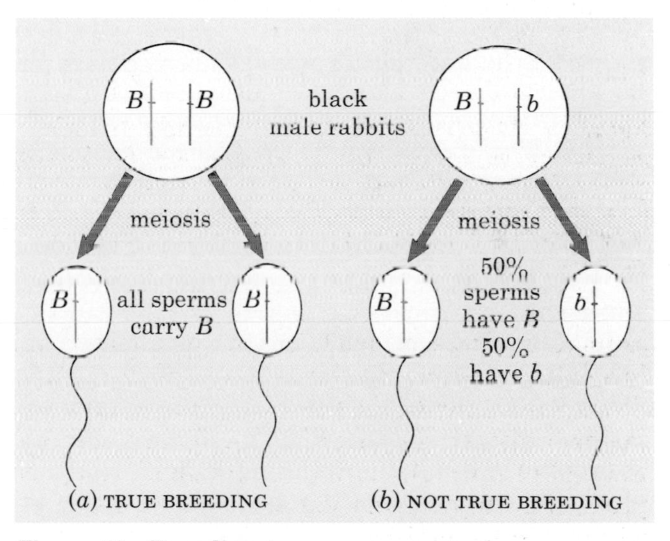

Figure 11 Breeding true

The black rabbit *BB* is called a true-breeding black and is said to be **homozygous** for black coat colour (*homo-* means 'the same'). If this rabbit mates with another black (*BB*) rabbit, all the babies will be black because all will receive a dominant allele for black fur. When all the offspring have the same characteristic as the parents, this is called 'breeding true' for this character.

When the *Bb* black rabbit produces gametes by meiosis, the chromosomes with the *B* alleles and the chromosomes with the *b* alleles will end up in different gametes. So 50 per cent of the sperm cells will carry *B* alleles and 50 per cent will carry *b* alleles (Figure 11*b*). Similarly, in the female, 50 per cent of the eggs will have a *B* allele and 50 per cent will have a *b* allele. If a *b* sperm fertilizes a *b* egg, the offspring, with two *b* alleles (*bb*), will be white. The black *Bb* rabbits are not true-breeding because they may produce some white babies as well as black ones. The *Bb* rabbits are called **heterozygous** (*hetero-* means 'different').

The black *BB* rabbits are homozygous dominant.

The white *bb* rabbits are homozygous recessive.

QUESTIONS

16 (a) Read question 15 again. Choose letters for the alleles for red hair and black hair and write down the combination of alleles for having red hair.

(b) Would you expect a red-haired couple to breed true?

(c) Could a black-haired couple have a red-haired baby?

17 Use the words **homozygous**, **heterozygous**, **dominant** and **recessive** (where suitable) to describe the following combinations of alleles: *Aa*, *AA*, *aa*.

18 A plant has two varieties, one with red petals and one with white petals. When these two plants are cross-pollinated, all the offspring have red petals. Which allele is dominant? Choose suitable letters to represent the two alleles.

Genotype and phenotype

The two kinds of black rabbit *BB* and *Bb* are said to have the same **phenotype**. This is because their coat colours look exactly the same. However, because they have different alleles for coat colour they are said to have different **genotypes**, i.e. different combinations of genes. One genotype is *BB* and the other is *Bb*.

You and your brother might both be brown-eyed phenotypes but your genotype could be *BB* and his could be *Bb*. You would be homozygous dominant for brown eyes; he would be heterozygous for eye colour.

The three to one ratio

Figure 12*a* shows the result of a mating between a true-breeding (homozygous) black mouse (*BB*), and a true-breeding (homozygous) brown mouse (*bb*). The illustration is greatly simplified because it shows only one pair of the 20 pairs of

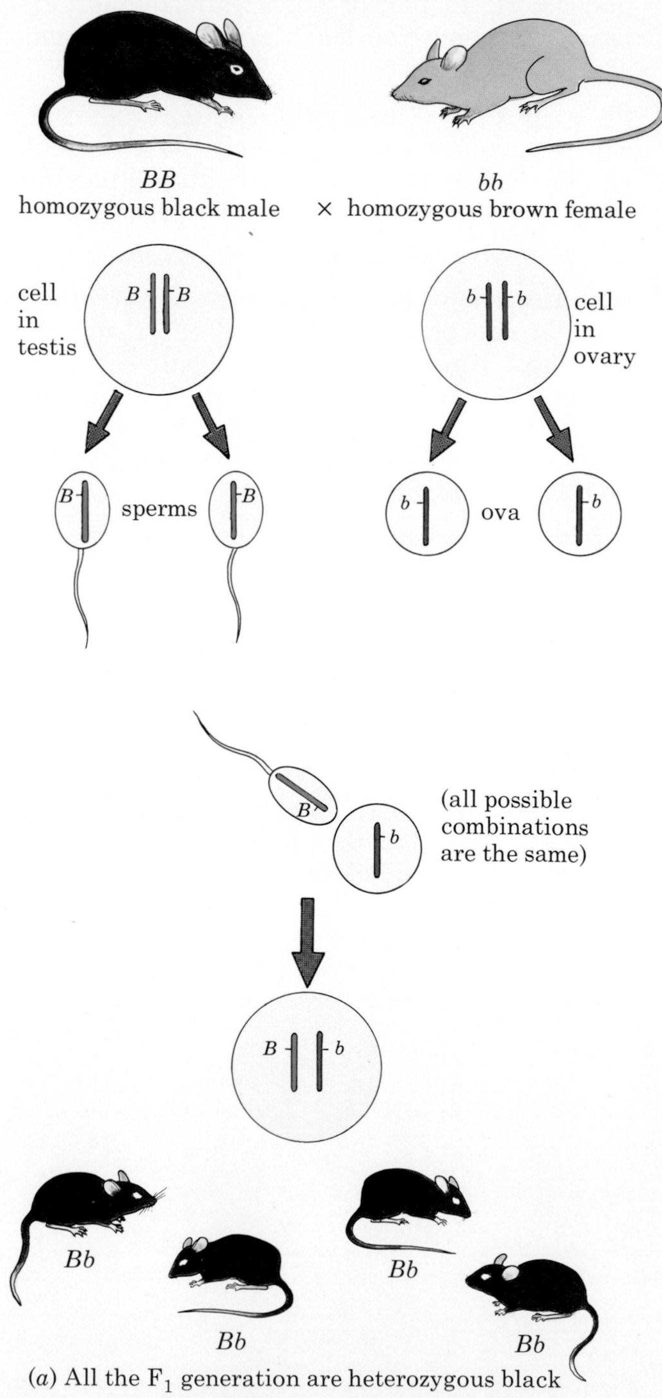

(*a*) All the F_1 generation are heterozygous black

Figure 12 Inheritance of coat colour in mice

mouse chromosomes and only one pair of alleles on the chromosomes.

Because black is dominant to brown, all the offspring from this mating will be black phenotypes, because they all receive a dominant allele for black fur from the father. Their genotypes, however, will be *Bb* because they all receive the recessive *b* allele from the mother. The offspring resulting from this first mating are called the **F_1 generation**.

Figure 12*b* shows what happens when these heterozygous, F_1 black mice are mated together to produce what is called the **F_2 generation**. Each sperm or ovum produced by meiosis can contain only one of the alleles for coat colour, either *B* or *b*. So there are two kinds of sperm cell, one kind with the *B* allele and one kind with the *b* allele. There are also two kinds of ovum with either *B* or *b* alleles. When fertilization occurs, there is no way of telling whether

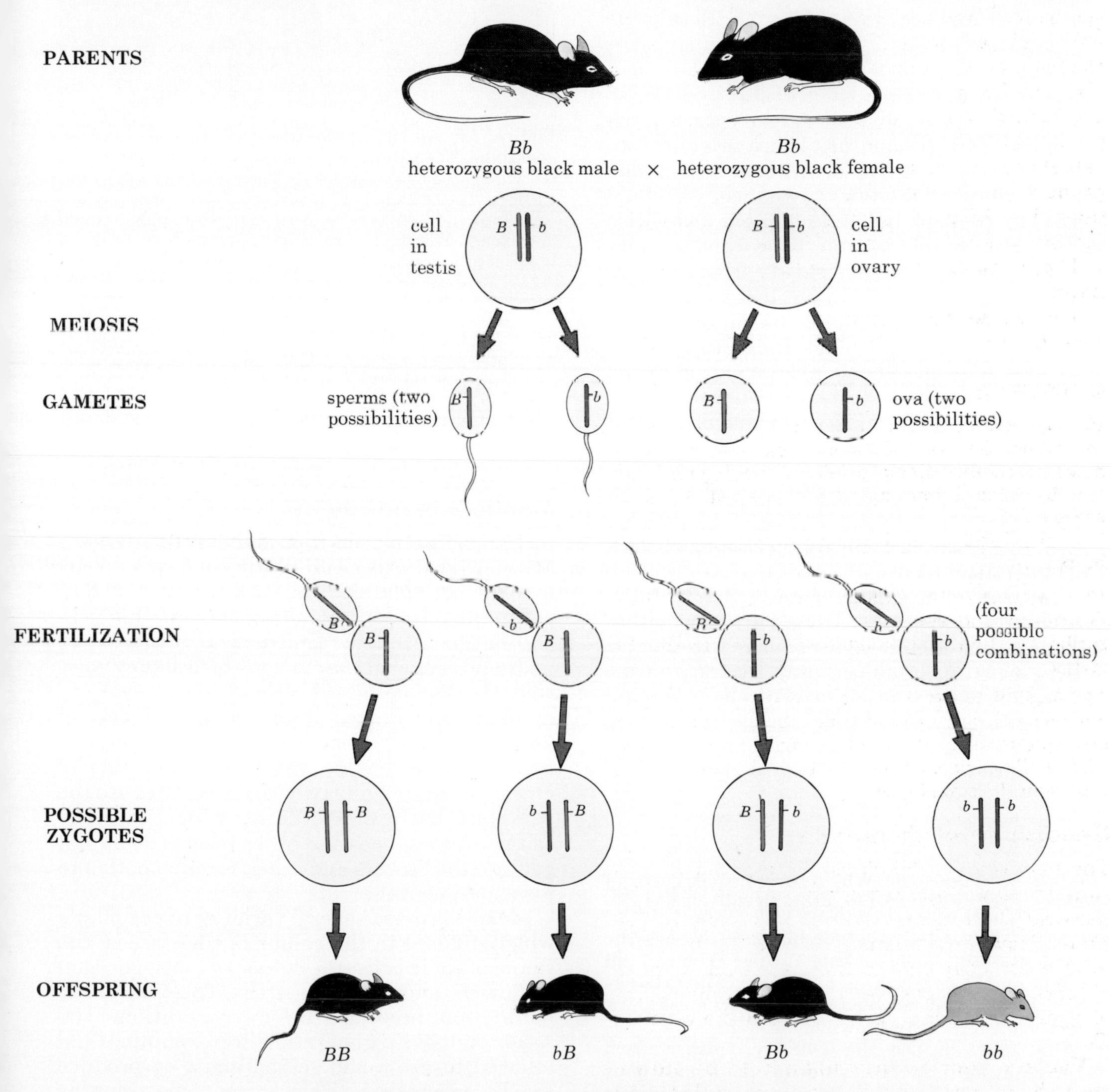

(*b*) The probable ratio of coat colours in the F_2 generation is 3 black: 1 brown a *b* or a *B* sperm will fertilize a *B* or a *b* ovum, so we have to look at all the possible combinations as follows:

(i) a *b* sperm fertilizes a *B* ovum. Result: *bB* zygote.

(ii) A *b* sperm fertilizes a *b* ovum. Result: *bb* zygote.

(iii) a *B* sperm fertilizes a *B* ovum. Result: *BB* zygote.

(iv) a *B* sperm fertilizes a *b* ovum. Result: *Bb* zygote.

There is no difference between *bB* and *Bb*, so there are three possible genotypes in the offspring—*BB*, *Bb* and *bb*. There are only two phenotypes—black (*BB* or *Bb*) and brown (*bb*). So, according to the laws of chance, we would expect three black babies and one brown. Mice usually have more than four babies and what we really expect is that the **ratio** (proportion) of black to brown babies will be close to 3:1.

If the mouse had 13 babies, you might expect 9 black and 4 brown, or 8 black and 5 brown. Even if she had 16 babies you would not expect to find exactly 12 black and 4 brown because whether a *B* or *b* sperm fertilizes a *B* or *b* ovum is a matter of chance. If you spun 10 coins, you would not expect to get exactly 5 heads and 5 tails. You would not be surprised at 6 heads and 4 tails or even 7 heads and 3 tails. In the same way, we would not be surprised at 14 black and 2 brown mice in a litter of 16.

To decide whether there really is a 3:1 ratio, we need a lot of results. These may come either from breeding the same pair of mice together for a year or so to produce many litters, or from mating 20 black and 20 brown mice, crossing the offspring, and adding up the number of black and brown babies in the F_2 families (see also Figure 13).

QUESTIONS

19 Look at Figure 12a. Why is there no possibility of getting a *BB* or a *bb* combination in the offspring?

20 You cannot tell whether black mice have the genotype *BB* or *Bb* by looking at them. You can tell the difference by a breeding experiment called a **back-cross**, i.e. a mating between either black mouse and a homozygous recessive brown mouse (*bb*).

What types of offspring would you expect from such a mating if the father was (a) the homozygous black mouse (*BB*), (b) the heterozygous black mouse (*Bb*)? (Work out what alleles will be in the gametes of the black and brown parents and see what possible combinations there are of these alleles combining at fertilization.)

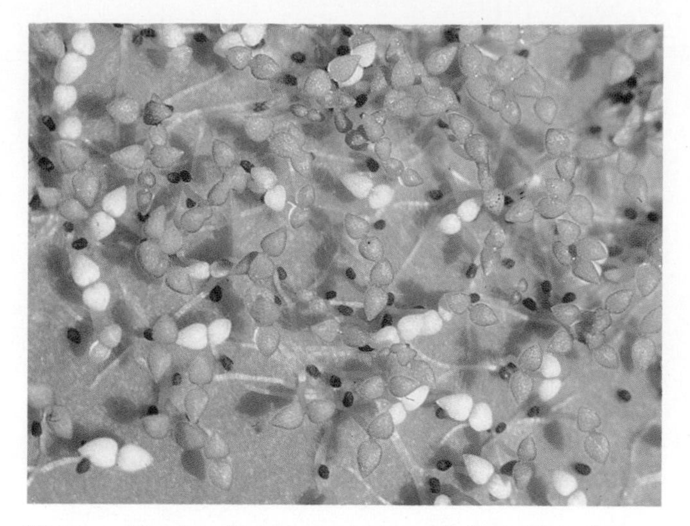

Figure 13 An F_2 generation of hybrid tobacco seedlings. The parents were heterozygous (*Cc*) for a gene (*C*) which produces chlorophyll. Their offspring will be *CC*, *Cc* or *cc*. The *cc* homozygous recessive plants can produce no chlorophyll. What ratio of green to white plants would you expect? What is the ratio in the picture? (There are 75 green seedlings present.)

21 In Figure 12*b* what proportion of the F_2 black mice are true-breeding?

Mendel's experiments

In the 1850s an Austrian monk called Gregor Mendel tried cross-pollinating different varieties of pea plants. The pea flower is self-pollinating (page 74) and to prevent this self-pollination, Mendel opened some of the flower buds and removed the stamens before they were ripe. He then collected pollen from one variety of plant and dusted it on to the stigmas of another variety from which the anthers had been removed. In this way he could be sure of cross-pollinating any two varieties. Because the pea plant has a closed flower which normally self-pollinates, bees and other insects could not get into the flowers and bring 'foreign' pollen to his experimental plants.

Mendel crossed several varieties of pea plants which differed in the colour of their seeds, the shape of their pods, the colour of pods, position of flowers and length of stem. The first cross-pollination produced an F_1 generation. The seeds of the F_1 generation were planted and allowed to grow and self-pollinate to produce the F_2 generation. (This is the same thing as mating the *Bb* mice together in Figure 12*b*.)

The following table shows some of Mendel's results:

		F_1	F_2	Ratio
yellow seeds	× green seeds	all yellow	6022 yellow 2001 green	3.01:1
green pods	× yellow pods	all green	428 green 152 yellow	2.82:1
tall stem	× short stem	all tall	787 tall 277 short	2.84:1

Notice that yellow seeds are dominant to green, green pods are dominant to yellow and tallness is dominant to shortness. Notice also that the results from the large number of offspring in the F_2 are nearly (but never exactly) 3:1. Figure 14 shows the result of a similar experiment using maize.

Figure 14 An F_2 maize cob. This is the F_2 offspring from a breeding experiment like Mendel's but using maize instead of peas. One of the alleles for colour gives yellow grains, the other gives dark grains. What was the colour of the seed which produced the plant with this cob?

From his results, Mendel deduced that each of the pea characteristics he studied was controlled by a pair of 'factors' (now called genes) and that only one of each pair of 'factors' could be present in a gamete. Since chromosomes were not discovered until 1882, these deductions were remarkable. Although Mendel published his results and conclusions in 1865, their importance was not realized until about 1900.

QUESTIONS

22 How would you try to cross two varieties of tulip (see page 73 for the flower structure)? Remember that tulip flowers are not 'closed' and are usually pollinated by insects.

23 Choose any one of Mendel's experiments from the table on this page and select letters for the genes. Using these letters, explain briefly (a) why all the F_1 generation were the same, and (b) why he obtained a 3:1 ratio in the F_2 generation.

Acquired and inherited characteristics

Your genes decide whether you have brown eyes. There is nothing you can do to change this because eye colour is an **inherited characteristic**. A fair-skinned person may be able to change the colour of his skin by exposing it to the sun, so getting a tan. The tan is an **acquired characteristic**. You cannot inherit a sun tan.

Many features in plants and animals are a mixture of acquired and inherited characteristics (Figure 15). For example, some fair-skinned people never go brown in the sun, they only become sun-burned. They have not inherited the genes for producing brown pigment in their skin. A fair-skinned person with the genes for producing pigment will go brown only if he exposes himself to sunlight. So his tan is a result of both inherited and acquired characteristics.

Height, physical strength and intelligence depend upon (a) **inherited** characteristics from the genes and (b) additional characteristics **acquired** during growth and development. Children who have the genes for tallness but are under-nourished will not grow to their full height. Also there is certainly more than one gene for each of these characteristics. There

Figure 15 Acquired characteristics. These apples have all been picked from different parts of the same tree. If all the apples have the same genotype, what environmental differences could have caused the variation in size?

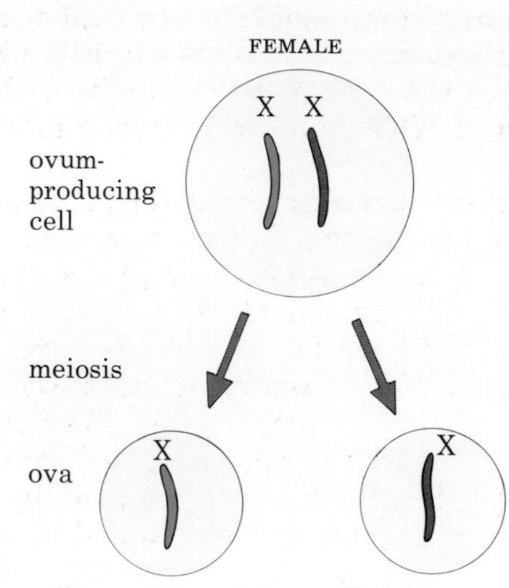

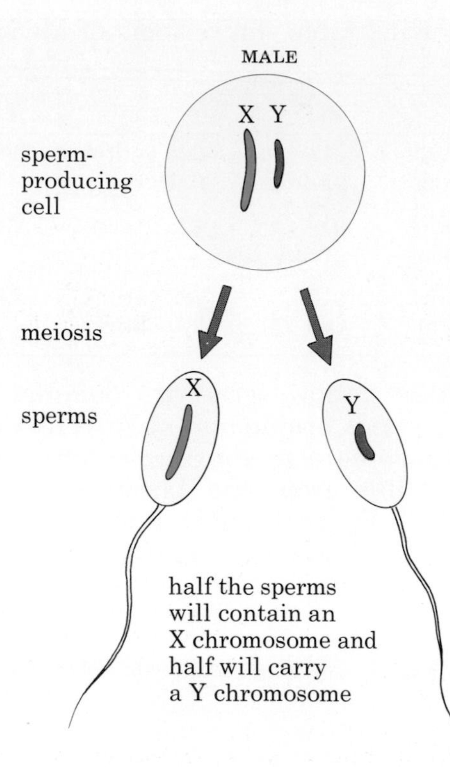

Figure 16 Determination of sex. Note that
(*i*) Only the X and Y chromosomes are shown
(*ii*) The Y chromosome is not smaller than the X in all animals
(*iii*) Details of meiosis have been omitted
(*iv*) In fact, four gametes are produced in each case, but two are sufficient to show the distribution of X and Y chromosomes

might be five pairs of alleles for height—(*Hh*), (*Tt*), (*Ll*), (*Bb*) and (*Gg*)—each dominant allele adding 4 cm to your height. If you inherited all ten dominant alleles (*HH, TT*, etc.) you could be 40 cm taller than a person who inherited all ten recessive alleles (*hh, tt*, etc.).

The actual number of genes which control height, intelligence and even the colour of hair and skin, is not known.

Sex determination

Whether you are a male or female depends on one particular pair of chromosomes called the sex chromosomes. In females, the two sex chromosomes, called the X chromosomes, are the same size as each other. In males, the two sex chromosomes are of different sizes. One corresponds to the female sex chromosomes and is called the X chromosome. The other is smaller and is called the Y chromosome. So the female genotype is XX and the male genotype is XY.

When meiosis takes place in the female's ovary, each ovum receives one of the X chromosomes, so all the ova are the same for this. Meiosis in the male's testes results in 50 per cent of the sperms getting an X chromosome and 50 per cent getting a Y chromosome (Figure 16). If an X sperm fertilizes the ovum, the zygote will be XX and grows into a girl. If a Y sperm fertilizes the ovum, the zygote will be XY and develops into a boy. There is an equal chance of an X or Y chromosome fertilizing an ovum, so the numbers of girl and boy babies are more or less the same.

QUESTIONS

24 Try to explain how fatness (overweight) might result from a mixture of inherited and acquired characteristics. Can you think of any evidence to support this idea? (Think of the people you know and their eating habits.)

25 Which of the following would you expect to be inherited and which acquired characteristics or a mixture of both: well-developed muscles, colour of hair, shape of nose, ability to read, rickets (page 94)?

Applied genetics

It is possible for biologists to use their knowledge of genetics to produce new varieties of plants and animals. For example, suppose one variety of wheat produces a lot of grain but is not resistant to a fungus disease. Another variety is resistant to the disease but has only a poor yield of grain. If these two varieties are cross-pollinated (Figure 17), the F_1 offspring should be disease-resistant and give a good yield of grain (assuming that the useful characteristics are controlled by dominant genes).

R represents a dominant allele for resistance to disease, and *r* is the recessive allele for poor

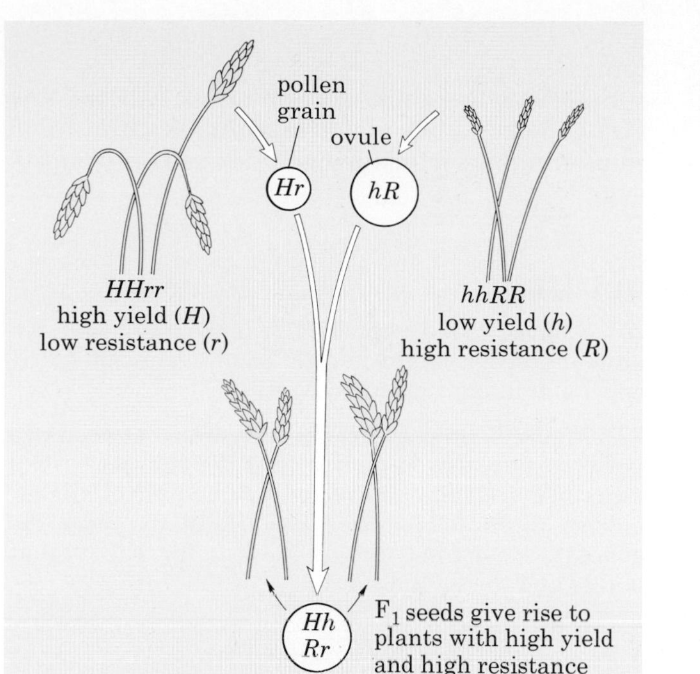

Figure 17 Combining useful characteristics

resistance. *H* is a dominant allele for high yield and *h* is the recessive allele for low yield. The high-yield/low-resistance variety (*HHrr*) is crossed with the low-yield/high-resistance variety (*hhRR*). Each pollen grain from the *HHrr* plant will contain one *H* and one *r* allele (*Hr*). Each ovule from the *hhRR* plant will contain an *h* and an *R* allele (*hR*). The seeds will therefore all be *HhRr*. The plants which grow from these seeds will have dominant alleles for both high yield and good disease resistance.

The offspring from crossing two varieties are called **hybrids**. If the F_1 hybrids from this cross bred true, they could give a new variety of disease-resisting, high-yielding wheat. As you learned on page 179, the F_1 generation from a cross does not necessarily breed true. The F_2 generation of wheat may contain

(a) high yield, disease resistant
(b) low yield, disease prone
(c) low yield, disease resistant } parental
(d) high yield, disease prone } types

This would not give such a successful crop as the F_1 plants.

With some commercial crops, the increased yield from the F_1 seed makes it worthwhile for the seedsman to make the cross and sell the seed to the growers. The hybrid corn (maize) grown in America is one example. The F_1 hybrid gives nearly twice the yield of the standard varieties.

In other cases it is possible to work out a cross-breeding programme to produce a hybrid which breeds true (Figure 18). If, instead of the *HhRr* in Figure 17, an *HHRR* could be produced, it would breed true.

An important part of any breeding programme is the selection of the desired varieties. The largest fruit on a tomato plant might be picked and its seeds planted next year. In the next generation, once again only seeds from the largest tomatoes are planted. Eventually it is possible to produce a true-breeding variety of tomato plant which forms large fruits. Figure 19 shows the result of such selective breeding.

The same principles can be applied to farm animals. Desirable characteristics, such as high milk yield and resistance to disease, may be combined. Stock breeders will select calves from cows which give large quantities of milk. These calves will be used for breeding stock to build a herd of high yielders. A characteristic such as milk yield is probably under the control of many genes. At each stage of selective breeding the farmer, in effect, is keeping the beneficial genes

(*a*) (*b*) (*c*) (*d*) (*e*)

Figure 18 The genetics of bread wheat. A primitive wheat (*a*) was crossed with wild grass (*b*) to produce a better-yielding hybrid wheat (*c*). The hybrid wheat (*c*) was crossed with another wild grass (*d*) to produce one of the varieties of wheat (*e*) which is used for making flour and bread.

Figure 19 Selective breeding in tomatoes. Different breeding programmes have selected different genes for fruit size, colour and shape. Similar processes have given rise to most of our cultivated plants and domesticated animals.

and discarding the less useful genes from his animals.

Selective breeding in farm stock can be slow and expensive because the animals often have small numbers of offspring and breed only once a year.

QUESTIONS

26 Suggest some good characteristics that an animal-breeder might try to combine in sheep by mating different varieties together.

27 Figure 17 on page 223 shows a cross between two varieties of plant. Assuming that the characteristics described are each controlled by one pair of alleles, choose letters to represent the alleles and write the genotype (combination of alleles) for the parent plants and the F_1 hybrid.

BASIC FACTS

- **In the nuclei of all cells there are thread-like structures called chromosomes.**
- **The chromosomes are in pairs; one of each pair comes from the male and one from the female parent.**
- **On these chromosomes are carried the genes.**
- **The genes are in pairs, because the chromosomes are in pairs.**
- **Each pair of genes controls a characteristic. These genes are called alleles.**
- **Although both alleles in a pair control the same characteristic, they do not necessarily have the same effect. For example, of a pair of alleles controlling fur colour, one may try to produce black fur and the other may try to produce white fur.**
- **Usually, one allele is dominant over the other, e.g. the allele (*B*) for black fur is dominant over the allele (*b*) for white fur.**
- **This means that a rabbit with the alleles *Bb* will be black even though it has an allele for white fur.**
- **Although *BB* rabbits and *Bb* rabbits are both black, only the *BB* rabbits will breed true.**
- ***Bb* black rabbits mated together are likely to have some white babies.**
- **The expectation is that, on average, there will be one white baby to every three blacks.**
- **Meiosis is the kind of cell division that leads to production of gametes.**
- **Only one of each chromosome pair goes into a gamete.**
- **A *Bb* rabbit would produce two kinds of gametes for coat colour; 50 per cent of the gametes would have the *B* allele and 50 per cent would have the *b* allele.**
- **Sex, in mammals, is determined by the X and Y chromosomes. Males are XY; females are XX.**
- **Your sex, eye colour and hair colour are controlled by genes. You cannot change these characteristics.**
- **Height, weight and intelligence are partly controlled by genes and partly by outside conditions. You may inherit the genes for growth to 6 ft (183 cm) but get only enough food to reach 5 ft 6 in. (168 cm).**

23 The theory of evolution and natural selection

There are hundreds of thousands of different kinds of animals and plants on the Earth. Have they always been there? If not, how did they get there? Have they always been the same or have they changed? The only honest answer to these questions is that we don't know and will probably never know for sure. Most biologists favour the theory of evolution but would change their minds if sufficient evidence was put forward in favour of a different theory. Scientific evidence can be found to support the theory of evolution but the evidence can be interpreted in other ways. 'Evolution' is a theory, not a fact.

THE THEORY OF EVOLUTION

The first life

The theory suggests that at one time there were no living organisms of any kind on the Earth, possibly because it was too hot. As the Earth cooled, the theory supposes that conditions were just right for certain kinds of chemical reaction to take place in the water. These chemical reactions might have produced compounds like amino acids and enzymes, which could make other reactions take place. If these chemicals somehow came together inside a membrane, they would form the first single-celled creatures, such as bacteria, which could feed, grow and reproduce. There is a great deal of argument among biologists about how this could have happened or whether it could have happened at all. It is possible to make amino acids (page 93) in laboratory experiments in the conditions suggested but it is impossible to be confident that these really were the conditions on the Earth millions of years ago.

Organisms became more numerous and more complicated

The first single-celled creatures might have given rise to many-celled creatures if the cells stuck together after cell division. The many-celled creatures could have started as a simple ball of cells (Figure 1). Some cells might have become specialised to carry food, conduct nerve impulses, or contract to produce movement. As a

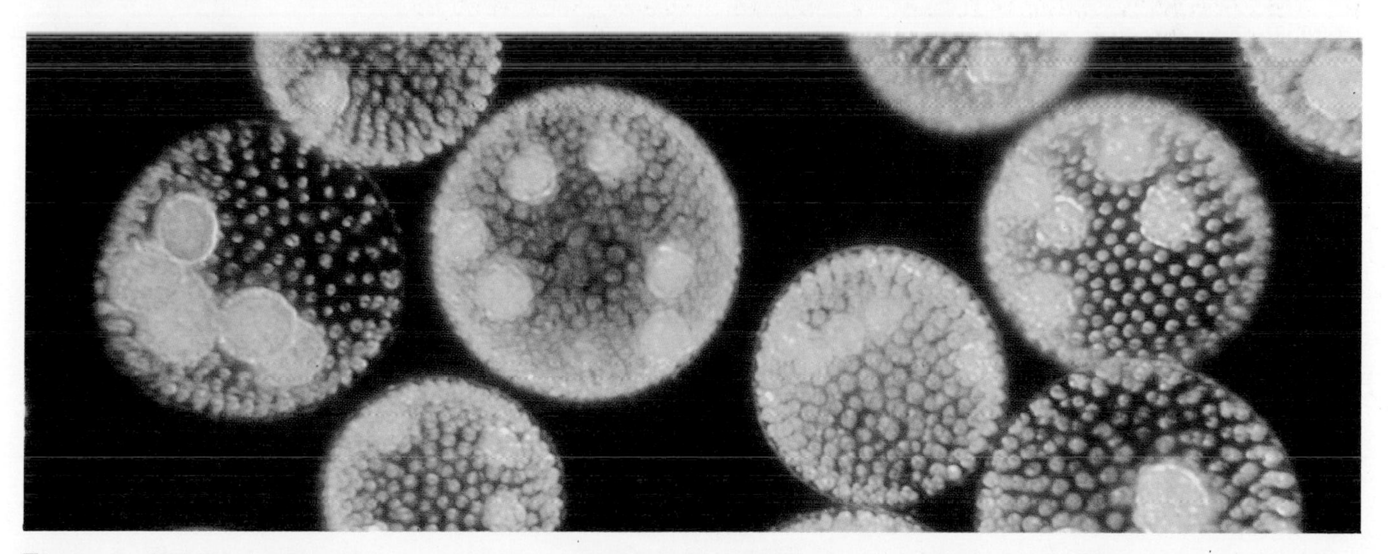

Figure 1 Volvox (×150). This is a simple, many-celled plant, in the form of a ball of cells. The earliest many-celled creatures might have been something like this. (The large green spheres in each organism are daughter colonies.)

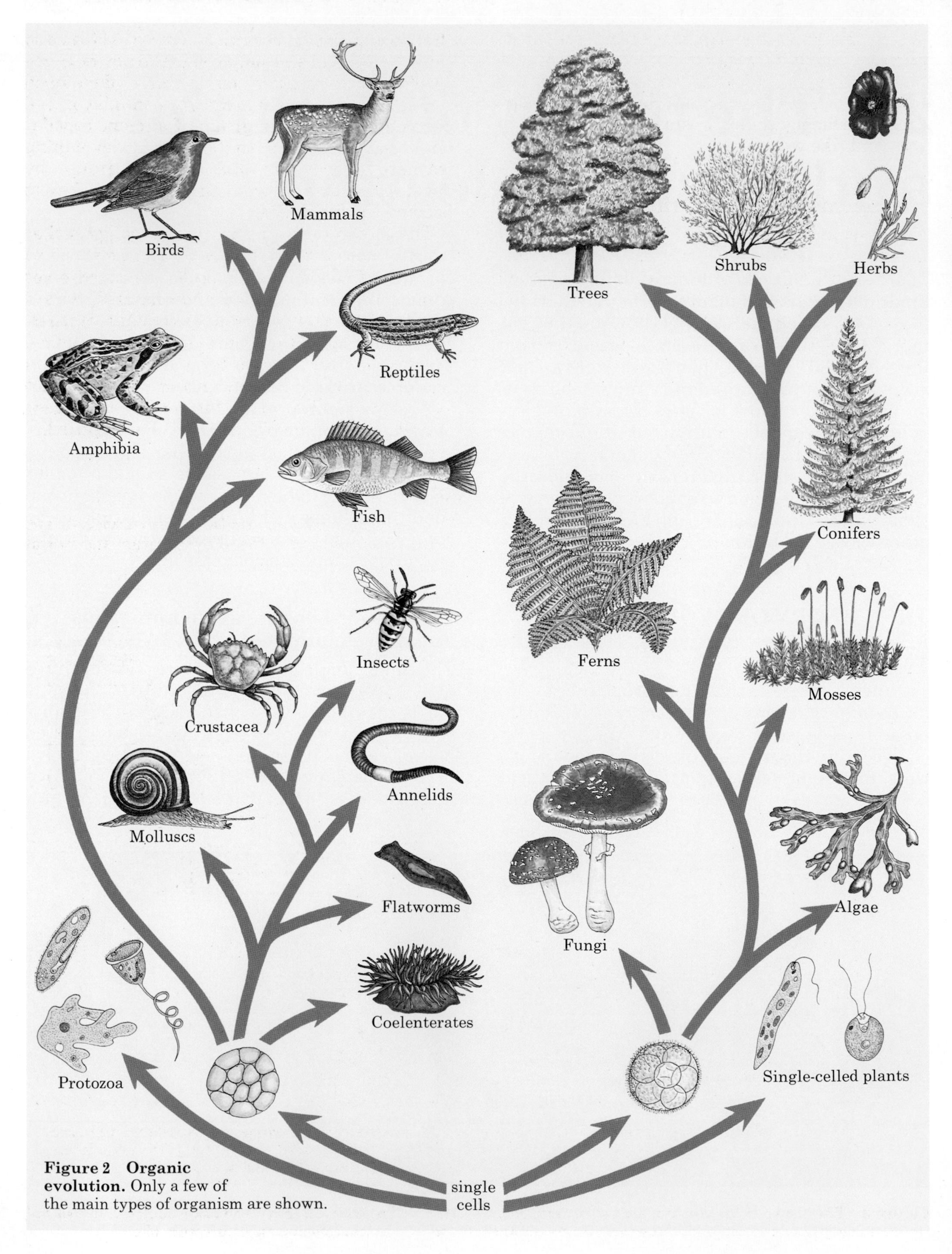

Figure 2 Organic evolution. Only a few of the main types of organism are shown.

result, the creatures would become more complicated and different forms would arise (Figure 2). Some could stick to rocks and produce tentacles like sea-anemones. Some might swim in the surface waters like shrimps or burrow in the sand like worms.

Organisms came on to the land

If evolution has occurred, it most likely started in the water. Gradually some plants and animals became able to survive on land for longer and longer periods. Some biologists think that, until green plants developed, there was no oxygen in the air. The plants' photosynthesis produced oxygen and this would have made it possible for animals to evolve.

Evolution of vertebrates

Biologists think that most evolution takes place as a result of a series of very small changes. So it would take millions of years for a fish-like creature to turn into a frog-like creature. Over a period of about 400 million years it is thought that some fish-like creatures developed legs and lungs and so became amphibia, like newts. Some amphibian ancestors could have developed scales and the ability to lay eggs on land and so become reptiles. One group of ancient reptiles might have given rise to the birds by developing feathers and wings and another group, by developing fur and producing milk, could have become the mammals.

The theory of evolution does not suggest that frogs turned into lizards or that lizards turned into birds. Frogs and lizards are themselves thought to be the products of millions of years of evolution from ancestral amphibia and reptiles. It is supposed that some of the ancestral amphibia gave rise to two main groups of descendants. One of these groups developed into primitive reptiles and the other continued evolving to become frogs and toads (Figure 3).

QUESTION

1 In Figure 2 why do you think that mammals and birds are drawn at the top of the diagram and worms and insects lower down?

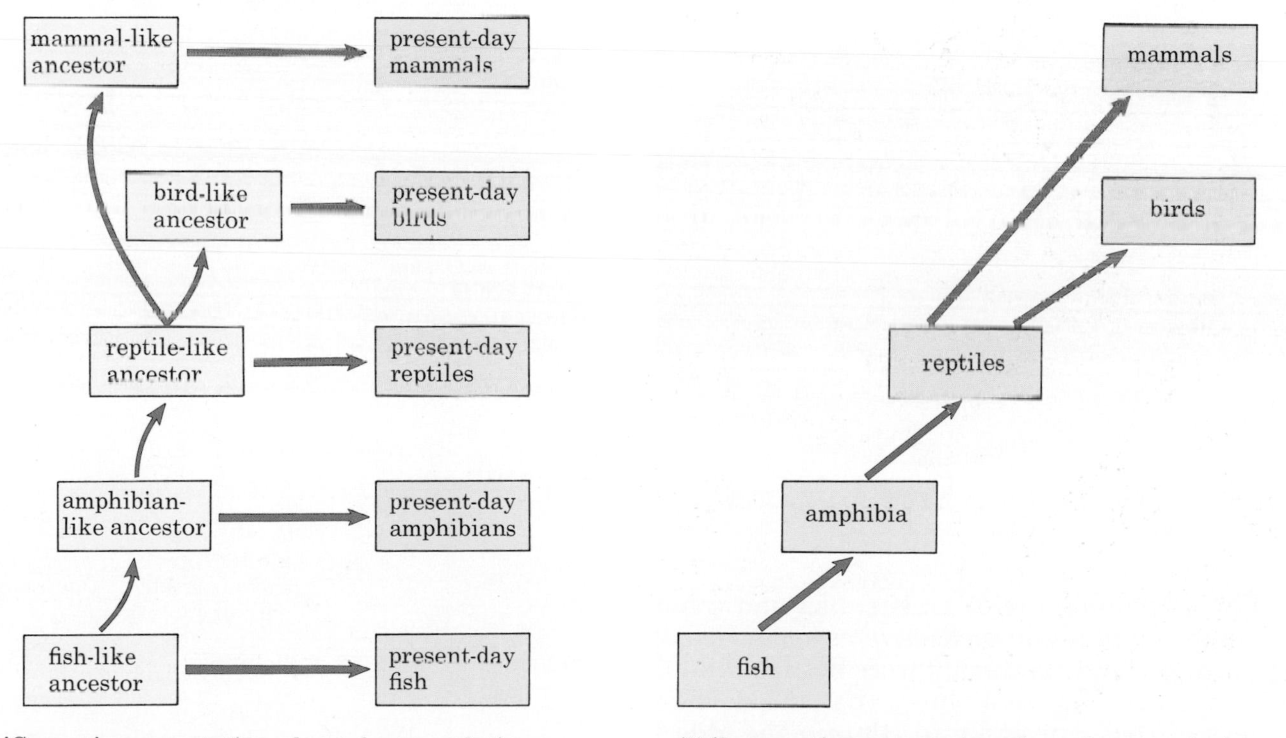

(*a*) 'Correct' representation of vertebrate evolution.

(*b*) 'Incorrect' representation of vertebrate evolution. It seems to suggest that present-day fish gave rise to present-day amphibia, and so on.

Figure 3 The theory of evolution. This says that fish and amphibia, for example, share a common ancestor. It does *not* say that fish give rise to amphibia. However, we do not know for sure what the common ancestors were like.

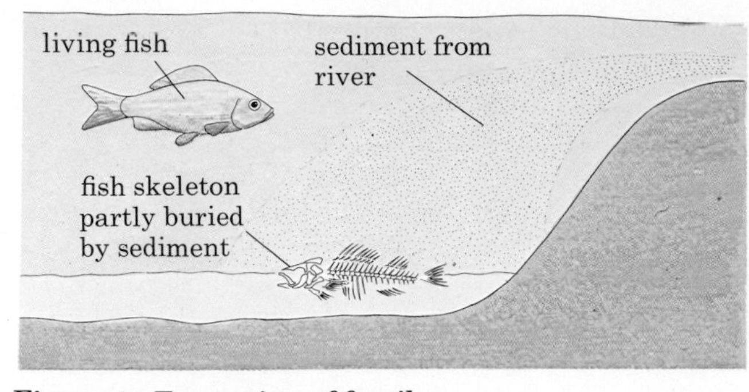

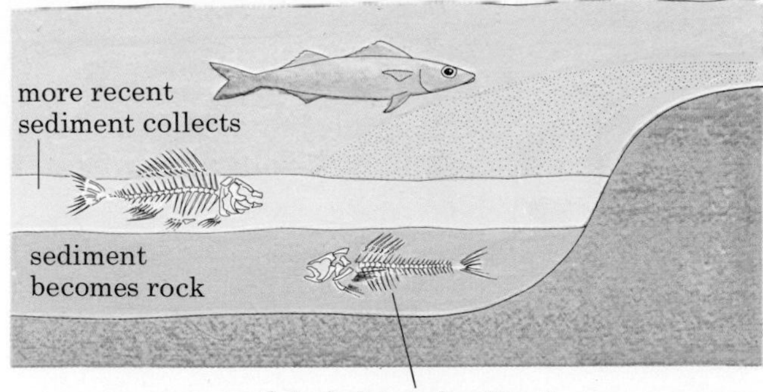

Figure 4 Formation of fossils

EVIDENCE FOR EVOLUTION

Fossils

Fossils are the remains of animals and plants preserved in certain kinds of rock. Figure 4 shows how the skeleton of a fish might become embedded in mud or sand which was settling down on the bottom of a lake. After a few million years, with the pressure of more layers building up on top of it, the mud or sand becomes rock. Massive earth movements may raise the rocks above the water and when they form cliffs or are quarried, the preserved skeletons can be found and chipped out of the rock (Figure 5).

Figure 5 The fossil remains of a fish (*Dapedius*) which lived 130 million years ago

The skeletons, shells, tree-trunks and other remains found in the rocks give us some idea of the animals and plants that were living millions of years ago. Figure 4 shows that the deepest layers of rock are likely to contain the oldest fossils. So by studying the fossil record, the theory of evolution can be put to the test.

Figure 6 shows the extent of the fossil remains of the five vertebrate classes in rock layers of increasing depth. The width of the red-brown bands represents the amount of fossil remains found. For example, 100 million years ago, it looks as if there were many more reptiles about than there are today. Also you will notice that fossils of mammals do not appear at all in rocks which are older than about 300 million years.

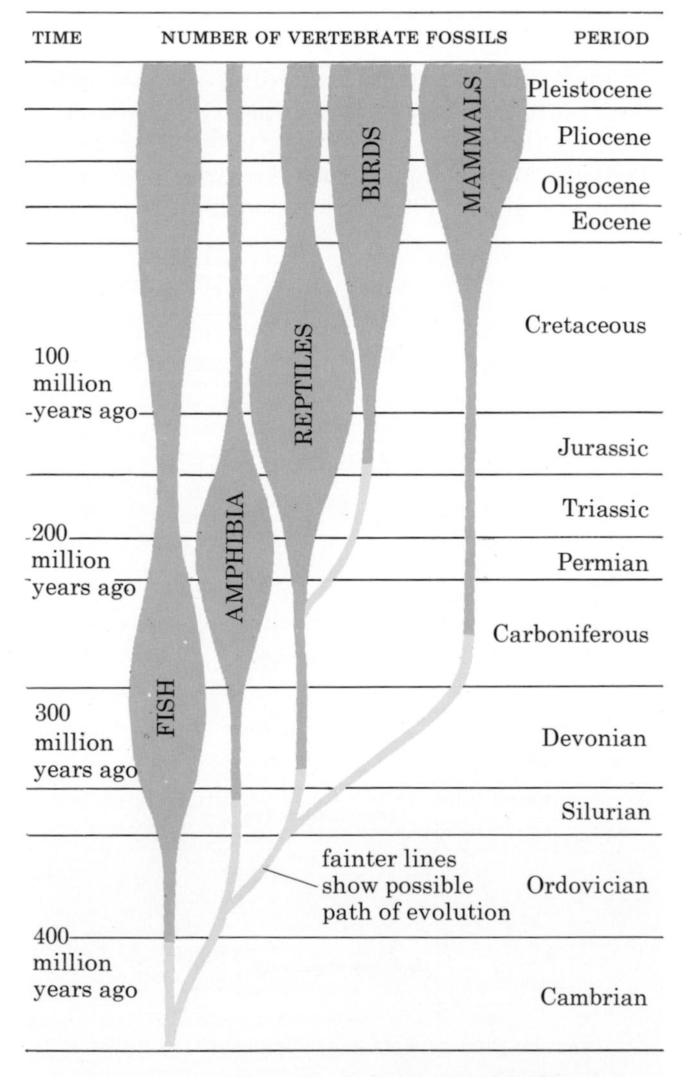

Figure 6 Numbers of vertebrate fossils found in rocks. (After Grove and Newell, *Animal Biology*, University Tutorial Press, 6th ed., 1961.)

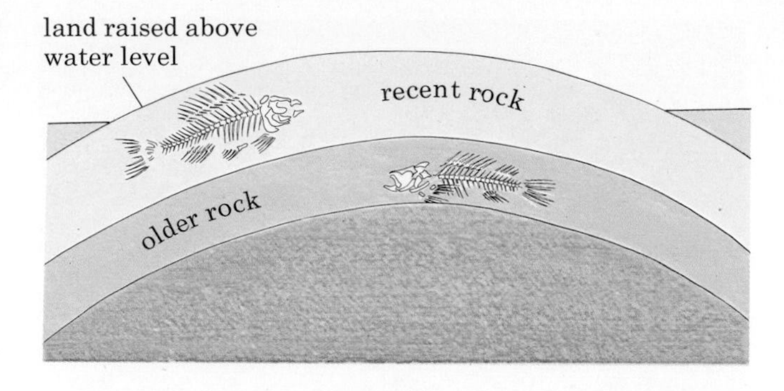

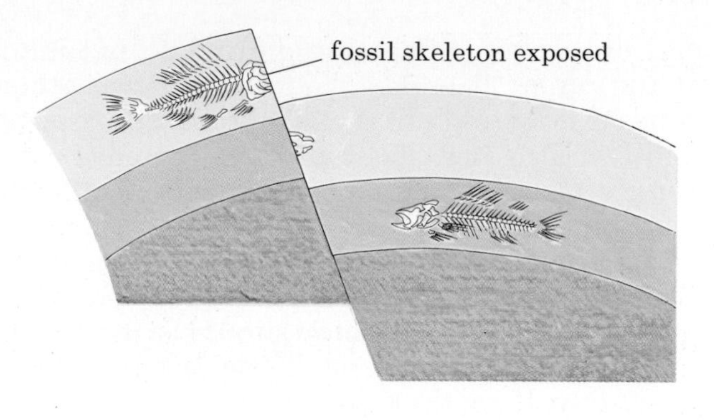

The fossil record appears to support the idea that fish-like creatures could have given rise to mammals because, 300 million years ago, there were plenty of fish but no mammals. Most of the organisms preserved in the fossil record appear to be different from the organisms we know today. They are obviously related but different in important ways. For example, Figure 7 shows a fish that lived 350 million years ago. It is clearly a fish, but quite different from any fish living today.

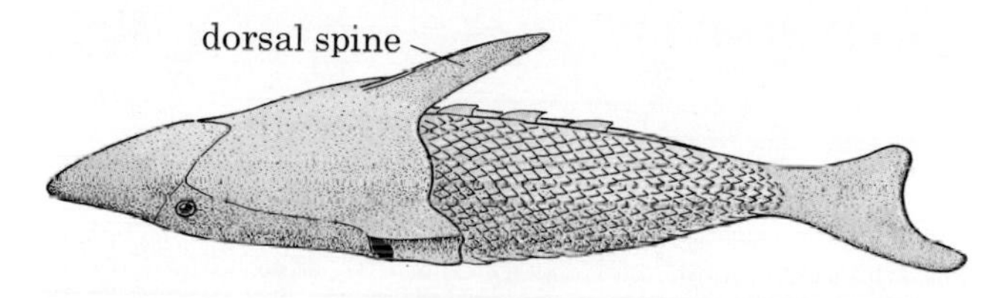

Figure 7 An extinct, 'armour-plated' fish of 350 million years ago (reconstructed from fossil remains)

So the fossil record shows that plants and animals have changed from one form to another over a long period—or does it?

Could it be that 350 million years ago there were all the fish, reptiles, amphibia, birds and mammals that we know today plus all those we know as fossils? Perhaps there were so few mammals at that time that we have not yet found any in the rocks of that period. In such a case, all that the fossil record would show is that some creatures have become more abundant, some have become scarce and some have died out altogether. This is a different interpretation of the fossil evidence and does not lend support to the theory of evolution.

Evidence from anatomy

Figure 8 shows the skeleton of the front limb of five types of vertebrate. Although the limbs have different functions such as grasping, flying, running and swimming, the arrangement and number of the bones is almost the same in all five. There is a single top bone (humerus), with a

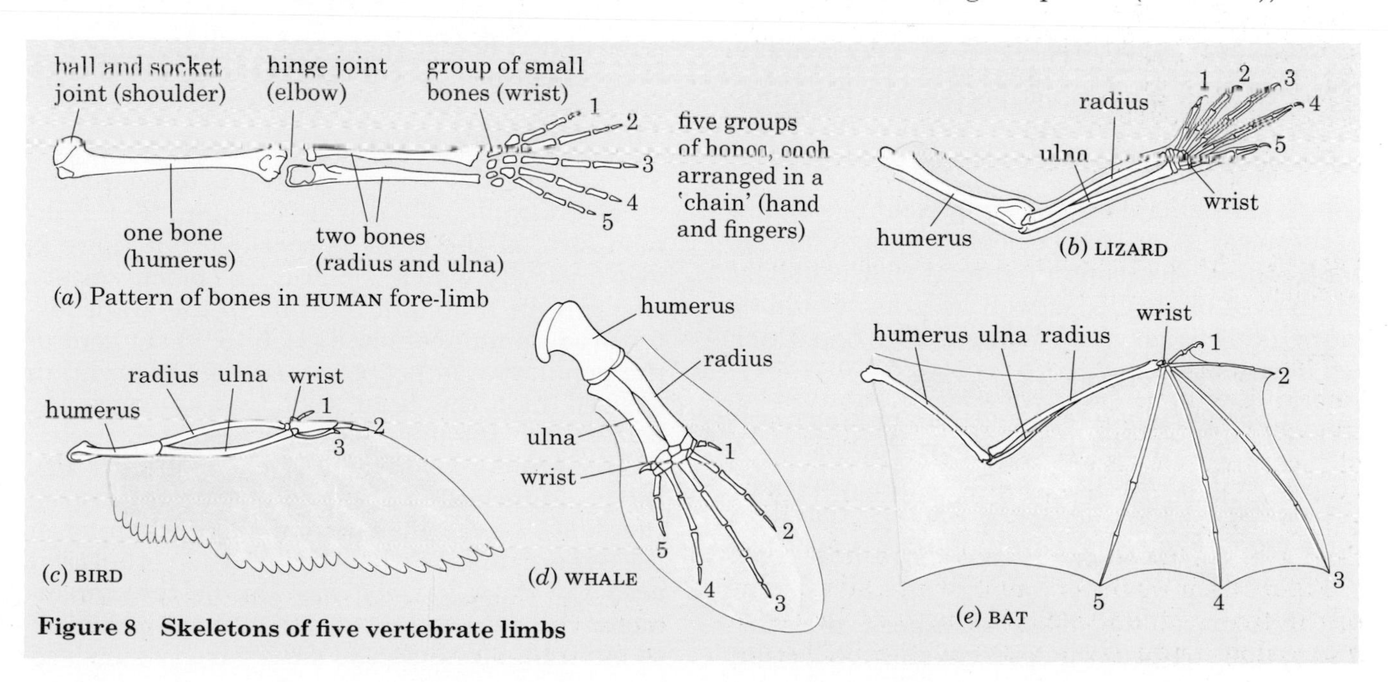

Figure 8 Skeletons of five vertebrate limbs

ball and socket joint at one end and a hinge joint at the other. It makes a joint with two other bones (radius and ulna) which join to a group of small wrist bones. The limb skeleton ends with five groups of bones (hand and fingers), although some of these groups are missing in the bird.

The argument for evolution says that, if these animals are not related, it seems very odd that such a similar limb skeleton should be used to do such different things as flying, running and swimming. If, on the other hand, all the animals came from the same ancestor, the ancestral skeleton could have changed by small stages in different ways in each group. So we would expect to find that the basic pattern of bones was the same in all these animals. There are many other examples of this kind of evidence among the vertebrate animals.

QUESTIONS

2 According to Figure 6 when were the amphibia most abundant?

3 What features of Figure 7 suggest that the fossil animal is related to present-day fishes?

NATURAL SELECTION

Theories of evolution have been put forward in various forms for hundreds of years. In 1858, Charles Darwin and Alfred Russel Wallace published a theory of evolution by natural selection which is still an acceptable theory today.

The theory of natural selection suggests that

(a) Individuals within a species are all slightly different from each other (Figure 9). These differences are called **variations**.

(b) If the climate or food supply changes, some of these variations may be better able to survive than others. A variety of animal that could eat the leaves of shrubs as well as grass would be more likely to survive a drought than one which fed only on grass.

(c) If one variety lives longer than others, it is also likely to leave behind more offspring. A mouse that lives for 12 months may have ten litters of five babies (50 in all). A mouse that lives for 6 months may have only four litters of five babies (20 in all).

(d) If some of the offspring inherit the variation that helped the parent survive better,

Figure 9 Variation. The garden tiger moths in this picture are all from the same family. There is a lot of variation in the pattern on the wings and abdomen.

they, too, will live longer and have more offspring.

(e) In time, this particular variety will outnumber and finally replace the original variety.

The peppered moth

The story of the peppered moth is often quoted as an example of natural selection. The common form of the moth is speckled but there is also a variety which is black. The black variety was rare in 1850, but by 1895 in the Manchester area its numbers had risen to 98 per cent of the population of peppered moths. Observation showed that the light variety was concealed better than the dark variety when they rested on tree-trunks covered with lichens (Figure 10). In the Manchester area, pollution had caused the death of the lichens and the darkening of the tree-trunks with soot. In this industrial area the dark variety was the better camouflaged (hidden) of the two and was not picked off so often by birds. So the dark variety survived

(*a*) A light coloured peppered moth and the dark variety are resting on a lichen-covered tree trunk. Which moth is better hidden?

(*b*) The two forms are resting on a tree trunk which has no lichen and whose bark is darkened with soot. Which form is more likely to be taken by a bird?

Figure 10 The peppered moth

better, left more offspring and nearly replaced the light form. (In fact, the story is more complicated than this and some of the evidence has been challenged.)

If a new variety does not breed with the original variety, it may go on changing its form and behaviour until it is quite unable to breed with the original type. If this were to happen, it would have become a new species. Although there are many examples of natural selection similar to that of the peppered moth, there is no example of a new species arising in this way. So although natural selection can, in theory, explain the gradual changes in evolution, there is no evidence that it has actually brought about an evolutionary change.

Inherited variations

These variations can arise in two ways: (i) new combinations of genes and (ii) mutation.

(i) New combinations of genes. The hybrid wheat (page 185) with the combination of genes for disease resistance and genes for good yield of seeds could have survival value. A severe attack of fungus disease could wipe out the original, non-resistant population. The small amount of seed produced by the low-yielding variety may not survive a very dry spring season. The disease-resistant, high-yielding variety would surpass both of the others in a few years if it bred true (page 179). Even if it doesn't breed true, it may still survive better if the other varieties are continually being selected out.

The offspring from crossing genotypes *Aa* and *Aa* (page 181 shows this for *Bb*) will always include some *aa* and *AA* genotypes, but if these are at a disadvantage, the *Aa* gene combination will survive better (Figure 11).

(ii) Mutation. Sometimes during cell division (page 175) a gene or a chromosome is not replicated properly (see page 176). The altered gene or chromosome is said to be a **mutation.** The colour of the black variety of the peppered moth results from the gene for wing colour mutating to one which produces black wings instead of speckled wings.

Sometimes a woman produces an ovum with 24 chromosomes instead of 23. This is a chromosome mutation. If the ovum is fertilized

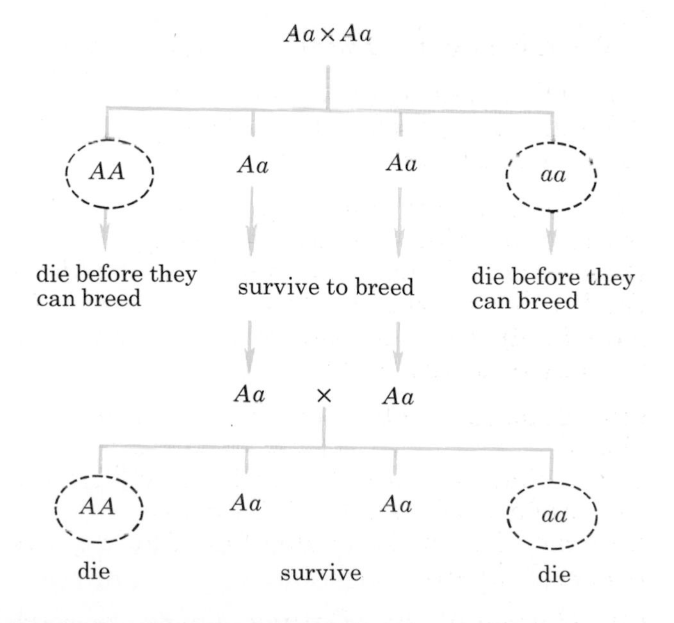

Figure 11 How a gene combination can be preserved by natural selection

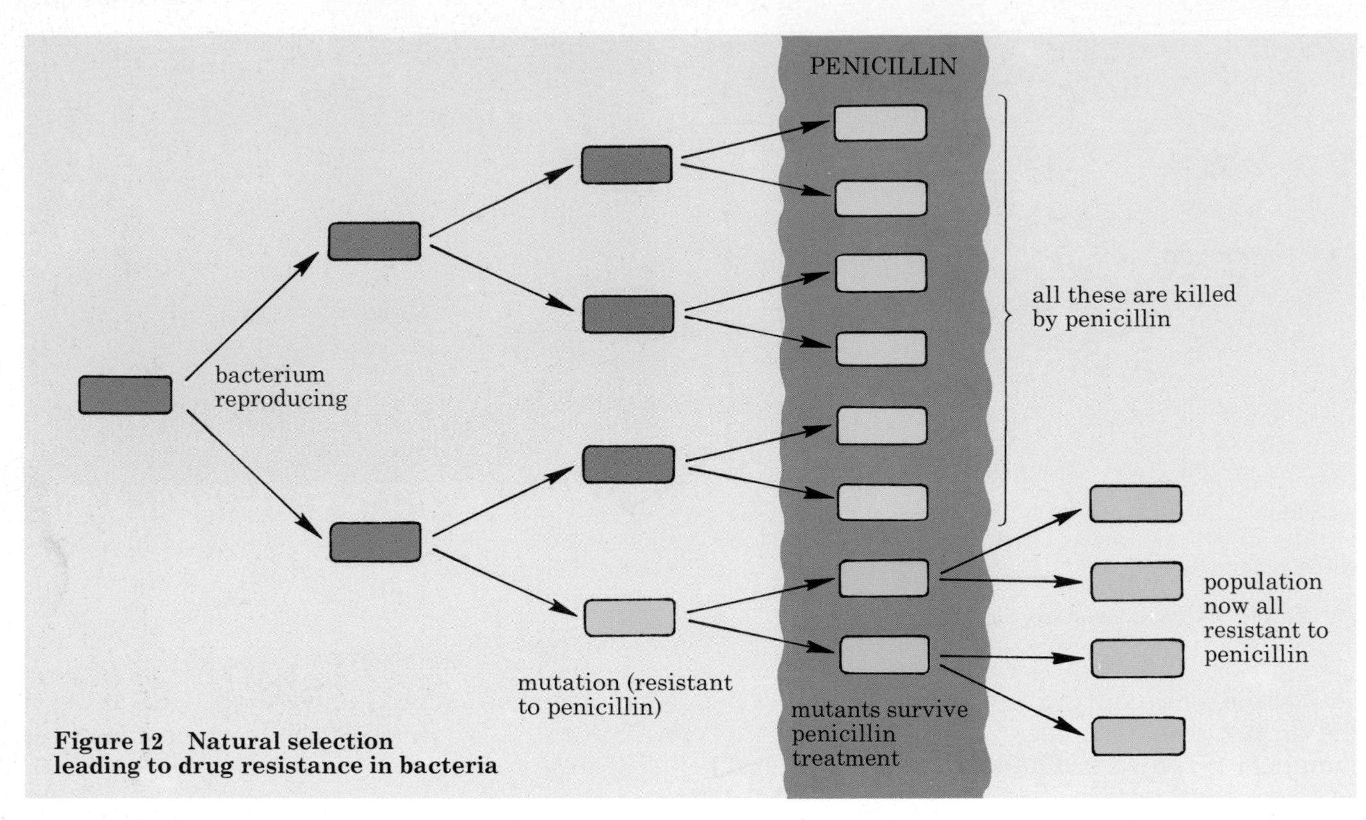

Figure 12 Natural selection leading to drug resistance in bacteria

by a normal sperm the zygote will contain 47 chromosomes instead of 46 and the individual who grows from it will suffer from a mental and physical deficiency called Down's syndrome (sometimes called 'Mongolism').

Most mutations of genes and chromosomes are harmful and the organisms with the mutation are likely to be eliminated by natural selection. Once in a while, however, a mutation is beneficial and gives the organism an improved chance of survival, e.g. the black variety of the peppered moth in industrial areas and the penicillin-resistant bacteria in Figure 12. The mutated gene is then passed on to the offspring.

QUESTIONS

4 What variations can you see in the moths of Figure 9?

5 Judging from Figure 12 why do you think it is not a good idea to take penicillin in low doses for trivial complaints?

BASIC FACTS

- **The theory of evolution tries to explain how present-day plants and animals came into existence.**
- **It supposes that simple living organisms, such as bacteria, were formed from non-living matter.**
- **These simple organisms changed over hundreds of millions of years to become many-celled and more complicated.**
- **They also become more varied, giving rise to many different groups of plants and animals.**
- **The evidence from the fossil record supports the theory of evolution because it shows animals and plants becoming more varied and more complicated as time goes on.**
- **The theory of natural selection suggests one way in which evolution could happen, namely:**
  (a) Plants and animals produce more offspring than can possibly survive.
  (b) Some of these offspring have small variations that help them to survive better.
  (c) Some of the useful variations are inherited by new generations.
  (d) In time, the variations build up until the organism is very different from its ancestors.
- **The two types of variation which can be inherited are caused by new gene combinations or changes in a chromosome or gene (mutation).**

SECTION FIVE

Characteristics of Living Things

Characteristics of living organisms

All living organisms, whether they are single-celled, many-celled, plants or animals, do the following things:

Feed They may take in solid food as animals do, or digest it first and absorb it later like fungi do, or build it up for themselves like plants do.

Breathe They take in oxygen and give out carbon dioxide. This exchange of gases takes place between the organism and the air or between the organism and water. The oxygen is used for respiration.

Respire That is, they break down food to obtain energy. Most organisms need oxygen for this.

Excrete Respiration and other chemical changes in the cells produce waste products such as carbon dioxide. Living organisms expel these substances from their bodies in various ways.

Grow Bacteria and single-celled creatures increase in size. Many-celled organisms increase the numbers of cells in their bodies, become more complicated and change their shape as well as increasing in size.

Reproduce Single-celled organisms and bacteria may simply keep dividing into two. Many-celled plants and animals may produce sexually or non-sexually.

Respond The whole animal or parts of plants respond to stimuli.

Move Most single-celled creatures and animals move about as a whole. Fungi and plants may make movements with parts of their bodies.

The next seven chapters describe some of these characteristics in different living things.

24 Living things

There are millions of different kinds of living things in the world. There are fish, frogs, birds, mosses, trees, jelly-fish and mushrooms. These are all quite different in appearance but because they are all living, they are alike in some very important ways; they breathe, feed, grow, reproduce and, in many cases, they make movements.

Most of this book deals with the activities which are common to all living things; how they get their food, what use they make of it, how they breathe, reproduce and so on. In this chapter, however, we look at how they are different from each other. This is best done if we try to sort out the enormous variety of living things into groups.

There are many ways in which living things could be grouped; by size, by colour or by the places where they live. The biologist looks for what he regards as important features which are shared by as large a group as possible. In some cases it is easy. Birds all have wings, beaks and feathers; there is rarely much doubt about whether something is a bird or not. In other cases it is not so easy. Some microscopic, singled-celled creatures are not obviously either plants or animals. As a result, people change their ideas from time to time about how living things should be grouped. New groupings are suggested and old ones are abandoned.

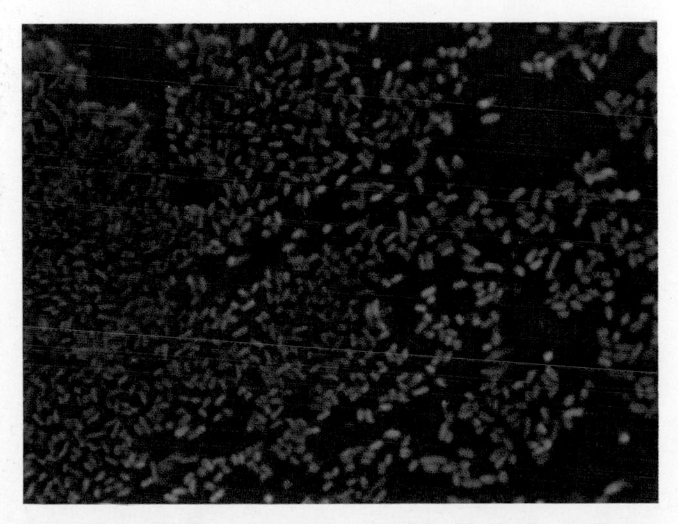

Figure 1 Cholera bacteria (× 800). These bacteria are growing safely on the surface of a special culture material.

The method of grouping organisms is called **classification** and one system of classifying living things is into five main groups:

Bacteria
Single-celled organisms
Fungi
Plants
Animals

Each of these main groups is divided into smaller sections, and these sections are subdivided. Animals can be divided into animals with backbones and animals without backbones. The animals with backbones can be sorted into different classes such as fish, reptiles and birds.

BACTERIA

These are tiny, 'single-celled' creatures, not more than about two thousandths of a millimetre (2 μm) long, and they can be seen only under a powerful microscope. They are everywhere: in the air, water, the soil and on or in the bodies of other living things. Most of them feed on the dead remains of other creatures and so they are the main agents of decay. Without bacteria, we might disappear under mountains of dead leaves, dead plants and dead bodies.

Some bacteria which get into the bodies of animals cause disease. Tuberculosis, typhoid and cholera are human diseases caused by bacteria. Figure 1 shows some bacteria magnified 800 times.

Each bacterium can reproduce by dividing into two. Since they can do this as often as once every 20 minutes, a population of bacteria can grow very rapidly.

Chapter 31 describes the bacteria more fully and considers their importance to man.

SINGLE-CELLED CREATURES (protista)

Most living things have bodies made up of thousands of cells (see page 2) but there are also large numbers of tiny creatures which consist of only a single cell. Most of them are microscopic but a few can just be seen as specks, with the naked eye (Figure 2).

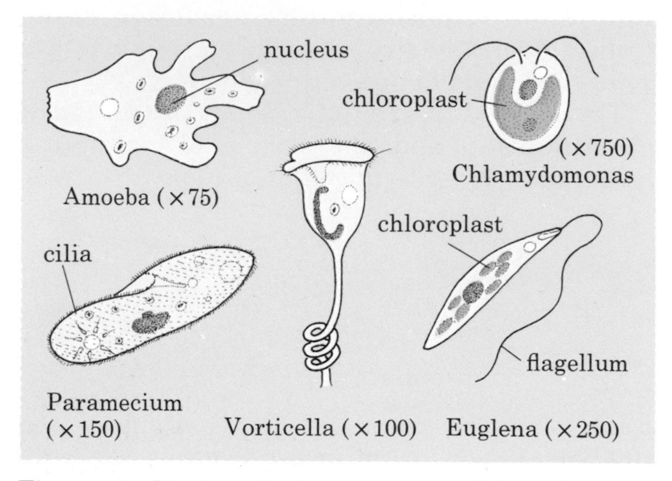

Figure 2 Single-celled organisms. Chlamydomonas and Euglena have chloroplasts and make their food by photosynthesis. The others ingest solid food such as bacteria or smaller protista.

They nearly all live in water; in the sea, rivers and lakes, or in the water in the soil, in puddles and in the body fluids of animals. The protista differ from bacteria because they have a definite nucleus (see page 4) in their cells and because they feed in a variety of different ways. Some take in solid food.

Amoeba (see also Figure 1 on page 208) moves by flowing along and feeds by picking up bacteria and other microscopic creatures as it goes. **Vorticella** makes a current of water by flicking its hair-like cilia (see page 6) and swallows the microscopic particles brought in by the current. **Euglena** can make its food in the same way as plants do because it has chlorophyll in its cell.

Many of the protista can move about by means of structures such as cilia or flagella. Cilia are like microscopic hairs which can flick to and fro. **Paramecium** is covered with cilia which all flick in a rhythmic pattern, driving the Paramecium through the water. Euglena has a single flagellum, like a whip, which draws it along by a lashing movement.

Most of the protista can reproduce by dividing into two cells (Figure 3). The protista can be subdivided into several classes, according to whether they have cilia or flagella for example.

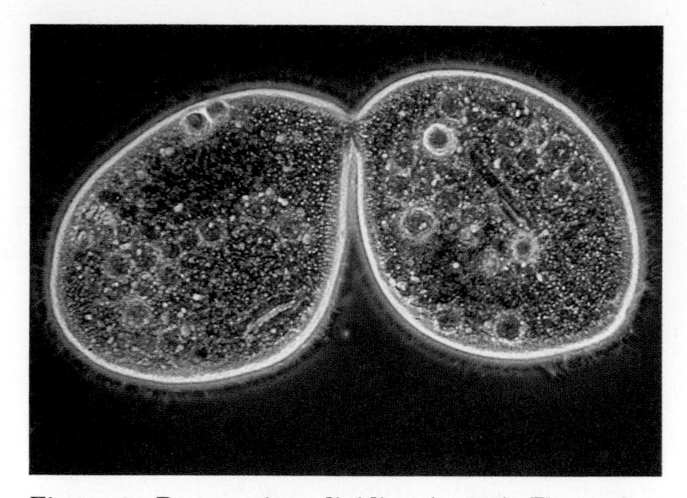

Figure 3 Paramecium dividing (× 300). The nucleus has already divided and now the cytoplasm is pinching off in the middle to produce two new organisms. The cilia can be seen on the outside of each Paramecium.

FUNGI

Familiar examples of fungi (singular=fungus) are mushrooms, toadstools, puff-balls, bracket fungi, moulds and mildews. They are all formed from fine thread-like structures called **hyphae** (pronounced 'high fee'). In moulds, these hyphae form a loose network over their food (Figure 11 on page 221), but in the larger fungi, the hyphae are packed together to form the structures we recognize as mushrooms or toadstools. Most fungi get their food from dead

Figure 4 Toadstools growing on a tree stump. The toadstools are the reproductive structures which produce spores. The feeding hyphae are inside the tree stump, digesting and absorbing the wood.

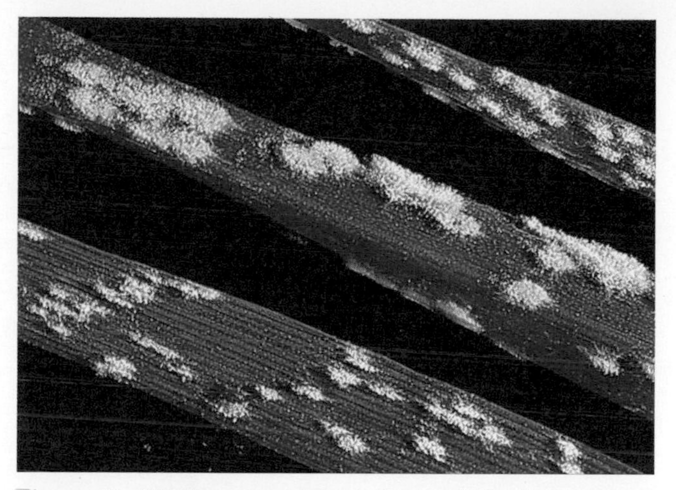

Figure 5 Mildew on wheat. Most of the hyphae are inside the leaves, digesting the cells, but some grow out and produce the powdery spores seen here. When the spores get blown on to a healthy wheat plant they will infect it.

and decaying material such as the rotting wood in tree-stumps (Figure 4), or the plant and animal remains in the soil. Some fungi are parasitic (page 208) and grow in living plants causing diseases such as mildew in important crop plants (Figure 5).

All fungi reproduce by forming **spores**. These are tiny cells, produced and dispersed in vast numbers (Figure 12 on page 221).

PLANTS

Plants make their food in their leaves. The green colouring matter, **chlorophyll**, in the leaves enables plants to absorb energy from sunlight and use it to combine carbon dioxide from the air, and water from the soil, to make sugar and all the other substances they need for food (page 24).

Subdivisions of the plant kingdom are:

- Algae
- Liverworts and Mosses
- Ferns
- Seed-bearing plants:
  - (a) Conifers
  - (b) Flowering plants:
    - (i) Monocots
    - (ii) Dicots

(See illustrations on next two pages.)

Algae

The algae are a strange mixture. Seaweeds (Figure 6) are the largest algae but they are quite simple structures with no proper roots, stems or leaves. Some of the simplest algae are the green filaments that form a slimy growth in ponds and ditches (Figure 7).

Liverworts and mosses

Liverworts are small green plants, usually growing close together in moist, shady places such as in the mouths of caves or on a stream bank near the water-line. Some look like small, overlapping strips of dark green leaf stuck to the rock or the bank (Figure 8). These liverworts have no stem and their 'roots' are only single cells like root hairs (page 68).

Figure 6 Seaweeds exposed at low tide. There are several types of seaweed here, belonging to the brown, red and green algae.

Figure 7 Fresh-water algae. The scummy-looking patches which can be seen on the surface of the lake are colonies of thread-like green algae.

Figure 8 A colony of liverworts. The small black knobs on the ends of the green stalks are the spore capsules (see also page 200, 'Liverworts').

ALGAE
Sea lettuce (×0.1)
Oar weed (×0.15)
Dulse (×0.3)
Bladder wrack (×0.3)
LIVERWORTS
Pellia (×2)
Lophocolea (×3)
Marchantia (×1.5)
FERNS
Spleenwort (×0.5)
Bracken (×0.1)
Hart's tongue (×0.3)
Male fern (×0.1)
Polypody (×0.3)
MOSSES
Funaria (×1)
Hypnum (×1.5)
Sphagnum (×0.8)
Polytrichum (×0.75)

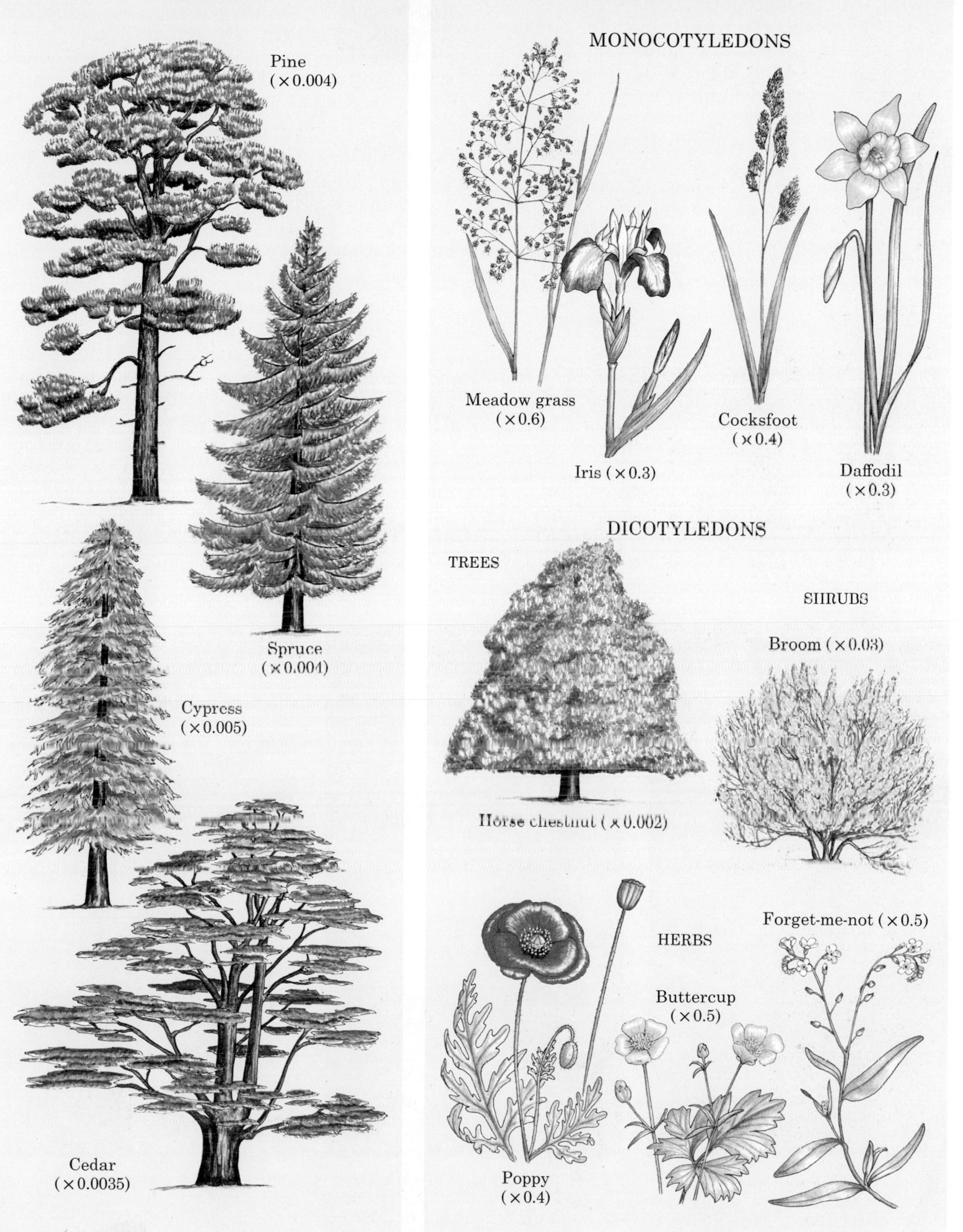
CONIFERS
Pine
(×0.004)
Spruce
(×0.004)
Cypress
(×0.005)
Cedar
(×0.0035)
FLOWERING PLANTS
MONOCOTYLEDONS
Meadow grass
(×0.6)
Iris (×0.3)
Cocksfoot
(×0.4)
Daffodil
(×0.3)
DICOTYLEDONS
TREES
Horse chestnut (×0.002)
SHRUBS
Broom (×0.03)
HERBS
Poppy
(×0.4)
Buttercup
(×0.5)
Forget-me-not (×0.5)

Mosses are familiar as green, cushion-like growths seen on walls, roofs and at the base of tree-trunks in woodland. Each 'cushion' is made up of hundreds of separate plants. Each moss plant has a slender stem with simple, tiny, overlapping leaves coming from it in a spiral arrangement. Neither the stem, nor the simple 'roots' have any vascular bundles (page 62) in them.

Mosses and liverworts produce microscopic spores, as do the fungi, and the capsules which contain the spores can often be seen sticking out of the tops of the moss plants (Figure 9). The capsules will break open, releasing the spores.

Ferns

Bracken is probably the most widespread and familiar type of fern. The male fern is also found in woods and gardens and the hart's tongue fern grows in damp woods and on shady banks (Figure 10).

Ferns have underground stems as well as roots and large well developed leaves. The roots and stem contain vascular bundles. The spores are produced in special structures under the leaves (Figure 11).

Seed-bearing plants

All the plants described so far reproduce by single-celled spores. The greatest number of plants, however, reproduce by seeds, which are made up of many cells and contain a tiny embryo plant (page 82) as well as a store of food for the early growth of the seedling.

(a) Conifers. These are seed-bearing trees but the seeds are produced in cones (Figure 12) and not in flowers. The leaves are usually small and often needle-like (Figure 13). Examples are pine, larch, spruce and cypress.

(b) Flowering plants. These produce their seeds from flowers (page 72). There are two main groups, monocots and dicots.

(*i*) Monocots. The name is short for monocotyledons, which means that the seeds of these plants contain only one cotyledon (see page 82). All the grasses including the cultivated grasses such as wheat, barley and maize are monocots. So are the plants which have bulbs or other underground storage organs; for example, daffodils, bluebells, tulips, crocuses, irises, lilies, onions and leeks. The leaves of monocots can usually be recognized because the veins run parallel to each other down the length of the leaf

Figure 9 Moss plants growing in a wood. The moss plants growing near the top of the picture have produced spore capsules.

Figure 10 Ferns. The two types of fern growing here are the hart's tongue and male fern. Some of the rocks have a covering of mosses and liverworts.

Figure 11 Hart's tongue fern. Each brown stripe on the underside of the leaves is made up of hundreds of spore-bearing structures.

Figure 12 Larch cones. The green cones are male and are producing pollen. The pink female cones face upwards and will produce the seeds.

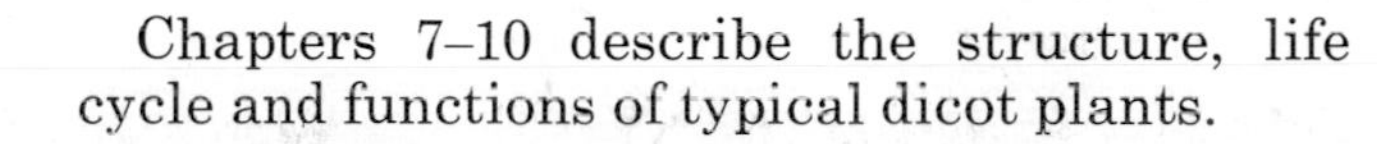

Figure 13 Pine needles and male cones. The needle-like shape of the leaves helps to reduce the rate of transpiration (page 66) in windy conditions.

and in many cases the leaves are narrow and long (Figure 14*a*).

(*ii*) Dicots. This is short for dicotyledons; the seed contains two cotyledons (Figure 2 on page 82). This group includes all the flowering plants not mentioned so far; the garden flowers and wild flowers (herbaceous dicots), shrubs and trees (woody dicots). Their leaves are broad and net-veined, that is, there is usually a large middle vein with other veins branching from it to make a visible network (Figure 14*b*).

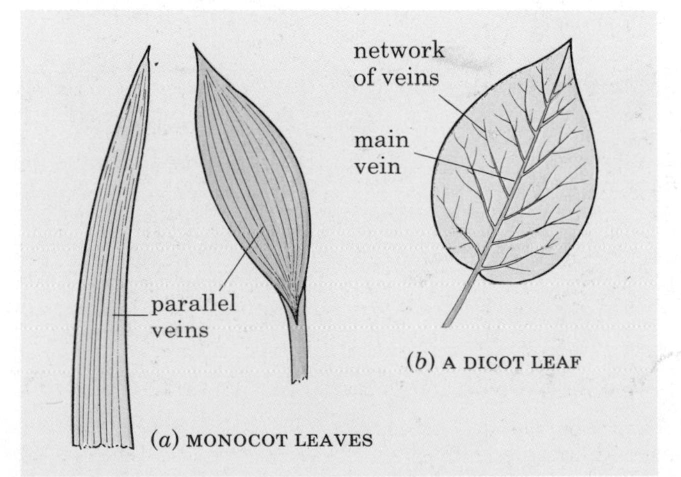

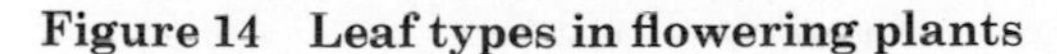

Figure 14 Leaf types in flowering plants

Chapters 7–10 describe the structure, life cycle and functions of typical dicot plants.

ANIMALS

All animals get their food by eating plants or other animals and most animals are able to move about quite freely. The animal kingdom is divided into several main groups, called **phyla** (pronounced 'filer'; singular—phylum), six of which are listed below:

- Coelenterates
- Flatworms
- Annelids
- Molluscs
- Arthropods
  - Crustacea
  - Insects
  - Arachnids
  - Myriapods
- Vertebrates
  - Fish
  - Amphibia
  - Reptiles
  - Birds
  - Mammals

(See illustrations on next two pages.)

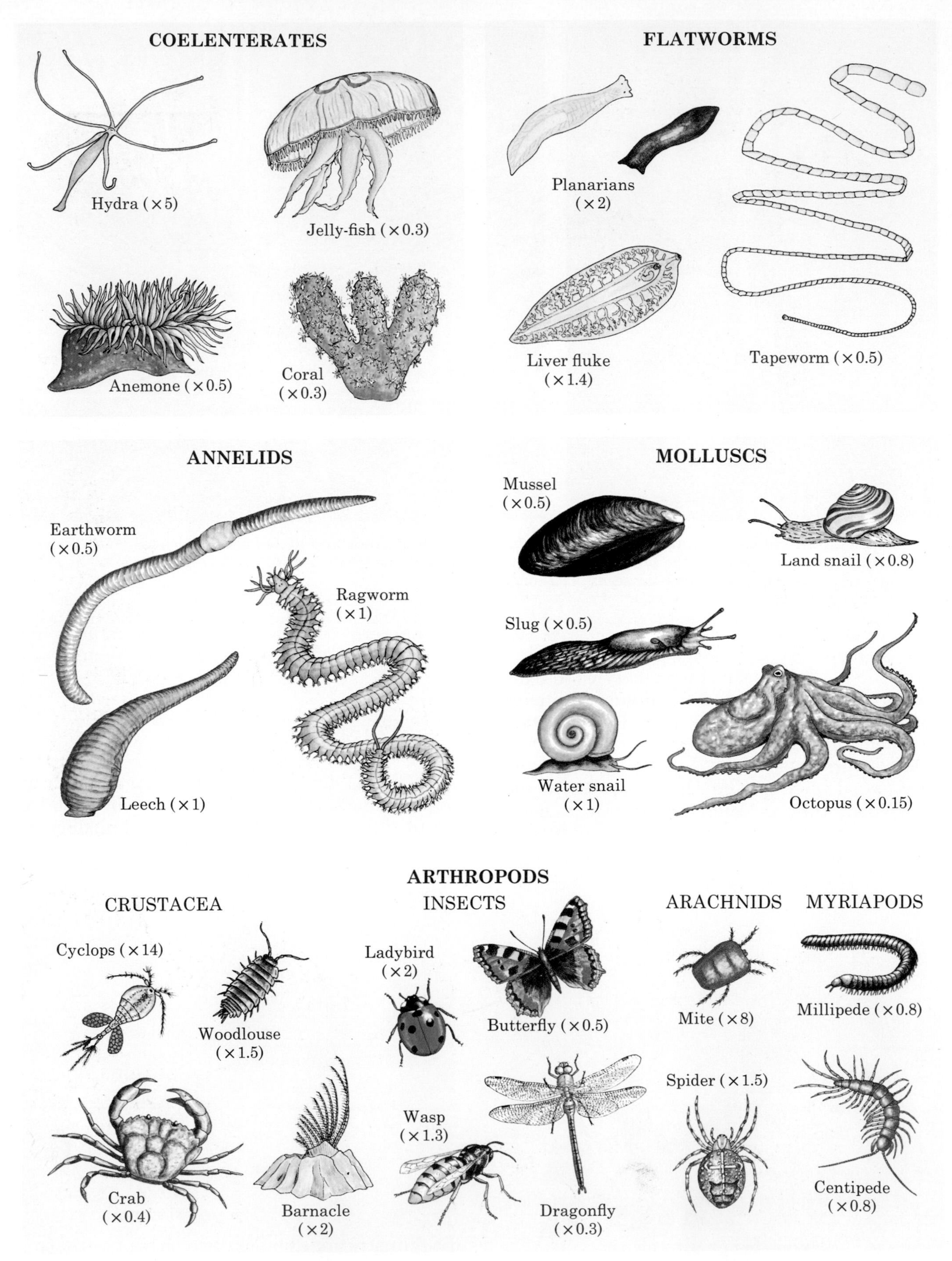
COELENTERATES
Hydra (×5)
Jelly-fish (×0.3)
Anemone (×0.5)
Coral
(×0.3)
FLATWORMS
Planarians
(×2)
Liver fluke
(×1.4)
Tapeworm (×0.5)
ANNELIDS
Earthworm
(×0.5)
Ragworm
(×1)
Leech (×1)
MOLLUSCS
Mussel
(×0.5)
Land snail (×0.8)
Slug (×0.5)
Water snail
(×1)
Octopus (×0.15)
ARTHROPODS
CRUSTACEA
Cyclops (×14)
Woodlouse
(×1.5)
Crab
(×0.4)
Barnacle
(×2)
INSECTS
Ladybird
(×2)
Butterfly (×0.5)
Wasp
(×1.3)
Dragonfly
(×0.3)
ARACHNIDS
Mite (×8)
Spider (×1.5)
MYRIAPODS
Millipede (×0.8)
Centipede
(×0.8)

FISH
Trout (×0.2)
Perch
(×0.2)
Plaice
(×0.14)
Dogfish (×0.13)
AMPHIBIA
Frog (×0.4)
Toad (×0.4)
Tadpoles (×0.8)
Newt (×0.6)
REPTILES
Slow-worm (×0.4)
Lizard (×0.5)
Snake (×0.18)
BIRDS
Kestrel
(×0.06)
Robin
(×0.25)
Mallard
(×0.1)
Kingfisher
(×0.2)
MAMMALS
Seal
(×0.03)
Mouse
(×0.4)
Bat (×0.2)
Deer (×0.03)
Squirrel
(×0.15)
Dog
(×0.03)

The vertebrates are animals which have a skull and a backbone (vertebral column). All the other groups are sometimes referred to as 'invertebrates', that is, animals without backbones. Although this is a convenient description, it is not really a proper subdivision of the animal kingdom.

Coelenterates

Coelenterates (pronounced 'sea-lenter-ates') are sea-anemones, corals and jelly-fish. They nearly all live in the sea and they have tentacles with sting cells. They feed on other animals by trapping them with their tentacles (Figure 10 on page 220).

Flatworms

The flatworms that live in ponds and streams are usually about a centimetre long or less. They are found under stones or under the floating leaves of pond plants and they glide about by means of the movements of the cilia which cover their bodies. Some members of the flatworm phylum are parasitic. Tapeworms live in the intestines of vertebrates and absorb the digested food there. Flukes are parasitic flatworms; some of them cause tropical diseases in man and others, such as the liver flukes, cause diseases in animals.

Annelids

Annelids are worms with tubular bodies divided up into segments (rings). They also have fine bristles which help them grip the soil or sand, or help them to swim.

Earthworms (Figure 8 on page 43) are annelids; other examples include lugworms and bristle-worms living on the sandy coasts between tide-marks.

Molluscs

Snails, slugs, limpets, mussels, oysters and octopuses are molluscs. Many of them have a shell outside or inside their bodies and they all have a 'foot'. This is the structure on which the snail creeps along or which holds the limpet on to a rock, or forms the tentacles of the octopus.

Arthropods

These animals have a hard, jointed skeleton like a suit of armour outside their bodies. They also have jointed legs and antennae ('feelers'). Their bodies are made up of segments and are usually clearly divided into three regions called **head**, **thorax** and **abdomen**.

(a) Crustacea include lobsters, shrimps, crabs, prawns, water fleas and woodlice. They all have more than six legs (e.g. crabs have ten).

(b) Insects have six legs and, in most cases, two pairs of wings. Flies, beetles, butterflies, fleas, wasps and earwigs are examples of insects. Insects such as fleas and lice are thought to have 'lost' their wings in the course of evolution.

(c) Arachnids are spiders, mites and scorpions. They have eight legs.

(d) Myriapods. The name means 'many feet' and the group consists of centipedes and millipedes.

Vertebrates

These animals all have a vertebral column ('backbone') (see page 148) and a skull. There are five classes:

(a) Fish, which live in water and have streamlined bodies covered with scales. They have fins and breathe by means of gills. Examples are herring, cod, haddock, trout, salmon and roach.

(b) Amphibia. These are frogs, toads and newts. They have four limbs and smooth moist skins without scales. They spend much of their time on land but they can live in water and most of them have to return to water to lay their eggs. On land they breathe with lungs. In water they can breathe through their skin.

(c) Reptiles also live on land but their skins are dry and covered with scales. They lay eggs which have leathery skins and they do not have to breed in water like the amphibians. The lizards have four limbs but snakes have no limbs.

(d) Birds. Birds are 'warm-blooded' vertebrates. They have feathers, wings and beaks and they reproduce by laying eggs.

(e) Mammals. The animals in this group of vertebrates are also 'warm-blooded' but their bodies are covered with hair. Their young are born 'alive', rather than hatched from eggs, and are suckled on milk. Some examples of mammals are dogs, cats, cows, whales, kangaroos, elephants and man.

BASIC FACTS

- **BACTERIA are microscopic organisms; they have no proper nucleus.**
- **PROTISTA are single-celled organisms containing a nucleus.**
- **FUNGI are made up of thread-like hyphae. They reproduce by spores.**
- **PLANTS make their food by photosynthesis.**

  (a) Algae are simple plants with few special organs. Most of them live in water.

  (b) Liverworts and mosses are land plants. Mosses have stems and leaves.

  (c) Ferns have well-developed stems, leaves and roots. They reproduce by spores.

  (d) Seed-bearing plants reproduce by seeds.
  (i) Conifers have no flowers; their seeds are not enclosed in an ovary.
  (ii) Flowering plants have flowers; their seeds are in an ovary which forms a fruit. Monocots have one cotyledon in the seed; dicots have two cotyledons in the seed.

- **ANIMALS get their food by eating plants or other animals.**

  (a) Coelenterates live mostly in the sea. They have tentacles and sting cells.

  (b) Flatworms have flat bodies; some have cilia. They live in fresh water or are parasitic.

  (c) Annelids have tubular bodies divided into segments.

  (d) Molluscs have an external shell and a creeping or holding 'foot'.

  (e) Arthropods have a hard exoskeleton and jointed legs.
  (i) Crustacea mostly live in water and have more than three pairs of legs.
  (ii) Insects mostly live on land and have wings and three pairs of legs.
  (iii) Arachnids have four pairs of legs and poisonous mouth parts.
  (iv) Myriapods have many pairs of legs.

  (f) Vertebrates have a spinal column and skull.
  (i) Fish have gills, fins and scales.
  (ii) Amphibia can breathe in air or in water.
  (iii) Reptiles are land animals; they lay eggs with leathery shells.
  (iv) Birds have feathers, beaks and wings; they are 'warm-blooded'.
  (v) Mammals have fur, and suckle their young; the young develop inside the mother.

25 Feeding

All living organisms need food. Some of this food is used to make new tissues for growth and replacement but most of it is used to provide energy (see page 14). Energy is obtained from food by breaking it down chemically to carbon dioxide and water.

Green plants

Green plants make their food from carbon dioxide and water by a process called photosynthesis. This is more fully explained on pages 24–32. Having made the food, the plants can then use it for energy or for growth.

Saprophytes

These are organisms which feed on dead and decaying matter. Often they release digestive enzymes into their food and absorb the soluble products back into their bodies. Bacteria and fungi are examples of saprophytes which digest dead wood, rotting vegetation or the humus in the soil (see pages 34 and 40).

Fungi are made up of fine threads called hyphae. In the mould fungi, these hyphae spread over organic matter such as dead wood or rotting vegetation which the fungus uses as a source of food (Figure 6 on page 35 and Figure 11 on page 221). Some of the hyphae grow into this food and produce enzymes which digest it. The soluble products are absorbed back into the hyphae and used for energy and growth.

Parasites

These are organisms which live on or in another organism (the **host**) and get their food from it. Fleas live on the bodies of mammals or birds and suck their blood. Tapeworms live inside the alimentary canals of vertebrates and absorb the digested food there. Parasites do not necessarily damage the host. A large number of bacteria live in our intestines and are not harmful. Indeed, some of them may be useful because they make a vitamin which we need. In this case they might be called symbiotic (page 39) rather than parasitic.

Protista

Some protista have chloroplasts (page 24) and make their food in a similar way to plants. Euglena (page 198) is an example. Others, such as Amoeba, Paramecium and Vorticella (page 198), take in solid food and digest it. The cilia on Vorticella and Paramecium create water currents which sweep small organisms, such as bacteria, towards the

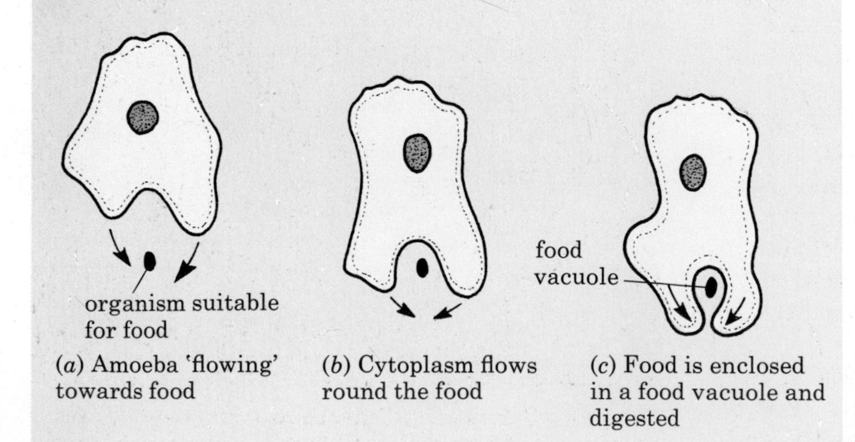

Figure 1 Amoeba ingesting food

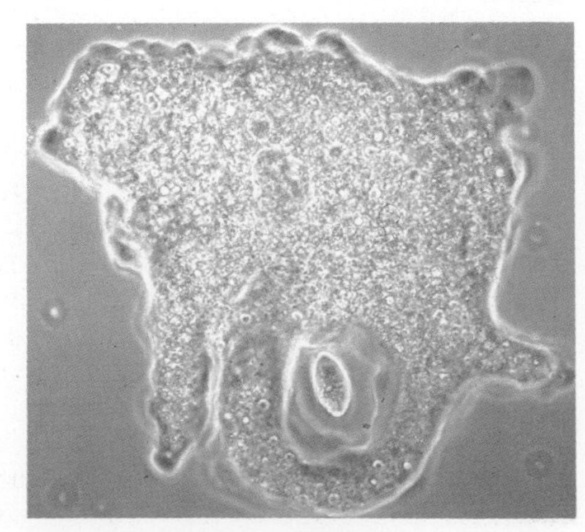

Figure 2 Amoeba feeding

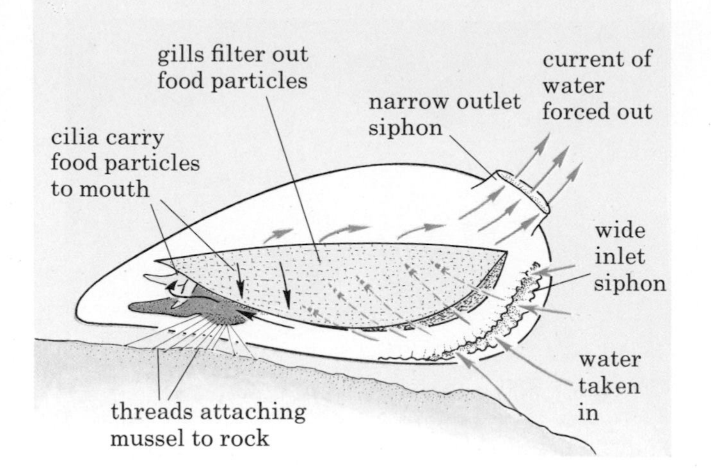

Figure 3 Filter feeding in a mussel

Figure 4 Mussels filter feeding. The water is going in through the opening fringed with 'tentacles' and coming out of the smooth-edged opening. If anything touches the tentacles, the mussel closes its shells at once. Some of the mussels have barnacles growing on them.

protistan. The food is then ingested into the protistan's cytoplasm as a food vacuole.

Amoeba has no cilia but ingests its food by 'flowing' round it (Figures 1 and 2) and trapping it in a food vacuole. The cytoplasm surrounding the food vacuole releases enzymes into it. These enzymes digest the food, and the soluble products are absorbed back into the cytoplasm. This method of feeding is similar to that of animals, i.e. ingestion, digestion, absorption.

Animals

Animals feed either on plants or on other animals. There is not much difference between green plants in the way they make their food, but there is a great variety of ways in which animals obtain their food.

Filter feeding. Many quite unrelated animals which live in water filter the water and ingest the particles which they extract from it.

The mosquito larva shown in Figure 5 on page 214 flicks its 'mouth brushes' and makes a current of water from which it extracts bacteria and other small organisms.

The mussel (Figures 3 and 4) has large gills covered with cilia. The beat of the cilia draws a current of water in between the shells and forces it through the gills. The gills filter out any food particles and the filtered water is expelled through the siphon. The cilia on the gills also carry the trapped food particles forward to the mouth.

These two creatures stay more or less in the same place and create a water current, but a great many animals swim through the water and filter it as they go.

As the herring swims along, it takes in water through its mouth and lets it out through the gill covers (see Figure 7 on page 215). This flow of water is partly for gaseous exchange as described on page 214, but the water is also filtered by gill rakers (Figure 5). These form a bony grid between the gills which allows water to pass out but traps small shrimps and other crustacea which are swimming in the surface waters of the sea.

Certain types of whale, enormous though they are, filter feed in a similar way.

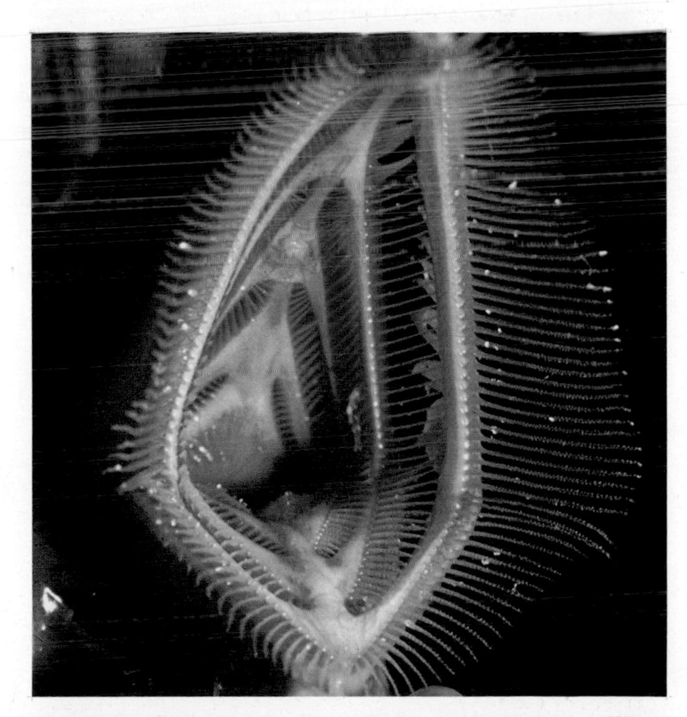

Figure 5 Gill rakers, dissected out of a herring. When water passes between the gill bars, food is trapped on the gill rakers.

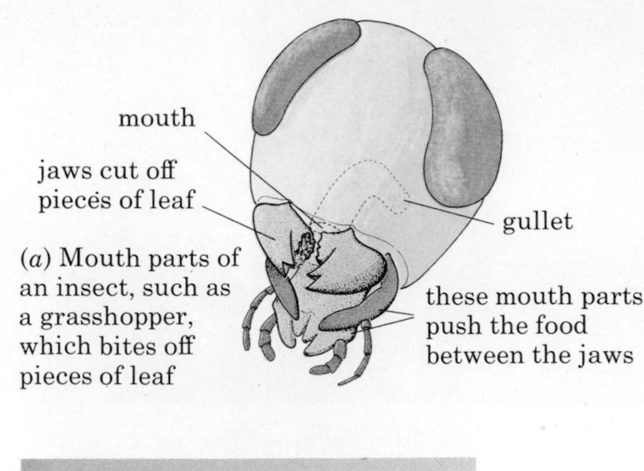

(*a*) Mouth parts of an insect, such as a grasshopper, which bites off pieces of leaf

(*b*) Butterfly feeding. The mouth parts form a tube, like a fine drinking straw, which probes into the flower to suck the nectar.

(*c*) Larva of the great diving beetle. This larva lives in fresh water and spears small fish or tadpoles with its jaws. It will change into a water beetle like the one in Figure 4 on page 214.

antenna
eye
tubular mouth parts

(*d*) Adult greenfly

food pumped into the alimentary canal
mouth parts piercing the leaf
food being sucked from phloem
vein in leaf
leaf section

(*e*) Section through leaf and head of greenfly to show how food is sucked from leaf

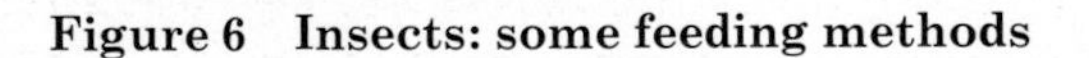

Figure 6 Insects: some feeding methods

Insects have a great variety of feeding methods. Grasshoppers, locusts and caterpillars have simple jaws (Figure 6*a*) with which they bite off pieces of leaf small enough to be ingested.

The butterfly's mouth parts form a long, fine tube which reaches into flowers and sucks the nectar (Figure 6*b*).

The greenfly also has tubular mouth parts but these are used to pierce through leaves and stems of plants and suck food from the phloem (Figures 6*d* and *e*).

The jaws of the water-beetle larva in Figure 6*c* clamp shut on passing tadpoles or small fish and inject them with enzymes. The enzymes digest the tissues of the prey and the digested fluid is sucked into the larva's alimentary canal.

The feeding method of the housefly is described on page 245.

Mammals. The way a carnivore, like a fox, uses its teeth to catch, kill and eat its prey has been described on page 98.

The teeth of a herbivorous mammal are different from the carnivore in many ways (Figure 7). The canine teeth are almost the same shape as the incisors and there is a big gap between the canines and the premolars. This enables the tongue to keep moving the food about in the mouth, to be ground up by the molars and premolars.

In sheep and cows, there are no top incisors. The grass is gripped between the bottom incisors

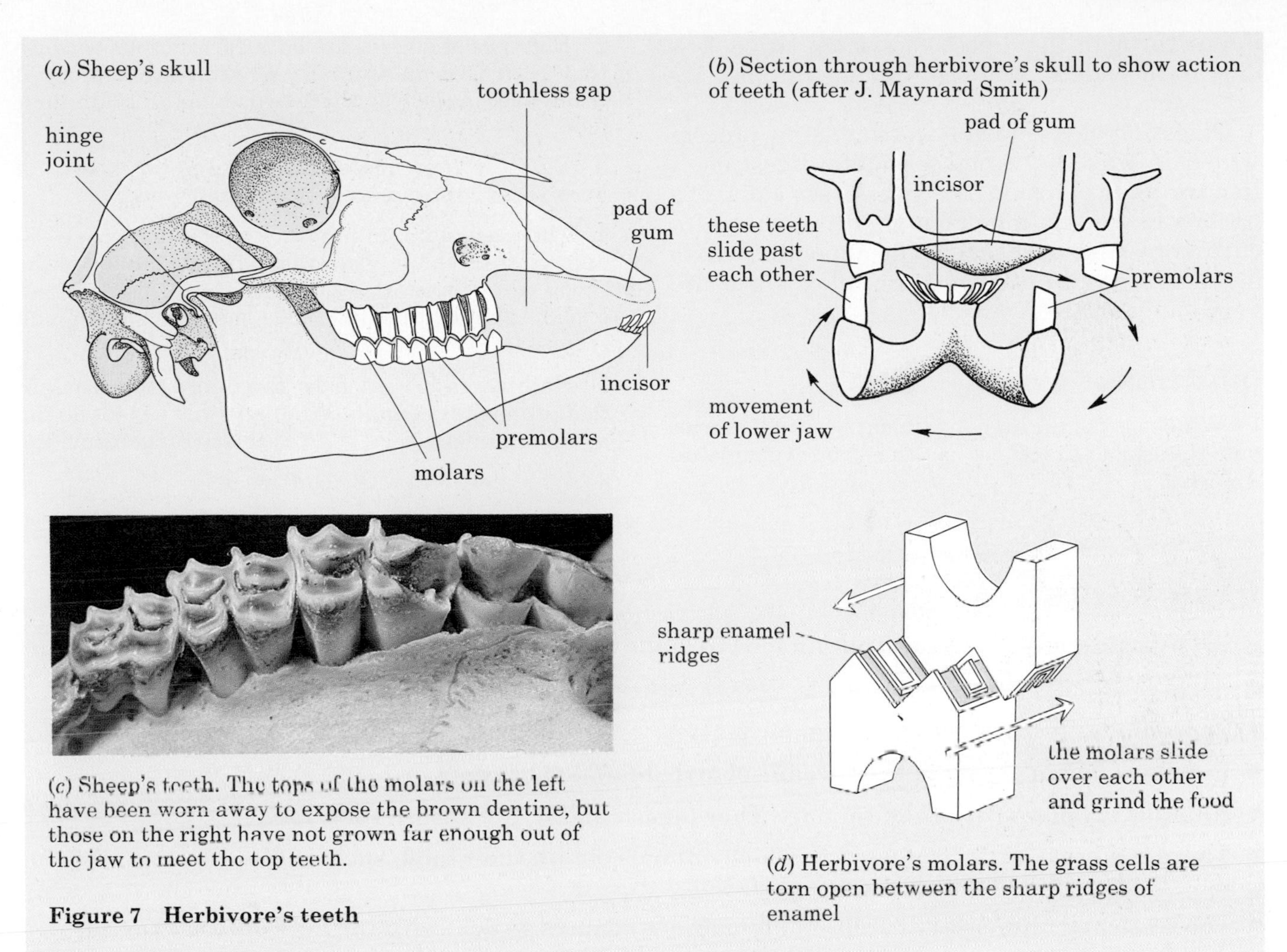

(c) Sheep's teeth. The tops of the molars on the left have been worn away to expose the brown dentine, but those on the right have not grown far enough out of the jaw to meet the top teeth.

(d) Herbivore's molars. The grass cells are torn open between the sharp ridges of enamel

Figure 7 Herbivore's teeth

and the gum of the top jaw. The bottom jaw moves from side to side and the premolar and molar teeth slide across each other, grinding the grass to a pulp (Figures 7*b* and *d*).

This action wears the teeth down so that the grinding surfaces of the top and bottom teeth fit together exactly (Figure 7*a*). The teeth continue to grow at the same rate as they wear down for the whole life of the animal.

In herbivores, the alimentary canal also is adapted to the diet of grass and leaves. Mammals cannot make an enzyme for digesting the cellulose of plant cell walls. They depend upon bacteria and protista living in their alimentary canals to do the digestion for them. These micro-organisms digest the cellulose slowly and, later on, the micro-organisms are digested by the herbivore. This is an example of symbiosis (page 39).

In cattle and sheep, there are several compartments to the stomach (Figure 8). The grass is chewed, swallowed and directed into the first compartment, called the **rumen**. In the rumen, the micro-organisms start to digest the grass. After a time the partly digested grass, called the **cud**, is sent back up the gullet to the mouth and chewed some more. When it is swallowed it goes, not to the rumen, but to the main stomach and so on its way through the rest of the alimentary canal.

Rabbits, at night, produce soft, moist faeces which they promptly eat. In this way, the grass

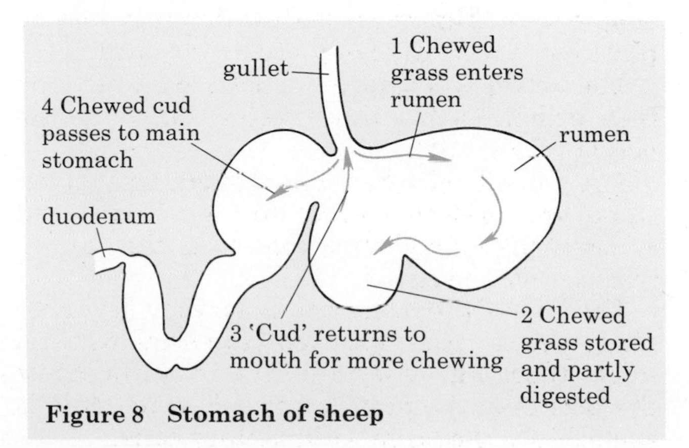

Figure 8 Stomach of sheep

is put through the digestive system twice. The dry, hard pellets produced during the day are not eaten.

Digestion of vegetation is a very slow process which is why the rabbit's double digestion is important. In addition, the alimentary canal of a herbivore is very much longer than that of a carnivore. Also the caecum and appendix, where much of the cellulose digestion takes place, are very much larger.

QUESTIONS

1 Which part of the human alimentary canal carries out the same functions as the food vacuole in Amoeba?

2 Herrings and foxes are both carnivorous because they both feed on animals. What are the essential differences in the way these two animals obtain their food?

3 Why do the molar and premolar teeth of herbivores continue to grow all their lives?

4 The premolars and molars of the dog (page 98) differ in their shape. The premolars and molars of the sheep are almost the same in shape. Suggest an explanation for this difference between the dog and sheep.

5 Herbivores benefit from the micro-organisms in their alimentary canals. What possible benefit do the micro-organisms receive from this arrangement?

6 What does 'chew the cud' mean?

BASIC FACTS

- **All living organisms need to obtain food to make new tissue and to provide energy.**
- **All green plants make their food by photosynthesis.**
- **Saprophytes feed on decaying organic matter.**
- **Parasites live in or on another organism and obtain food from it.**
- **Animals eat plants or other animals. They ingest, digest and absorb their food.**
- **There is a great variety of ways in which animals obtain their food, such as filter feeding, chasing and catching prey or chewing up vegetation.**
- **The teeth and alimentary canals of animals are adapted to deal with the kind of food they eat.**

26 Breathing

In order to get energy from their food, animals and plants have to break it down to carbon dioxide and water by the chemical process called respiration (page 14). One form of respiration, aerobic respiration, uses oxygen for this process. The oxygen is obtained by gaseous exchange (page 125) with the air or water around the organism. The same process gets rid of carbon dioxide produced by respiration.

In some animals, gaseous exchange is helped by a process of ventilation (page 123) but many plants and animals depend on diffusion alone.

Very small organisms, like bacteria and protista, have no need of ventilation. The distances over which the oxygen and carbon dioxide have to travel in these creatures are so small that diffusion is fast enough to meet their needs (Figure 1).

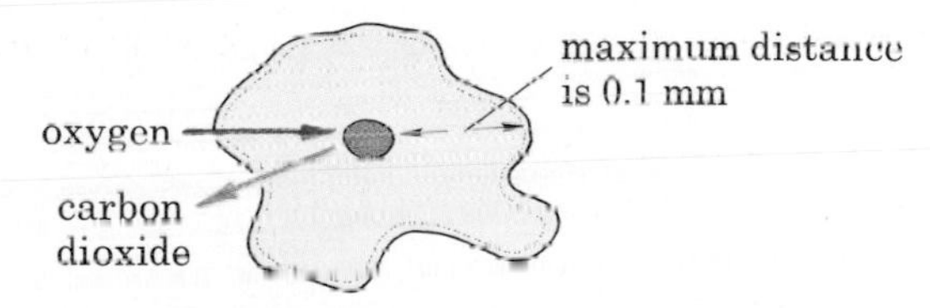

Figure 1 The distance is so small in Amoeba that diffusion is a rapid enough process for the cell's needs

Larger animals need special organs for gaseous exchange and some method of ventilating these organs.

BREATHING IN AIR

Mammals

All mammals breathe by means of lungs. The method of ventilation and gaseous exchange in man is described on pages 123–125 and this is similar in all mammals.

Amphibia and reptiles

Frogs can absorb oxygen and give out carbon dioxide through their moist skin. They also use the lining of the mouth for gaseous exchange. Gulping movements of the throat and the floor of the mouth force air in and out of the mouth through the nostrils (Figure 5 on page 239). Frogs also have lungs, but no ribs or diaphragm. The gulping action also fills the lungs when the frog is using them. Unlike mammals there are no regular breathing movements. Frogs ventilate their lungs only as the need arises.

Reptiles have lungs but no diaphragm and are thought to fill their lungs by movements of the ribs, in a similar way to mammals.

Insects

Insects have a branching system of tubes, called **tracheae** (pronounced 'tray-key-ee') running through their bodies (Figure 2). These tubes

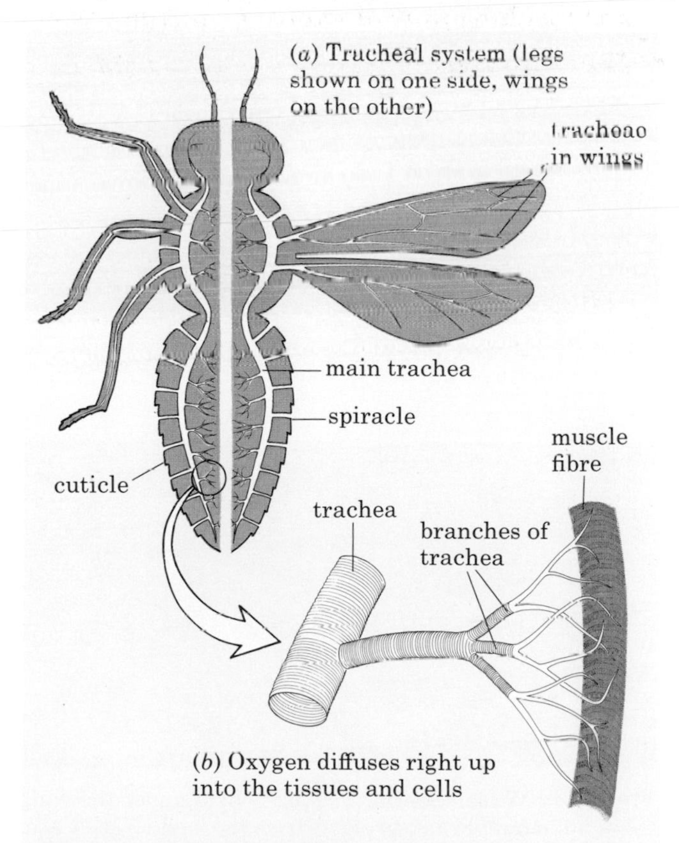

Figure 2 Breathing system of an insect

open through the cuticle (see page 226) by holes called **spiracles**. Oxygen diffuses in through the spiracles, into the tracheae and all the way up to the cells and tissues which are using up oxygen. In a similar way, carbon dioxide diffuses out.

In many insects diffusion is fast enough to meet their needs, but in actively flying insects there is usually a method of ventilation. This consists of compressing the abdomen, so partly squashing the tracheae and forcing air out of them. When the pressure is released, air is sucked back into the tracheae. If you watch a wasp or bee when it has alighted, you will see the pumping movements of the abdomen.

BREATHING IN WATER

Insects

The larval stages of some insects live in water and have gills. These are plate-like or thread-like structures sticking out from the body. There are branches of the tracheal system in the gills and these absorb oxygen from the water (Figure 3).

Some insects breathe air even though they live in water. The mosquito larva in Figure 5 is exchanging oxygen and carbon dioxide with the air by means of a breathing tube which is connected to the tracheal system.

The adult water beetle holds a store of air under its wing cases and uses it during its dives (Figure 4). From time to time it goes to the surface to replace the oxygen and get rid of the carbon dioxide in this store.

Fish

Fish absorb dissolved oxygen from the water by means of gills. The herring has four gills on each side of the head underneath a bony gill cover (Figure 5 on page 233). Each gill consists of a curved bar with gill filaments sticking out from one side and gill rakers projecting forwards from the other side (Figure 6). The gill filaments are branched structures with blood capillaries running in them. The branched filaments present a very large surface to the water for absorbing oxygen into the blood and getting rid of carbon dioxide.

By movements of the floor of the mouth and the gill covers, a fish takes water in through the mouth, and forces it between the gills and out

Figure 3 Mayfly larva. The seven pairs of gills flick to and fro creating a current of water over them.

Figure 4 Water beetle. The air is held under the wing cases but a bubble also projects from the back. Notice also the fringe of hairs on the third legs. These legs act like oars to drive the beetle through the water.

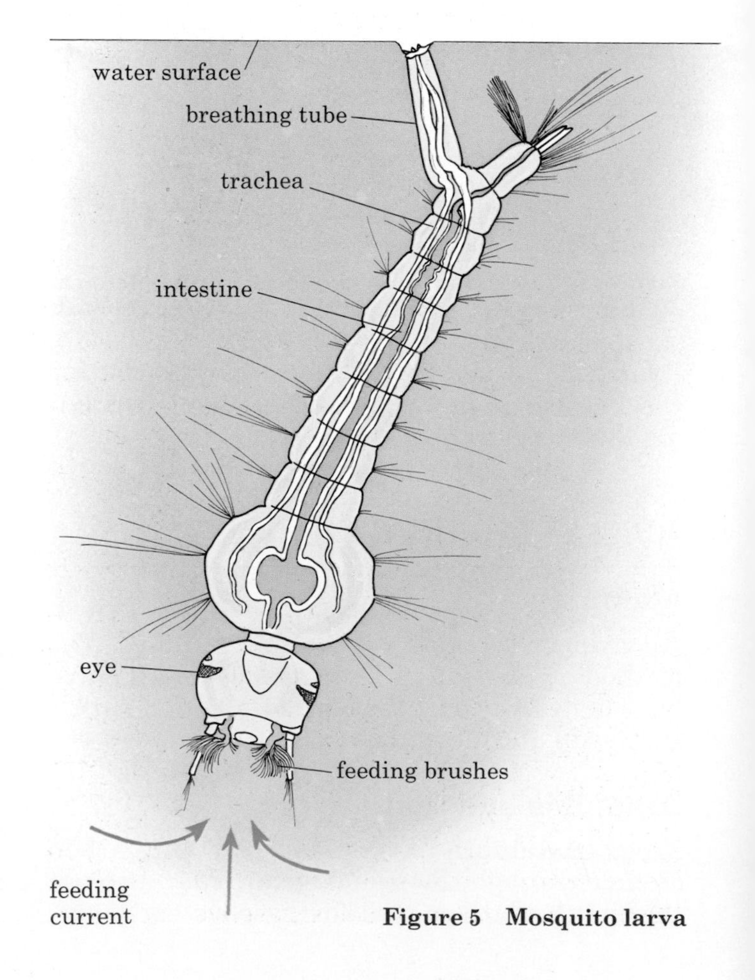

Figure 5 Mosquito larva

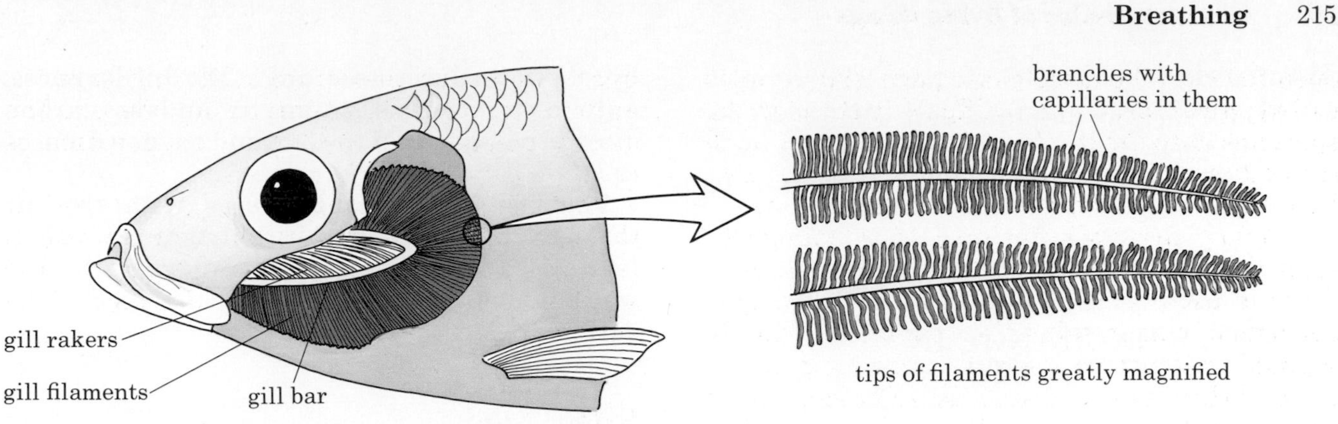

Figure 6 Herring gills. (The gill cover has been cut away to show the gills.)

through a gap between the gill cover and its body (Figure 7). This method of ventilation changes the water in contact with the gills, so bringing fresh supplies of oxygen and carrying away carbon dioxide.

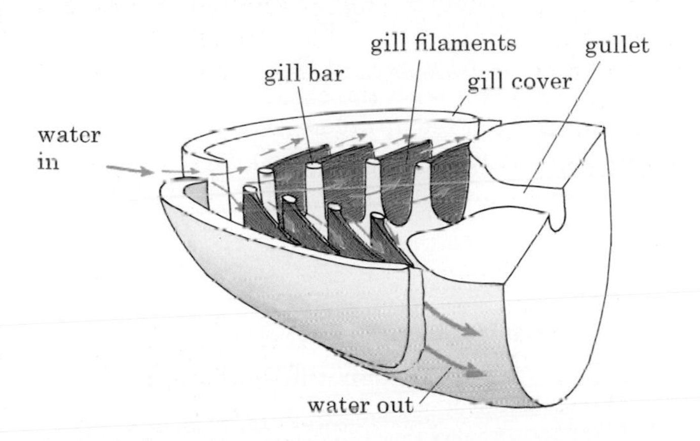

Figure 7 Diagram to show ventilation of gills

The lungs of mammals and gills of fish both show the basic requirements of an efficient structure for gaseous exchange. These structures need a large surface area. The gills achieve this by having hundreds of filaments, with many branches on each filament. In the lungs, it is the millions of alveoli (page 122) which give a large surface area.

In both gills and lungs, there is a dense network of capillaries which enable rapid exchange of oxygen and carbon dioxide between the blood and the water or air. The capillaries are separated from the water or air by only a thin epithelium (single layer of cells), so that diffusion of the gases is as rapid as possible.

Amphibia

The larvae of frogs (Figure 10 on page 230), toads and newts have external gills at first (Figure 8, and Figure 10 on page 230). These are little more than branched blood vessels covered with a thin epithelium, sticking out from the side of the body. Later on, the tadpole uses internal gills rather like those of a fish. The adult frog in water can absorb dissolved oxygen through its skin. When it is at the surface, the frog can breathe air through its nostrils, which are just above the water-level (Figure 9).

Figure 8 Newt larva with external gills

Figure 9 Head of frog partly submerged

Mammals

Mammals such as dolphins and whales, which spend all their lives in water, still breathe air at the surface and use their lungs for gaseous exchange.

GASEOUS EXCHANGE IN FLOWERING PLANTS

Plants do not have special breathing organs. Because they do not move about, like animals, they do not use up oxygen or produce carbon dioxide very rapidly. Their leaves present a very large surface area to the air, and diffusion of oxygen and carbon dioxide through the stomata into the inter-cellular spaces (page 60) is fast enough for the respiration going on in their cells. Because most plant leaves are thin, the distances for diffusion are also very short (see also pages 25 and 61).

In daylight, some of the oxygen produced by photosynthesis is used in respiration and the carbon dioxide released from respiration is used by photosynthesis (see page 24). In darkness, only respiration is going on and so carbon dioxide passes out of the leaf and oxygen diffuses in.

Plant stems exchange gases with the air through their stomata or structures called lenticels. Roots use oxygen dissolved in the soil water which they absorb.

QUESTIONS

1 What are the essential differences in the way a cell in the body of a mammal receives its oxygen, and the way a cell in an insect's body receives its oxygen?

2 On page 213 the text suggests that frogs ventilate their lungs only as the need arises. In what conditions would you expect a frog to be ventilating its lungs?

3 The mosquito larva has only one spiracle. Where do you think this is?

4 Ventilation of lungs or gills brings fresh air or water into contact with the breathing organs. How does this help to speed up diffusion of oxygen into the blood? (Page 18 may provide some ideas.)

BASIC FACTS

- **Diffusion is rapid enough to supply microscopic organisms with the oxygen they need.**
- **All the vertebrate animals except fish have lungs, but they are ventilated in different ways.**
- **Insects breathe by means of a branching system of tracheae.**
- **Fish force water over their gills, which absorb the dissolved oxygen.**
- **All breathing organs have a large surface, thin epithelium and many capillaries.**
- **Frogs can absorb oxygen by means of their mouth lining, lungs or skin.**
- **Plants exchange carbon dioxide and oxygen with the air by diffusion through their stems, roots and leaves.**

27 Reproduction

No plant or animal can live for ever, but part of it lives on in its offspring. Offspring are produced by the process of reproduction. This process may be **sexual** or **asexual** (see below), but in either case it results in the continuation of the species.

SEXUAL REPRODUCTION

The following statements apply equally to plants and animals. Sexual reproduction involves the production of sex cells. These sex cells are called **gametes** and they are made in reproductive organs. The process of cell division which produces the gametes is called **meiosis** and is described on page 177. In sexual reproduction, the male and female gametes have to come together and **fuse**, that is, their cytoplasm and nuclei join together to form a single cell called a **zygote**. The zygote then grows into a new individual (Figure 1 on page 138).

Sexual reproduction in flowering plants

The male gamete is a cell in the pollen grain. The female gamete is an egg-cell in the ovule. The process which brings the male gamete within reach of the female gamete (i.e. from stamen to stigma) is called **pollination**. The pollen grain grows a microscopic tube which carries the male gamete the last few millimetres to reach the female gamete for fertilization. The zygote then grows to form the seed. These processes are all described in more detail on pages 72–79.

Sexual reproduction in animals

The male gamete is the sperm; the female gamete is the ovum. The male and female animals are usually attracted towards each other by patterns of behaviour called **courtship**. The process of mating places the sperms as close as possible to the ovum. The sperms then swim to the ovum and fertilize it. The zygote grows into an embryo. Sexual reproduction in man is described more fully on pages 138–147.

In both plants and animals, the male gamete is microscopic in size and mobile (i.e. can move from one place to another). The sperms swim to the ovum and the pollen cell moves down the pollen tube (Figure 1). The female gametes are always larger than the male gametes and are not mobile. Pollination in the seed-bearing plants and mating in animals are the processes which bring the male and female gametes close together. Fertilization is the fusion of the gametes, and as a result the zygote undergoes rapid cell division to produce an embryo.

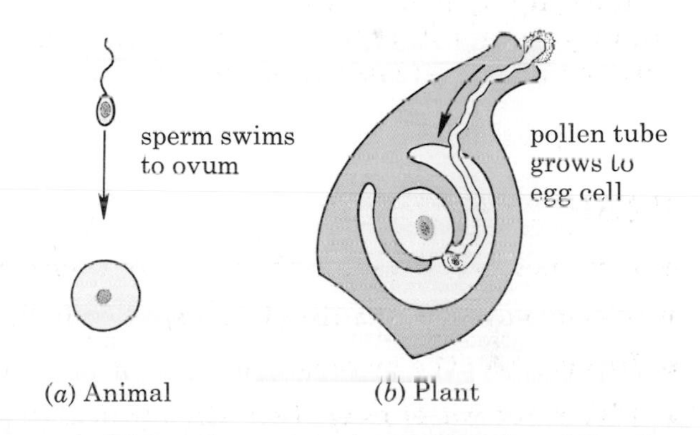

Figure 1 The male gamete is small and mobile; the female gamete is larger

External fertilization. In many animals which live or breed in water, the female sheds her unfertilized eggs into the water and the male releases sperms over them. The eggs are fertilized outside the body of the female and so the process is described as external fertilization. For fertilization to be successful, the eggs and sperms must be released at the same time and close to each other. This usually results from a behaviour pattern in which the male and female are first attracted towards each other and then stimulate each other to produce gametes.

Figure 2 Male three-spined stickleback showing the breeding colour. A one-week-old baby can also be seen in the picture.

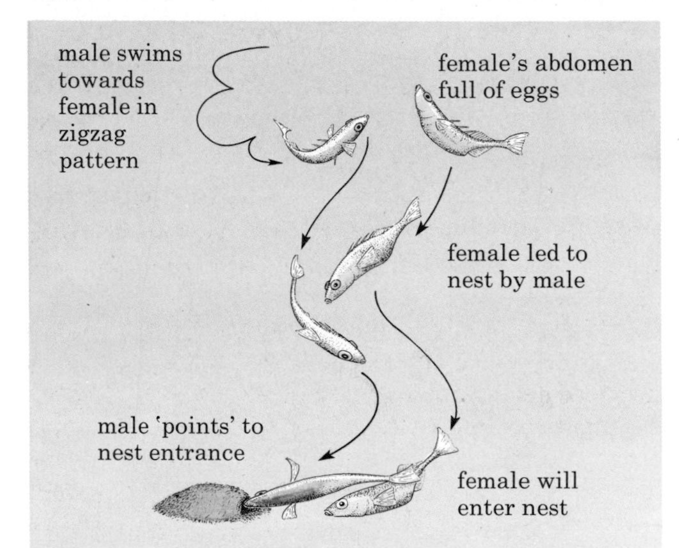

Figure 3 Courtship in the three-spined stickleback (from N. Tinbergen, *The Study of Instinct*, O.U.P., 1951)

Figure 4 Sticklebacks at the nest. The male's movements at the nest entrance induce the female to enter and lay eggs. Notice her abdomen swollen with eggs.

The male stickleback, in the breeding season, develops a red belly and blue eyes (Figure 2). These colour changes seem to keep other males at bay. The male digs a small hollow at the bottom of the pond and roofs it over with pieces of vegetation. When a mature female enters the male's territory, she is attracted by his bright colours. The male's next behaviour is triggered by the sight of her abdomen, swollen with eggs. He swims towards her and then down to the nest. She follows him and enters the nest (Figures 3 and 4). The male taps the female's tail with his snout and this stimulates her to lay eggs. When she leaves the nest, the male enters and sheds sperms on the eggs. The nest and the courtship behaviour ensure that eggs and sperms are shed in the same place and at the same time.

Frogs and toads, in the spring, migrate to ponds. Here the male climbs on the back of the female and is carried about by her till she lays her eggs (Figure 5). When the male feels the eggs being laid, he releases his sperms which fertilize the eggs. This is external fertilization, but the male's behaviour in remaining attached to the female makes sure that the sperms are released at the same time as the eggs.

Figure 5 Toads pairing. The male clings to the female's back and releases his sperms as she lays the eggs.

Parental care is the way in which animals look after their offspring. In British frogs and toads there is no parental care; both parents leave the eggs to develop entirely on their own. The male stickleback, on the other hand, guards the developing eggs by driving off intruders and fans water through the nest with his fins to improve the oxygen supply.

Figure 6 Parental care in birds. The sedge warbler has brought a caterpillar to feed the chicks in the nest. The bright colour on the inside of their beaks stimulates the parents to put food into them.

Figure 7 Large numbers of offspring. One method of survival is to produce a large number of eggs, as in this egg-rope of the perch. Many of the young fish will die but enough will survive to continue the species.

In many birds and mammals, the parental care is even greater (see page 144). The young are reared in a nest, fed and kept warm usually by both parents (Figure 6).

There is often a connection between the amount of parental care and the number of eggs produced. Where there is little parental care, there is usually a large number of eggs (Figure 7). Because the parents do not protect the eggs and young or bring them food, many offspring will die. The large number of eggs, however, ensures that some offspring will survive to maturity. The parental care shown by birds and mammals gives the young a better chance of survival and they have small numbers of offspring.

Internal fertilization. In reptiles, birds and mammals, the eggs are fertilized by the male placing sperms inside the body of the female. This is internal fertilization and it requires a behaviour pattern which brings male and female together at the right time so that sperms are released when the eggs are mature. The first step is **pair formation**, which ensures that the male and female stay together at least for the breeding period. There is usually a courtship pattern which brings the pair together in the first place, stimulates them to mate, and keeps them together while the young are reared. In birds, the male often feeds the female as part of the courtship ritual (Figure 8).

The reproductive organs have to develop in such a way that internal fertilization can occur. This usually means that the male animal has developed a penis which can be inserted into the female's reproductive organs in order to deposit the sperms.

Internal fertilization is essential for animals living on land. External fertilization here is impossible because the eggs and sperms would dry up when exposed to the air. In reptiles and birds, the egg is prevented from drying up by having a shell, so the sperms have to get to the egg before the shell is put on it. This means that internal fertilization is necessary. In mammals, the egg develops inside the body of the female after internal fertilization.

Other land-dwelling animals such as insects, spiders and snails rely on methods of internal fertilization for their reproduction.

Figure 8 Courtship. The female herring gull begs for food, which the male regurgitates in response to her pecking at his beak. This behaviour helps to establish and maintain the pair bond in many species.

QUESTIONS

1 Draw up a table with three columns. Head the second column 'Flowering plants' and the third column 'Mammals' as shown below. In the first column write (a) Male reproductive organs, (b) Female reproductive organs, (c) Male gamete, (d) Female gamete, (e) Place where fertilization occurs, (f) Zygote grows into . . .

Now complete the other two columns.

	Flowering plants	Mammals
(a) Male reproductive organs		
(b) Female reproductive organs		
(c) Male gamete etc.		

2 Which of the following animals would you expect to have external, and which internal fertilization: butterfly, mussel, trout, sparrow, earthworm? In each case give the reason for your decision.

3 Why do you think courtship behaviour is necessary for the success of both internal and external fertilization?

4 In what ways is the human reproductive system adapted for (a) internal fertilization, (b) internal development of the embryo?

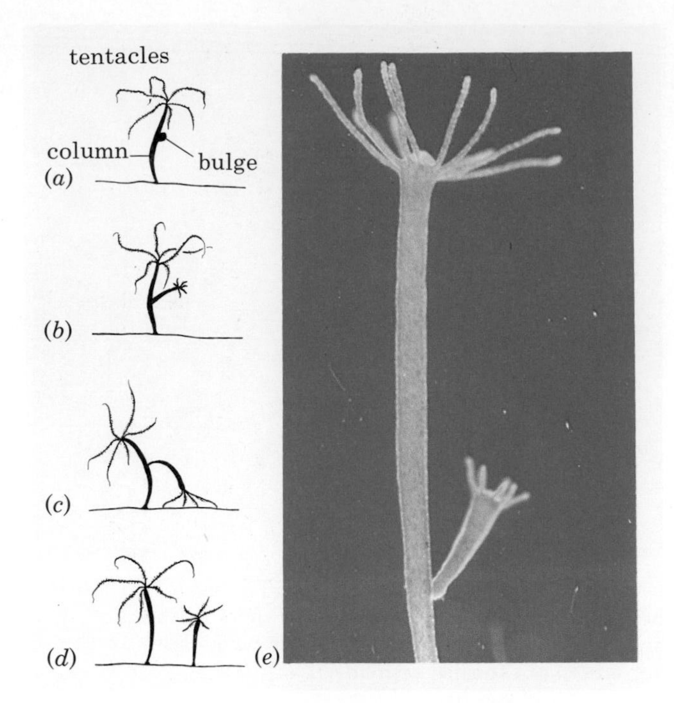

Figure 10 Asexual reproduction in Hydra
(*a*) A group of cells on the column start dividing rapidly and produce a bulge
(*b*) The bulge develops tentacles
(*c*) The daughter Hydra pulls itself off the parent
(*d*) and becomes an independent animal
(*e*) Hydra with bud

ASEXUAL REPRODUCTION

Asexual means 'without sex' and this method of reproduction does not involve gametes. In the single-celled protista or in bacteria, the cell simply divides into two and each new cell becomes an independent organism (Figure 9, and Figure 3 on page 198). In more complex plants and animals, a small part of the organism may grow and develop into a separate individual. For example, a small piece of stem planted in the soil may form roots and grow into a complete plant.

Asexual reproduction in animals

A small number of invertebrate animals are able to reproduce asexually.

Hydra is a small animal, 5–10 mm long, which lives in ponds attached to pond weed. It traps small animals with its tentacles, swallows and digests them. Hydra reproduces sexually by releasing its male and female gametes into the water but it also has an asexual method which is shown in Figure 10.

Asexual reproduction in a fungus

Mucor is an example of a mould fungus. Fungi have sexual and asexual methods of reproduction. In the asexual method, they produce

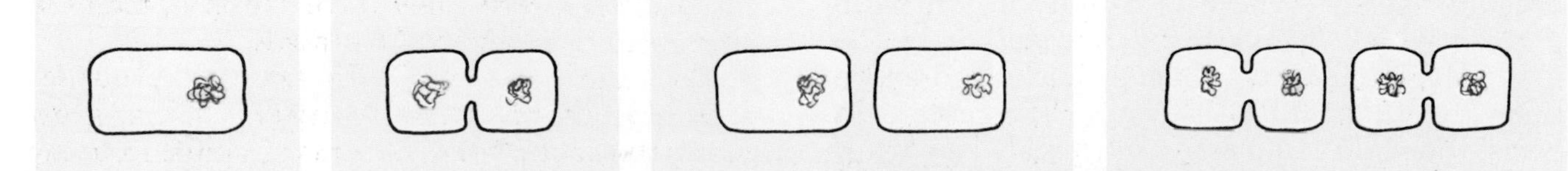

(*a*) Bacterial cell (*b*) Chromosome replicates (*c*) Cell divides (*d*) Each cell divides again

Figure 9 Bacteria reproducing asexually by cell division. Some bacteria can reproduce every 20 minutes.

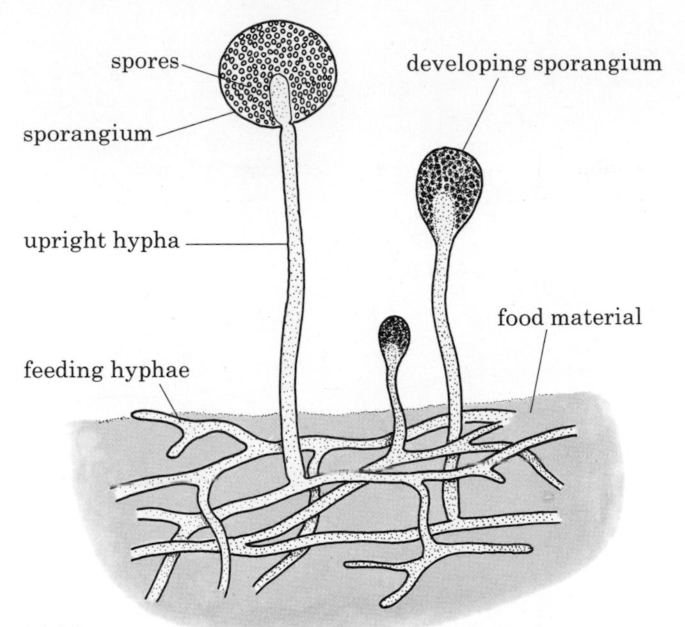

(*a*) Mucor

(*b*) A mould fungus, similar to Mucor, growing on a carrot. The black knobs are the sporangia. The white threads are the hyphae.

Figure 11 Asexual reproduction in mould fungi

single-celled **spores** which are formed in structures called **sporangia**. These microscopic spores are widely scattered and grow into new individuals by cell division.

Fungi are made up of fine threads called **hyphae** which grow into the organic matter which the fungus uses as a source of food. The hyphae are microscopic tubes with an outer wall, central vacuoles and a lining of cytoplasm which contains many nuclei. The tips of the hyphae produce enzymes which digest the food.

Figure 12 Puff-ball. When a raindrop hits the ripe fungus, a cloud of spores is shot out.

The moulds also produce upright hyphae. In Mucor, the ends of these upright hyphae swell up, and the contents undergo rapid cell division to produce a large number of single-celled spores contained in a sporangium (Figures 11*a* and *b*). When the sporangium breaks open, the spores are carried either in air currents or on the feet of insects, or are splashed about by raindrops. If a spore reaches a suitable place for growth, a hypha grows out of it and eventually forms a new mould colony.

The gills on the underside of a mushroom or toadstool (Figure 6 on page 35) produce spores. Puff-balls release clouds of spores (Figure 12).

Since no gametes are involved and no fertilization takes place, this reproduction is asexual. Most fungi do have methods of sexual reproduction in addition to the asexual process.

Asexual reproduction in flowering plants (vegetative propagation)

Although all flowering plants reproduce sexually (that is why they have flowers), many of them also have asexual methods.

The strawberry plant, for example, produces long side shoots called **runners**. These grow along the ground and, at some distance from the parent, take root and produce a complete new plant. When the runner dies, the new plant is quite independent of the parent. One parent

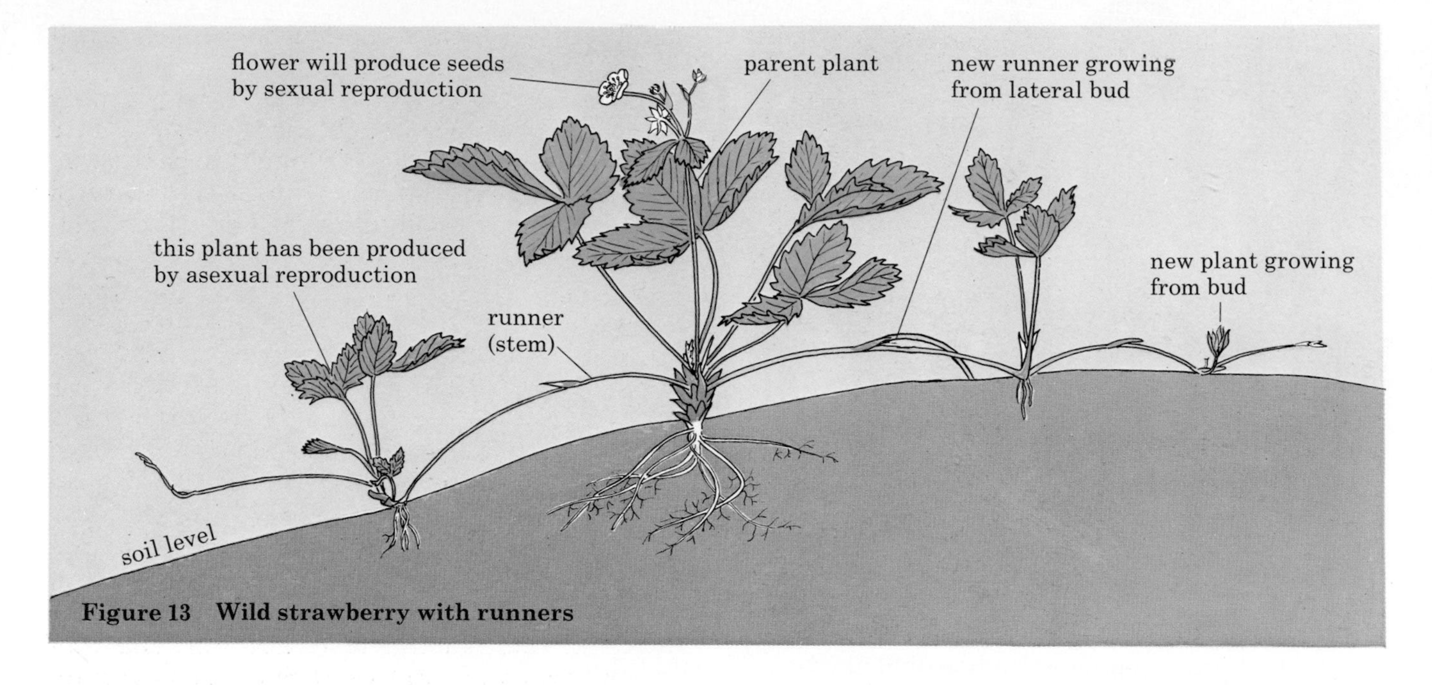

Figure 13 Wild strawberry with runners

plant may produce several daughter plants in this way. Figure 13 shows this method of asexual reproduction or **vegetative propagation** as it is sometimes called.

In the potato plant, vegetative propagation is combined with a method of food storage. Shoots grow from the lateral buds on the lowest part of the potato plant's stem. These shoots grow down in the soil instead of producing leafy branches. Food made in the leaves of the parent plant travels down the stem into these underground branches which swell up to form potatoes (Figure 14). The potato is an underground stem called a **tuber**. It is swollen with stored food,

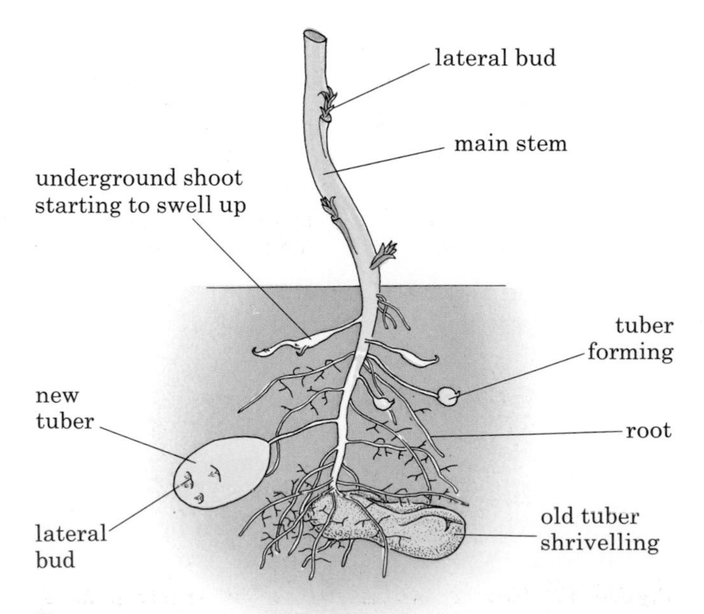

Figure 14 Stem tubers (potatoes)

mainly starch. Because it is a stem, the potato tuber has lateral buds (sometimes called 'eyes'). When the parent plant dies, the buds on the tuber sprout and produce shoots and roots which form new plants (Figure 15). When the

Figure 15 Potato tuber sprouting. Two shoots are growing from the left and five or six from the right.

buds sprout, they can use the large supply of food stored in the tuber and so they grow far more rapidly than plants growing from seeds.

Daffodil, tulip and onion bulbs are examples of a method of asexual reproduction as well as food storage. The bulbs are formed by closely packed leaves, swollen with stored food. Asexual reproduction occurs when the buds in the bulb grow and produce new plants.

Taking cuttings (Figure 16) is an example of a method by which humans make plants reproduce asexually. It is known as artificial vegetative propagation.

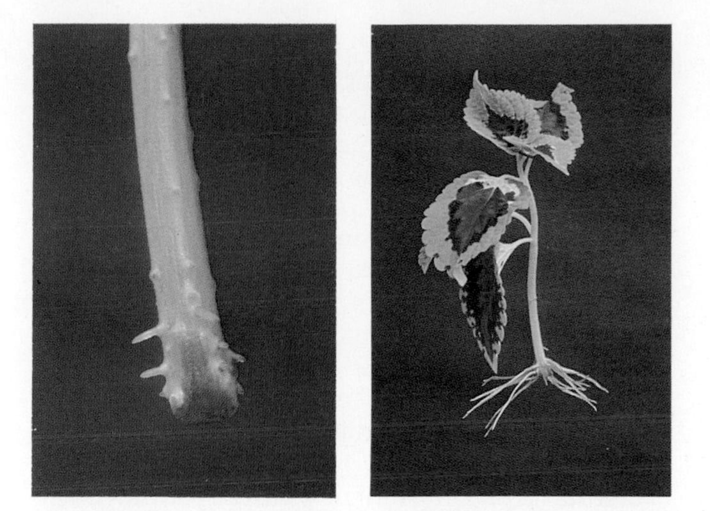

Figure 16 Cuttings. *Left*, in some plants when a branch is cut off and placed in the soil, roots will grow from the cut end of the stem so that the cut branch becomes a new plant. *Right*, the cut stem of this Coleus plant was placed in water and has formed roots in a week or two.

QUESTIONS

5 (a) In what ways does asexual reproduction in Mucor differ from asexual reproduction in flowering plants?

(b) How does a spore differ from a seed? (See page 82 for seed structure.)

6 Why do you think gardeners heap soil round the base of potato plants early in the growing season?

7 Gardeners will sometimes cut a large potato into two before planting, in the hope of getting more plants. How would the gardener decide where to make the cut?

8 In the growing season, what plant structures will be produced by a strawberry plant as a result of (a) asexual reproduction, (b) sexual reproduction?

SEXUAL AND ASEXUAL REPRODUCTION COMPARED

Variation

Sexual reproduction allows the formation of new varieties of plants or animals by combining the genes (page 175) of both parents. For example, plant A may produce a large number of seeds but have a thin stalk that is easily blown over in the wind. Plant B (a variety of the same species as A) has a short strong stalk but produces only a few seeds. In sexual reproduction between plants A and B, the pollen from B may bring the genes for a strong stalk to the ovule of A, which has the genes for producing many seeds. This new combination of genes will give plants with strong stalks and many seeds (Figure 17). These plants should survive longer and leave behind more offspring and so replace the parental types. This is thought to be the way evolution takes place. There is the possibility, of course, of the combination of 'few seeds' and 'thin stalk', but these plants will not survive for long.

Asexual reproduction rules out any possibility of new combinations like this. The new organism is produced from one parent's body cells and not from gametes, so it has exactly the same genes as its parent. This means that there is little chance of evolving new, more successful varieties. From our point of view, however, it is a great advantage to be able to preserve the qualities we like in our fruit, flowers and vegetables, by asexual reproduction.

Sexual reproduction and selection can produce a variety of potato with many large tubers. When these tubers are planted all the plants grown from them by vegetative propagation have the same genes and, therefore, keep the same desirable qualities as the parent. If the plants were grown from seed, only some of them would have the useful characteristics of the parents.

Food supply

In asexual reproduction in plants, the offspring remains attached to the parent (e.g. strawberry) while it grows, or is supplied with a large reserve

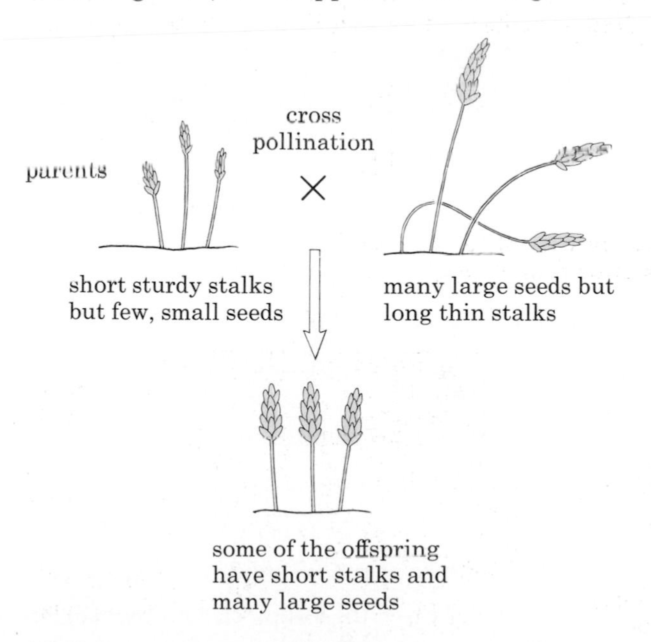

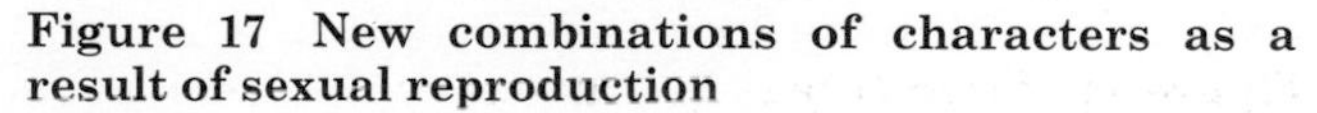

Figure 17 New combinations of characters as a result of sexual reproduction

Figure 18 Vegetative propagation in daffodils. The daffodil plants are growing in clumps. Each clump is probably derived by vegetative propagation from a single bulb over a few years.

of food (e.g. potato). In both cases the plant has the advantage of an abundant food supply until it is well developed and can make its own food by photosynthesis. The seeds produced by sexual reproduction always have some food reserve, but if they are to travel any distance from the parent (see page 80), they cannot be large and heavy, so the amount of food in the seed is rather small.

Dispersal

In the process of vegetative propagation, the offspring have to be produced close to the parent because they are formed from a bud on the parent's stem. This has the disadvantage that they are not likely to spread to a new environment by this means. Since all the plants with vegetative methods of reproduction also have flowers and seeds, this is not a serious drawback. Vegetative propagation does result in the formation of dense colonies because the daughter plants are produced close together. These colonies help to keep out competitors for light and water (Figure 18).

Sexual reproduction in plants results in the production of seeds which may be dispersed over long distances (page 80). This has the advantage that new areas, at a distance from the parents, can be colonized by the species. It is, however, a 'wasteful' method because so many seeds will land in places where germination and growth are not possible.

QUESTIONS

9 How would you attempt (a) to produce a strawberry plant which had bigger and better-tasting strawberries (see page 185), (b) to reproduce this plant so that all the offspring had these qualities?

10 Make a list of the advantages of asexual reproduction (vegetative propagation) in flowering plants.

BASIC FACTS

- **Sexual reproduction involves the male and female sex cells (gametes) joining together.**
- **The male gamete is small and mobile. The female gamete is larger and not often mobile.**
- **The male gamete of an animal is a sperm. The male gamete of a flowering plant is the pollen nucleus.**
- **The female gamete of an animal is an ovum. The female gamete of a flowering plant is an egg cell in an ovule.**
- **Fish and frogs have external fertilization. The sperm is placed on the eggs after they are laid.**
- **Reptiles, birds and mammals have internal fertilization. The sperm is placed in the female's body to fertilize the eggs.**
- **Asexual reproduction does not involve gametes.**
- **Fungi, liverworts, mosses and ferns reproduce asexually by single-celled spores.**
- **Many flowering plants reproduce asexually by vegetative propagation.**
- **Sexual reproduction can produce new varieties of plants and animals.**
- **Asexual reproduction keeps the characteristics of the organism the same from one generation to the next.**

28 Growth and change

GROWTH

Most living organisms start their life as a single cell (zygote, see page 138) too small to be seen with the naked eye. This cell divides many times (by mitosis, see page 175) to produce an organism made up of thousands or millions of cells. Not only does this cell division increase the size of the organism, it also makes it far more complicated than it was to start with. The new cells become specialized and form tissues and organs (see page 7). So growth involves an increase in size and weight and also an increase in complexity.

Cell division on its own does not always produce growth. The fertilized frog's egg in Figure 1*a* divides into hundreds of cells but the embryo in Figure 1*g* is no bigger than the single cell from which it came. It consists of hundreds of tinier cells. As the cells continue to divide, the tissues they produce fold or roll up to form organs. They tuck in at the front to form a mouth and alimentary canal. Along the top, they roll up to form a tube which will later become the spinal cord and brain (Figure 2).

In plants, there are special groups of rapidly dividing cells in the tip of the root or the shoot. Some of these cells, when they stop dividing, start to increase in size because their vacuoles enlarge and stretch the cell walls (see pages 5 and 64). As a result of hundreds of these cells all getting larger, the root grows down into the soil and the shoot grows upwards.

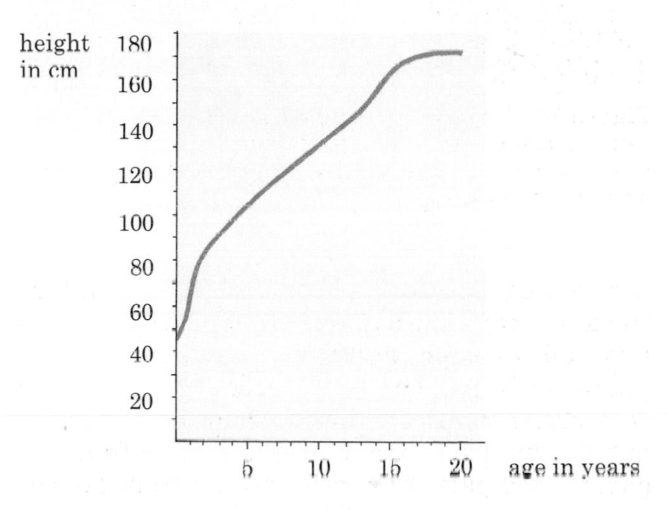

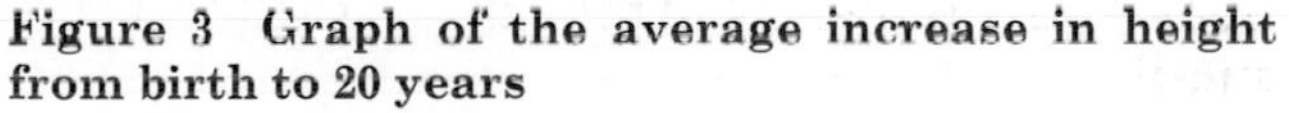
Figure 3 Graph of the average increase in height from birth to 20 years

In most organisms, growth is a gradual process, though the rate of growth may vary at different ages. Figure 3 shows that your growth in height is rapid for the first five years and slows down in the next five years. Between the ages of 10 and 17 (adolescence) there is an

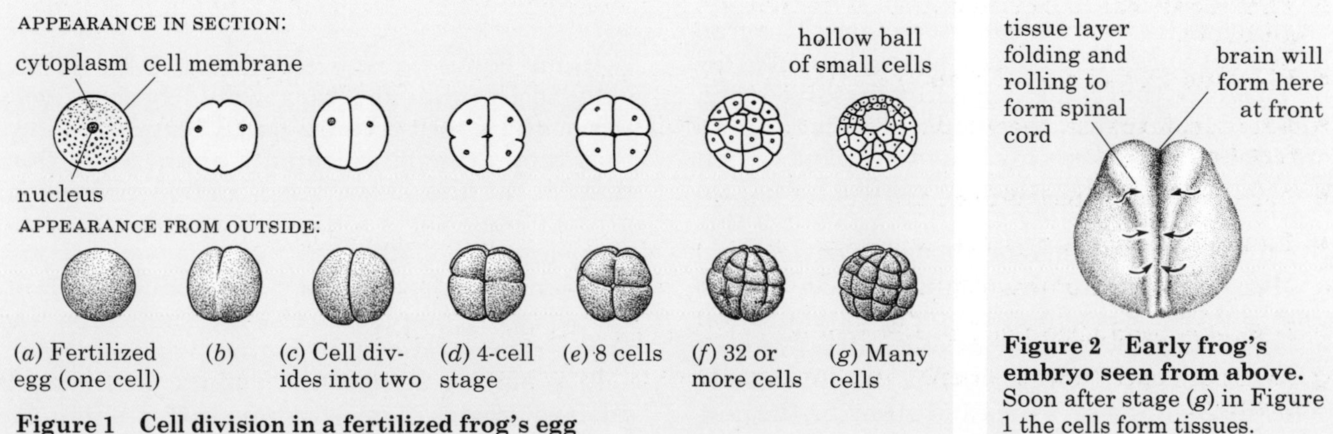

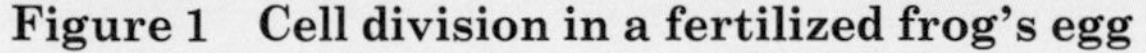
Figure 1 Cell division in a fertilized frog's egg

Figure 2 Early frog's embryo seen from above. Soon after stage (*g*) in Figure 1 the cells form tissues.

Figure 4 Moulting in insects. The locust has wriggled out of its old cuticle which remains attached to the branch. The wings will expand to full size before the cuticle hardens.

increase in the growth rate, but after the age of 20 you will probably not grow at all.

Some fish go on growing all their lives but the growth rate gets slower as they get older.

In the arthropods (page 206) growth in size takes place in spurts. The hard shell, called the **exoskeleton** or **cuticle**, covering the body of a crustacean like a crab, cannot expand. The crab has to get out of this cuticle before he can grow in size. So the shell is partly dissolved and softened from inside, then it splits open, the crab crawls out and allows his body to expand. When the body is expanded, the new cuticle forms and hardens on the outside and the crab stays this size until the next time he sheds his cuticle.

Although an insect's cuticle is much thinner than the crab's it can still not be extended. So insects also grow by a series of moults. Figure 4 shows a locust emerging from its old cuticle before expanding its wings and allowing the new cuticle to harden.

QUESTIONS

1 Some of the cells in Figure 1*g* will become skin, some will become nerves and muscles. What kind of changes must take place in one of the cells for it to become a muscle cell?

2 According to the graph in Figure 3, what is the average increase in height (a) from the age of one to five years, (b) from five to ten years?

CHANGE IN SHAPE

Growth involves an increase in size and complexity, but it also results in a change of shape. The tadpole in Figure 10*g* (page 230) is a very different shape from the adult frog in Figure 10*k*.

The growth of a human baby to an adult involves a considerable change in shape as well as an increase in size. Your head does not grow so fast as your limbs and trunk, so your proportions change as shown in Figure 5.

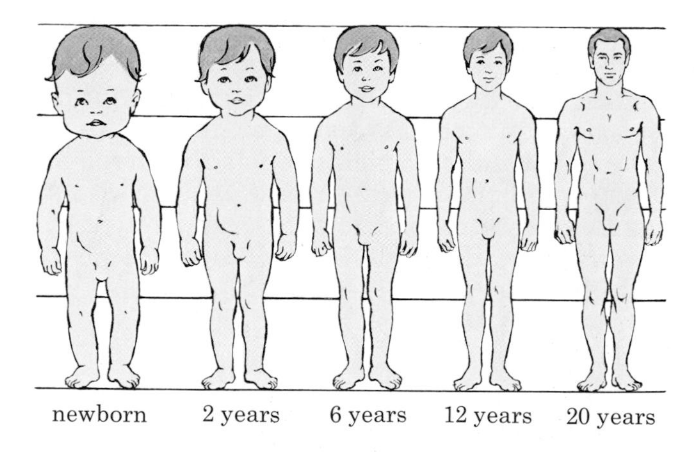

Figure 5 Human growth. All the figures are drawn to the same height to show how the body proportions change with age (after C. M. Jackson).

If our bones were to grow simply by adding cells to the outside, they would become very thick and short (Figure 6*a*). Figure 6*b* shows that they grow more rapidly at the ends than they do at the sides. Also, in the process of growth, some parts actually have to be dissolved away and remodelled.

When a seed germinates to produce a plant, the embryo completely changes its shape and proportions as shown in Figure 3 on page 83.

In some organisms, the change in shape at different stages of growth is so extreme that the young form of the creature seems to be totally

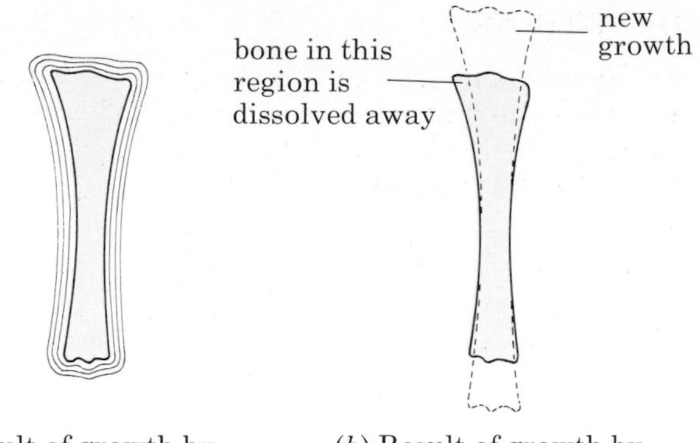

(*a*) Result of growth by adding layers of cells equally all over

(*b*) Result of growth by adding most new cells at the ends

Figure 6 Growth of limb bones

unlike the adult form. The caterpillar growing into a butterfly or the tadpole into a frog involves drastic changes like this. The process is called **metamorphosis**.

QUESTION

3 From Figure 5 you will see that a baby's head occupies over one-quarter of the length of its body.

(a) What fraction of the body does the head occupy in an adult?

(b) How does the proportion of the legs to body change between birth and 20 years?

METAMORPHOSIS

Insects

All insects hatch from eggs. The maggot, grub or caterpillar which comes out of the egg (Figures 7*a* and 9*b*) is called a **larva**. This larva feeds almost non-stop, grows very rapidly and moults its cuticle several times, but without greatly changing its form. Then it stops feeding and goes into a resting stage called a **pupa**. At this time most of the structures of the larva are digested away and replaced by the structures of the adult. When the pupal stage moults and the adult pulls itself out (Figure 7*d*), it has a totally different form from the larva. For example, the adult has wings; the larva has none. The adult butterfly has tube-like mouth parts for sucking nectar from flowers; the larva has simple jaws for biting off pieces of leaf (Figure 7*b*). The larva has prolegs (Figure 8*a*) with hooks for gripping on to the plant where it is feeding; these are not present in the adult.

Figures 8 and 9 show the metamorphosis of a butterfly and a housefly. The adult insects continue to feed, but do not grow or moult any more. They mate and then the female lays eggs, usually on the food material that the larvae will eat; leaves for caterpillars and decaying material for housefly maggots.

(*a*) Butterfly larva hatching. Caterpillar emerging from its egg. Later on it will eat the egg shell.

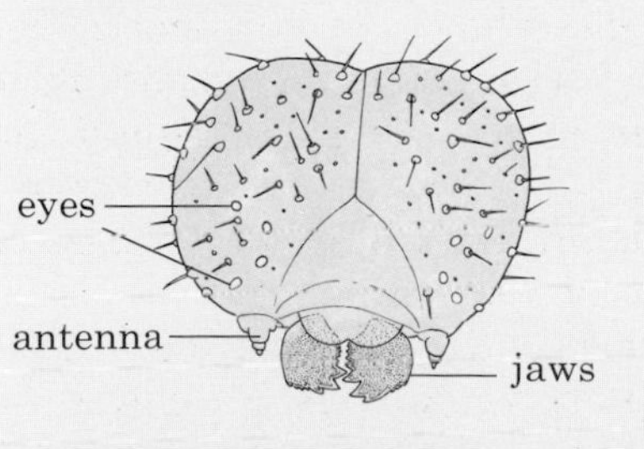

(*b*) Head of caterpillar

(*c*) Head of butterfly

(*d*) Metamorphosis. After a few weeks the pupa's cuticle splits and the butterfly comes out.

Figure 7 The change from caterpillar to butterfly

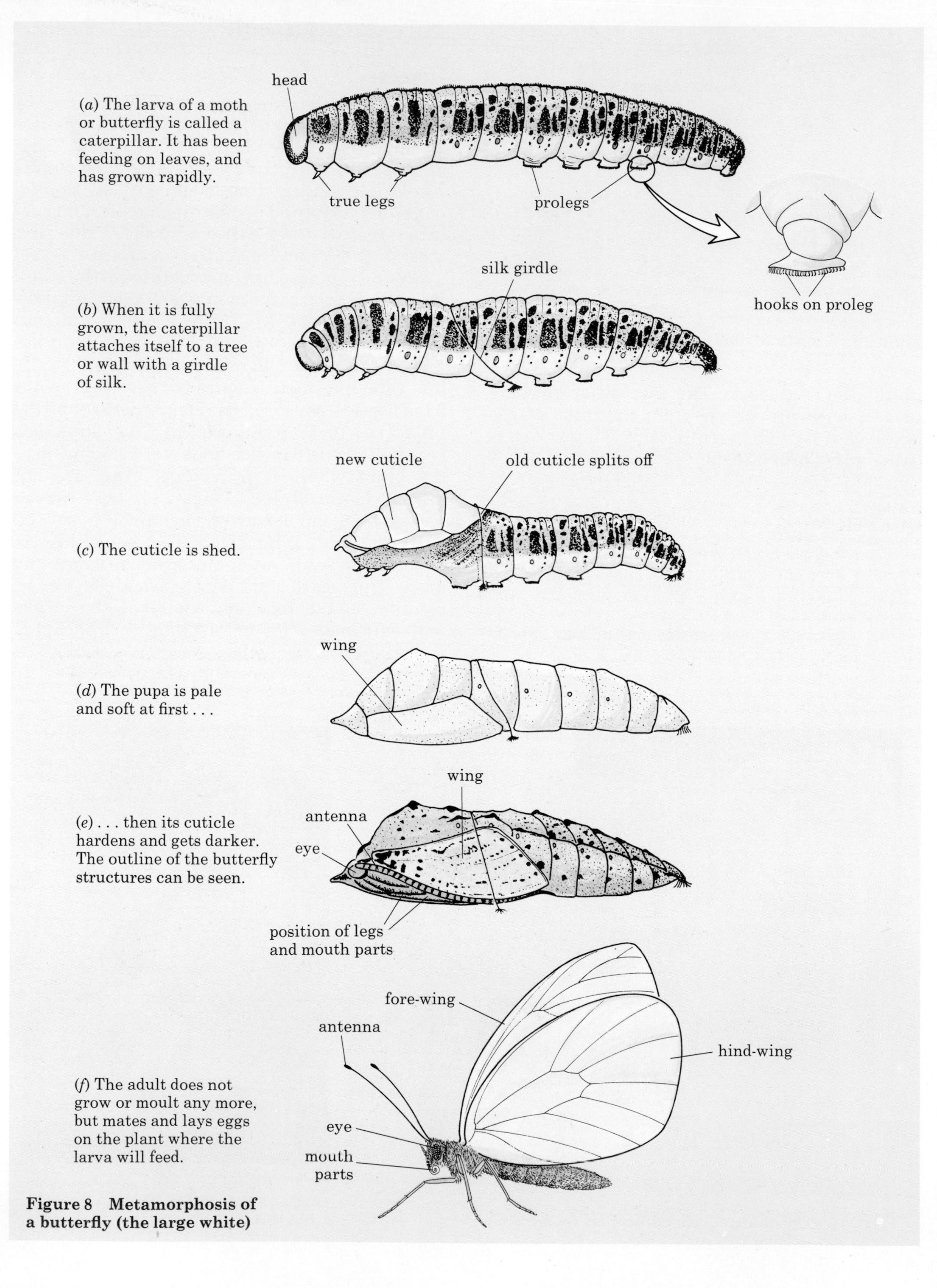

(*a*) The larva of a moth or butterfly is called a caterpillar. It has been feeding on leaves, and has grown rapidly.

(*b*) When it is fully grown, the caterpillar attaches itself to a tree or wall with a girdle of silk.

(*c*) The cuticle is shed.

(*d*) The pupa is pale and soft at first . . .

(*e*) . . . then its cuticle hardens and gets darker. The outline of the butterfly structures can be seen.

(*f*) The adult does not grow or moult any more, but mates and lays eggs on the plant where the larva will feed.

Figure 8 Metamorphosis of a butterfly (the large white)

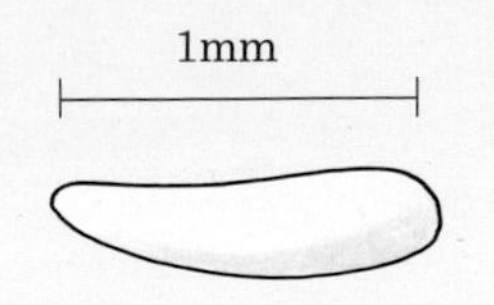

(*a*) Egg of housefly. The housefly lays its eggs in decaying matter such as rotting food in dustbins.

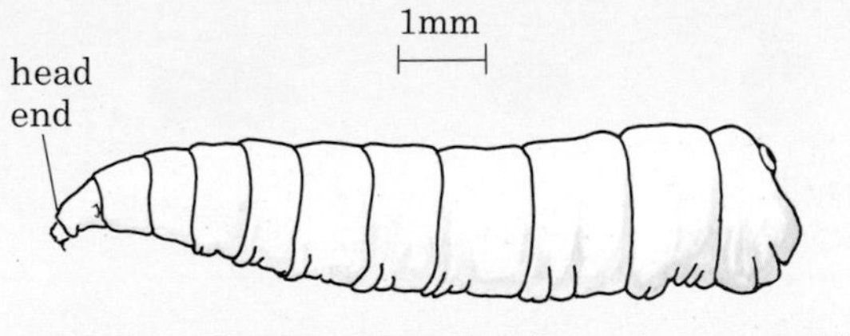

(*b*) Larva. The larva hatches from the egg and feeds on the decaying matter, burrowing through it with wriggling movements. It grows to about 1 cm in 5 days.

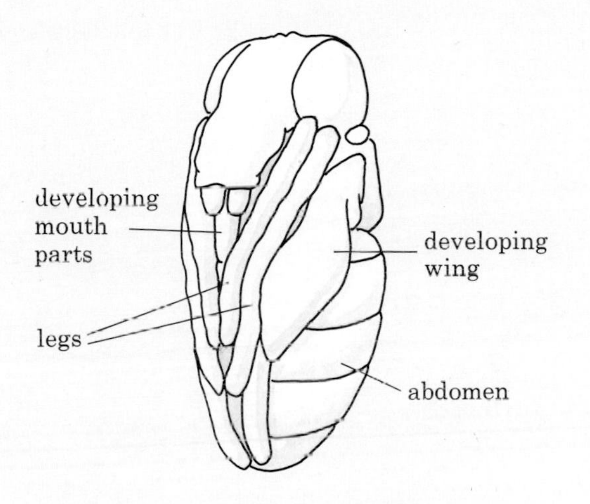

(*c*) Pupa. Then it changes to a pupa which develops inside a pupal case into the adult form.

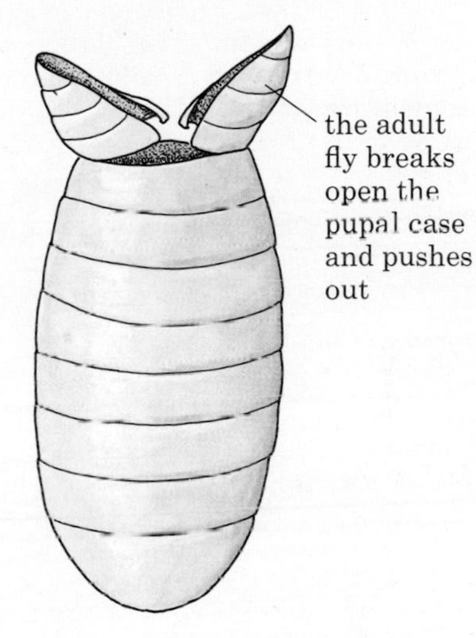

(*d*) Pupal case. When the adult is fully developed, it pushes its way out of the pupal case, its wings expand and it flies away.

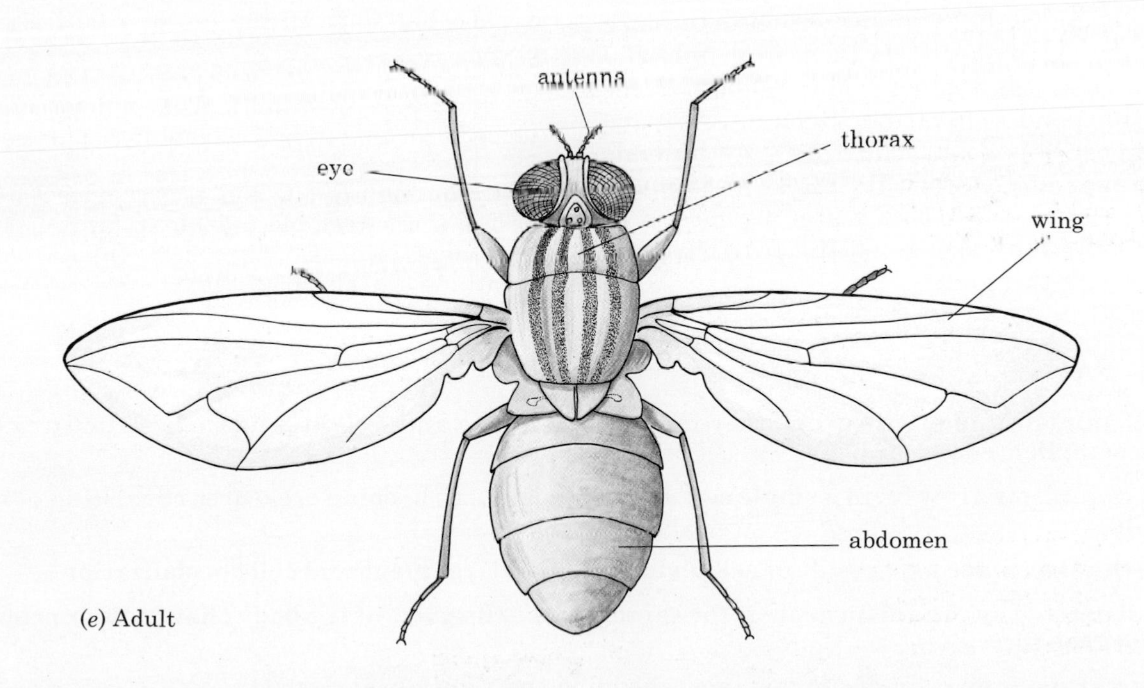

(*e*) Adult

Figure 9 Metamorphosis of the housefly (after Hewitt).

(*a*) A ball of cells as in Figure 1*g* . . .

(*b*) . . . grows longer . . .

(*c*) . . . and develops a head and body.

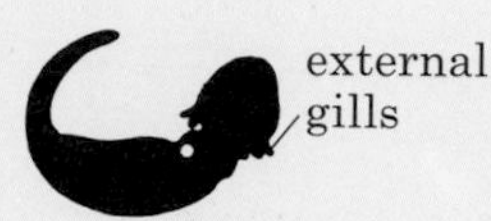

(*d*) The tail and the external gills appear.

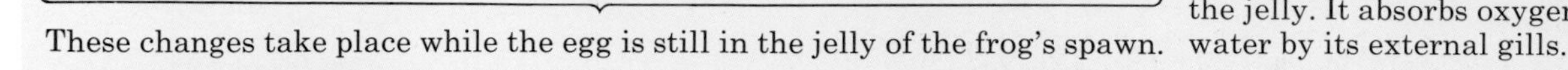

These changes take place while the egg is still in the jelly of the frog's spawn.

(*e*) 2 days. The tadpole wriggles out of the jelly. It absorbs oxygen from the water by its external gills.

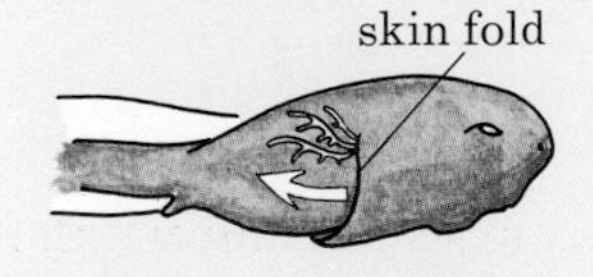

(*f*) 6 days. A fold of skin grows over the external gills and covers them.

(*g*) 3 weeks. The external gills have gone and the tadpole now breathes with internal gills, like a fish. It swims by wriggling movements.

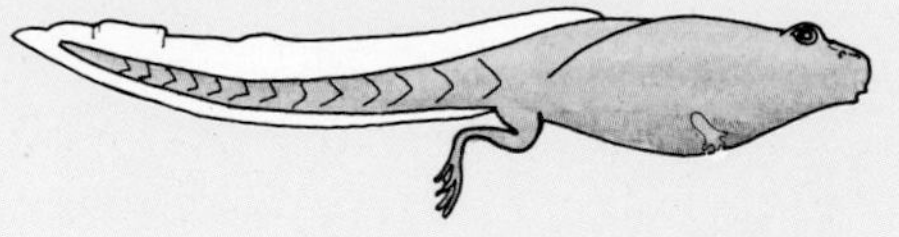

(*h*) 10–11 weeks. The hind legs appear first.

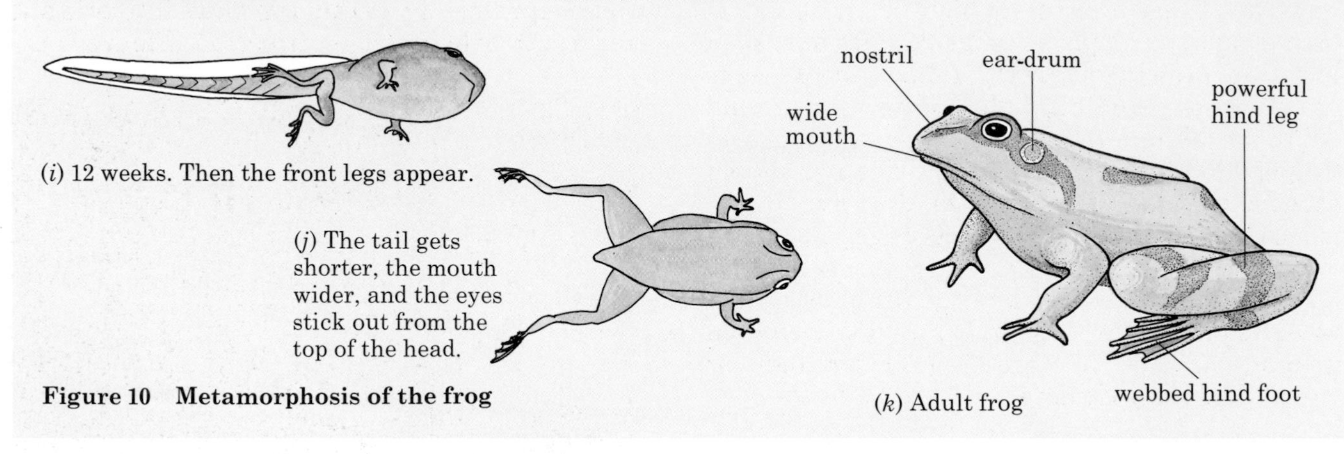

(*i*) 12 weeks. Then the front legs appear.

(*j*) The tail gets shorter, the mouth wider, and the eyes stick out from the top of the head.

(*k*) Adult frog

Figure 10 Metamorphosis of the frog

Amphibia

Figure 10 shows the changes which take place during the metamorphosis of a frog. Internal changes also take place. The tadpole feeds mainly on pond vegetation; its long intestine helps digest this food. The frog eats animals such as flies and worms; its intestine is much shorter. Tadpoles breathe water by means of gills, but these disappear as the frog develops lungs for breathing air.

QUESTIONS

4 At what stage in the life of a butterfly does it grow?

5 Study Figures 9*b* and *e*. What changes have taken place in turning from a maggot into a housefly?

6 Study Figures 10*g* and *k*. How does a frog differ from a three-week-old tadpole in (a) its methods of movement (see also page 232), (b) its method of breathing (see also page 215)?

BASIC FACTS

- **When animals and plants grow, they increase their size and weight, and their structures become more complicated.**
- **Most organisms grow from a single cell, by cell division, to become creatures consisting of millions of cells.**
- **Growth takes place as a result of cell division, cell enlargement and cell specialization.**
- **Not all parts of an organism grow at the same rate, so the parts of the body change their proportions during growth.**
- **The arthropods grow in distinct stages, shedding their cuticle at each stage.**
- **In the insects and some amphibia, there are drastic changes in body structure during growth. This change is called metamorphosis.**

29 Movement and locomotion

Locomotion refers to the movement of an organism from place to place. Plants do not show locomotion; they remain in one place for their whole life. They do make movements, however, of parts of their bodies. These movements, such as tropisms (page 86), are also seen in the opening up of flowers, folding up of leaves at night or climbing plants twining round supports. When these movements are photographed by 'time-lapse' photography and shown speeded up, they reveal plants as active organisms, even though they do not move about as a whole.

Most protista and animals move about a great deal. They do this to search for food, to avoid danger, build nests, defend their territory or move to a better climate.

PROTISTA

Euglena and its relatives have a long flagellum (Figure 1). Wave-like ripples pass down this flagellum and make the Euglena rotate and move forward. Paramecium and many similar single-celled creatures move by means of hundreds of cilia covering their bodies. The cilia make rhythmic lashing movements, like thousands of tiny oars rowing, and drive the organism through the water.

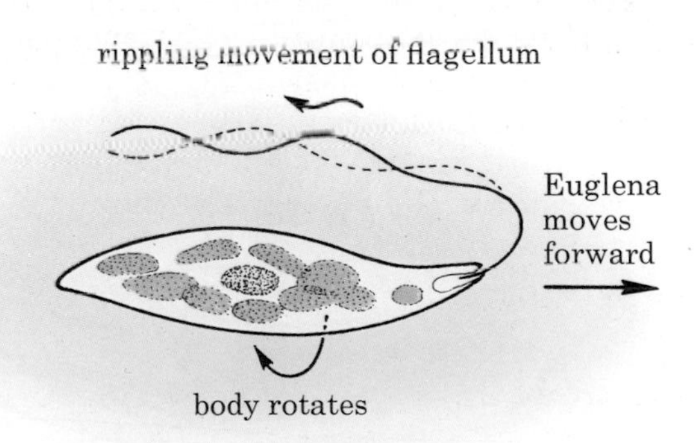

Figure 1 Euglena moving

Amoeba's method of locomotion is unusual. It has no fixed shape and when a part of its cell membrane bulges out, the cytoplasm flows into it (Figure 2). If a bulge forms at a different place on the surface, the cytoplasm flows into this one and so Amoeba changes direction (Figure 2*e*).

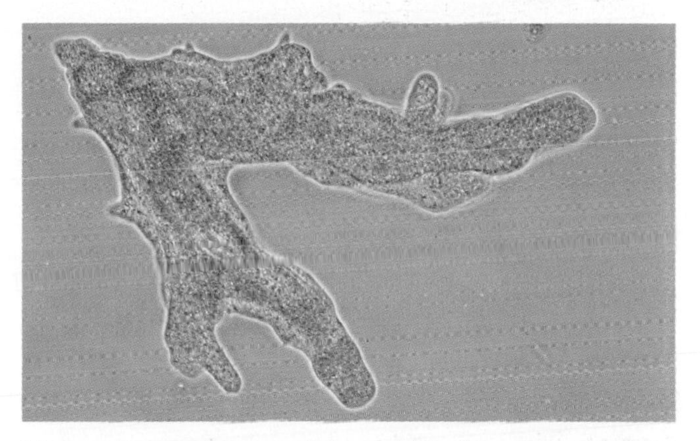

(*a*) Amoeba moving from left to right. Cytoplasm will flow into one of these 'arms' and the other will be withdrawn.

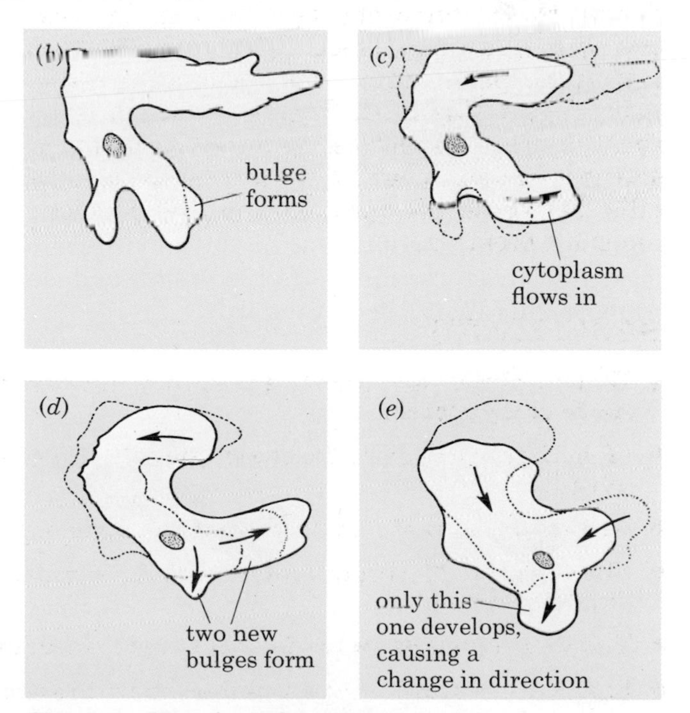

Figure 2 How Amoeba moves

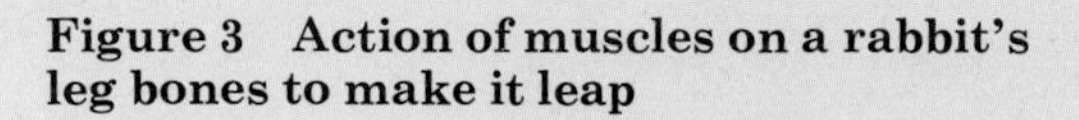

Figure 3 Action of muscles on a rabbit's leg bones to make it leap

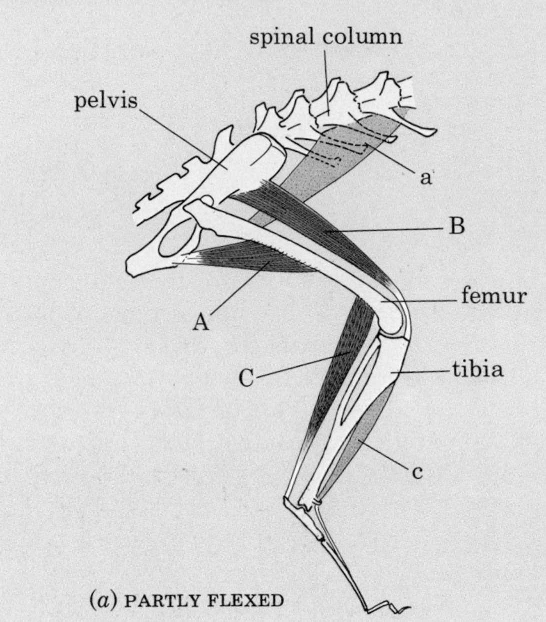

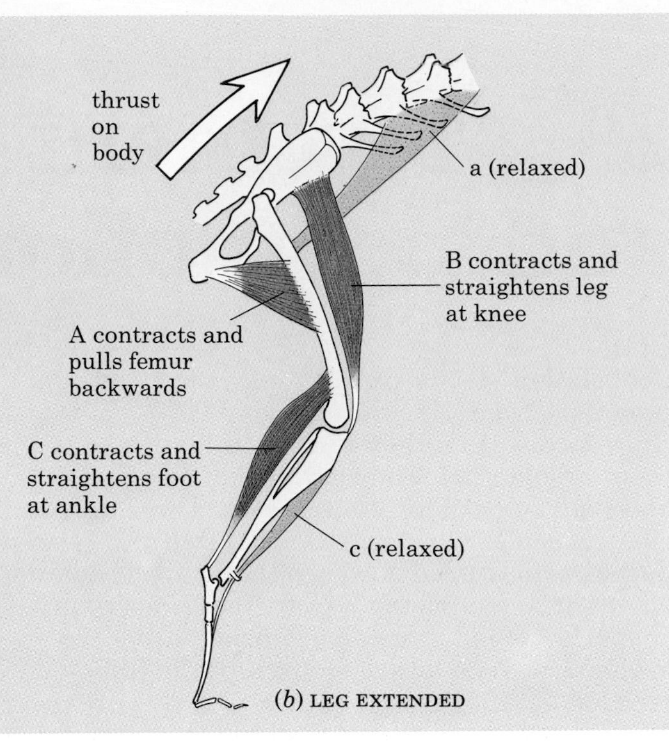

VERTEBRATES

Vertebrates move by making their muscles contract and pull on the bones of their skeletons. This process is described more fully on page 151.

Mammals

Figure 3 shows how some of the muscles in a rabbit's hind leg would extend the leg and thrust the animal forward and upward in a leaping movement. In Figure 3*b*, muscles A, B and C contract at the same time, straightening the hind leg and pulling it backwards. The upwards and forward thrust of the leg on the pelvic girdle is transmitted to the spinal column and so propels the rabbit's body forwards.

Frog

Muscular action, similar to the rabbit's, would produce the leaping movement of a frog as shown in Figure 4. The frog's hind legs are long and powerful, which is useful both for leaping and the swimming movements it makes when in the water.

Fish

The fish has no limbs (Figure 5*a*), but when the muscles in its body contract they make the tail move sideways. The muscles on opposite sides of the body contract alternately to send the tail first to one side and then to the other (Figure 5*b*). As the tail moves, it pushes sideways and backwards on the water. The sideways move-

Figure 4 Frog leaping (from James Gray, *How Animals Move*, Cambridge University Press, 1953).

(*a*) At rest; hind legs flexed

(*b*) Hind legs extended, pushing frog forward and upwards

(*c*) Front legs being extended; beginning to draw up the hind legs

(*d*) Front legs being extended to take first shock of landing; hind legs drawn up

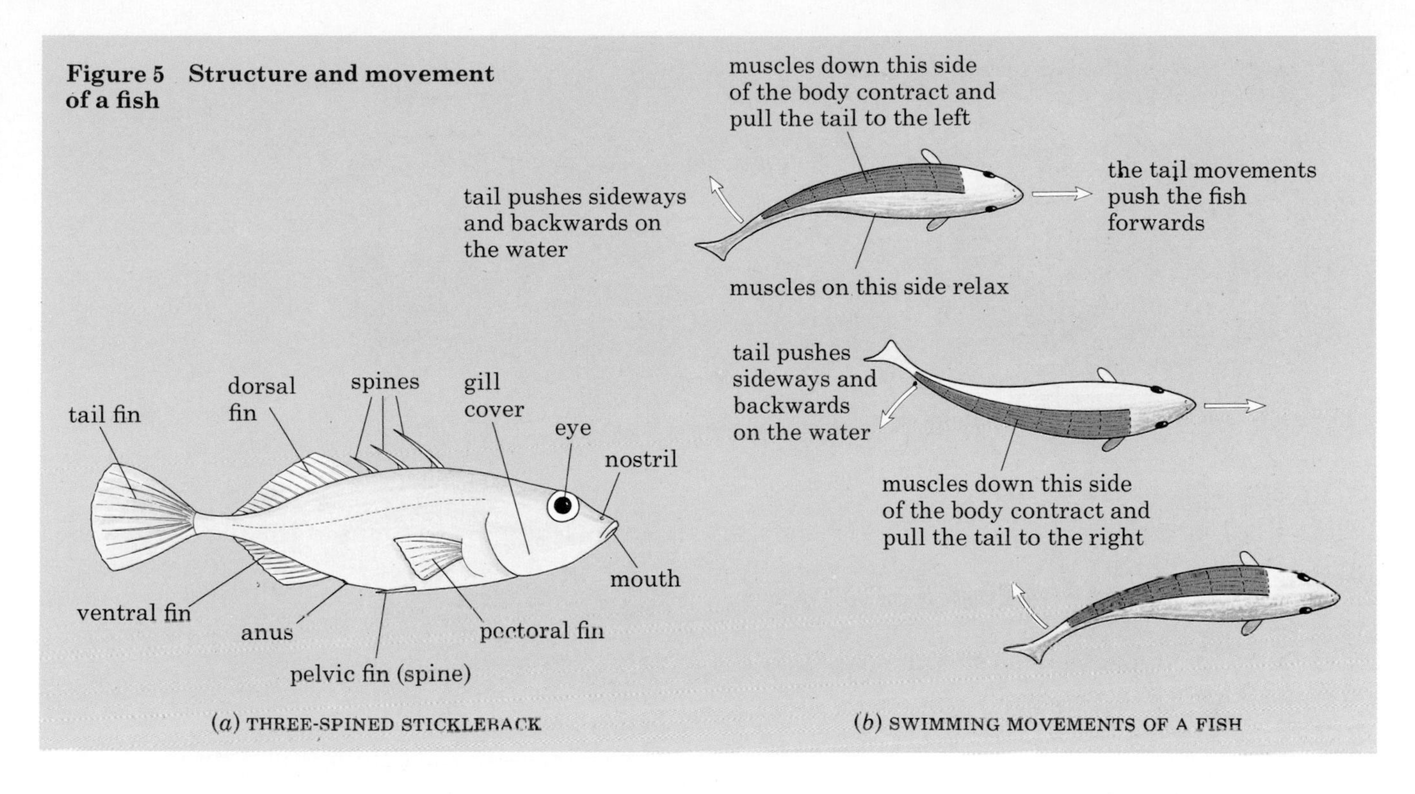

Figure 5 Structure and movement of a fish

(*a*) THREE-SPINED STICKLEBACK (*b*) SWIMMING MOVEMENTS OF A FISH

ments to left and right cancel each other out, and the backwards thrust of the tail on the water drives the fish forward.

The fins do not propel the fish, but help it to steer a course. The streamlined shape of the fish offers the least possible resistance to forward movement through the water.

The fish may, therefore, be said to be adapted to movement in water by having a streamlined shape, powerful body muscles to move its tail and fins to help it steer.

Many fish also have an air bladder in their bodies which makes them buoyant, so that they don't sink down every time they stop swimming.

Birds

In mammals, the muscles for locomotion act mainly on the hind limbs. In birds, they act on the fore-limbs which are the wings. One set of muscles pulls the wing down and an antagonistic set pulls the wings up. (Antagonistic muscles have opposite effects on the same limb, as explained on page 151.)

Figure 6 shows the skeleton of a bird and you will see that the breastbone has a deep keel projecting down. This provides a large surface for the attachment of the powerful flight muscles that act on the wing. The **coracoid** bone acts as a strut between the breastbone and the spine so that when the flight muscles contract, they pull the wing bone down rather than pull the breastbone up. Figure 7 shows this part of the bird's skeleton as seen from in front. You can see how the flight muscles are attached to the breastbone at one end and the humerus at the other.

When the large flight muscle contracts

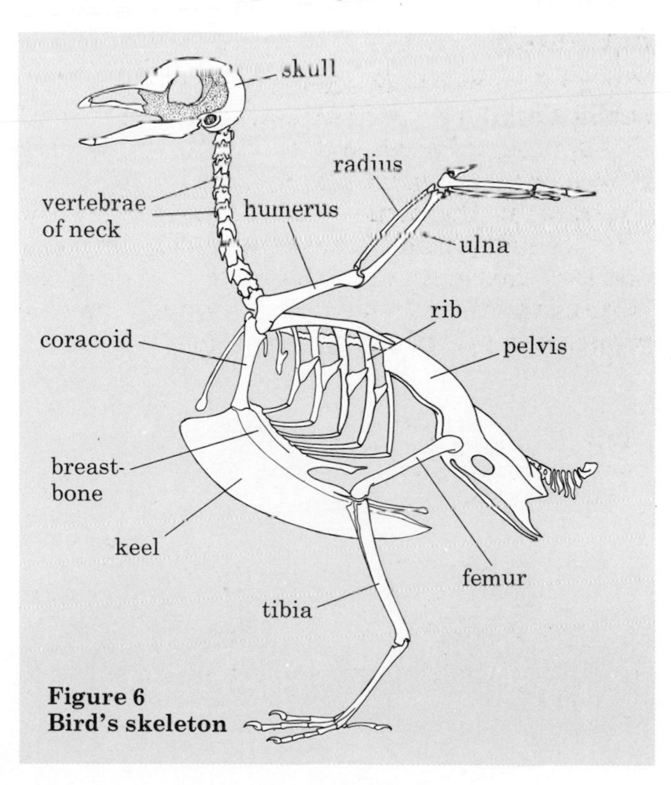

Figure 6 Bird's skeleton

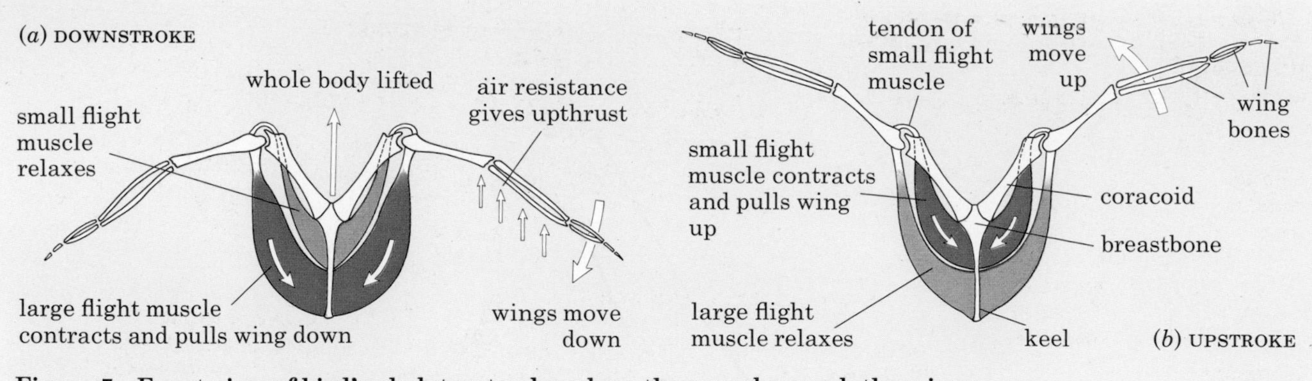

Figure 7 Front view of bird's skeleton to show how the muscles work the wings

(Figure 7*a*), it pulls the wing down. The small flight muscle when it contracts (Figure 7*b*), pulls the wing up, because its tendon runs over a groove in the coracoid, like a pulley, and attaches to the upper side of the humerus. So the large and small flight muscles are antagonistic to each other. When they contract alternately, they pull the wing up and down.

During the upstroke, the wing is bent at the wrist and so it does not offer much air resistance (Figure 8*a*). In the downstroke the wing is spread out and pushes downwards on the air. The air resistance causes an upthrust on the wing and so lifts the bird up. Figure 8*c* also shows how the flight feathers overlap in such a way as to let the air pass between them during the upstroke but not during the downstroke.

A bird is adapted to flight in the following ways:
(a) Its front legs are modified to form wings.
(b) Its flight feathers are light but present a large air-resistant surface.
(c) It has powerful flight muscles.
(d) The deep keel provides a large surface for attachment of these muscles.
(e) Its limb bones are hollow and light.
(f) Its skeleton forms a rigid framework to stand up to the contractions of the flight muscles.

ARTHROPODS

The muscles of the vertebrate animals are attached to the outside of their skeletons. Their skeletons are on the inside of their bodies and

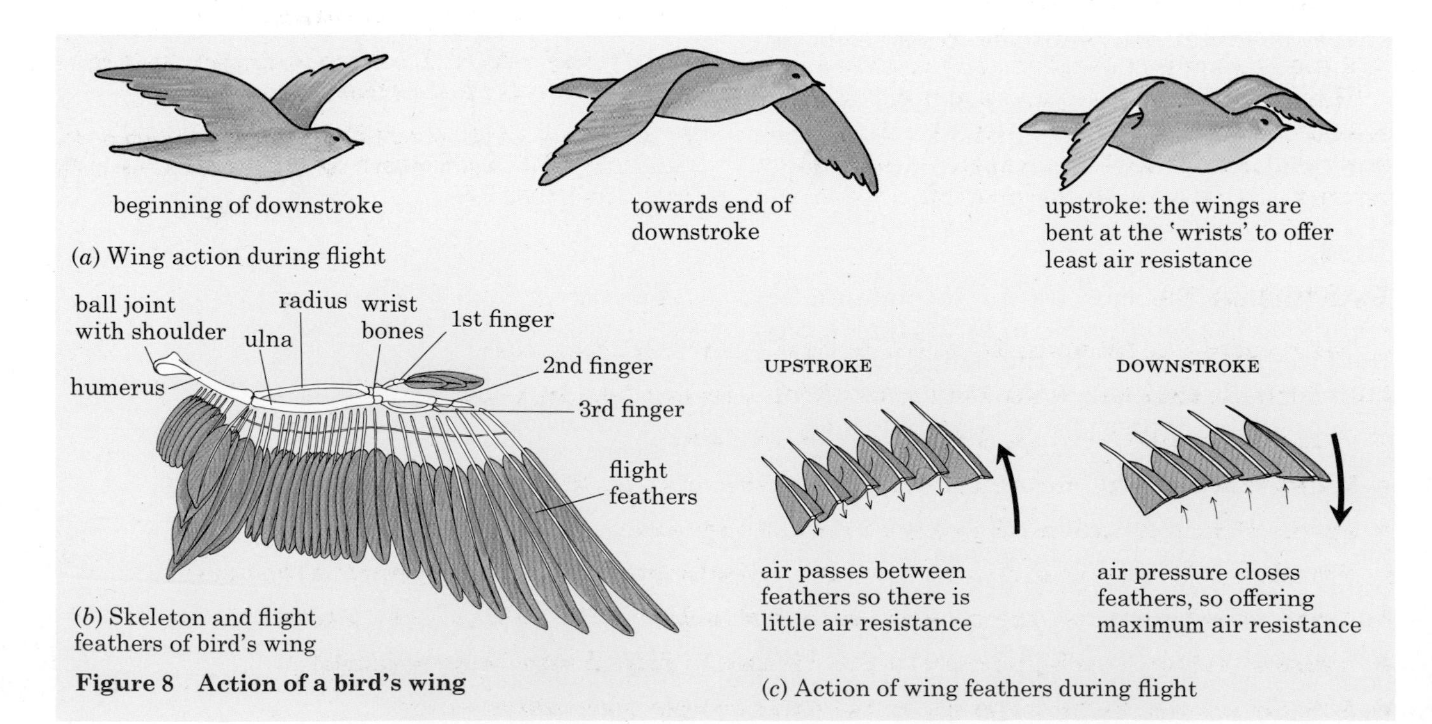

Figure 8 Action of a bird's wing

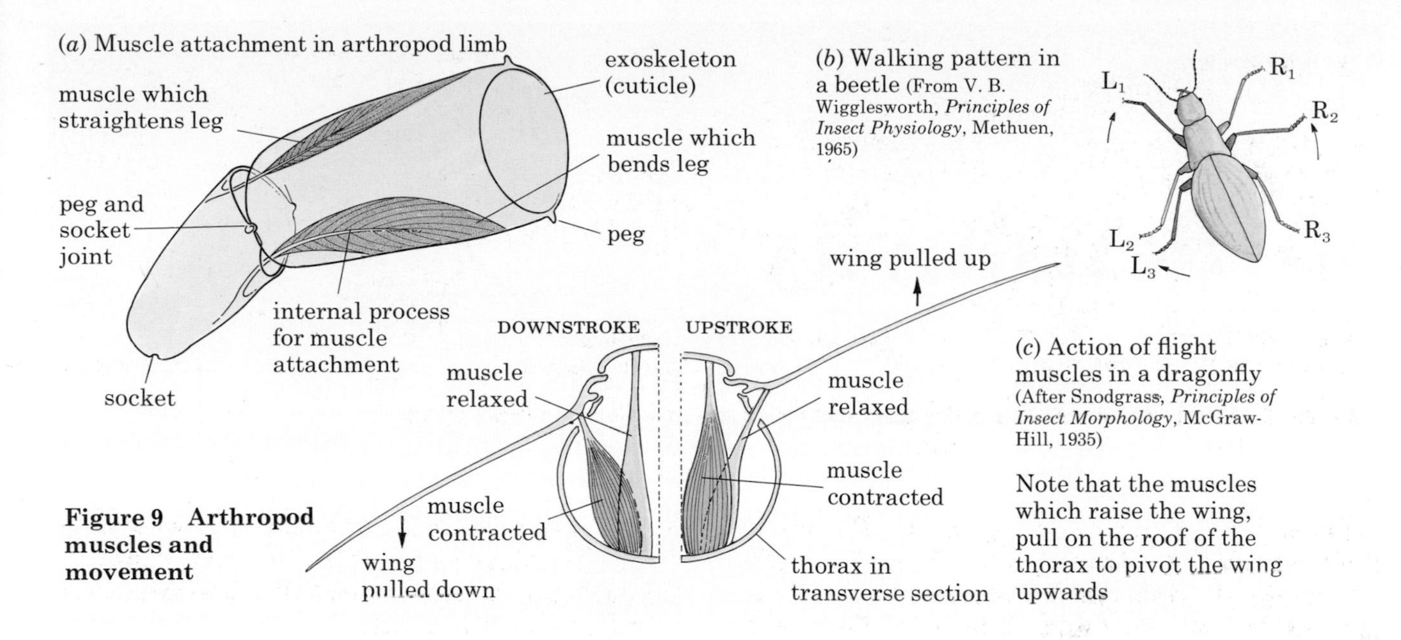

Figure 9 Arthropod muscles and movement

are referred to as **endoskeletons**. Arthropods (e.g. crustacea and insects) have **exoskeletons**, which consist of a hard outer layer called the **cuticle**. This cuticle covers the entire body but, at the joints, the cuticle is flexible enough to permit movement.

Any muscle which produces movement is attached to a point on the inside of the skeleton. Figure 9*a* shows how this arrangement works. An insect's leg is like a series of rigid tubes, with each tube attached to the next by a hinge joint. When the top muscle of Figure 9*a* contracts, it pulls the two limb sections into line. Contraction of the bottom muscle flexes the limb by bending the left-hand section down at the joint.

Figure 9*b* shows how a beetle uses its six legs to walk. The muscles inside legs L_1, R_2 and L_3 are contracting to lift the legs clear of the ground and swing them forwards, while the muscles of R_1, L_2 and R_3 are contracting to swing the legs backwards, so pushing the insect forward. The insect is supported on the three legs which are propelling him forward, while the other three legs are swinging forward ready for the next step.

Figure 9*c* shows how the antagonistic flight muscles inside the thorax of a dragonfly contract to pull the wings up and down.

QUESTIONS

1 Why do you think that muscle C in Figure 8 is bigger than its antagonistic muscle c?

2 Why do you think it is more important for a fish to be streamlined than a mammal like a dog?

3 If there were no coracoid bone in a bird's skeleton, what would happen when the flight muscles contracted (Figure 7*a*)?

BASIC FACTS

- **Nearly all living things make movements with parts of their body.**
- **Protista and animals move their whole bodies around (locomotion).**
- **Protista use mainly cilia and flagella to move about.**
- **Plants make growth movements (tropisms) and move their leaves and flower petals.**
- **Animals' muscles pull on their skeletons and so produce movement.**
- **Vertebrates have endoskeletons with muscles attached to the outer surface of the bones.**
- **Arthropods have exoskeletons with muscles attached to the inside of the cuticle.**
- **A bird's skeleton and flight muscles are adapted in special ways to produce flight.**
- **A fish is adapted by its shape and muscle arrangement to swim in water.**

30 Sensitivity

If you push a brick hard enough, it might move, but if you just touch it lightly, nothing will happen. Most animals and many plants, however, do respond to a light touch. The touch is called a **stimulus** and it is followed by an action called a **response**. Living organisms will respond to a stimulus (plural = stimuli); non-living matter will not. The ability to respond to a stimulus is called **sensitivity**.

A stimulus could be, for example, a touch, a noise, a passing shadow or a change in temperature. The response is usually a movement of some kind. A fly will respond to the movement of your approaching hand by flying away before you can swat it. The stimulus might have been the sight of the moving hand, the shadow it cast or even the air disturbance that it caused.

Figure 1 An insect-eating plant. The sundew plant grows in damp swampy places and gets part of its food by trapping insects. Insects can be seen on three of the leaves. Once the insect is trapped on the leaf, the sticky tentacles bend towards it, digest it with their enzymes, and then absorb the liquid products of digestion. The insects probably supply the plant with nitrates, which are lacking in the soil.

PLANTS

Although plants do not respond by moving their whole bodies, parts of them do respond to stimuli. The tropisms described on pages 86–89 are good examples of responses made by the growing parts of shoots or roots to the stimuli of light or gravity.

Some plants respond to touch. A runner bean shoot that is stimulated by being touched repeatedly on one side for an hour or two will curve towards the stimulus. As a result, the bean shoot twines round a string or a stake as it grows upwards. The insect-eating plant, sundew, has sticky tentacles on its leaves. When the tentacles are touched by an insect walking on the leaf, they curve over the insect, trap it and slowly digest it (Figure 1).

QUESTIONS

1 On page 154 some examples of possible stimuli were given. Suggest two or three other possible kinds of stimuli.

2 What might be the stimulus, and what is the response, when a sundew leaf (Figure 1) traps an insect?

ANIMALS

Obvious responses by animals are such things as moving towards food or away from danger, but among the protista and invertebrate animals there are many examples of non-directional responses; the animal moves about at random until it escapes from an unpleasant stimulus.

Non-directional responses

If Paramecium, observed on a microscope slide, swims into an obstacle in its path, the contact acts as a stimulus. The Paramecium responds by reversing the beat of its cilia and moving

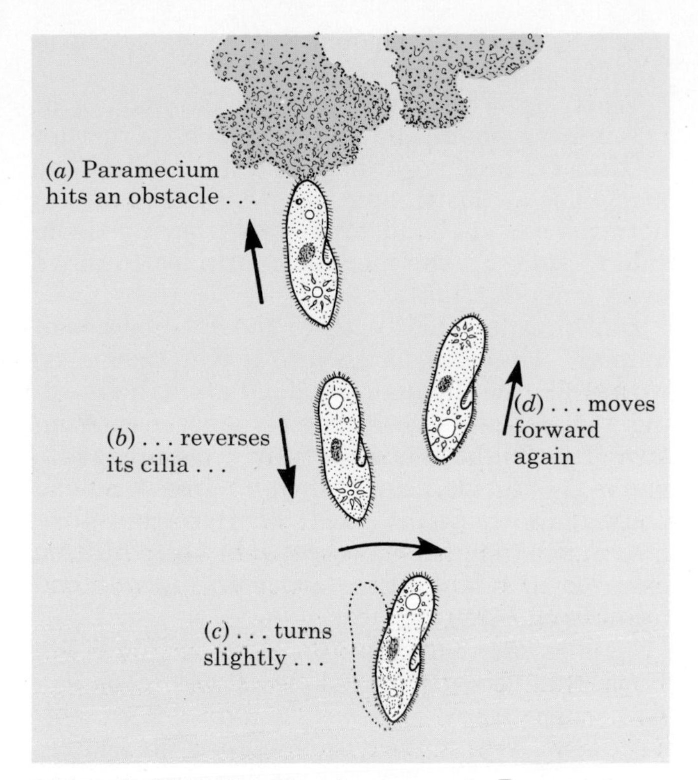

Figure 2 Non-directional response in Paramecium

backwards, away from the obstacle (Figure 2). Then Paramecium turns through a small angle and moves forward again. The direction of turning is not related to the direction of the obstacle; it could be a turn to the left or to the right. If the Paramecium bumps into the obstacle again on its forward path, it simply repeats the 'reverse–turn–forward' pattern. By a series of random turns, it will sooner or later get past the obstacle. In its normal environment, Paramecium's reactions are probably more varied than this simple response.

Figure 3 shows a choice chamber. The air on the left side will be moister (more humid) because of the water in the lower compartment.

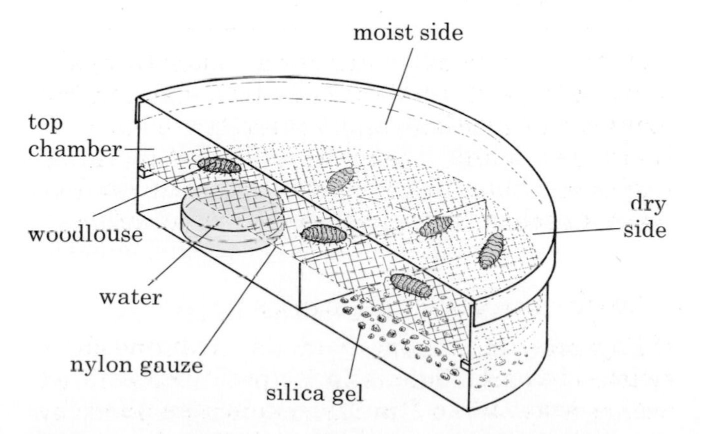

Figure 3 A choice chamber (only one half shown).

The air on the right will be dryer because the silica gel absorbs water vapour. Some woodlice are kept in a dry container for 3 hours. If ten of these woodlice are then placed in the top compartment of the choice chamber, they will move about at random. On the dry side, they will move rapidly but on the moist side they will move slowly and, sooner or later, they will stop.

It looks as if they have moved deliberately towards the moist side but they have not. Their final positions result from the fact that the stimulus of dry air makes them move rapidly in any direction. When a movement happens to bring them by chance into the moist air, they stop moving.

If you do an experiment like this, you must take certain precautions if your results are to be reliable.

(a) You must make sure that there is no stimulus other than moisture (humidity) which could affect the woodlice. If light was coming from the right side it could be said that the woodlice were responding to the light by moving away from it, rather than responding to the humidity. The light would have to come from above and not from one side. Alternatively you could repeat the experiment several times with light coming from each side in turn to show it made no difference to the results, or you could cover the choice chamber with a box so that it was in total darkness and lift the box at one minute intervals to see where the woodlice were.

(b) You must use at least ten woodlice. If you only had five and they ended up with two on the dry side and three on the moist side, this is just as likely to happen by chance. For the results to be significant, the woodlice must end up with at least nine on the moist side and one or none on the dry side. Even 8:2 would not be accepted as very much better than chance.

(c) You must do the experiment several times. The more often you repeat the experiment, and the more often you get a 9:1 or 10:0 distribution of woodlice, the more confident you can be that the animals are responding to humidity.

QUESTIONS

3 In the choice chamber experiment, what differences, other than humidity and light, might affect the final distribution of the woodlice?

4 What would be a good control experiment (see page 17) for the choice chamber experiment to show that it was humidity and not some other stimulus that affected the woodlice? What results would you expect in the control experiment?

5 How would you design an experiment to see if woodlice responded to the stimulus of light?

Directional responses

In these responses, the direction of the response is related to the direction of the stimulus. Tropisms are directional responses made by plant organs. A shoot is positively phototropic; it grows towards the source of light. The most obvious responses of animals are directional; towards food, towards a mate, away from an enemy.

A simple directional response can be shown with blowfly larvae (maggots). In Figure 4 the maggot is moving away from the light source at A. Then light A is switched off and light B is switched on. The maggots change direction and move away from light B.

Close study of their movement shows that as they wriggle forwards, the maggots swing their head end from side to side. There is a group of cells in the head which are sensitive to light. If these cells are stimulated by light coming from the right, they send off nerve impulses to the maggot's body which make it swing its head more violently to the left.

When light B is switched on, the maggot at position *d* responds by swinging its head further to the left and so altering course. When the maggot is at positions *a*, *b* or *c*, light A affects both sides of its head equally as it swings from side to side. So the maggot continues to move away from the light.

Blowflies lay their eggs in the dead bodies of animals. The eggs hatch into larvae (maggots) which burrow through the dead animal, digesting and eating its flesh. The response of moving away from light occurs in fully grown maggots and makes them move downwards from the dead body and burrow into the soil. Here they turn into pupae (page 227) protected by the soil from changes of temperature, from drying out and from being eaten by birds.

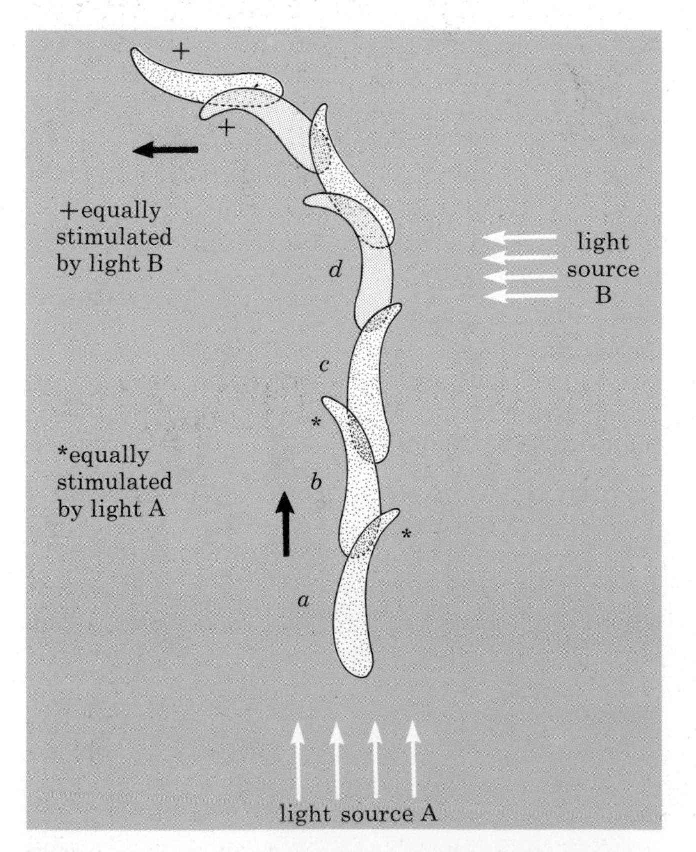

Figure 4 Directional response of blowfly larva (after Fraenkel and Gunn, *The Orientation of Animals*, Oxford, 1940).

QUESTIONS

6 In what way do shoots show a directional response to a stimulus other than light?

7 What directional responses do wasps appear to make (a) when they pester you at a picnic, (b) when they try to escape from a room?

Sense organs

Before an animal can respond to a stimulus, it must be able to detect it. This detection is done by sensory organs. Sometimes these are sensory cells all over the bodies of the animals. Some of these cells detect changes in light, some detect changes in temperature and some respond to chemicals or touch. In the vertebrates, the skin contains some sensory cells of this kind (page 154), but there are also special sense organs where the sensitive cells are packed closely together in one place and the stimulus is directed on to them. In our eyes (page 156) the retina consists of thousands of closely packed cells, all able to respond to light. The stimulus of light is directed on to the sensitive cells by the cornea and lens.

The cochlea (page 160) contains thousands of cells which detect vibrations. The ear-drum and ossicles direct the stimulus on to the sensitive cells.

Figures 5–10 show parts of the sensory equipment of various animals. Although some parts of plants, such as the growing points, are more sensitive to stimuli than others, there are no special sense organs.

Figure 5 Common frog. The special sense organs of animals are concentrated on the head. The frog's circular ear drum is seen behind and below the eye, and its nostril in front of the eye. The nostrils can be closed when the frog is under the water.

Figure 6 Hare. The long ears help to pick up and locate sound vibrations. The eyes at the side of the head give the hare good vision all around.

Figure 7 Grass snake; sense of smell. The flicking tongue picks up chemicals in the air and carries them to a sense organ in the roof of the mouth which 'tastes' them.

Figure 8 Long-eared bat. The bat gives out high-pitched sounds which are reflected back, from its prey and from obstacles, to its ears and sensitive patches on its face. By timing these echoes the bat can judge its distance from the obstacle or prey.

Figure 9 Tawny owl. The owl's eyes point forwards and help it to judge distances accurately (see page 158) and so capture prey. The large size of the eyes helps to pick up what little light there is at night time.

Figure 10 Gerbils. The whiskers on the face affect nerve endings in the skin. The slightest movement of the whiskers, even those caused by air movements, will cause nerve impulses to be sent to the brain.

QUESTIONS

8 What human sense organs can detect chemicals? What sensations do we experience when our chemical senses receive a stimulus?

9 Why do you think that most animals have their main sense organs on their heads?

10 Why do you think your voice sounds different when you hear it played back on a tape?

BASIC FACTS

- **Living organisms respond to stimuli. They are said to be sensitive.**
- **Stimuli are such things as touch, heat, cold, and light acting on the organism.**
- **Animals have sense organs for detecting these stimuli.**
- **When animals make non-directional responses, they move at random until they escape from an unpleasant stimulus and then they slow down or stop moving.**
- **A directional response is made towards or away from a stimulus.**
- **Plants do not move about, but their roots and shoots make directional responses to the stimuli of light and gravity.**
- **When designing experiments to show responses of animals, (a) be sure that the animal is receiving only one kind of stimulus, (b) use a sufficient number of animals and (c) repeat the experiment several times.**

SECTION SIX

Health

31 Infectious diseases

Bacteria, protista, viruses and some fungi are micro-organisms, which may cause disease when they get into the bodies of plants and animals. These micro-organisms can reproduce inside the bodies of their hosts. So even a small number getting in to start with can give rise to a large population inside the body and so cause illness. The symptoms of illness result from the damage the micro-organisms do to cells and the poisonous substances they produce.

MICRO-ORGANISMS WHICH CAUSE DISEASE

Bacteria

Bacteria (singular = bacterium) are single-celled organisms about one thousandth of a millimetre long. Figure 1 shows the general structure of a bacterium; Figure 2 shows the various shapes and sizes of some bacteria harmful to man.

Only a small proportion of bacteria are harmful. Most of them are harmless or helpful. They occur everywhere and are especially abundant in the soil where they bring about decay and so replace the mineral salts needed by plants (see the nitrogen cycle, page 36). These bacteria are saprophytic (i.e. feed on decaying matter, page 34). The bacteria which invade the bodies of other organisms are parasitic or, in some cases, symbiotic (i.e. harmless or beneficial, page 38). Bacteria which cause illness are sometimes called **pathogenic** bacteria.

Bacteria reproduce by cell division, sometimes as often as every 20 minutes. So in a few hours, a small number of bacteria invading the body can multiply to tens of thousands.

When a large number of bacteria become established in the body, they cause the symptoms of disease; high temperature, skin rash, sore throat, or diarrhoea and vomiting, for example. These symptoms are often the result of poisonous substances, called **toxins**, which the bacteria produce.

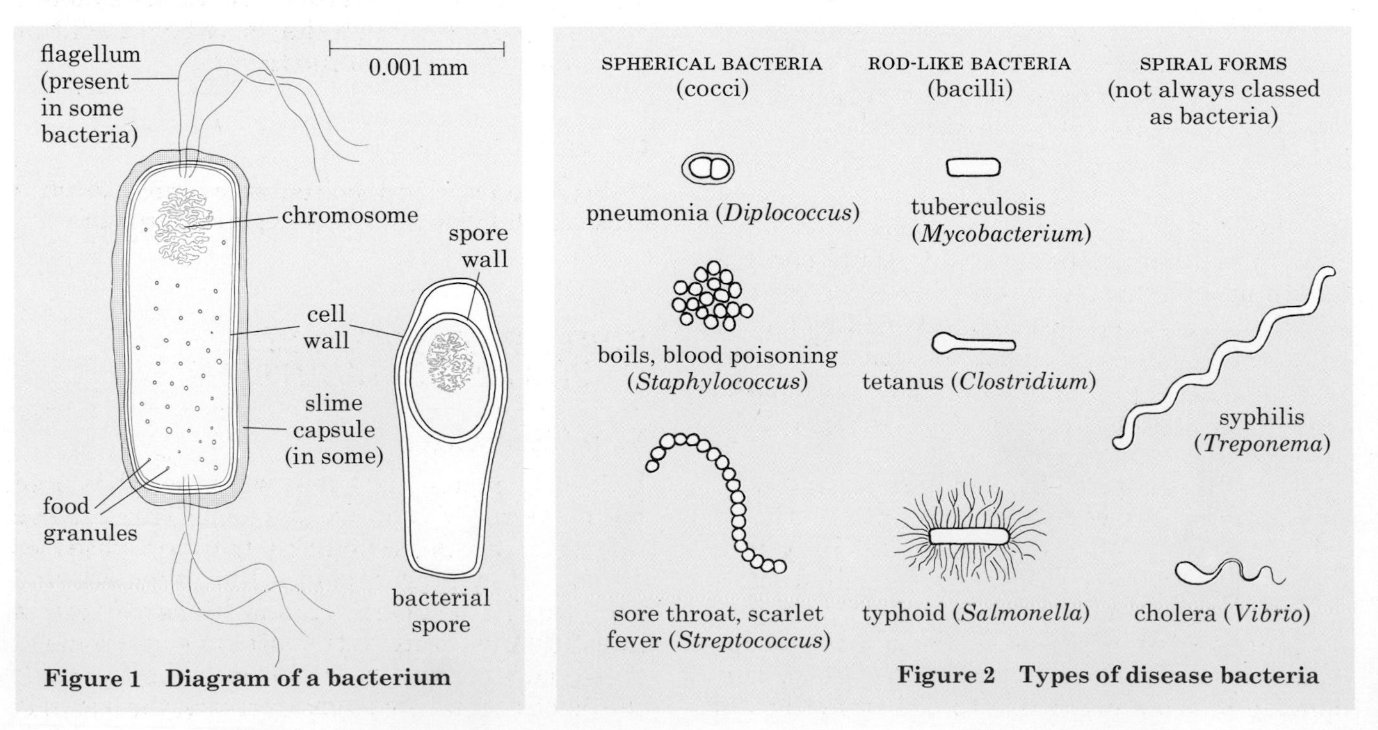

Figure 1 Diagram of a bacterium

Figure 2 Types of disease bacteria

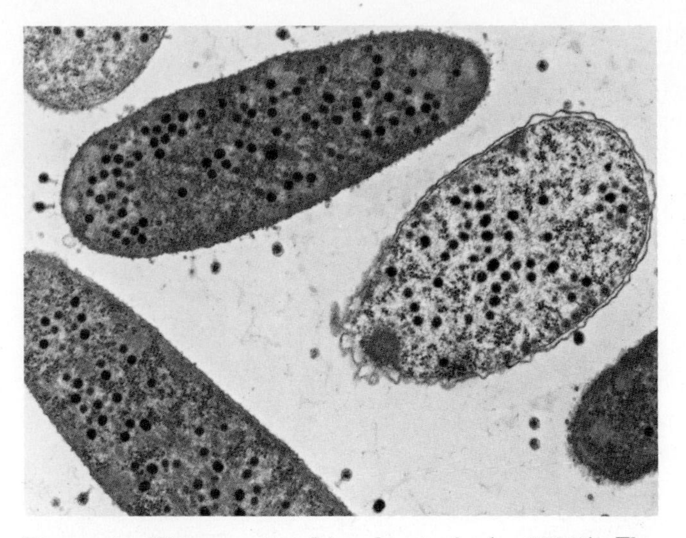

Figure 3 Viruses attacking bacteria (× 18 000). The photograph shows all or part of five bacteria, each containing many viruses which show up as dark blobs.

Some bacteria can form **spores** (Figure 1). The cytoplasm of the bacterium shrinks up to one end of the cell and is enclosed in an extra wall. Spore-forming bacteria are dangerous because the spores can often survive temperatures of 100 °C which kill other forms of bacteria.

Some human diseases caused by bacteria are pneumonia, tuberculosis, typhoid, cholera and syphilis. The diseases may be spread in one or more of the following ways: by contact, in food or drinking water, by houseflies or droplets of moisture in the air.

QUESTION

1 If ten bacteria get into your body and reproduce every 20 minutes, how many will there be in four hours, assuming that none die or are killed?

Viruses

These are even smaller than bacteria, perhaps ten thousand times smaller. There are viruses which can invade bacterial cells and kill them (Figure 3). Viruses cause diseases as mild as the common cold or as serious as polio. Influenza, chicken pox and measles are all caused by viruses.

The virus is not a cell; it can hardly be called a living organism because it does not breathe or grow or feed. It does reproduce, but only in a living cell. When a virus gets into a cell, it takes over the cytoplasm of the cell and turns it into thousands of new viruses. These will escape from the killed cell and invade other cells. This damage to cells gives rise to the symptoms of disease. The common cold virus invades the cells lining the nose and throat and causes them to break down, so giving you a sore throat and runny nose.

All viruses are harmful and can cause diseases in both plants and animals.

QUESTION

2 In what ways do viruses differ from bacteria?

Fungi

A small number of fungi cause mild skin diseases or allergic reactions in man. 'Ringworm' and 'athlete's foot' (tinea) are common examples of skin diseases. Allergies similar to 'hay fever' are caused by inhaling fungus spores from the air. Some fungi are parasites of plants and can cause serious outbreaks of disease in crops. Potato blight, mildew (Figure 5 on page 199) and 'rust' disease of wheat are just three examples of fungus diseases which can ruin crops.

Protista

In tropical countries, there are some single celled organisms which cause disease. Malaria and sleeping sickness are caused by single-celled parasites in the blood. There is also a kind of Amoeba which causes dysentery when it gets into the alimentary canal. In Britain and most other temperate countries these diseases do not normally occur, though they may be brought in by people coming from the tropics.

QUESTION

3 Why would it be incorrect to say that diseases caused by fungi are not very important to man?

SPREAD OF DISEASE: CAUSE AND PREVENTION

Drinking water

If disease bacteria get into water supplies used for drinking, hundreds of people can become infected. This is particularly true with diseases of the alimentary canal, like typhoid and cholera. The sick person has millions of bacteria infesting the lining of the intestine and some of these bacteria will pass out with the faeces. If the faeces get into streams or rivers, the bacteria

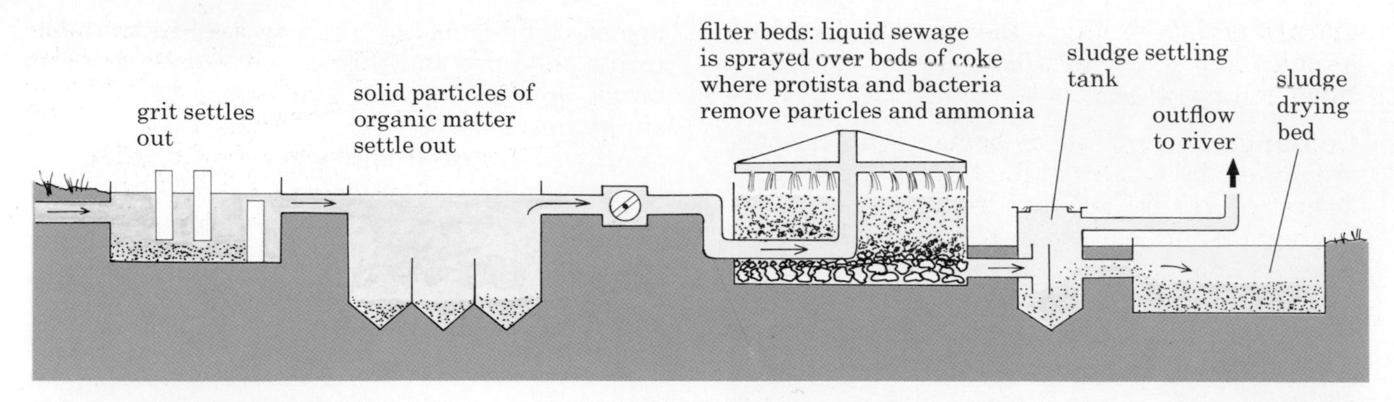

Figure 4 One kind of sewage treatment, the biological filter (from Mackean and Jones, *Introduction to Human and Social Biology*, Murray, 1975).

may be carried into reservoirs of water used for drinking. Even if faeces are left on the soil or buried, rain-water may wash the bacteria into a nearby stream.

To prevent this method of infection, faeces have to be made harmless, and drinking water must be purified.

Sewage treatment. The faeces are broken down at the sewage works by saprophytic bacteria. These bacteria are eaten by protista, like Vorticella and Amoeba, and the solid sludge which settles out contains mostly dead bacteria and protista. The sludge is removed and dried and can be used as a farm fertilizer, while the remaining liquid, now free from harmful bacteria, can be discharged into a river (Figure 4).

Water purification. Water to be used for drinking and washing is filtered through beds of gravel and sand which remove most of the bacteria. To make sure that any remaining bacteria are killed, the water is exposed for a time to the poisonous gas, chlorine. This is called **chlorination**.

QUESTION

4 When earthquakes occur, one of the dangers is that sewage pipes and water pipes may be cracked by the earth movements. Why should this be a cause for concern?

Food

The bacteria most likely to get into food are the same as or similar to those which might be in water. These are the ones which cause diseases of the alimentary canal such as typhoid and food poisoning. The bacteria are present in the faeces of infected people and may find their way into food from the unwashed hands of the sufferer. When patients are recovering from one of these diseases, they may feel quite well but bacteria are still present in their faeces. If they don't wash their hands thoroughly after going to the lavatory, they may have small numbers of bacteria on their fingers. If they then handle food, the bacteria may be transferred to the food. When this food is eaten by healthy people, the bacteria will multiply in their bodies and give them the disease.

People working in food shops, kitchens and food-processing factories could infect thousands of other people in this way if they were careless about their personal cleanliness (Figure 5).

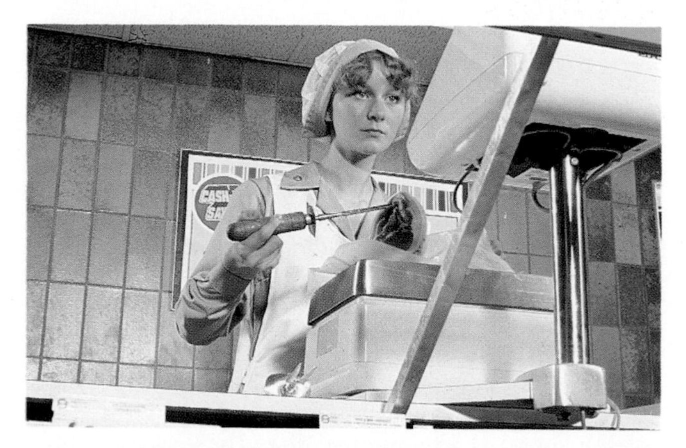

Figure 5 Hygienic handling of food. A shop assistant avoids handling meat with her fingers.

Some forms of food poisoning result from poisons (toxins) that are produced by bacteria which get into food. Cooking kills the bacteria in the food but does not destroy the toxins which cause the illness. Only one form of this kind of food poisoning, called **botulism**, is dangerous and it is also very rare.

Personal hygiene. People who handle food must wash their hands before touching the food, and always after going to the lavatory. Lavatory handles and doorknobs may have bacteria on them, left by other people.

Cooking. The spread of bacteria can also be prevented by cooking food before eating it. Bacteria are killed by high temperatures. Frying, baking, boiling or roasting will destroy any bacteria (but not their spores) which may be in the food. The high temperatures used in canning factories also destroy any bacteria in the food being canned.

Refrigeration and **deep freezing** do not kill bacteria. The low temperature only slows down their rate of reproduction. If harmful bacteria are present in refrigerated food, the low temperature will stop them multiplying and so their numbers may not be sufficient to cause disease. If the refrigerated food is not cooked properly but only warmed up, any bacteria present will multiply rapidly and produce a population large enough to cause infection.

Cooking, canning and refrigerating, which kill or slow down the reproduction of disease bacteria in food, have the same effect on saprophytic bacteria which would cause the food to decay. So these measures help to preserve food as well as make it safe to eat.

QUESTIONS

5 If people are not sure whether water is drinkable, they often boil it first. Why does this make it safe to drink?

6 An electrical fault puts your refrigerator out of action. It cannot be repaired for the next two days. How would you store (a) the eggs, (b) the bacon, (c) the cooked beef, and (d) an unopened pack of butter that are in it? Explain your actions.

7 Before people are employed to work in a canteen, they often have to undergo a medical examination. What do you think this examination tries to find out?

Houseflies

Flies walk about on food. They place their mouth parts on it and pump saliva on to the food. Then they suck up the digested food as a liquid. This would not matter much if flies fed only on clean food, but they also visit decaying food or human faeces. Here they may pick up bacteria on their feet or their mouth parts. They then alight on our food and the bacteria on their bodies are transferred to the food. Figure 6 shows the many ways in which this can happen.

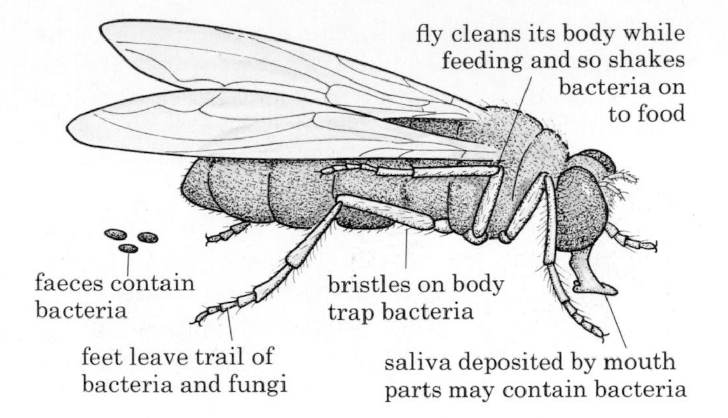

Figure 6 Spread of disease organisms by a housefly (from *Introduction to Human and Social Biology*).

Methods of prevention are to

(a) keep all unwrapped food in fly-proof containers such as refrigerators or larders,
(b) enclose all food waste in fly-proof dustbins so that the flies cannot pick up bacteria,
(c) never leave human faeces where flies can reach it (if faeces cannot be flushed into the sewage system, they must be buried),
(d) destroy houseflies, wherever possible, in the places where they breed, such as rubbish tips and manure heaps.

QUESTION

8 Although wasps often walk on our food, they are not suspected of spreading disease. Why do you think they are less harmful in this respect than houseflies?

Droplet infection

When we sneeze, cough, laugh, speak or merely breathe out, we send a fine spray of liquid drops into the air. These droplets are so tiny that

Figure 7 Protection of food on display. The glass barrier stops customers from touching the food and helps stop droplets from coughs and sneezes falling on the food.

they remain floating in the air for a long time and may be breathed in by other people or fall on to exposed food (Figure 7). If the droplets contain viruses or bacteria, they may cause disease when they are inhaled or eaten with food.

Virus diseases like colds, 'flu, measles and chicken pox are spread in this way, as are the bacteria (streptococci) which cause sore throats. When the water in the droplets evaporates, the bacteria often die as they dry out, but the viruses remain infectious, floating in the air for a long time.

In buses, trains, cinemas and discothèques, the air is warm and moist, and full of droplets which remain floating for a long time. These are places where you are likely to pick up one of these infections.

Wearing a gauze mask over the mouth and nose reduces the chances of either inhaling or exhaling the droplets but this precaution is usually taken only in hospitals. By covering your mouth and nose with a handkerchief when you cough or sneeze, you can greatly reduce the number of infected droplets which you expel into the air.

QUESTIONS

9 Very often, soon after the beginning of a new school term, many people suffer from colds. Suggest one reason why this happens.

10 Why do you think surgeons and nurses wear gauze face masks when they are doing an operation?

11 Although your resistance to disease may be lowered by getting cold and wet, you cannot 'catch a cold' in this way. Explain why not.

Contagious diseases

These are diseases which can be spread by direct contact with infected persons or by contact with their clothing, bed-linen or towels. The skin diseases caused by fungi (page 243) are very contagious. The sexually transmitted diseases are spread by sexual contact with an infected person (page 256).

Syphilis is a sexually transmitted disease. It causes ulcers on the reproductive organs at first, but if it is not treated the infection spreads to the whole body. The bacteria can also get across the placenta and infect an unborn child.

The disease can be cured with antibiotics provided it is treated in the early stages. The surest way of avoiding it is not to have sexual intercourse with an infected person. The early symptoms are often not noticed and any ulcers on the reproductive organs heal in about six weeks, so there is no easy way of recognizing an infected person. People such as prostitutes, who are known to have sexual intercourse with a large number of other people, are the most likely to be infected.

QUESTION

12 If you borrow another person's towel or running shoes, what risk are you taking?

RESISTANCE TO DISEASE

Natural barriers

Although there may be many bacteria living on the surface of the skin, the outer layer of the epidermis (page 134) seems to act as a barrier which stops them getting into the body. If the skin is cut or damaged, the bacteria may then get into the deeper tissues and cause infection.

The sweat glands and sebaceous glands produce substances which kill bacteria and keep down their numbers on the skin. Tears contain an enzyme called **lysozyme** which dissolves the cell wall of some bacteria and so protects the eyes from infection.

The acid conditions in the stomach destroy most of the bacteria which may be taken in with the food. The moist lining of the nasal passages traps many bacteria, so does the mucus produced by the lining of the trachea and bronchi. The ciliated cells (page 6) of these organs carry the trapped bacteria away from the lungs.

When bacteria get through these barriers, the body has two more lines of defence – the white cells and the antibodies. The way these work is described on page 120.

Immunity

When you catch a disease, your blood system makes antibodies which destroy the bacteria or viruses and so help you to recover. These antibodies remain for a time in your blood and make you **immune** to the disease. That is, you are most unlikely to get the same disease again, at least for a time. The immunity which you get as a result of having a disease is called **acquired immunity** (see page 120).

Immunity may last for a lifetime, as in the case of measles, or for only a few weeks as with influenza. Acquired immunity works only for

Figure 8 Immunization. The children are being inoculated against tuberculosis in a refugee camp.

the disease from which you have recovered; antibodies against diphtheria will not protect you against polio. Diseases such as the common cold and influenza produce no lasting immunity, possibly because there are so many different varieties of the virus which causes the disease. The antibody for one variety is no good against a different variety.

Immunization

Although it is a good thing to acquire immunity as a result of having a disease, it is not worth running the risk of dying from polio in order to get this immunity. Immunization is a way of making your body produce antibodies without you having to endure the dangers and discomforts of the disease.

Diphtheria. To protect you against this disease, you are injected with the toxin produced by the bacteria which cause it. The toxin is first made harmless so that you have only mild symptoms of illness, but your blood makes antibodies that protect you against the real toxin if the bacteria should ever get into your body.

Tuberculosis. To protect you against tuberculosis, you are injected with the BCG **vaccine** (pronounced 'vaxeen'). A vaccine is a collection of living bacteria or viruses which have been grown in artificial conditions in a laboratory for so long that they have become harmless. When the BCG bacteria are injected into you, they may cause slight symptoms, such as a rise in temperature, but your blood system will make antibodies against the bacteria. These antibodies will protect you against the real tuberculosis bacteria if they should ever get into your body.

Poliomyelitis. In most cases, the virus of this disease affects only the throat and intestine, producing the symptoms of a mild intestinal disorder or perhaps no symptoms at all. In about 1 per cent of cases the virus gets to the nervous system and causes temporary, or more rarely permanent, paralysis of some muscles. You can be immunized against polio either by a vaccine prepared from a weakened form of the living virus taken by mouth, or by an injection of the killed virus.

QUESTIONS

13 How might a harmful bacterium be destroyed or removed by the body if it arrived (a) on the cornea, (b) on the hand, (c) in a bronchus, (d) in the stomach?

14 Revise pages 120 and 121, and then explain briefly why immunization against diphtheria does not protect you against polio as well.

15 Even if there have been no cases of diphtheria in a country for many years, children may still be immunized against it. What do you think is the point of this?

BASIC FACTS

- **Infectious diseases of animals and plants are caused by viruses, bacteria, protista and fungi.**
- **All these grow or reproduce inside the bodies of their hosts and cause a build-up of infection.**
- **The micro-organisms are spread from diseased to healthy individuals by various methods: (a) in water, (b) in food, (c) by flies (or other insects), (d) by droplets and (e) by contact.**
- **The spread of disease can be prevented once we find out how it is spread.**
- **Our bodies have natural barriers like skin, mucus and cilia which keep most bacteria out.**
- **White cells and antibodies attack any bacteria and viruses which do get into the body.**
- **When we have recovered from a disease, there are antibodies in our blood which give us some immunity from further attacks of that disease.**
- **Immunity can be acquired by being injected with a harmless form of the bacterium or virus.**

32 Personal health

A character in a play once said, 'Everything I enjoy is either illegal, immoral or fattening.' It is indeed true that quite a number of human pleasures can be harmful to health. It is better, however, to study the facts as far as we know them, rather than to preach about health. The facts will allow us to make a conscious choice about our life style and not smoke, drink or eat our way to an early grave simply because we are ignorant of what we are doing.

We rarely realize the value of feeling physically fit until we experience a spell of ill-health. Some illnesses are unavoidable or accidental. It is not your fault if you suffer from hay fever or catch the 'flu during an epidemic, but some kinds of heart disease, lung cancer and venereal disease are illnesses that people bring upon themselves.

The information which follows is not trying to turn you into a 'fitness freak' but to present some facts so that, if you enjoy life, you can take steps to enjoy it for as long as possible, in a state of physical and mental well-being.

CORONARY HEART DISEASE

This is one of the commonest causes of death in countries with a high standard of living. It is caused by the formation of fatty deposits in the lining of the arteries. The arterial lining also grows thicker. These two changes may cause a blockage of the coronary arteries which supply the heart muscle (Figure 4*a* on page 113). When these vessels become blocked, the heart muscle does not get enough oxygen or glucose and without these, the heart cannot pump properly and it may stop altogether. This is usually what is meant by a 'heart attack'.

One man in five is likely to have symptoms of coronary heart disease before the age of 65. In about 25 per cent of these cases, the first sign will be collapse and death. Some people are more 'at risk' than others because they genetically inherit a tendency towards certain diseases, including heart disease. You may know of old people who boast of drinking, smoking and over-eating all their lives with no ill-effects, but this proves nothing about the effects of such a life style on other people. Evidence has been collected scientifically over many years, from a large number of people. It shows clearly that whether you are genetically a 'high risk' or a 'low risk' person, you can greatly reduce your chances of early death from coronary heart disease by

(a) taking regular exercise,
(b) keeping your body weight at a reasonable level and
(c) not smoking.

These precautions will help to protect you not only against heart attacks but against many other crippling illnesses such as bronchitis.

QUESTIONS

1 What is the function of the coronary arteries?

2 What three things can you do to protect yourself against coronary heart disease?

EXERCISE

In 1973, a group of 16 882 men between the ages of 40 and 64 and with 'office jobs' took part in a study on the possible effects of exercise on heart disease. They kept records of their leisure activities that involved exercise. In the following years, it was found that those men who suffered heart attacks had taken less exercise than those who were still free from heart disease. Light exercise such as car-polishing, housework or lawn-mowing appeared to be of little benefit. The kind of exercise needed was vigorous activity such as running, cycling, playing football or tennis and had to be kept up for at least 30 minutes at a time.

It is not known why exercise helps to reduce heart attacks. The improved blood circulation resulting from exercise could stop the fatty substances from settling down in the lining of the arteries, or the arteries might grow wider and produce more branches.

There is evidence to suggest that vigorous exercise taken during the years of adolescence has a long-term effect on the efficiency of the heart. Steady work or exercise for periods of about an hour, which raises the heart rate to about 140 beats per minute, helps to develop a heart with large ventricles. Vigorous exercise or heavy work, which gets the pulse rate up to about 180 per minute, causes the heart muscle to thicken up and its blood supply improves.

Exercise increases the flow of blood through the muscles and so helps to remove waste products. Contraction of the body muscles during exercise squeezes the veins and lymphatics (page 117) and so helps to return blood and lymph to the heart. The improved appetite resulting from exercise probably helps digestion. Exercise also helps to keep your body weight down, but the best way to do this is to avoid over-eating.

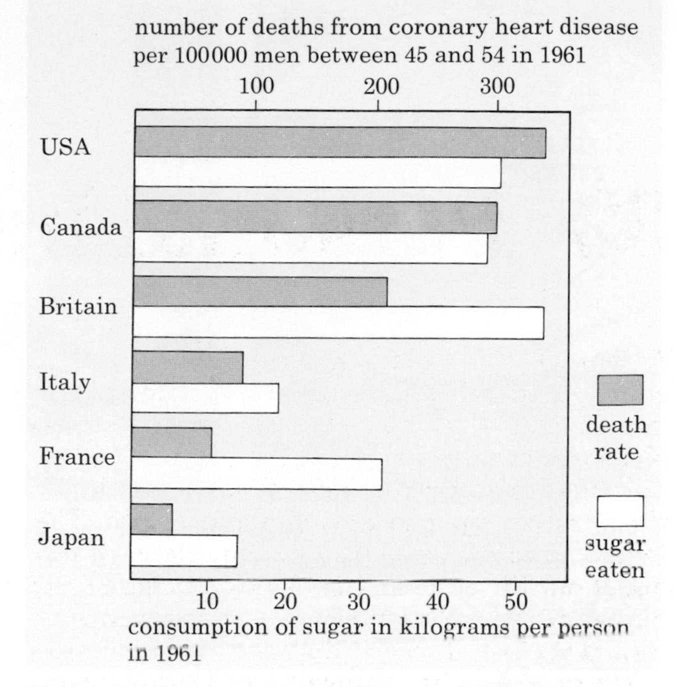

Figure 1 Sugar consumption and heart disease. Generally speaking, the greater the sugar consumption the higher the death rate from coronary heart disease, but this relationship does not prove that sugar *causes* heart disease. (From J. K. Brierley, *Biology and the Social Crisis*, Heinemann, 1967, and *The Sunday Times*, 1964.)

QUESTIONS

3 For exercise to be of long-term benefit, it needs to be vigorous enough to make you out of breath and raise your heart beat to 120 or more per minute. What regular exercise do you take which does these things?

4 What are the effects of exercise on the circulatory system?

DIET

Chapter 11 explained the need for a balanced diet; protein for building and replacing the tissues, carbohydrates for energy, and vitamins for chemical reactions in the cells.

In Western society, it is unlikely that any of these food items will be missing from the diet. It is much more likely that we will damage our health by eating too much of some kinds of food, or simply eat too much of everything.

Too much sugar

The white sugar from the sugar bowl is called refined sugar. Some kinds of 'brown' sugar consist only of white sugar with a little molasses added to darken it. Sugar comes from sugar-cane or sugar-beet. It is made by the plants along with a large number of other substances. If we ate sugar-cane or sugar-beet, we would probably not do ourselves much harm. It is the purified (refined) sugar which damages our health.

There is plenty of evidence to show that sugar in the mouth is an important cause of tooth decay but refined sugar also affects us in many other ways. It is a very concentrated source of energy. You can absorb a lot of sugar from biscuits, ice-cream, sweets, soft drinks, tinned fruits and sweet tea without ever feeling 'full up', so you tend to take in more sugar than your body needs. A high intake of refined sugar, therefore, causes people to become overweight, and this in turn leads to other forms of illness as described below.

There is a connection between sugar intake and heart disease. In the last 200 years, the average sugar consumption in Britain has increased from 2 kg per person to 55 kg per person each year. Figure 1 shows the sugar consumption in different countries and the death rate from heart attacks. This does not prove that sugar *causes* heart disease, but it makes us suspect that excessive intake of sugar does contribute to heart attacks. Similar charts

Figure 2 Margarine from sunflower oil. Notice that the cholesterol level is low and the unsaturated fatty acids (polyunsaturates) are high.

Figure 3 A wise choice of diet. Try to explain why the items of food in this picture have been selected to represent a good diet. Consider the fibre and vitamin content, the amount of refined sugar, the number of calories and the types of fat and protein.

could show a connection between heart disease and the sale of television sets, but it is not suggested that watching television is the cause of heart attacks. Eating a lot of sugar, driving cars and watching television are characteristics of a wealthy society. The members of a wealthy society tend to eat too much (especially sugar and fat) and take too little exercise (car-driving and television-watching). It is probably a combination of these, together with smoking, which increases the likelihood of heart disease.

Too much fat

The fatty layer which forms in the lining of arteries and leads to coronary heart disease contains fats and a substance called cholesterol (coal-ester-oll). The more fat and cholesterol you have in your blood, the more likely you are to suffer a coronary heart attack. Many doctors and scientists think that if you eat too much fat, you raise the level of fats and cholesterol in the blood and so put yourself at risk. There is still some argument about this. Some scientists think that eating a large amount of fat does not necessarily increase the cholesterol level in the blood. Until more is known, it seems to be a good idea to keep a low level of fats in your diet.

The culprits appear to be animal fats; that is, butter, cream, some kinds of cheese, egg yolk and the fat present in meat. These fats are digested to give what are called **saturated** fatty acids (because of the structure of their molecules). Many of the fats and oils from plants, such as the oil from sunflower seeds, contain **unsaturated** fatty acids (Figure 2). These are thought to be less likely to cause fatty deposits in the arteries. For this reason, it seems to be better to fry food in vegetable oil (e.g. sunflower, soya, corn) and to use margarine from certain vegetable oils rather than butter.

Too little fibre

We eat too many processed foods, such as sugar, which have been purified from their vegetable sources, and too much white bread, from which the bran has been removed. Unprocessed foods such as potatoes, vegetables and fruit contain a large amount of cell walls which we call **dietary fibre** (also called 'roughage'). Although we may not be able to digest the cell walls ourselves, there are bacteria in our intestines which can do so and we get the benefit from the digested products. Apart from fibre being a source of food, there is evidence to show that it has other highly beneficial effects. It prevents constipation and probably other disorders and diseases of the intestine, including cancer. Eating a diet with a lot of fibre makes you feel 'full up' and so stops you from over-eating. A 100-gram portion of boiled potato provides only 80 kilocalories (kcal, page 258). (A potato about the size of an egg weighs 50–70 grams.) You could feel quite full after eating 300 grams of potatoes but would take in only 240 kcal. A 100-gram portion of milk chocolate will give you 580 kcal but it is not filling. So a high fibre diet helps to keep your weight down without leaving you feeling hungry all the time (Figure 3).

Too much of everything

If you eat more food than your body requires for its energy needs or for building tissues, you are likely to store the surplus as fat and so become overweight (Figure 4). An overweight person is much more likely to suffer from high blood pressure, coronary heart disease and diabetes than a person whose weight is about right. Being fat also makes you less willing to take exercise because you have to carry the extra weight around.

Whether you put on weight or not depends to some extent on genetics. You may inherit the tendency to get fat. Some people seem able to 'burn off' their excess food as heat and never get fat, no matter how much they eat. You can't change your genetics but you can avoid putting on too much weight by controlling your diet. This does not necessarily mean eating less but simply eating differently. Avoid sugar and all processed food with a high sugar level, such as sweets, cakes and biscuits, and include more vegetables, fruit and bread in your diet. Your teeth, waistline, intestines and health in general will benefit from such a change in diet.

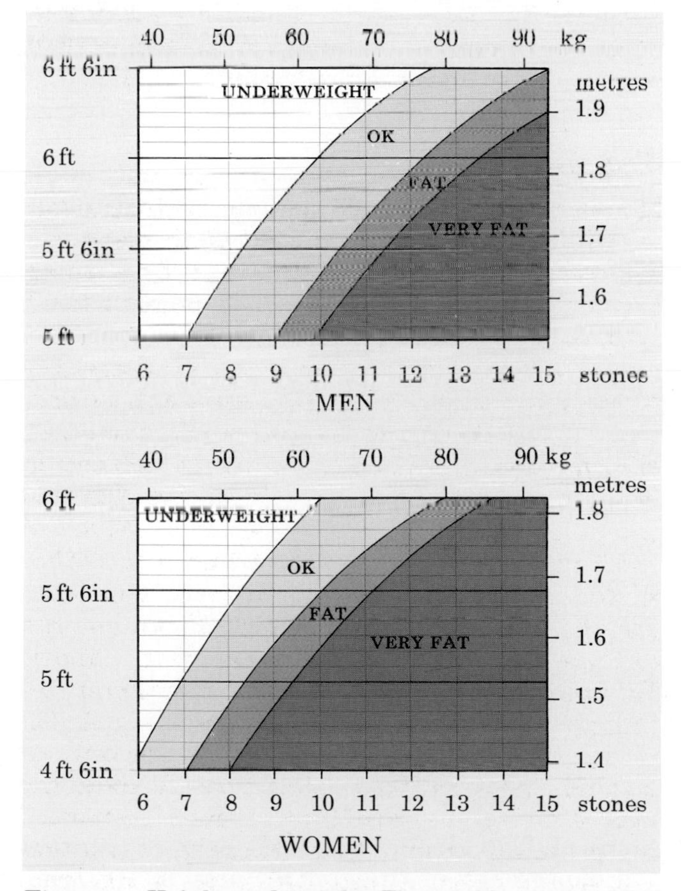

Figure 4 Height and weight. These tables are intended for adults who have reached their full height. They would find their height on the left-hand scale and look along the line till they reached their weight on the bottom scale. (By permission of The Health Education Council, London.)

QUESTIONS

5 (a) If you feel 'peckish' between meals, why is it better to eat an apple than a bar of chocolate?

(b) If you are going to do a long-distance walk, why is it better to take chocolate bars than apples?

6 100 grams of boiled potato will give you 80 kcal, but 100 grams of chips give you 236 kcal. Why do you think there is such a big difference?

7 For what reason might a person who has had a heart attack refuse cream in his coffee?

8 Why should a 'high fibre' diet help to stop you putting on weight?

MOOD-INFLUENCING DRUGS

Any substance used in medicine to help our bodies fight illness or disease is called a **drug**. One group of drugs helps to control pain and relieve feelings of distress. These are the mood-influencing drugs.

Events in your life may make you feel excited, depressed, anxious or angry. All these sensations must arise from changes taking place in your nervous and endocrine systems (page 164). The chemical substance adrenalin (page 170), when released into your blood, makes you feel tense and anxious or excited. In a similar way, it is thought that different chemicals produced by nerve endings in your brain give rise to most of your emotional sensations.

It is not always easy to be sure which is cause and which is effect. Feelings of anxiety may cause the production of adrenalin, or it may be that adrenalin causes feelings of anxiety. Nevertheless, it is known that swallowing or injecting certain substances can give rise to distinct changes of mood. It is often not known how these substances produce their effect. Even the method of action of alcohol, one of the oldest known mood-influencing drugs, is not known.

If these drugs are used wisely and under medical supervision, they can be very helpful. A person who feels depressed to the point of wanting to commit suicide may be able to lead a normal life with the aid of an anti-depressant drug which removes the sensation of depression. However, if drugs are used for trivial reasons, to produce sensations of excitement or calm, they may be extremely dangerous because they can cause **tolerance** and **dependence**.

Tolerance

This means that if the substance is taken over a long period, the dosage has to keep increasing in order to have the same effect. The continuing use of barbiturates in sleeping pills may require the dose to increase from one to two or three tablets in order to get to sleep. People who drink alcohol in order to relieve anxiety may find that they have to keep increasing their intake to reach the desired state. If the dosage continues to increase it will become so large that it causes death.

Dependence

This is the term used to describe the condition in which the user cannot do without the substance. Sometimes a distinction is made between emotional and physical dependence. A person with emotional dependence may feel a craving for the substance, may be bad-tempered, anxious or depressed without it, and may commit crimes in order to obtain it. Cigarette-smoking is one example of emotional dependence. Physical dependence involves the same experiences but in addition there are physical symptoms, called **withdrawal symptoms**, when the substance is withheld. These may be nausea, vomiting, diarrhoea, muscular pain, uncontrollable shaking and hallucinations. Physical dependence is sometimes called **addiction**.

Not everyone who takes a mood-influencing drug develops tolerance or becomes dependent on it. There are millions of people who can take alcoholic drinks in moderation with no obvious physical or mental damage. Those who become dependent cannot drink in moderation; their bodies seem to develop a need for permanently high levels of alcohol and the dependent person (an **alcoholic**) gets withdrawal symptoms if alcohol is withheld.

Physical or emotional dependence is a very distressing state. Getting hold of the substance becomes the centre of the addict's life, and he loses interest in his person, his job and his family. Because the substances he needs cannot be obtained legally or because he needs the money to buy them, he resorts to criminal activities. Cures are slow, difficult and usually unpleasant. There is no way of telling in advance which person will become dependent and which will not. Dependence is much more likely with some drugs than with others, and these are, therefore, prescribed with great caution. Experimenting with drugs for the sake of emotional excitement is extremely unwise. Some people may become dependent on almost any substance which gives them a conscious sensation.

Some of the mood-influencing drugs will now be considered.

Stimulants

Caffeine. This is the active substance present in tea, coffee and cocoa. It acts on the nervous system and increases wakefulness and reduces sensations of fatigue. In moderation, it seems to have no harmful effects or to build up tolerance. In excess it may cause tension and anxiety, hand-tremor, over-excitability and sleeplessness. It seems reasonable to assume that there is some emotional dependence involved in tea- and coffee-drinking.

Amphetamines. Drugs in this class make people feel alert and they reduce fatigue. Often they increase confidence but they also cause a reduction of accuracy. They increase the heart rate and give a sensation of nervousness but this is followed by feelings of depression. Unlike caffeine, the users can become tolerant of and dependent on amphetamines. At one time they were used to relieve depression, but their value for this is now doubtful and there are better drugs for this condition.

Amphetamines can cause dangerously high blood pressure if taken by athletes in an attempt to improve performance and they are quite useless for helping examination candidates because although they increase confidence, they also reduce accuracy.

Cocaine. This is extracted from the leaves of the coca plant which grows in South America. Natives of this region chew the leaves which are said to give them relief from fatigue and hunger, but it also results in dependence. Pure cocaine, taken as snuff or eaten, gives a temporary sensation of excitement followed by depression and listlessness. Prolonged use constricts the arteries and causes mental disorders. Cocaine was once used as a local anaesthetic but it has now been replaced by less poisonous compounds.

Depressants

Alcohol. The alcohol in wines, beer and spirits is a depressant of the central nervous system. Small amounts give a sense of well-being, with a release from anxiety. However, this is accompanied by a fall-off in performance in any activity requiring skill. It also gives a misleading sense of confidence in spite of the fact that one's judgement is clouded. The drunken driver usually thinks he is driving extremely well.

Alcohol causes vaso-dilation in the skin, giving a sensation of warmth but in fact leading to a greater loss of body heat (see page 136). A concentration of 500 mg of alcohol in 100 cm^3 of blood results in unconsciousness. More than this will cause death because it stops the breathing centre in the brain.

Some people build up a tolerance to alcohol and this may lead to both emotional and physical dependence (alcoholism). The way alcohol acts on the nervous system is not known, but if taken in excess for a long time, it causes damage to the brain and the liver which cannot be cured.

Barbiturates. These form one class of the drugs which depress the central nervous system and can be used as anaesthetics, sleeping pills or sedatives (to calm you down) according to the dosage. The concentration used for sedatives also reduces anxiety or aggressiveness but makes the user feel drowsy and reduces his mental powers, reaction time and co-ordination. Overdoses cause death by stopping the breathing centre in the brain. The combination of alcohol and barbiturates is particularly dangerous for this reason. Patients develop tolerance to barbiturates and easily become dependent with regular dosage.

Analgesics

Analgesics (Ann-al-jee-ziks) are drugs which relieve pain. Although the cause of pain may start in the body, the sensation of pain occurs in the brain. Depressants such as alcohol and barbiturates have analgesic effects because they alter the brain's reaction to pain.

Aspirin. This is a mild analgesic, particularly useful for relief of pain resulting from inflammation of tissues. It is also used to lower the body temperature during a fever. The side-effects with small irregular doses are mild but about 1 in 15 of aspirin users suffer from indigestion, irritation and possibly bleeding of the stomach lining if aspirin is taken frequently. Tolerance and dependence do not seem to be serious problems for most users. World consumption of aspirin is thousands of tonnes per year.

Paracetamol and phenacetin are mild analgesics which do not cause indigestion or gastric bleeding, but phenacetin has been suspected of causing serious kidney disorders.

Narcotic analgesics. Morphine, codeine and heroin are narcotics made from opium. Morphine and heroin relieve severe pain and produce a feeling of well-being and freedom from anxiety. They can both lead to tolerance and physical dependence within weeks and so they are prescribed with caution. Their illegal use has terrible effects on the unfortunate addict.

Codeine is a less effective analgesic but does not lead to dependence so easily as morphine. It is still addictive if used in large enough doses.

Tranquillizers. These substances are used to relieve tension and anxiety. They do not make people so drowsy as barbiturate sedatives do and they do not usually lead to tolerance and dependence. Some tranquillizers have been extremely valuable in treating severe mental illnesses such as schizophrenia and mania. Many thousands of mental patients have been enabled to leave hospital and live normal lives as a result of using the tranquillizing drug **chlorpromazine**.

Nowadays, tranquillizers are prescribed in their millions for the relief of anxiety and sleeplessness. Some people think that it is reasonable to use drugs to relieve acute anxiety and stress. Others think that these drugs are being used merely to escape the stresses of everyday life that could be overcome with will-power and determination. Some degree of anxiety is probably needed for mental and physical activity. These activities are unlikely to be very effective in someone who takes tranquillizers every time a problem crops up. Moreover, evidence is accumulating to show that some tranquillizers are severely addictive if used over long periods.

Hallucinogens

Cannabis and other extracts of Indian hemp are chewed or smoked to produce a sense of well-being, detachment and sometimes hallucinations. There does not seem to be much evidence of tolerance or emotional and physical dependence, but unstable individuals looking for more 'exciting' experiences are thought to be likely to move on to the 'hard drugs' such as morphine and heroin.

In the last 10 years, studies have shown that smoking cannabis can be more harmful than smoking cigarettes. The smoke from cannabis is held in the lungs for a longer period than cigarette smoke. As a result, five times more carbon monoxide enters the blood and four times as much tar remains in the lungs. Regular smokers of cannabis appear not only to damage their lungs but to have less effective immune systems and impaired brain function. The changes taking place in the lungs are similar to those which occur in cigarette smokers. Studies over a longer period will reveal whether these changes lead to cancer.

It has taken a great many years of careful study and scientific research to collect the evidence which now shows tobacco smoke to be so harmful. It would be irresponsible to legalize the smoking of another substance which might, after the same intensive study, turn out to be just as harmful.

LSD (lysergic acid). This substance has little medical use. It has been studied because it can cause symptoms similar to the mental disorder called schizophrenia. The sensations after taking LSD will vary according to the person concerned. Some people report sensations of happiness and heightened awareness, visual distortions and hallucinations. Others experience fearfulness, depression and terrifying illusions. Some individuals may commit violent and pointless acts when the drug is active and may experience the terrifying symptoms at intervals long after the initial dose has worn off. The possible benefits of new sensations are greatly outweighed by the chance of temporary or permanent mental damage.

QUESTIONS

9 What is the difference between (a) becoming tolerant of a drug and (b) becoming dependent on a drug? Which of these do you think is meant by being 'hooked' on a drug?

10 Why are amphetamine stimulants unsuitable for improving performance in (a) athletics, (b) examinations?

11 Why should drinking alcohol cause you to lose heat but make you 'feel' warm?

12 Why is it dangerous to take alcoholic drinks before driving?

13 Why is it dangerous to take an overdose of a depressant drug?

14 What do you think is the difference between an anaesthetic, a sedative and an analgesic?

15 If morphine and heroin make an addict 'feel good', why can he not keep taking a steady low dose to stay in this 'happy' state?

16 Apart from its immediate effect, what is the long-term danger of taking LSD?

SMOKING AND HEALTH

About 300 chemical compounds have been found in tobacco smoke. Of these, nicotine seems to have most effect on the nervous system. It stimulates some types of synapse (page 165), increases blood pressure and heart rate by the production of adrenalin (page 170), causes vasodilation (page 136) in the muscles and vasoconstriction in the skin. It is not clear how these changes produce the pleasure derived from smoking. Some regular smokers would claim that smoking calms their 'nerves'; others claim that smoking stimulates them.

The short-term effects of smoking cause the bronchioles (page 123) to constrict and the cilia lining the air passages (page 6) to stop beating. The smoke also makes the lining produce more mucus. The long-term effects may take many years to develop but they are severe, disabling and often lethal.

Emphysema

Emphysema (pronounced em-fi-seamer) is a breakdown of the alveoli (page 123). The action of one or more of the substances in tobacco smoke weakens the walls of the alveoli. The irritant substances in the smoke cause a 'smokers' cough' and the coughing bursts some of the weakened alveoli. In time, the absorbing surface of the lungs is greatly reduced (Figure 5). Then the smoker cannot oxygenate his blood properly and the least exertion makes him breathless and exhausted.

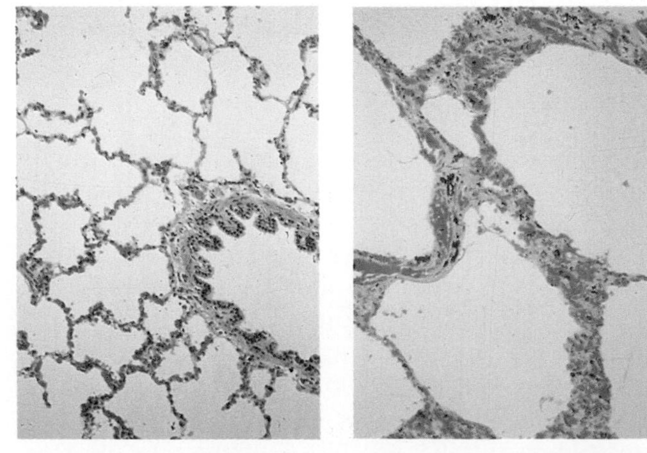

(*a*) Alveoli of normal lung (*b*) Lung with emphysema

Figure 5 Thin sections of lung tissue (× 100). In (*a*) there are about 30 thin-walled alveoli (and a small bronchiole). In (*b*) there are only about 6 alveoli. They present a smaller surface area and a thicker lining for the oxygen to diffuse through to reach the blood capillaries.

Chronic bronchitis

The smoke stops the cilia in the air passages from beating and so the irritant substances in the smoke and the excess mucus collect in the bronchi. This leads to the inflammation known as **bronchitis**. Over 95 per cent of people suffering from bronchitis are smokers and they have a 20 times greater chance of dying from bronchitis than non-smokers.

Heart disease

Coronary heart disease is the leading cause of death in most developed countries. It results from a blockage of the coronary arteries by fatty deposits. This reduces the supply of oxygenated blood to the heart muscle and sooner or later leads to heart failure (see page 248). High blood pressure, diets with too much animal fat, and lack of exercise are also thought to be causes of heart attack, but about a quarter of all deaths due to coronary heart disease are caused by smoking (Figure 6).

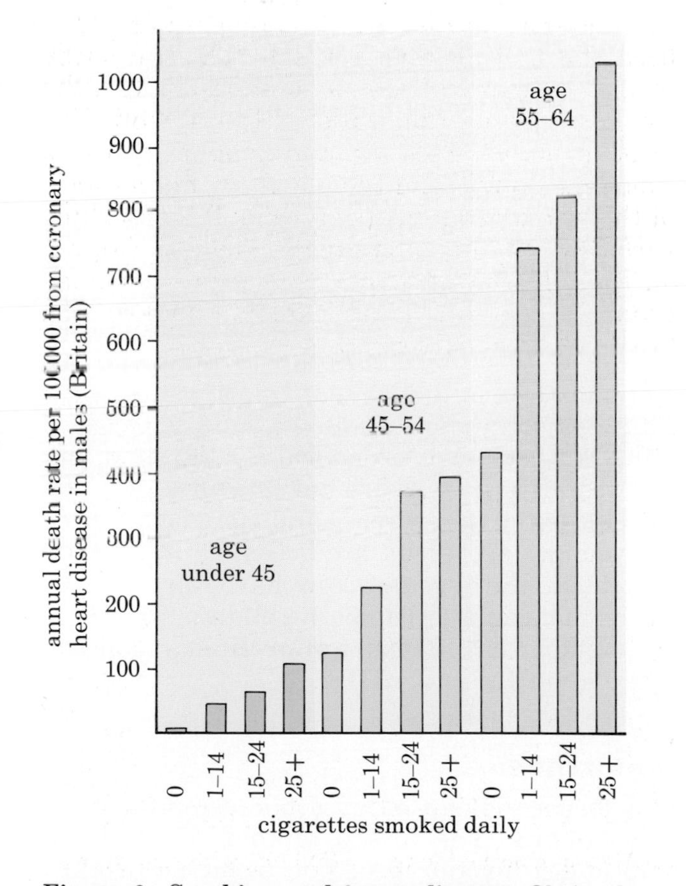

Figure 6 Smoking and heart disease. Obviously, as you get older you are more likely to die from a heart attack, but notice that in any age group the more you smoke, the higher your chances of dying from heart disease. (From *Smoking or Health: a report of the Royal College of Physicians*, Pitman Medical Publishing Co. Ltd.)

The nicotine and carbon monoxide from cigarette smoke increase the tendency for the blood to clot and so block the coronary arteries, already partly blocked by fatty deposits. The carbon monoxide increases the rate at which the fatty material is deposited in the arteries.

Lung cancer

Although all forms of air pollution are likely to increase the chances of lung cancer, many scientific studies show, beyond all reasonable doubt, that the vast increase in lung cancer (4000 per cent in the last century) is almost entirely due to cigarette-smoking (Figure 7).

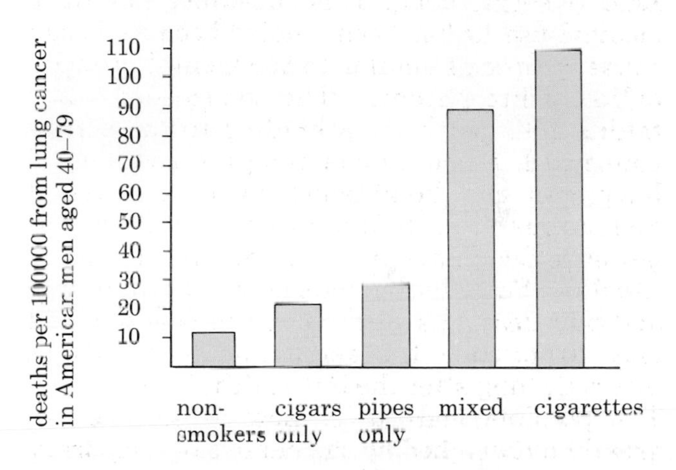

Figure 7 Smoking and lung cancer. Cigar and pipe smokers are probably at less risk because they often do not inhale. But notice that their death rate from lung cancer is still twice that of non-smokers. (From *Smoking and Health Now: a report of the Royal College of Physicians*, Pitman Medical Publishing Co. Ltd.)

There are at least 17 substances in tobacco smoke known to cause cancer in experimental animals, and it is now thought that 90 per cent of lung cancer is caused by smoking. The table below shows the relationship between smoking and the risk of developing lung cancer.

Number of cigarettes per day	Increased risk of lung cancer
1–14	×8
15–24	×13
25+	×25

Other risks

About 95 per cent of patients with disease of the leg arteries are cigarette-smokers. This condition is the most frequent cause of leg amputations.

Figure 8 Cigarette smoke can harm the unborn baby. Pregnant women who smoke may have smaller babies and a higher chance of miscarriage or stillbirth.

Strokes due to arterial disease in the brain are more frequent in smokers.

Cancer of the bladder, ulcers in the stomach and duodenum, tooth decay, gum disease and tuberculosis all occur more frequently in smokers.

Babies born to women who smoke during pregnancy are smaller than average, probably as a result of reduced oxygen supply caused by the carbon monoxide in the blood. In smokers, there is twice the frequency of miscarriages, a 50 per cent higher still-birth rate and a 26 per cent higher death rate of babies (Figure 8).

In 1976 two famous doctors predicted that one in every three smokers will die as a result of their smoking habits. Those who do not die at an early age will probably be seriously disabled by one of the conditions described above.

Reducing the risks

By giving up smoking, a person who smokes up to 20 cigarettes a day will, after 10 years, be at no greater risk than a non-smoker of the same age. The pipe- or cigar-smoker, provided he does not inhale, is at less risk than a cigarette-smoker but still at greater risk than a non-smoker. The risk of disease or death is also reduced by changing to low-tar cigarettes, leaving longer stubs, inhaling less and taking fewer puffs.

QUESTIONS

17 What are (a) the immediate effects and (b) the long-term effects of tobacco smoke on the trachea, bronchi and lungs?

18 Why does a regular smoker get out of breath sooner than a non-smoker of similar age and build?

19 If you smoke 20 cigarettes a day, by how much are your chances of getting lung cancer increased?

20 Apart from lung cancer, what other diseases are caused by smoking?

STDs (Sexually Transmitted Diseases)

Sexually transmitted diseases (STDs) are caught by having sexual intercourse with an infected person. The bacteria which cause two of these diseases can live only in warm moist conditions and so they do not survive long outside the body. For this reason, they can be caught only by direct sexual contact. It is not easy to avoid catching colds or 'flu because the germs may be floating about in the air. It is quite possible to avoid catching STDs by not having sexual contact with an infected person.

Since the symptoms of the disease are often not obvious, it is difficult to recognize an infected person. So the disease must be avoided by not having sexual intercourse with a person who *might* have the disease. Such persons are (a) prostitutes, who have sexual intercourse with men for money, (b) other people who are thought to have had sexual relationships with many others ('slept around'), and (c) casual acquaintances whose general background and past sexual activities are not known.

STDs are on the increase and more people from the younger age group are becoming infected. If untreated, the diseases can have very serious effects and can affect unborn children. There are three important STDs, **gonorrhoea**, (gone-er-rear), **syphilis** and **AIDS**.

Gonorrhoea

This disease is caused by a bacterium called a **gonococcus**. The first symptoms in men are pain and a discharge of pus from the urethra (page 139). In women, there may be similar symptoms or no symptoms at all. In men the disease leads to a blockage of the urethra and to sterility (inability to reproduce). In a woman the disease can be passed to her child during birth.

The bacteria in the vagina invade the infant's eyes and cause blindness.

The disease can be cured with penicillin but some strains of the gonococcus have become resistant to this antibiotic. There is no immunity to gonorrhoea; having had the disease once does not prevent you catching it again.

Syphilis

This is caused by a germ called a **spirochaete** (spiro-keet). In the first stage of the disease, a lump or ulcer appears on the penis or the vulva one week to three months after being infected. This ulcer usually heals without any treatment after about six weeks. By this time the spirochaetes have entered the body and may affect any tissue or organ. There may be a skin rash, a high temperature and swollen lymph nodes (page 118), but the symptoms are variable and the infected person may appear to be in good health for many years. However, if the disease is not treated in its early stages, the spirochaetes will eventually cause inflammation or ulcers almost anywhere in the body and do permanent damage to the blood vessels, heart or brain, leading to paralysis and insanity.

In a pregnant woman, the spirochaetes can get across the placenta (page 142) and infect the embryo. Penicillin will cure syphilis but unless it is used in the early stages of the disease, the spirochaetes may do permanent damage.

If a person suspects that he or she has caught an STD, treatment must be sought at once, and anybody who has had sexual contact with him/her must also get treatment. There is no point in one partner being cured if the other is still infected.

AIDS (Acquired Immune Deficiency Syndrome)

AIDS is caused by a virus called **HIV** (Human Immunodeficiency Virus). The disease is caught mainly through sexual contact, though it has also been transmitted by infected blood, e.g. in transfusions. The virus suppresses natural immunity, so the patients are at risk from a variety of serious illnesses such as rare forms of cancer and pneumonia. It may take several years for the symptoms to develop.

The drugs 'zidovudine' and 'acyclovir' may control the disease to a limited extent but they do not cure it.

QUESTIONS

21 Why is it often difficult to recognize the symptoms of STDs?

22 How can a baby become infected with an STD?

23 (a) How can syphilis or gonorrhoea be cured?
(b) What are the difficulties in bringing about a cure?

BASIC FACTS

- **Coronary heart disease results from a blockage of the arteries supplying the heart muscle.**
- **You can reduce the chances of early death from heart attack by (a) taking exercise, (b) not smoking, (c) keeping your weight down.**
- **Natural foods are better than processed foods because they contain (a) more fibre, (b) less sugar.**
- **Too much refined sugar causes tooth decay, fatness and possibly heart disease.**
- **Too much animal fat in the diet may contribute to heart disease.**
- **Being too fat makes you more liable to get heart disease and diabetes.**
- **Mood-influencing drugs are valuable for treating mental disorders but dangerous if misused.**
- **'Tolerance' means that you have to keep taking greater doses to achieve the same effect.**
- **'Dependence' (addiction) means that you feel physically ill and mentally disturbed if you do not take the drug.**
- **Smoking can cause emphysema, bronchitis, heart disease, lung cancer and strokes and it can affect unborn babies.**
- **STDs are caught through sexual intercourse with an infected person.**
- **Syphilis and gonorrhoea can be cured with antibiotics if treated at an early stage.**

GLOSSARY OF SCIENTIFIC TERMS

Acid. A sharp-tasting chemical, often a liquid. Some acids can dissolve metals and turn them into soluble salts. Nitric acid acts on copper and turns it into copper nitrate, which dissolves to form a blue-coloured solution. Acids of plants and animals (amino acids, fatty acids) are weaker and don't dissolve metals. Amino acids and fatty acids are organic acids. Hydrochloric, sulphuric and nitric acids are called mineral acids or inorganic acids.

Agar. A clear jelly extracted from one kind of seaweed. On its own it will not support the growth of bacteria or fungi, but will do so if food substances (e.g. potato juice or Bovril) are dissolved in it. Agar with different kinds of food dissolved in it is used to grow different kinds of micro-organism.

Alcohol. Usually a liquid. There are many kinds of alcohol but the commonest is ethanol (or ethyl alcohol) which occurs in wines, spirits, beer, etc. It is produced by fermentation of sugar. Ethanol vaporizes quickly and easily catches fire.

Alkali. The opposite of an acid. An alkali can neutralize an acid and so remove its acid properties. Sodium hydroxide is an alkali. It neutralizes hydrochloric acid to form a salt, sodium chloride.

$$Na\boxed{OH + H}Cl \rightarrow \underset{\text{salt}}{NaCl} + \underset{\text{water}}{H_2O}$$

Amino acids. Organic acids whose molecules contain nitrogen in an amino ($—NH_2$) group.

$$H - \overset{\displaystyle NH_2}{\underset{\displaystyle H}{C}} - C \begin{matrix} \nearrow O \\ \searrow OH \end{matrix}$$

is a simple amino acid called glycine. Proteins are made up of long chains of amino acids joined together. When proteins are digested, the amino acids are set free.

Atom. The smallest possible particle of an element. Even a microscopic piece of iron would be made up of millions of iron atoms. When we write formulae, the letters represent atoms. So H_2O for water means an atom of oxygen joined to two atoms of hydrogen.

Calorie. Just as a centimetre is a unit of length, a calorie is a unit of heat or energy. It is the amount of heat that would raise the temperature of one gram of water one degree Celsius. The energy value of food is measured in kilocalories (kcal). 1000 calories = 1 kilocalorie. In scientific studies, calories have been replaced by joules. For the energy in food, however, calories are still used. (1 calorie = 4.2 joules.)

Capillary attraction. The tendency of water to fill small spaces is called capillary attraction. If a narrow bore tube is placed in water, the water will rise up it for several centimetres. In a similar way, water will creep into the spaces between the fibres in a piece of blotting paper or between the particles in soil.

Carbon. A black, solid non-metal which occurs as charcoal or soot, for example. Its atoms are able to combine together to make ring or chain molecules (see 'Glucose'). These molecules make up most of the chemicals of living organisms (see 'Organic'). One of the simplest compounds of carbon is carbon dioxide (CO_2).

Carbon dioxide. A gas which forms 0.03 per cent (by volume) of the air. It is produced when carbon-containing substances burn ($C + O_2 \rightarrow CO_2$). It is also produced by the respiration of plants and animals. It is taken up by green plants to make food during photosynthesis.

Catalyst. A substance which makes a chemical reaction go faster but does not get used up in the reaction. Platinum is a catalyst which speeds up the rate at which nitrogen and hydrogen combine to form ammonia, but does not get used up. Enzymes are catalysts for chemical reactions inside living cells.

Caustic. A caustic substance can damage the skin and clothing and therefore should be handled with great care.

Cellulose. A chemical which makes up plant cell walls. It occurs in paper because this is made from wood, and in the clear plastic material called Cellophane. The cellulose molecule is made up of about a thousand glucose molecules joined end to end to form a long chain:

Compound. Two or more elements joined together form a compound. Carbon dioxide, CO_2, is a compound of carbon and oxygen. Potassium nitrate, KNO_3, is a compound of potassium, nitrogen and oxygen.

Cubic centimetre (cm^3). This is a unit of volume. A tea-cup holds about 200 cm^3 liquid. One thousand cubic centimetres are called a cubic decimetre (dm^3) but this volume is also called a litre. Some measuring instruments are marked in millilitres (ml). A millilitre is a thousandth of a litre and therefore the same volume as a cubic centimetre. So 1 cm^3 = 1 ml.

Dissolve. A substance which mixes with a liquid and seems to 'disappear' in the liquid is said to dissolve. Sugar dissolves in water to make a solution.

Element. An element is a substance which cannot be broken down into anything else. Sulphur is a non-metallic element. Iron is a metallic element. Oxygen and nitrogen are gaseous elements. Water (H_2O) is not an element because it can be broken down into hydrogen and oxygen.

Energy. This can be heat, movement, light, electricity, etc. Anything which can be harnessed to do some kind of work is energy. Food consists of substances containing chemical energy. When food is turned into carbon dioxide and water by respiration, energy is released to do work such as making muscles contract.

Expand. If a metal rod is heated strongly it gets longer. It is said to have expanded. If air is heated, it will expand and take up more space. If air, or any gas, is heated in a closed container which will not allow the gas to expand, the gas pressure will rise instead.

Fatty acids. Organic acids containing carbon, hydrogen and oxygen only.

```
     H   H   H
     |   |   |        O
H -- C - C - C - C ≤
     |   |   |        OH
     H   H   H
```

is butanoic acid. The $—C\begin{smallmatrix}\nearrow O\\ \searrow OH\end{smallmatrix}$ group makes it acid. Fats are made up of various kinds of fatty acid combined with glycerol.

Filtrate. The clear solution which passes through a filter; e.g. if a mixture of copper sulphate solution and sand is filtered, the blue copper sulphate solution which passes through the filter paper is called the filtrate.

Formula. A way of showing the chemical composition of a substance. Letters are chosen to represent elements, and numbers show how many atoms of each element are present. The letter for carbon is C and for oxygen is O. A molecule of carbon dioxide is one atom of carbon joined to two atoms of oxygen and the formula is CO_2. There are more elements than letters in the alphabet, so some of the elements have two letters, e.g. Mg for magnesium. Other elements have letters standing for the latin name, e.g. sodium is Na (= natrium).

Glucose. One kind of sugar. Its formula is $C_6H_{12}O_6$ and the atoms are arranged something like this:

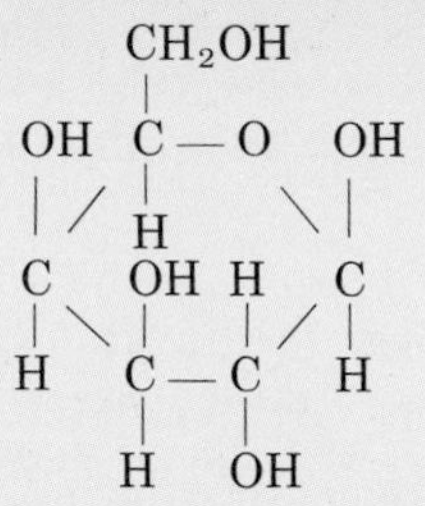

It is represented in this book by

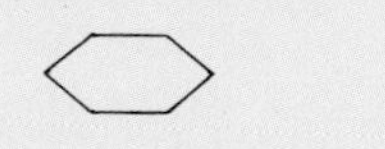

Glycerol. An organic compound containing carbon, hydrogen and oxygen. Its formula is:

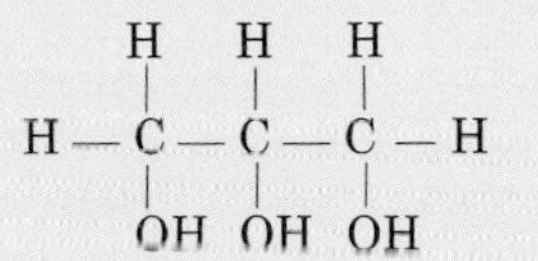

Each —OH group can combine with a fatty acid and so make a fat.

Gram (g). A unit of weight in the metric system.

A penny weighs 3½ grams.
A pack of butter is 225 grams.
1000 grams is a kilogram (kg).
One thousandth of a gram is a milligram (mg).

Hydrogen. Hydrogen is a gas which burns very readily. It is present in only tiny amounts in the air but forms part of many compounds such as water (H_2O), and organic compounds like carbohydrates (e.g. $C_6H_{12}O_6$ glucose), fats and proteins.

Inorganic. Substances like iron, salt, oxygen and carbon dioxide are inorganic. They do not have to come from a living organism. Salt is in the sea, iron is part of a mineral in the ground, oxygen is in the air. Inorganic substances can be made by industrial processes or extracted from minerals.

Insoluble. An insoluble substance is one which will not dissolve. Sugar is soluble in water but insoluble in petrol.

Lime water. A weak solution of lime (calcium hydroxide) in water. When carbon dioxide bubbles through this solution, it reacts with the calcium hydroxide to form calcium carbonate (chalk) which is insoluble and forms a cloudy suspension. This makes lime water a good test for carbon dioxide.

$$Ca(OH)_2 + CO_2 \rightarrow CaCO_3 + H_2O$$

Maltose. A sugar which has the same formula as sucrose $C_{12}H_{22}O_{11}$. It is formed when starch is broken down by enzyme action. The malt for making beer and whisky contains maltose.

Molecule. The smallest amount of a substance which you can have. For example, the water molecule is H_2O, that is, two atoms of hydrogen joined to one atom of oxygen. A drop of water consists of countless millions of molecules of H_2O moving about in all directions and with a lot of space between them.

Organic. This usually refers to a substance produced by a living organism. Organic chemicals are things like carbohydrates, protein and fat. They have very large molecules and are often insoluble in water. Inorganic chemicals are usually simple substances like sodium chloride (salt) or carbon dioxide (CO_2).

```
     H   H   H   H
     |   |   |   |        O
H -- C - C - C - C - C ≤
     |   |   |   |        OH
     H   H   H   H
```

molecule of a fatty acid
C_4H_9COOH (organic)

$$O = C = O$$

molecule of carbon dioxide
CO_2 (inorganic)

Oxygen. Oxygen is a gas which makes up about 20 per cent (by volume) of the air. It combines with other substances and oxidizes them, sometimes producing heat and light energy. In plants and animals it combines with food to release energy.

Peptide. A peptide is smaller than a protein. A dipeptide is made from two amino acids joined together. A polypeptide consists of many amino acids but is still not large enough to be called a protein. When proteins are digested they are first broken down to peptides.

Permeable. Allows liquids or gases to pass through. A cotton shirt is permeable to rain but a PVC mackintosh is impermeable. Plant cell walls are permeable to water and dissolved substances.

pH. This is a measure of how acid or how alkaline a substance is. A pH of 7 is neutral. A pH in the range 8–11 is alkaline, while those in the 6–2 range are acid; pH 6 is slightly acid, and pH 2 is very acid.

PIDCP. The initials of an organic chemical called phospho-indo-dichloro-phenol. It changes from blue to colourless in the presence of certain chemicals, including vitamin C.

Pigment. A chemical which has a colour. Haemoglobin in blood is a red pigment; chlorophyll in leaves is a green pigment. A black pigment called melanin may give a dark colour to human skin, hair and eyes.

Pipette. A glass tube designed to deliver controlled amounts of liquid. A bulb pipette has a plastic squeezer on one end so that it can deliver a drop at a time. A graduated pipette has marks on the side to show how much liquid has run out.

Proteins. Organic chemicals with large molecules containing carbon, hydrogen, oxygen and nitrogen (and usually sulphur). Proteins are made up of long chains of amino acids, often a hundred or more. The long chain is twisted and folded in a way which gives the protein a special shape (see 'Enzymes' on page 13). Enzymes and the structures in cells are made mostly of protein.

Reaction (chemical). A change which takes place when certain chemicals meet or are acted on by heat or light. The change results in the production of new substances. When paper burns, a reaction is taking place between the paper and the oxygen in the air.

Salt. A salt is a compound formed from an acid and a metal. Salts have double-barrelled names like sodium chloride, (NaCl) and potassium nitrate (KNO_3). The first name is usually a metal and the second name is the acid. Potassium (K) is a metal, and the nitrate (NO_3) comes from nitric acid (HNO_3).

Sodium hydrogencarbonate. At one time this was called sodium bicarbonate. It is a salt which is used to make carbon dioxide in experiments. Its formula is $NaHCO_3$.

Sodium hydroxide (NaOH). An alkali with caustic properties, i.e. its solution will dissolve flesh, wood and fabrics.

Soluble. A soluble substance is one which will dissolve in a liquid. Sugar is soluble in water.

Solution. When something like sugar or salt dissolves in water it forms a solution: The molecules of the solid become evenly spread through the liquid.

Volume. The amount of space something takes up, or the amount of space inside it. A milk bottle has an internal volume of one pint. Your lungs have a volume of about 5 litres; they can hold up to 5 litres of air. This cube has a volume of 8 cubic centimetres (8 cm^3).

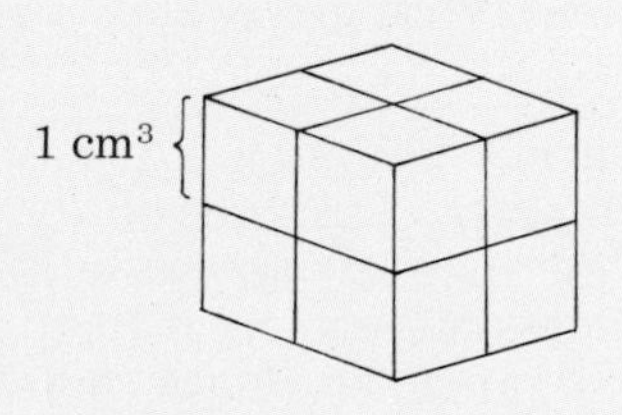

REAGENTS

Benedict's solution. Dissolve 170 g sodium citrate and 100 g sodium carbonate in 800 cm^3 distilled water. Add a solution of copper sulphate made from 17 g copper sulphate in 200 cm^3 distilled water.

Biuret test. 10% sodium hydroxide; 1% copper sulphate.

Iodine solution. 1 g iodine and 1 g potassium iodide are ground in a mortar with distilled water. Make up to 100 cm^3 and dilute 5 cm^3 of this solution with 100 cm^3 water for experiments.

Methylene blue. 1% solution.

PIDCP. (phenol-indo-2,6-dichloro-phenol). 0.1% solution in distilled water.

Pyrogallic acid in sodium hydroxide. Place 1 g solid pyrogallic acid in each flask and add 10 cm^3 10% sodium hydroxide solution just before setting up the experiment.

Solutions for tasting:

sweet; 5% sucrose solution.

sour; 0.5% citric acid or 5 cm^3 pure lemon juice in 20 cm^3 water.

salt; 2% sodium chloride solution.

bitter; 1 cm^3 tincture of quinine in 100 cm^3 water or boil 3 g dried hops in 200 cm^3 water for 30 minutes. Strain the mixture and make up to 200 cm^3.

Water cultures. 2 g calcium nitrate, 0.5 g each of potassium nitrate, magnesium sulphate and potassium phosphate in 2 litres of distilled water, plus a few drops of iron(II) chloride solution.

No nitrate; use potassium and calcium chlorides instead of the nitrates.

No calcium; use potassium nitrate instead of calcium nitrate.

No phosphate; use potassium sulphate instead of the phosphate.

INDEX

(Page numbers in bold type show where a subject is introduced or most fully explained)

PHOTO CREDITS

Thanks are due to the following copyright owners for permission to reproduce their photographs. The abbreviations used are *t* top, *c* centre, *b* bottom, *l* left, *r* right.

Heather Angel 35, 39, 51, 76*tr*, 80*br*, 144*l*, 199*bl,br*, 203*l*, 209*t*, 210*l*, 219*tr*, 221*t,b*, 239*tl,cl,br*

Ardea: 49*tr*, 202*bc*; Elizabeth Burgess 203*r*; Werner Curth 239*bl*; Bob Gibbons 76*br*, Ake Lindau 49*br*; John Mason 214*b*; P. Morris 50*tl*; J. Swedberg 55*l*, A. R. Weaving 99*t*

Biophoto Associates: Dr. Leedale, Leeds Dental School 99*br*, 101*r*, 197

Barnaby's Picture Library 52, 198*br*

University of Basel: M. Meader, M. Wurtz & F. Traub 243

Frank V. Blackburn 195

Brian Bracegirdle 3*tl*, *tr*, 62*t*, 63, 64*r*, 116, 134*l,r*, 170, 178

Bruce Coleman: Gene Ahrens 67*r*, 74*r*; S. C. Bisserot 209*cr*; Nicholas Brown 214*t*; Jane Burton 33*br*, 80*bl*, 83, 144*r*, 173, 215*r*, 218*tl*, 227*r*, 236; J. & A. Clare 219*tl*; Eric Crichton 40*r*; Inigo Everson 54; Udo Hirsch 219*b*; Hans Reinhard 57, 218*r*, 239*tr*

Bundeszentrale für gesundheitliche Aufklärung 91

Cornelsen-Velhagen & Klasing 156

Ken Coton 185

Gene Cox 23*tl,tr*, 59, 61, 101*l*, 106, 107, 123, 129*l*, 141, 168

Farmers Weekly 50*bl*

Finefare Ltd 244, 245

Glasshouse Crops Research Institute 31, 32

Colin Green 34

Griffin and George, Gerrard Biological Centre 169

Susan Griggs Agency Ltd 241

Philip Harris Biological Ltd 10, 62*b*, 111, 115, 125, 175, 183*l*; J. K.Burrass 186; M. J. D. Hirons 192; J. H. Kugler 3*bl,br*, 129*r*, 140, 254*l,r*; M. I. Walker 208

Health Education Authority: Sidney Harris 250*r*

ICI Plant Protection Division 50*cr,br*

Institute of Geological Sciences 190

Dr J. E. Jackson, East Malling Research Station 183*r*

Howard Jay 76*tl*

Andrew Johnson 99*bl*

Anthony Langham 95

Le Roye Productions 164

Ian Mackean 145*r*

Malmberg Educational 1

Leo Mason 137, 152

Maternity Centre Association, New York 143

National Institute of Agricultural Botany 199*tl*

Natural History Photographic Agency: Stephen Dalton 43*l*; E. A. Janes 49*bl*; M. Tweedie 193*l,r*; K. G. Preston-Mafham 202*bl*

Organon Laboratories Ltd 142

Oxford Scientific Films 33*bl*, 43*r*, 145, 187, 202*br*, 210*r*, 215*l*, 218*bl*, 220, 226, 227*l*

Picturepoint: M. I. Walker 231

Popperfoto 247

Rank Organisation 149, 150*l,r*

Royal Society for the Protection of Birds: Michael Richards 55*r*

Spectrum 130, 154

Swiss League Against Cancer 256

Syndication International 56

John Topham Picture Library: Parkhouse 224

M. I. Walker 198*tr*

The following photographs are by the author: 14, 20*l,r*, 22, 38, 40*l*, 42, 64*l*, 73*tl,cl*, 74*tl,cl,bl*, 75*tr,br*, 78*tr,bl,bc,br*, 79, 80*bc*, 84, 87, 88, 182, 199*bc*, 209*b*, 211, 222, 223*l,r*, 250*r*

Cover picture **Naturfotoarchiv:** Hans Pfletschinger

Title page picture **Ardea:** J. A. Bailey